D1568511
5/12
Tredyffrin Public Library
582 Upper Gulph Road
Strafford, Pennsylvania 19087-2052
610-688-7092
Chester County Library System
www.ccls.org
www.tredyffrinlibraries.org

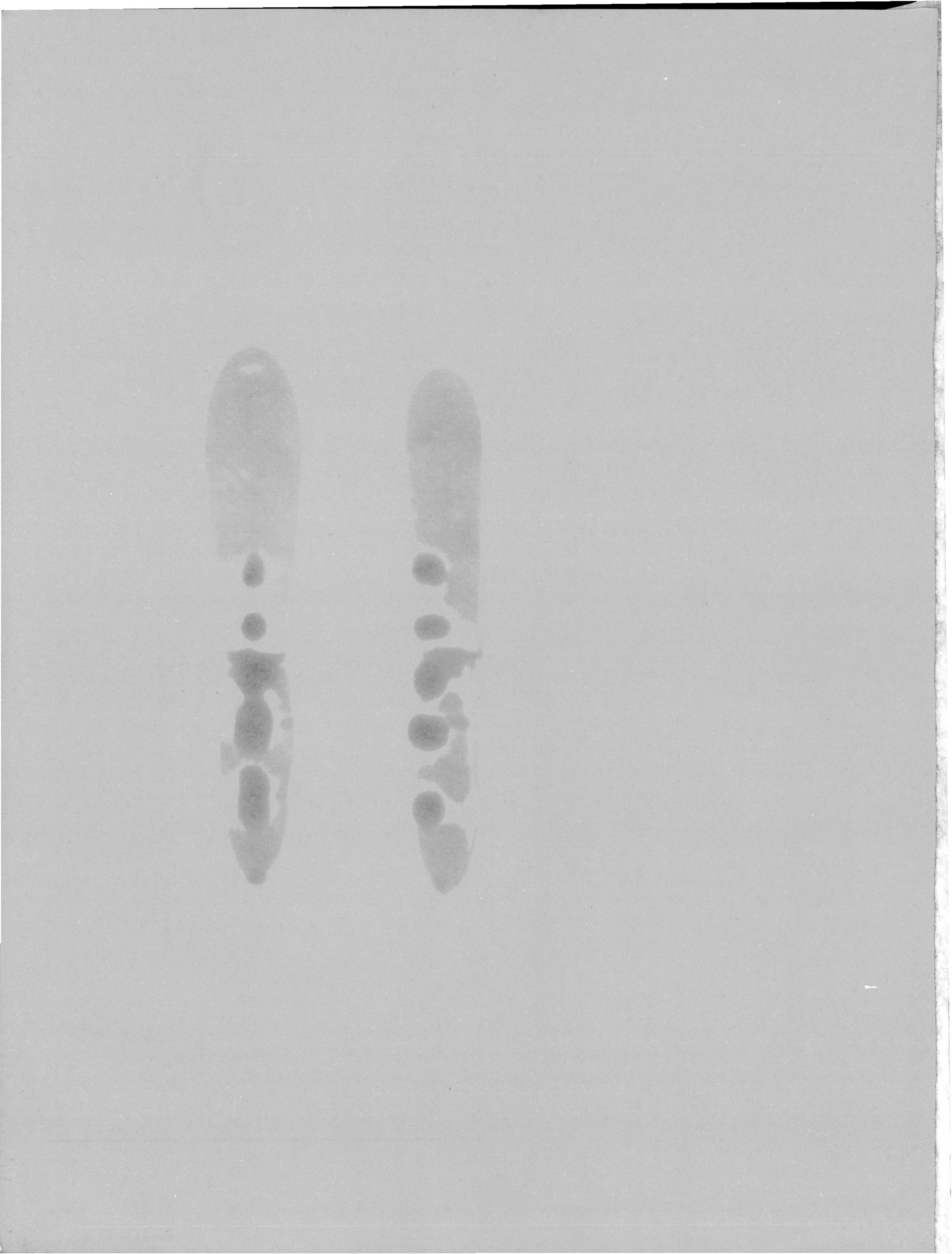

A Glorious Enterprise

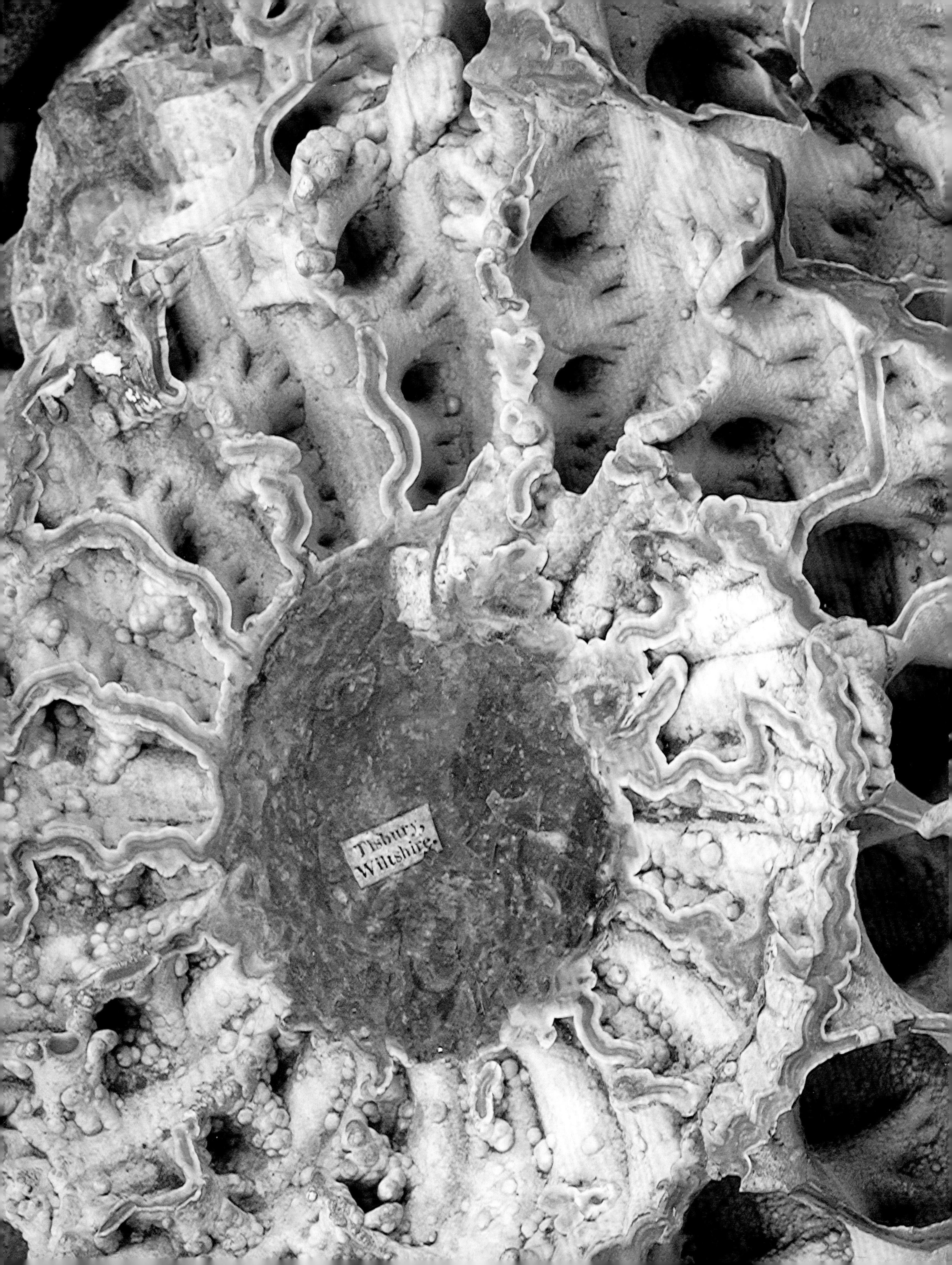
Tisbury,
Wiltshire.

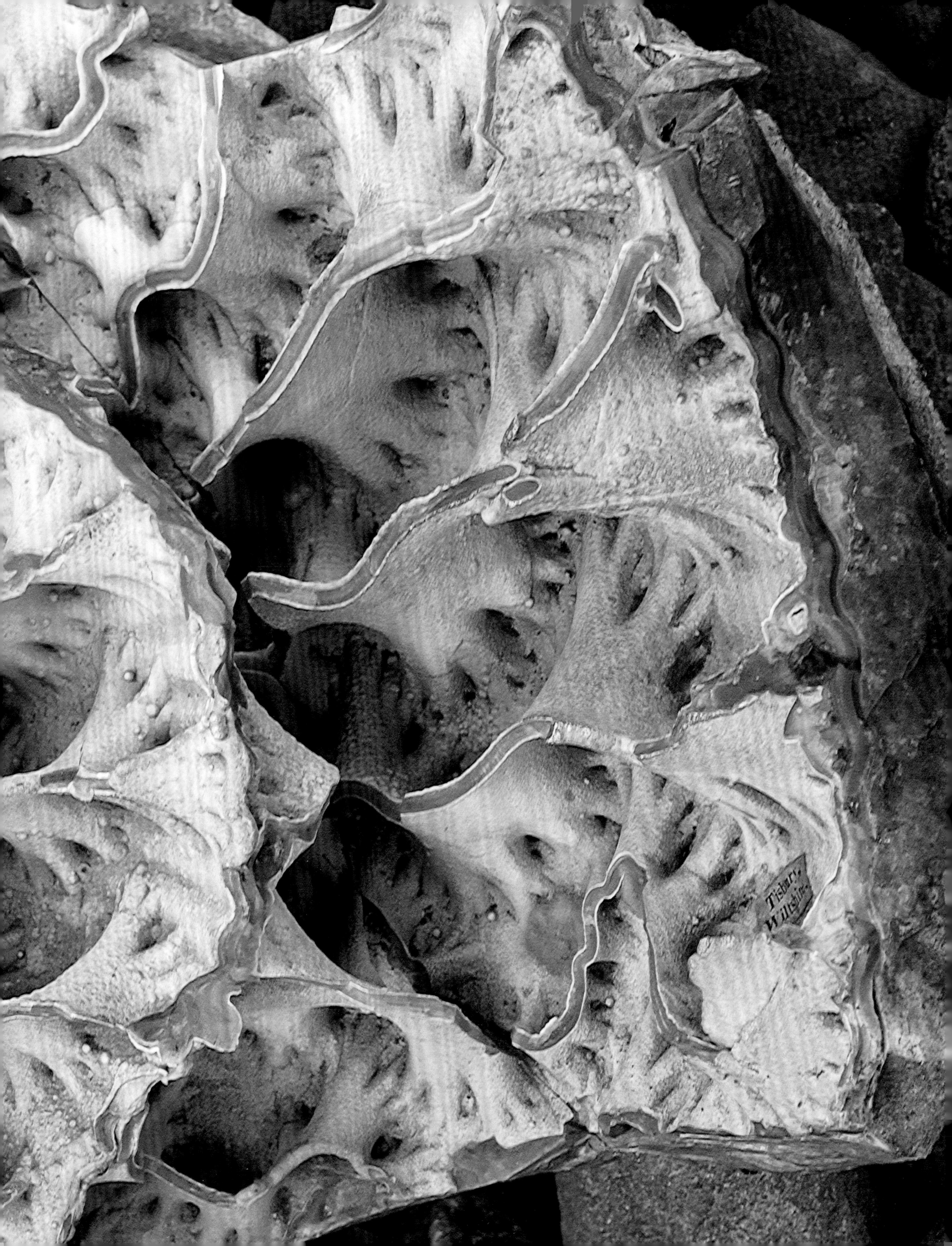
Tisbury,
Wiltshire.

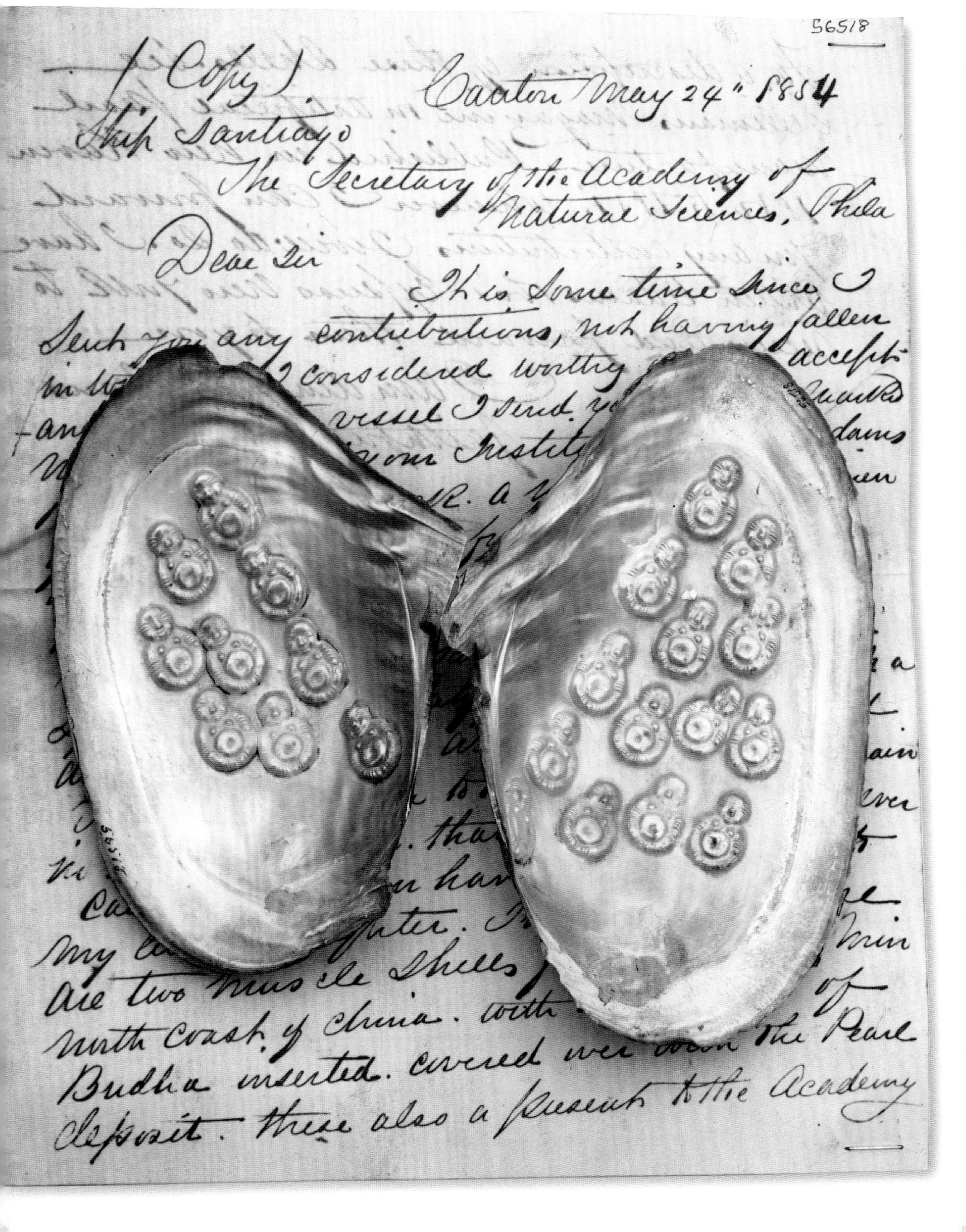

56518

(Copy). Canton May 24th 1854

Ship Santiago

The Secretary of the Academy of
Natural Sciences, Phila

Dear Sir

It is some time since I
sent you any contributions, not having fallen
in [illegible] I considered worthy [illegible] accept
an[illegible] vessel I send. y[illegible]
[illegible] your Instit[illegible]

[illegible]

my [illegible]pter. [illegible]
are two muscle shells [illegible]
north coast of China. with [illegible] of
Budha inserted. covered over with the Pearl
deposit. these also a present to the Academy

A Glorious

Enterprise

The Academy of Natural Sciences of Philadelphia and the Making of American Science

Robert McCracken Peck and Patricia Tyson Stroud
Photographs by Rosamond Purcell

PENN

University of Pennsylvania Press
Philadelphia

1

2

3

4

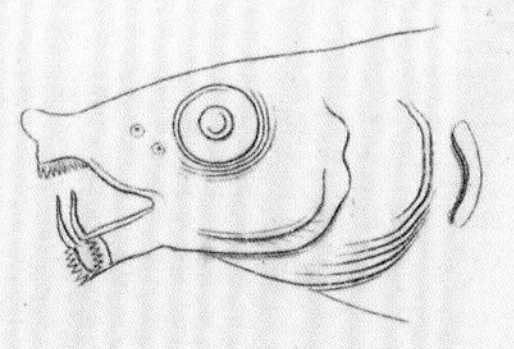

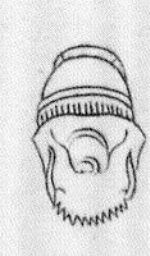

Prevalet pinx.

Mlle Louvier sc.

531

Contents

Preface

A Glorious Enterprise is a book about the extraordinary people who conceived, built, and continue to shape America's oldest continuously operating natural history museum. The title might also describe the creation of the book itself, for this has been a project many years in the making and has allowed its authors an opportunity to explore one of the world's great centers for scientific research.

Surprisingly, except for a short, anecdotal account published at the time of the Academy's centennial and a smattering of articles and book chapters written over time, there has never been a formal history of the venerable institution where so many fields of natural science were given their start in America. The challenge and opportunity to tell such a story was at once daunting and exhilarating.

The history of the world, wrote Thomas Carlyle, is but the biography of great men. The same can be said for an institution. During its two centuries of growth, the Academy of Natural Sciences has had more than its share of brilliant, quirky, courageous, genial (and not so genial), heroic, funny, insightful, and generally intriguing characters at its core. Collectively, these men and women have pioneered the study of natural science in America, amassed some eighteen million specimens, and created one of the greatest natural history libraries in the world. In pursuing their individual and collective passions, they have explored the planet, risked life and limb, and shared their findings with countless millions of people. This is the story we tell.

Having coedited four exhibition catalogues and the Academy's members' magazine *Frontiers* from 1979 to 1982, we felt that our research and writing styles were sufficiently complementary to work together as coauthors of this book. Our independent publications, including nine biographies and dozens of articles on people associated with the Academy or natural history in America, gave us a working background for the history we had long wanted to write. The Academy's bicentennial seemed the ideal occasion to produce a comprehensive history—not only for internal use, but to share with the world at large. So, with the blessing of the Academy's administration, we commenced the active work of gathering information and structuring our narrative.

While many authors come to their subjects late, we have known the Academy from the inside out for almost one-sixth of its existence. For much of that time we have been active participants in one part or another of its many-faceted operations. This firsthand experience helped us appreciate and interpret the copious amount of information we gleaned from the Academy's archives and from the living memories of the members, volunteers, and staff who generously shared their stories for this project. All were essential in writing this book.

We divided the chapters evenly, allowing each of us to pursue in depth particular themes, time periods, and personalities. As each chapter took shape, we exchanged our drafts, working closely to flesh out and refine them in ways that would ensure their accuracy and interest. We wanted each chapter to be able to stand alone but also fit well with the others in advancing the story as a whole. After our drafts were refined and polished, we sent them to internal and external experts for vetting before submitting them to the University of Pennsylvania Press for a professional peer review.

The book is comprehensive but by no means exhaustive. Many people whose contributions to the Academy and to science were as significant as those we included, for one reason or another did not make it into the pages of *A Glorious Enterprise*. For some, we simply couldn't find enough information about them. For others, we decided that their stories were too similar to ones we were telling elsewhere. In all cases, we focused on people who we believe best exemplify the spirit of the Academy and the times in which they lived. Sometimes, in lieu of full coverage, we augmented our narrative with notes that give depth and breadth to the story. Names, dates, and explanations that did not fit naturally into the text, but that we considered too important to omit, found their way into the notes. Similarly, we relegated some of our research—the names of presidents, trustees, and Academy medalists—to the back of the book. We also included a time line that provides a quick overview of the events described in more detail in the text.

To give the book a strong visual dimension, we combined historical photographs and other illustrations from the Academy's archives and library and other sources with contemporary photographs that capture the scope of the Academy's remarkable collections. For this, we were fortunate to secure the enthusiastic participation of Rosamond Purcell, whose three decades of work in natural history collections around the world has established her as the dean of natural-object, natural-light photographers. During three weeklong visits, she applied her unique vision to hundreds of specimens drawn from every part of the museum, creating masterful images that both record the subjects as they are and suggest the many layers of interpretation that they have and will continue to receive. In addition to featuring her photographs within the chapters, we have used the pages between chapters to highlight others that offer unexpected views of selected items from the Academy's collections.

We hope that *A Glorious Enterprise* will serve as a balanced and accurate beginning on which future historians will continue to build as the Academy flourishes in its third century and in its new affiliation with Drexel University. It would not have been possible to write it without a great deal of help from many, many people. While we reserve our detailed acknowledgments for that section of the book, we want to recognize three individuals here.

Eileen Mathias, reference librarian of the Academy, shared much useful information, based on her many years of work at the Academy. More important, she served as our illustration guru, scanning, labeling, and organizing the more than three hundred images we culled from the archives for reproduction in this book. There is no doubt that without her countless hours of hard work, this publication would be far less glorious than it is.

Clare Flemming, the Brooke Dolan Archivist of the Academy, was always cheerful and professional in her responses to our endless requests, which made working in the archives as productive as it was pleasant. To Clare and all of the Academy archivists and librarians who amassed and organized the Academy's irreplaceable records over the years, we are extremely grateful.

Finally, we wish to acknowledge the generous patron who made publishing this book possible. The intellectual, emotional, and financial support provided by Robert L. McNeil Jr. was critical to the book's existence. We had hoped to share with him the first copy of *A Glorious Enterprise*. Instead we can only thank him for the confidence he put in us and hope that the book we have written—the last of many he supported during his philanthropic life—is one he would have enjoyed. Bob felt strongly that an institution with as long and distinguished a history as the Academy of Natural Sciences deserves a book worthy of its accomplishments. We have tried our best to create such a book.

RMP

PTS

A Glorious Enterprise

Chapter 1

A Gathering of Gentlemen: The Founding and Early Years

This institution is conceived for the purposes of rational, free, literary and scientific conversation.... We meet also to compare the advancement of the sciences in the rest of the world with our own.... We are lovers of science.
—Contributions toward a Plan and Regulations, 1812

Philadelphia had been a thriving center of early science and medicine for decades when on Saturday evening, 25 January 1812, a small group of enthusiastic amateurs decided to pool their interest and knowledge in natural science. They met at a private residence on the northwest corner of High (Market) and Second Streets (figure 1.2).[1] At a subsequent meeting (21 March 1812), the title "Academy of Natural Sciences of Philadelphia" was employed for the first time, and it was decided that the origin of the Academy should be assigned that date.[2]

Several blocks away from their meeting room, the Academy founders had an imposing precedent in the American Philosophical Society (APS). Founded by Benjamin Franklin more than sixty years earlier, in 1743, the institution aimed to promote "the dissemination of useful knowledge." Thomas Jefferson had been the society's president during both his terms in the White House. John Bartram, a fellow founder with Franklin, had long ago established his famous botanical garden on the banks of the Schuylkill River a few miles outside Philadelphia. His son William, whose book *Travels through North & South Carolina, Georgia, East & West Florida* (1791) had influenced the English Romantic poets William Wordsworth and Samuel Taylor Coleridge, still lived and worked at the garden.

In Pennsylvania's state house (now Independence Hall) was Charles Willson Peale's museum, where the multitalented artist and naturalist housed his voluminous collections (figure 1.3). High on the walls of the Long Room hung Peale's portraits of famous men of the time, representing all disciplines of science and art, as well as political and military leaders. Numerous American Indian artifacts, many collected by Meriwether Lewis and William Clark on their pioneering expedition to the Pacific (1804–1806), were artfully displayed. But pride of place was given to natural history, with unusual live and stuffed birds and animals, shells, insects, minerals, and fossil bones, including those of a giant mastodon that Peale had unearthed in upstate New York in 1801. Many of the bird and mammal specimens were shown with painted backgrounds representing their habitats. Peale and his artistic sons rendered these in a way that anticipated the diorama exhibits of the twentieth century. At different times, the museum showed monkeys dressed in human clothes and "a small calf of very singular formation, and presenting various points of similitude with the human figure," which could be seen "on application."[3] Also present was a five-legged, two-tailed cow that produced milk for the Peale family for years before it died and joined the other stuffed animals on permanent display.[4]

Because no national museum existed at the time, Peale's Philadelphia museum served as the federal government's repository for natural history specimens collected on government-sponsored expeditions. Jefferson had sent Peale numerous bones

1.1 The first ballot box of the Academy of Natural Sciences of Philadelphia. ANSP Archives coll. 115. Purcell photograph.

1.2 *Second Street North from Market Street with Christ Church* by William Birch, from *The City of Philadelphia in the State of Pennsylvania North America as it appeared in the Year 1800.* Hand-colored lithograph. The Library Company of Philadelphia. The Academy's first meeting rooms were on the second floor of the center building, with the steeple of Christ Church behind it.

and skins of newly discovered western animals from the Lewis and Clark Expedition as well as a live prairie dog that, Peale reported, grew sleepy as winter approached. More spectacular were the two grizzly bear cubs from Zebulon Pike's western journey (1805) (figure 1.4). The pair, docile at first, grew fierce as they aged, and Peale, of necessity, shot them and stuffed their skins for display. That spring of 1812, a butcher shop advertised in a local newspaper "the meat of a remarkably fine Missouri or Grisly Bear . . . procured by Major Pike."[5] Nothing was wasted in those days of fantastic new discoveries.

The founders of the protean Academy of Natural Sciences sought to distinguish their activities from the more popular, less scientific aspects of Peale's commercial enterprise. It has been suggested that the Academy's founders formed their society because they had been excluded from the American Philosophical Society for social reasons, the APS being only for the well connected and socially prominent.[6] But there was a more fundamental difference between the members of the two societies. Whereas the members of the APS were established Americans, eminent in their fields, having been elected because of their individual accomplishments in many disciplines, the focus of Academy members was entirely on natural history. Three Academy founders were immigrants, with European scientific backgrounds, pursuing new ventures in America, and the other four were established businessmen but not of the socially elite. The exception was Thomas Say, son of Benjamin Say, a distinguished Philadelphia doctor who served in the Pennsylvania Senate and the U.S. Congress.[7]

The founders were nevertheless cultured, well educated, and broad-minded men who quoted Plato and Franklin in their initial documents and stated as their manifesto words of the preeminent mystical theologian and writer François de Salignac de la Mothe Fénelon (1651–1715): "Inasmuch as I prefer the liberal interests of my country to my own sordid or narrow interests, so do I prefer the universal interests of all mankind to any invidious interest of my own country."[8] This quote from Fénelon was used to reinforce the founders' conviction that "we are all human beings essentially natives of one common world . . . [and that] we adhere to the spirit which dictates: 'Peace on Earth and Good will to all men.'"[9]

Because they thought that Galileo was burned alive for promulgating laws of nature and theories of nature's phenomena,

FACING PAGE 1.3 *The Artist in His Museum* by Charles Willson Peale. Oil on canvas. Pennsylvania Academy of the Fine Arts. Gift of Mrs. Sarah Harrison (The Joseph H. Harrison Jr. Collection).

the founders stated that "All party-cant-words, polemics and controversy, religious, national and political, are by consequence excluded [from] our conversation in the Society."[10]

Of the seven men who united to create the Academy, Dr. Gerard Troost (1776–1850) was the most accomplished naturalist (figure 1.5). A mineralogist and former pharmacist from Le Havre, France, he had been a pupil of the great French scientist the Abbé René-Just Haüy. Troost had earned a doctor of medicine degree at Leyden and a master's degree in pharmacy from the University of Amsterdam in 1801, and in 1810 he had been elected a correspondent of the Natural History Museum in Paris.[11] His employment in America at the time, producing alum (a white mineral salt used in medicine and dyeing), earned him a very modest living. The other founders included Nicholas S. Parmentier (1776–1835), a Frenchman and a former officer in Napoleon's army, who emigrated to the United States in 1805 and engaged in the distilling and manufacture of spermaceti (whale) oil; "Dr." Camillus MacMahon Mann, a physician described in the *Minutes* as an "Irish patriot and refugee," who had spent time in France, where he had scientific friends; Jacob Gilliams, a second-generation dentist whose father had treated George Washington during his residence in Philadelphia as the first U.S. president; John Shinn Jr., a manufacturing chemist from New Jersey; John Speakman, a Quaker druggist; and Thomas Say, Speakman's business partner, who soon became the nation's leading authority on insects and shells (figure 1.6).

Say was unable to be present at the first several meetings, but at the session of 21 March, it was decided that, because he was part of their group, he should be included as one of the founders. An early Academy historian, Edward J. Nolan, notes: "It may be claimed that the continued existence of the Academy was in great measure due to [Say's] devotion, and the dignity he was able to give the proceedings by the high character of his scientific work."[12] Portraits of Say, Speakman, Gilliams, and Troost hang today in the Academy's library.

The young society soon rented meeting rooms above a milliner's shop at 94 North Second Street. These included a "conversation hall," a reading room, and a place to deposit scientific specimens. The natural history collections at first encompassed a few insects and shells, several stuffed birds, and "a fine herbarium collected in the vicinity of Paris by Dr. Mann." It also included some birds from Thomas Say, a group of shells gathered by Dr. John Barnes, a herbarium belonging to Nicholas Parmentier, and crystals from Gerard Troost.[13] A few months later, in the summer of 1812, the Academy acquired a collection of two thousand minerals purchased by Speakman from Dr. Adam Seybert for $750. This acquisition enabled Troost soon afterward to deliver a course of lectures on mineralogy to the members.[14] In

1.4 *Missouri Bear, Ursus horribilus*: Ord. *Specimen col[lecte]d by Lt. Pike, presented to C. W. Peale* by Titian Peale. Watercolor. American Philosophical Society.

1.5 *Gerard Troost* (1776–1850) by Charles Willson Peale, [1824]. Oil on canvas. ANSP Library coll. 286.

FACING PAGE 1.6 Members' signatures from the Academy of Natural Sciences of Philadelphia Constitutional Act, Foundation Meeting, 1812. ANSP Archives coll. 527.

We therefore pledge ourselves to each other to bear one another out through all the difficulties, responsability before the world and expences of the establishment of the Academy of Natural Sciences, and the seven primary and foundation members constitute themselves and are constituted accordingly Comittee and Board, of Regulations, Management, and Direction.

Thomas Say

N. S. Parmantier

Camillus McMahon. Secretary.

J. Gilliams

G. Troost

John Speakman

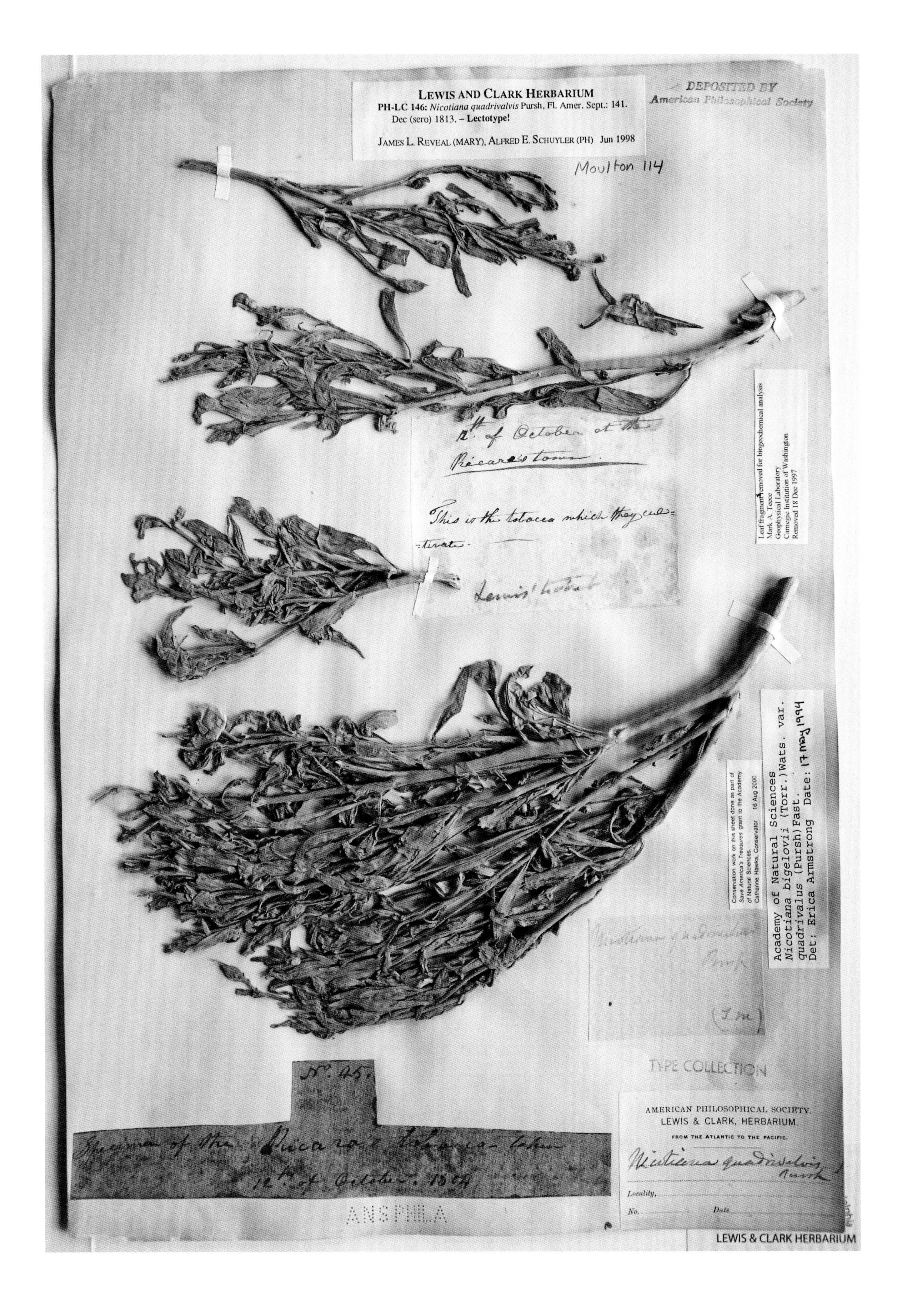

LEWIS AND CLARK HERBARIUM
PH-LC 146: *Nicotiana quadrivalvis* Pursh, Fl. Amer. Sept.: 141. Dec (sero) 1813. – Lectotype!
JAMES L. REVEAL (MARY), ALFRED E. SCHUYLER (PH) Jun 1998
DEPOSITED BY
American Philosophical Society
Moulton 114
12th of October at the Ricara's town.
This is the tobacco which they cul-tivate.
Leaf fragment removed for biogeochemical analysis
Mark A. Teece
Geophysical Laboratory
Carnegie Institution of Washington
Removed 18 Dec 1997
Conservation work on this sheet done as part of Save America's Treasures grant to the Academy of Natural Sciences.
Catharine Hawks, Conservator 16 Aug 2000
Academy of Natural Sciences
Nicotiana bigelovii (Torr.) Wats. var.
quadrivalvus (Pursh) East.
Det: Erica Armstrong Date: 17 May 1994
TYPE COLLECTION
AMERICAN PHILOSOPHICAL SOCIETY.
LEWIS & CLARK, HERBARIUM.
FROM THE ATLANTIC TO THE PACIFIC.
Nicotiana quadrivalvis Pursh
Locality,
No.
Date
No. 45.
Specimen of the Ricara's tobacco taken 12th of October. 1804
ANSPHILA
LEWIS & CLARK HERBARIUM

1815, to accommodate a new set of botanical lectures, a room was granted to Academy members in the hall of the Agricultural Society since the Academy had no room in which to hold lectures.[15]

Despite the facts that these lectures were well attended and the Academy was able to repay Speakman for the mineral purchase, a shortage of money plagued the society. By the end of 1815, the Academy was in debt for $1,000. The rental of rooms and the occasional purchase of specimens exceeded the income from dues and lectures.[16]

Aside from the Academy's educational aims, it also had a moralistic one. "In this large city," an early document states, "the young men generally have few modes of social recreation when their daily business is over except frequenting the theatres at a considerable expense, taverns, gaming houses, dancing rooms or places of a more degrading atmosphere, where they can gain no improvement but are sure to acquire habits injurious to the spirit of a good citizen."[17] Therefore, to be a member of the Academy would be an uplifting and edifying experience.

The founding members' consensus that their society should be "perpetually exclusive of political, religious and national partialities," as these were adverse "to the interests of Science,"[18] was an important injunction at a time when politics was often controversial. Jefferson's Republican administration from 1800 to 1808 had been fraught with contention because of its advocacy of states' rights over the New England Federalists' preference for a more centralized government. At the time of the Academy's founding, James Madison, another Virginian sympathetic to the economics of the slave-holding South, was president and continuing many of Jefferson's policies.

Science reflected a decidedly democratic element in this early period: anyone could participate in the activity of collecting, describing, and gathering facts about elements of the natural world. The founders' stated objective was the diffusion of knowledge, not the advancement of science, which would come later, and relations with those involved were friendly and sharing.[19] Under the heading "Conversation in the Society," one of the Academy's founding documents clearly states that the members intended to admit only "persons of gentlemanly manners 'genteel and gentle' [and that] it cannot enter into our contemplation to suffer the presence or entrance of persons of a contrary deportment."[20]

Despite being prohibited, politics must have entered at least some Academy discussions in the spring of 1812, when international tensions reached a crisis. Merchants, sea captains, and the U.S. government had had enough of Great Britain seizing American ships and impressing their sailors, and because Napoleon Bonaparte had banned English commerce everywhere on the coast of Europe, England refused to allow Americans to trade where its own merchants could not go. The American political scene was close to the boiling point over this issue when, on 7 April 1812, a Philadelphia newspaper announced that the American government had imposed "an embargo on all the ships and vessels in the ports and harbors of the United States for a limited time."[21] There were many who believed that this decision, surely a prelude to war, foreshadowed the ruin of the American economy.

In such a turbulent time, it must have been difficult, if not impossible, for the Academy's members to pursue their research in natural history without at least thinking about the political implications of their activities. According to historian John C. Greene: "There was progress in American science in the Jeffersonian era—in establishing an institutional base, assimilating European developments, and exploring a continent—but it took place in a context of flamboyant patriotism, political and religious controversy, and practical concern with commerce and industry that alternately inspired and distorted scientific development."[22]

Patriotism was an important force propelling the advance of science in America generally, and it may have been the central influence on the Academy's founding. First and foremost was the drive by Americans to name and describe the species of their own land, examples of which were being discovered almost daily. Alexander Wilson, a zealously "American" Scottish immigrant and early Academy member, voiced this sentiment when he railed against "the reproach of being obliged to apply to Europe for an account and description of the productions of our own country."[23] Wilson probably had in mind the plants collected by John and William Bartram, many of which were described abroad, and certainly, had he known of them (he died in 1813), also those collected by Lewis and Clark, which were described and published in England in 1814 by the German botanist Frederick Pursh (1774–1820) (figure 1.7).

Despite their focus on encouraging American science, the Academy founders knew that if they were to be successful, they would need to enlist foreign assistance in establishing what they hoped would become America's own clearinghouse for new discoveries. They also recognized the importance of having Old World specimens to compare with those from America. Thus they sought new members and international connections for the fledgling institution.

1.7 Indian Tobacco (*Nicotiana quadrivalvis*) collected by Lewis and Clark on their expedition to the Pacific Ocean (1804–6). ANSP Botany Department, PH-LC 146, on deposit by the American Philosophical Society, Philadelphia. Purcell photograph.

Lewis and Clark's Plants

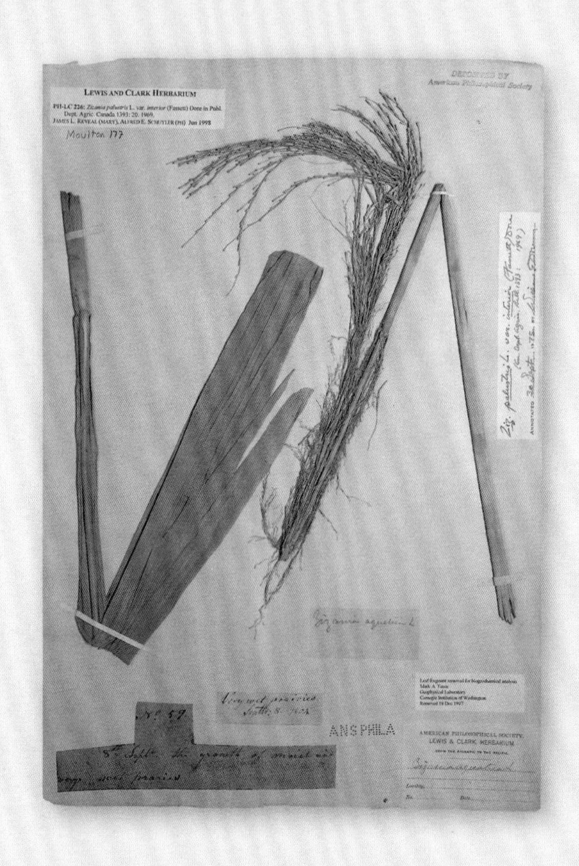

ABOVE Wild rice (*Zizania palustris*) collected by Meriwether Lewis near the future Nebraska–South Dakota state line, September 1804. ANSP Botany Department, PH-LC 226, on deposit by the American Philosophical Society, Philadelphia. Note the purple tag at the bottom of the herbarium sheet, with Lewis's handwriting.

FACING PAGE Arrow-leaf balsamroot (*Balsamorhiza sagittata*) collected by Meriwether Lewis along the Columbia River in future Washington State, April 1806. ANSP Botany Department, PH-LC 36, on deposit by the American Philosophical Society, Philadelphia.

Meriwether Lewis and William Clark, as charged by Thomas Jefferson, led an expedition from 1804 to 1806 to find what they could between the Mississippi River and the Pacific Ocean. On their return trip east, they brought with them, among other things, hundreds of specimens of plants. The long journey of these specimens continued when parts of the collection were carried to England (where a few still reside) and others to Washington, Boston, and other cities in the United States. A century passed before the collection was brought together in Philadelphia in an institution that did not exist at the time of the expedition.

Lewis collected almost all the plants himself, pressing them flat and drying them between sheets of blotting paper. At the expedition's terminus, bundles of plants were sent from St. Louis to Thomas Jefferson in Washington, who, in turn, sent them to the American Philosophical Society in Philadelphia, where they were distributed for study. Among those who received them was the horticulturist Frederick Pursh, who worked for William Hamilton, a wealthy landowner in the suburbs of Philadelphia. Pursh helped himself to some samples, presumably without permission, and left for England, where in 1814 he published the first treatise on North American plants.

While most of the specimens were held in storage at the American Philosophical Society by 1818, Pursh's plants found their way into the collections of Aylmer Lambert in London. When Lambert died in 1842, his collection of thousands of plant specimens was auctioned off, and at the end of the day, a visiting Academy member named Edward Tuckerman, later president of Amherst College, bought a few of the "dregs." In 1856 he gave his prize to the Academy, where the specimens were kept for forty years until curator Thomas Meehan recognized them as Lewis's collections. Meehan, on a hunch suggested by a Harvard botanist in 1897, went to the American Philosophical Society and found the rest of the expedition's specimens in the attic, still as they had been bundled by Lewis.

Meehan's search at the end of the nineteenth century accidentally rejoined the widely traveled plants of Lewis and Clark, but it was not until later in the twentieth century that the society's plants and Tuckerman's gift were reassembled as a single collection again, missing just ten plants that remain in England today. Now more than two centuries after their gathering by Lewis, 222 specimens are housed in controlled environmental conditions in the Academy's herbarium, the botanical treasures from an iconic American journey.

—Rick McCourt
Associate Curator
Department of Botany

—Earle Spamer
Reference Archivist
American Philosophical Society

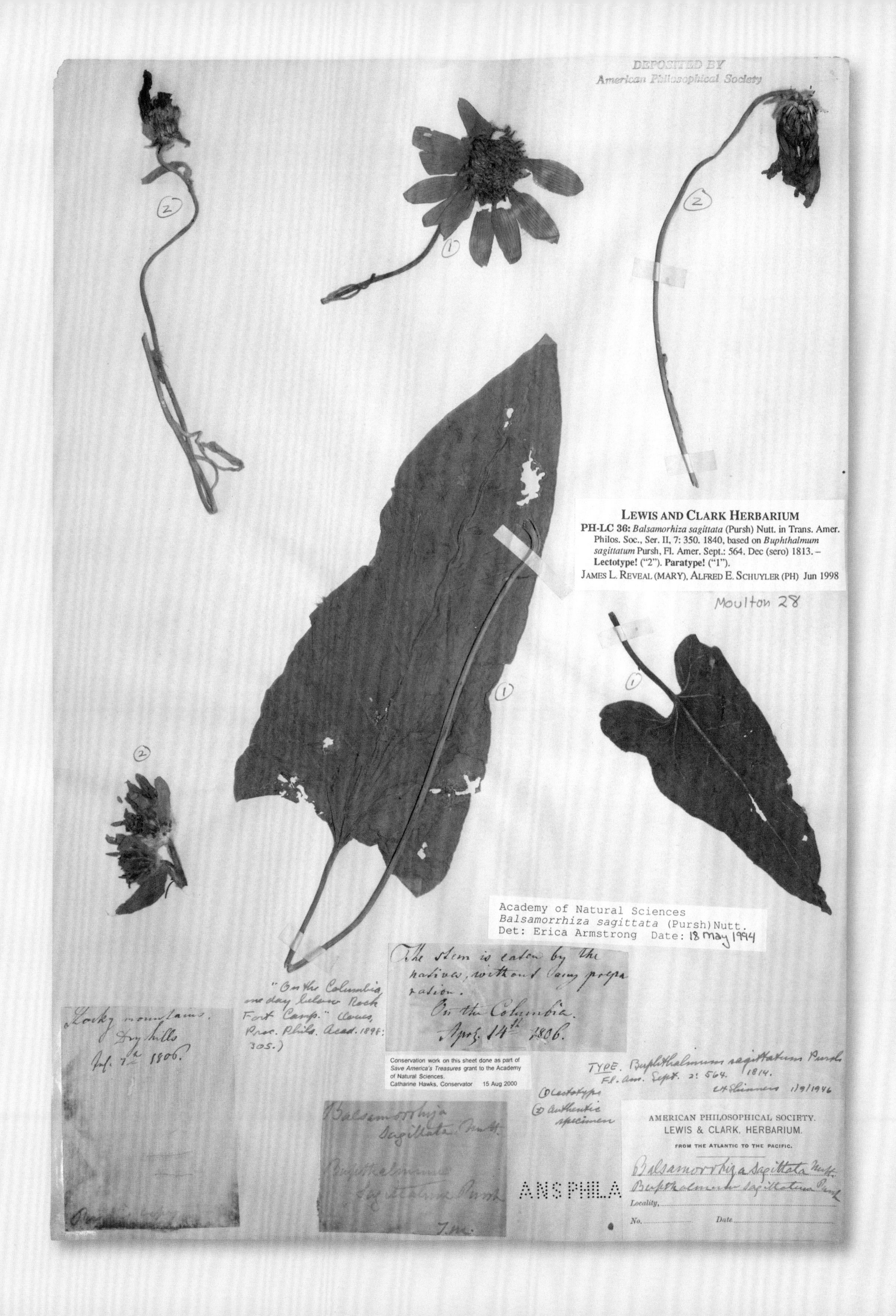

DEPOSITED BY
American Philosophical Society
LEWIS AND CLARK HERBARIUM
PH-LC 36: Balsamorhiza sagittata (Pursh) Nutt. in Trans. Amer. Philos. Soc., Ser. II, 7: 350. 1840, based on Buphthalmum sagittatum Pursh, Fl. Amer. Sept.: 564. Dec (sero) 1813. – Lectotype! ("2"). Paratype! ("1").
JAMES L. REVEAL (MARY), ALFRED E. SCHUYLER (PH) Jun 1998
Moulton 28
Academy of Natural Sciences
Balsamorrhiza sagittata (Pursh)Nutt.
Det: Erica Armstrong Date: 18 May 1994
The stem is eaten by the natives, without any preparation.
On the Columbia.
Aprl. 14th 1806.
"On the Columbia one day below Rock Fort Camp." Coues, Proc. Phila. Acad. 1898: 305.)
Rocky mountains.
Dry hills
Jul. 7th 1806.
Conservation work on this sheet done as part of Save America's Treasures grant to the Academy of Natural Sciences.
Catharine Hawks, Conservator 15 Aug 2000
TYPE. Buphthalmum sagittatum Pursh
Fl. Am. Sept. 2: 564. 1814.
①Lectotype
② authentic specimen
Balsamorrhiza sagittata Nutt.
Buphthalmum sagittatum Pursh
ANSPHILA
AMERICAN PHILOSOPHICAL SOCIETY.
LEWIS & CLARK, HERBARIUM.
FROM THE ATLANTIC TO THE PACIFIC.
Balsamorrhiza sagittata Nutt.
Buphthalmum sagittatum Pursh
Locality,
No.
Date

Both resident and corresponding members (those who did not live in Philadelphia or nearby and therefore could not be active in the Academy's affairs) were proposed at every meeting. They were approved or disapproved by secret ballot (see figure 1.1).

The announcements sent to those elected to corresponding membership employed a flattering rhetoric designed to open the door wide to further correspondence and involvement with the society.

Positive responses were usually prompt and often accompanied by specimens or inscribed copies of the new member's publications. During the Academy's first three years, fifty-one correspondents were elected, eleven of them from foreign countries, and by 1817 the number had reached one hundred.[24] Thus began the building of the Academy membership, with many members of considerable distinction, and of a library that in its field was soon second to none.

Early members included Alexander Wilson, the father of American ornithology; William Bartram (figure 1.8), the country's best-known botanist and Say's great-uncle;[25] and Daniel Henry Drake of Cincinnati, who founded a science museum in that city and employed John James Audubon when Audubon was a struggling artist, yet to publish his great bird paintings, to prepare animal specimens for display. Among European members with strong links to America were the Abbé José Correa da Serra, Portuguese minister to the United States and a naturalist of international reputation; the French botanist François André Michaux, "the Naturalist of the forests" as he is described in the *Minutes*; and another Frenchman, the educator William Phiquepal.

Phiquepal's most important contribution to the Academy's history occurred at the first meeting he attended, in early June 1812, when he proposed William Maclure for membership.[26] A wealthy Scotsman and mineralogist, Maclure became the philanthropic mainstay of the society, enabling it to advance to a position of world prominence in many fields of natural history. At the time of his election, he was in France and would not return to America for four years. Soon after his return, however, he was elected president of the Academy, a post he held from 1817 until his death in 1840.

Born in Ayr, Scotland, in 1763, by middle age Maclure had amassed a fortune through textile manufacture and then turned his considerable abilities to the study of geology, in both Europe and America. He lived in Philadelphia in the late 1790s, became a citizen, and numbered among his friends Thomas Jefferson, Benjamin Henry Latrobe, Joseph Priestley, Benjamin Smith Barton, and other American scholars and scientists.[27] His geological studies in Europe, after his retirement in 1797, secured his European reputation. On his return to America, he made a geological survey of the entire eastern United States, which resulted in a pioneering paper on the subject that was read before the American Philosophical Society in 1809. Among many historians of geology, this work earned him the title "father of American geology."

Maclure benefited the Academy financially by supporting the institution and by funding exploratory journeys to collect specimens, an endeavor that became an Academy hallmark. A precedent for such activities was set in April 1812, several years before Maclure appeared at the institution, when the founders met for "a mineralogical excursion to the mines of Perkiomen," with the treasurer "requested to engage a suitable Carriage at common expense."[28]

Maclure's donations to the society's library were immense. Soon after his election as Academy president in 1817, he gave the new institution a large number of valuable books, including nearly fifteen hundred volumes on subjects covering natural

1.8 *William Bartram* (1739–1823) by Charles Willson Peale, 1808. Oil on canvas. Independence National Historical Park Collection.

history, antiquities, fine arts, travel, and voyages of exploration. Few institutions in America could boast of such an extensive library.[29]

No one benefitted more from Maclure's largesse than Say, who became the father of American entomology (the study of insects) and of American conchology (the study of shells). A few others had preceded him in these disciplines, but he was the first to study a wide range of indigenous insects and shells and to describe them in scientific form. John Bartram, Say's great-grandfather, had not known Latin well enough to describe his insect discoveries, although he was able to puzzle out the descriptions published by others. In 1813, Say read to his Academy colleagues an original essay introducing the science of entomology, in which he "endeavoured to defend it against the aspersions cast upon it by some writers and against the ridicule of the inconsiderate."[30]

Shortly after the Academy's founding, its members prepared a request for the newspapers addressed to "farmers, planters and owners of land throughout the United States," in an effort to procure information and samples of rocks, minerals, soil, and other materials. These samples were to be forwarded to "the house of Messieurs Speakman and Say, Market Street, corner of Second Street in Philadelphia, [where] the said pieces shall be analyzed by able mineralogists free of all expense." The advertisement continued that in this effort to locate and identify those specimens, it "would be beneficial to the country at large to bring forth the resources that God has given us within ourselves to enable us to withstand the deprivation of foreign trade and render ourselves as independent as possible of other countries and governments. . . . Printers throughout the United States, favourable to the interests of Science and the prosperity of our country are invited to give the above advertisement occasional insertion as convenient."[31]

The War of 1812 had helped to foster this patriotic sentiment since the embargo had temporarily canceled exchanges from abroad. The reference to "God" may have been included to refute the charges of atheism that had been leveled against Academy members for excluding religion from their discussions.[32]

Whether or not this solicitation was effective, donations from other sources began to augment the collections given by the earliest members. As soon as the shipping embargo of 1812 was lifted, donations of shells, insects, birds, reptiles, amphibians, and mammals arrived from all parts of the globe. After a letter was received from the Linnaean Society of New England addressed to "sea captains and others relative to the procuring and the mode of preparing articles for the Museum," a three-man committee of the Academy prepared a similar document.[33] This request reaped a rich harvest. The *Minutes* list, among other acquisitions, a box of insects from China, "a small flying-fish" from a Captain Kitty, the skin of a bird of paradise from the East Indies, a group of tarantulas and their nests from Jamaica, a robe made from the skins of diver ducks and edged with ermine from the northwest coast of America, and a Captain Craycroft's collection of shells from the West Indies.[34]

The Academy's growth was such that by 1815 it was running out of space in which to house its collections. Jacob Gilliams, one of the founders, proposed building a hall behind his father's dwelling on Arch Street, between Front and Second Streets. At the meeting of 18 April 1815, "a Committee of Messieurs [William] Strickland, [Reuben] Haines, [Thomas] Say, and [John] Randolph propose that the building for the new Academy accommodations be built as proposed by Mr. Gilliams and rented for $200 per annum." The Academy moved into its new quarters the following August.[35]

In addition to the Academy's collections, correspondence with other scientific institutions around the world drew the society more and more into the mainstream of international science. Mineral specimens arrived from Alexandre Brongniart in Paris, seeds were received from the botanical garden in Calcutta, one hundred specimens of shells were given by M. J. O'Kelly of Ireland, and a request for an exchange of natural history objects arrived from Charles von Schreiber, director of the natural history cabinet of the Austrian emperor.[36]

The Academy now had an impressive international membership, but the most important European contact, aside from William Maclure, was Charles-Alexandre Lesueur, a brilliant French artist and naturalist brought to the United States by Maclure in 1816 (figure 1.9). Born at Le Havre and the son of a naval officer, Lesueur at age twenty-three had joined a three-and-a-half-year scientific expedition sent by Napoleon to explore and chart the coasts of New Holland (Australia) and Van Diemen's Land (Tasmania).[37] On the journey, Lesueur collected many animal species new to science and made hundreds of drawings of these discoveries. With his friend and fellow traveler François Peron, who wrote the descriptions, he planned to publish his findings. But tragedy intervened when Peron died from tuberculosis in 1810. Grief-stricken, Lesueur was unable to continue with their joint project. At this time he met William Maclure, who was then living in Paris. Maclure was planning a trip to the West Indies and enlisted Lesueur's services as artist. The two men reached

1.10 *Thomas Jefferson* (1743–1826) by Thomas Sully, 1821. Oil on canvas. American Philosophical Society. Gift of William Short, 1830.

the United States in the spring of 1816 after having toured the islands of the Caribbean and many eastern states. They settled in Philadelphia that fall. Lesueur was welcomed by the scientific community and elected a member of the Academy of Natural Sciences in December.

The arrival of Lesueur was an unexpected boon to the Academy. Not only did he add greatly to the intellectual and artistic activity of the institution, but his wide circle of friends and correspondents in France helped give the Academy an even more highly visible presence internationally. Lesueur soon proposed several eminent French naturalists as Academy members, all of whom were elected correspondents of the society.[38] Two years later, in 1818, the list of new members sponsored by Lesueur was even more impressive: Baron Georges Cuvier (1769–1832), founder of the studies of comparative anatomy and paleontology; Antoine Laurent de Jussieu (1748–1836), whose *Genera Plantorum* (1789) formed the basis of modern botanical classification; Chevalier Jean-Baptiste Pierre de Lamarck (1744–1829), best known for his distinction of vertebrate from invertebrate animals and his theory of evolution that anticipated the work of Charles Darwin; and Etienne Geoffroy de St. Hilaire (1772–1844), who had been with Napoleon on his expedition to Egypt, where he made numerous natural history discoveries. He became director of the Muséum national d'histoire naturelle in Paris. These scientific luminaries were willing to exchange knowledge and specimens with their American colleagues, which helped to open important European channels for all the Academy's scientists.[39]

By 1817, it was evident that in order to record and disseminate information about the discoveries of its members and promote

1.9 *Charles-Alexandre Lesueur* (1778–1846) by Charles Willson Peale, 1818. Oil on canvas. ANSP Library coll. 286. Lesueur, a French artist and naturalist, is identified as such by the pen he is holding and the bottled eel specimen beside him. The eel attribute reflects the fact that Lesueur was the first naturalist to study marine life of the American Great Lakes.

an exchange of scientific ideas, both in America and abroad, the Academy would need to publish its own journal. Maclure, the greatest champion of this idea, believed so strongly in public education that he bought a printing press, which he set up in his own house, to assist the Academy with its publications. It was probably also Maclure's idea to keep the *Journal* inexpensive so that it could be more widely available. In line with this philosophy, members of the Academy's publishing committee did the printing themselves for the first few years.

The financial assistance of Maclure and the scientific zeal of Say, who chaired the committee, were two of the most important factors in keeping the publication alive in its early days. According to Academy historian Maurice Phillips, "considering the fact that the *Journal* contributed so much to the success and sound establishment of the Academy at this time, the strong support of these two gentlemen may even have been responsible for the actual survival of the Society."[40]

The mandate of the *Journal* was to communicate to the public "such facts and observations as, having appeared interesting to them, are likely to be interesting to other friends of science."[41] Facts and observations were part of the Enlightenment philosophy that characterized early American science. Benjamin Franklin had laid the foundation for this practical approach with his inventions, and Thomas Jefferson had expanded it with his farming innovations, weather charts, experiments in horticulture, interest in fossils, and systematic excavation of an Indian mound on his property, the earliest archaeological dig in the United States that was carefully carried out. In a letter of 1787, Jefferson wrote that he wished that those who explored the West for fossil remains would make "exact descriptions" of what they saw "without forming any theories. The moment a person forms a theory, his imagination sees in every object, only the traits which favor that theory. But it is too early to form theories on those antiquities. We must wait with patience until more facts are collected."[42] When he sent Captains Lewis and Clark on their expedition to the Pacific Ocean, Jefferson hoped that Lewis would find a living mastodon similar to the fossil bones Charles Willson Peale had uncovered in New York in 1801.

Jefferson's years of interest in science had, of course, been inspiring for Academy members. It was thus reason for great pride when, in the spring of 1818, the former president accepted their invitation to become a corresponding member of the society (figure 1.10). Jefferson mentioned that he was "particularly gratified by the perusal of the journal you have been so kind as to send me; in which I find many distinguished papers on subjects of much interest."[43]

The *Journal* was intended to be international in scope, based on the premise that mammals, reptiles, birds, fish, shells, insects, plants, and minerals were all part of a worldwide study and were of significance to scientists everywhere. New World flora and fauna were of special interest abroad as in America. An article on Rocky Mountain sheep, first seen by Lewis and Clark on their famous journey, was particularly pertinent to the Academy.[44] These "white buffaloes," as the Indians called them, were of as much curiosity to Europeans as to Americans because hardly any white men had yet seen them. Academy member George Ord, who had never been west, used Lewis and Clark's specimens deposited in Peale's museum as the basis for his description of these stunning animals. Lewis had died under mysterious circumstances in 1809, only three years after returning from the Pacific, and he had been too occupied as governor of the Missouri Territory to write up his discoveries for publication. At the time, no one was as qualified as Ord to do so. Ord, a "cabinet naturalist" (one who did not go on expeditions), was emerging as a "natural historian" on the scientific scene.

When Say wrote an article describing new species of American mollusks, he dealt with a science as yet unexamined in the New World. He stated in a paper he read to his colleagues in the summer of 1817 that although his list of specimens was not considerable, "it may nevertheless form the commencement of a complete account of our crustaceous animals—a very imperfect one it is true, but it may be considered of some importance, in as much as the errors which may be discovered in it, will, by being corrected by competent naturalists, introduce us to a more perfect knowledge of these curious depurators of the ocean."[45] In speaking of depurators, or agents that purify, Say was an early observer of the importance of these creatures to water quality, an anticipation of an Academy priority in the second half of the twentieth century and into the twenty-first (see chapter 15).

Lesueur provided illustrations for Say's papers, as he did for those of many other writers in the *Journal*. The first volume had twenty-three plates, all by Lesueur. His drawings for this and the following four volumes of the *Journal* gave artistic distinction to the project. Although Lesueur's earliest drawings were engraved on copper, he was experimenting with the new technique of lithography, which he had probably learned in France. The Philadelphia artist Bass Otis is credited with the first lithograph in America, but a magazine article on Otis from July 1818 refers to Lesueur, "whose exquisite designs are well known to men of science and arts here, [and he] has procured some [lithographic] stones, such as are used at Paris[,] . . . and is proceeding with the experiment, we hope successfully; for in truth, it is an

experiment in which the whole circle of science and literature is very much interested."[46]

Lesueur did not publish his lithographs of fish until 1822, but he may have made them much earlier, because his *Journal* article concerning a certain specimen of bass, which he illustrated with lithographs, involved a fish he had caught six years before in Lake Erie.[47] These pictures appear to have been among the first lithographic book illustrations published in the United States (figure 1.11).[48]

During the Academy's early years, so much of the present United States was still unexplored wilderness that exciting new discoveries were made all the time. In a paper submitted to the *Journal* in 1817, the botanist Thomas Nuttall (1786–1859), an Englishman so absorbed in science that, "protected by his own innocence, [he] wandered unharmed from one peril to the next,"[49] braving many harrowing adventures and illnesses to gather unknown flora in large areas of the country, described a plant he thought was a new genus. He expressed the primitive state of geographical knowledge when he said that though the plant was as yet known only from a small part of North America, it "may probably form a numerous genus, whenever the great plains of *California*, the *Columbia*, the *Missouri* and the *Arkansa* shall be explored."[50]

Say's article on the "Hessian fly" and the parasitic insect that feeds on it would be important for American agriculture. After his minute description of this insect in the first number of the *Journal,* he wrote: "This well known destroyer of wheat has received the name of 'Hessian Fly,' in consequence of an erroneous supposition, that it was imported in some straw with the Hessian troops during the revolutionary war. But the truth is, it is absolutely unknown in Europe, and is a species entirely new to the systems—being now for the first time described."[51] This article was of particular interest to Jefferson, who in 1791, under the auspices of the American Philosophical Society, took a journey in the company of James Madison to upstate New York and New England to study the Hessian fly because of the damage it was then doing to wheat crops.[52]

In addition to its domestic audience, the Academy's *Journal* articles were of considerable interest to both institutions and individuals abroad. In London, the Royal Society, the Linnaean Society, and the British Museum received copies, as did the eminent naturalists Sir Joseph Banks and Sir James E. Smith. The Cork Institution, the Dublin Society, and the Royal Irish Academy were among the recipients in Ireland, and in France many leading savants were on the subscription list, Alexandre Brongniart, Georges Cuvier, Anselm Gaetan Desmarest, and

1.11 *Cichla Aenea* (a bass) by Charles-Alexandre Lesueur. This is one of the first lithographs published in America. ANSP *Journal*, 1822, ANSP Library.

Jean-Baptiste Pierre de Lamarck among them. In acknowledging with enthusiasm the receipt of the *Journal*, the French naturalist Bernard Lacépède wrote that the Academy was setting an example by encouraging the study of natural science that would be followed in both North and South America.[53]

George Ord, writing from Paris to Reuben Haines, corresponding secretary of the Academy, to request copies of the *Journal* for the Jardin des Plantes, alluded to the resistance to scientific study in the United States, and for that very reason he stressed the importance of publishing the *Journal*: "Our little Academy, although undervalued by some at home, yet bears a good reputation abroad, particularly in Paris; and all this has been owing to its Journal." Continuing in his somewhat stilted style and throwing in a bit of Shakespeare, Ord concluded, "You may meet, and specify, and confabulate, and elect members, and enlarge your collections, and beautify your domicile, but if you publish not, down goes your character to the tomb of the Capulets."[54]

Description, nomenclature, and classification characterized Academy discussions and *Journal* papers in the early years. Expanding their international collections and exchanging their publications for European articles helped Academy scientists to see minute distinctions in species and better understand their place in nature. The beginning of the century had seen the romantic, poetic pantheism of Alexander Wilson and William Bartram, but by the time of the Academy's founding, and in the subsequent quarter century or so, the systematic studies of men such as Say, Lesueur, Ord, and their contemporaries were of primary importance in piecing together an understanding of New World flora and fauna.

In December 1820, Thomas Nuttall, who had contributed to this process with the publication of his highly acclaimed book *Genera of North American Plants* (1818), reflected on the limits of scientific knowledge at the time in a paper he read to Academy members and later submitted to the *Journal*. In his essay, Nuttall discussed the ancient face of North America, saying that the valley of the Mississippi River showed throughout its extent evidence of a primeval ocean in the fossilized remains of marine creatures, which he had seen during his travels there. He found that the antiquity of these animals, which were apparently created before any other living things, was incomprehensible and observed that what had caused the drawing away of "these mighty waters, and the consequent elevation of the land" was a subject equally mysterious. He concluded that at the time it was sufficient to mark the different stages of the ocean flowing back, "so as to connect our remarks and render them intelligible to those who wish to follow us in the course of observation."[55]

Although some theoretical and speculative articles, like Nuttall's, appeared in the Academy's *Journal*, papers describing and naming new species were its primary focus. Publishing a detailed description of and assigning a name to a species established its place in the structure of nature. Priority was strictly adhered to. In a paper submitted to the *Journal* in 1818, the eccentric naturalist Constantine Samuel Rafinesque (1783–1840), who often rushed his discoveries into print without a thorough review of the existing literature, named as new a species that already had another name. This error was compounded when it was discovered that he had already published his "discovery" elsewhere. Several months later, the *Journal* included a retraction of Rafinesque's article, explaining that multiple names and duplicate publications on the same subject were contrary to the rules of nomenclature and thus ran counter to the Academy's guidelines for publication.[56]

Rafinesque's papers were never accepted again by the Academy, which is unfortunate, because despite his eccentricities —it was once claimed that he had described seven species of lightning—Rafinesque's work is seen by modern scholars to have been touched by genius. The historian Henry Savage Jr. described him as "a tragic figure, too erratic, undisciplined, and unreliable in his scientific work to be fully accepted by the scientific community of his day."[57] Yet A. Hunter Dupree, in his biography of the botanist Asa Gray (1810–1888), states that "some of Rafinesque's views have given him some fame as a precursor of Darwin."[58]

The novelist James Fenimore Cooper (1789–1851) satirized the early nineteenth-century idea of a naturalist in *The Prairie* (1827) with his eccentric character Dr. Obed Battius, who is thought to be a composite of Rafinesque and Nuttall. In the novel, Dr. Battius gives a Latin name to every bird, animal, and flower he encounters. At one point in the story, he is terrified by the sound at dusk of what he thinks is an undescribed bat, and he immediately names the creature *Vespertilio horribilis*, only to find, to his chagrin, that this great unknown is none other than his own mule—*Asinus domesticus*.[59]

In December 1817, Maclure invited Say, Ord, and Titian Peale to accompany him on a sailing excursion to the islands off the coast of Georgia and to Spanish-held Florida. Say may have suggested the eighteen-year-old Peale, since the young man had

1.12 *Crustacea* (crabs) by Charles-Alexandre Lesueur, illustration for Thomas Say's "An Account of the Crustacea of the United States," ANSP *Journal*, 1817. Copperplate. ANSP Archives coll. QH1 A16. Purcell photograph.

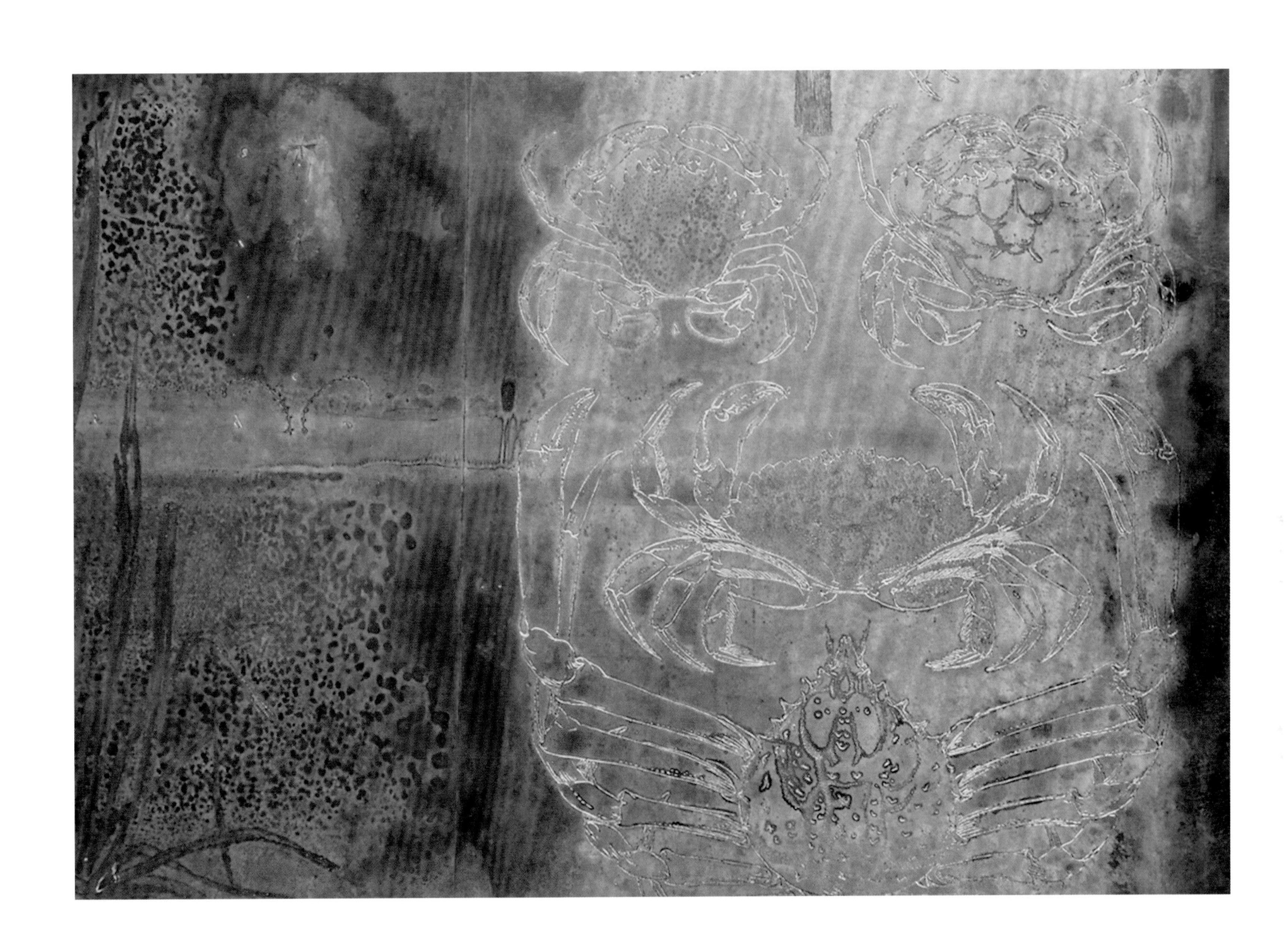

been working for a year on illustrations for the first issue of Say's *American Entomology* (not fully published until 1824). He knew Peale to be well trained in collecting and preserving specimens, skills that would be highly useful during the trip.

Although Maclure had good contacts in Spain, having visited the country on many occasions and owning land there, in the prevailing political climate of the time, his feat of obtaining a letter of safe conduct from the Spanish king, Ferdinand VII, was quite an accomplishment. Relations between Spain and the United States had been strained ever since the British returned Florida to Spain in 1783. When the War of 1812 seemed imminent, the American government had asked Spain's permission to occupy East Florida, to prevent the British from establishing a base of operations there, but because Spain was Britain's European ally, the request was refused.

In the spring of 1812, United States forces seized the Spanish town of Fernandina on Amelia Island, just off the Florida coast, and pressed southward to St. Augustine on the mainland. In the face of Spanish outrage, the American government officially denied its action, but American frontiersmen from border towns in Georgia continued fighting Spanish-incited Indians caught in the crossfire between the two nations. These southern border areas were a hotbed of trouble, and late 1817 was an inauspicious time to launch an expedition. But Maclure had often been present in places of political turmoil—Paris during the French Revolution, for example—and his motives for the excursion were probably as political as they were scientific. Maclure's friend and lifelong correspondent, the American diplomat George William Erving, who had been sent to Spain to resolve the Florida question three years earlier, may have asked for Maclure's assistance in assessing the situation firsthand. Over a year later, in 1819, unable to maintain its hegemony, Spain sold Florida to the United States.[60]

Maclure and Say traveled by carriage to Charleston, South Carolina, where Maclure sold his carriage and horses. They then took a steamboat to Savannah and met Ord and Peale, newly arrived by packet from Philadelphia. Maclure chartered a thirty-ton sloop, and the party soon cast off, accompanied by Maclure's servant, three sailors, and Ord's hunting dogs. The plan was to visit the Sea Islands off Georgia and then travel by water into Florida's interior. Say wrote to Jacob Gilliams, his fellow founder at the Academy, that they expected "to ascend as far as convenient the river St. Johns, pursuing pretty much the track of [William] Bartram my excellent & ingenious relative; but whether or not we shall go further than he did will entirely depend on circumstances."[61] Circumstances referred to the "Indian problem," something William Bartram had not had to contend with during his travels in Florida from 1773 to 1777.[62]

Although Say made many discoveries and collected numerous species of crustacea, sponges, and gorgonia (horny corals), as well as a few fish for Lesueur, he and the others were disappointed that the expedition had to be abridged. He was especially sorry not to have been in Florida during a season auspicious for insects, his principal interest. He was also disgusted with the treatment the Indians were receiving from the United States. "Thus, in consequence of this most cruel & inhuman war that our government is unrighteously & unconstitutionally waging against these poor wretches whom we call savages," he complained to his friend and fellow entomologist John Melsheimer, "our voyage of discovery was rendered abortive, as we were not in Florida at the season we wished, the Spring. We therefore obtained but very few Insects & these few of little consequence—My discoveries were principally in the Crustacea" (figure 1.12).[63] Say read several papers on American crustaceans to Academy members soon after his return. These papers were duly published in the *Journal* over the next few years, as were several others resulting from the Florida expedition. Included were essays by Ord on birds and mammals (figure 1.13) and several on fish by Lesueur, who was working on a book about North American fishes.[64] As the biologist and former president of the Academy, Dr. Thomas Peter Bennett, has written: "The result of the 1817 Florida Expedition was an institutional paradigm for natural history research in the European tradition, formalized in America. The model was collection, study, presentation, publication, and curation. The Academy now made this possible."[65]

PTS

1.13 *Neotoma floridana* (Florida Woodrat) by Charles-Alexandre Lesueur. The specimen was collected on the Academy's Florida expedition of 1818. Illustration for ANSP *Journal*, 1825. Copperplate. ANSP Archives coll. QH1 A16. Purcell photograph.

Notes

1 The first meeting of the Academy of Natural Sciences of Philadelphia was held at the house of John Speakman, with the next two get-togethers at Mercer's Cake Shop on High Street near the corner of Franklin Place.

2 Edward J. Nolan, M.D., *A Short History of the Academy of Natural Sciences of Philadelphia* (Philadelphia: Academy of Natural Sciences, 1909), 6.

3 *Poulson's American Daily Advertiser*, 2 January 1811.

4 Charles Coleman Sellers, *Mr. Peale's Museum* (New York: W. W. Norton, 1980), 90.

5 *Poulson's American Daily Advertiser*, 6 May 1812.

6 For a discussion of the elite in the city's early institutions, see Simon Baatz, "Patronage, Science, and Ideology in an American City: Patrician Philadelphia, 1800–1860," Ph.D. dissertation, University of Pennsylvania, 1986.

7 Patricia Tyson Stroud, *Thomas Say: New World Naturalist* (Philadelphia: University of Pennsylvania Press, 1992), 14.

8 Quoted in ANSP Archives, coll. 502, folder 5, #8.

9 ANSP coll. 502.

10 Ibid.

11 A. Goldstein, "Gerard Troost and His Collection," *Mineralogical Record* 15 (1984): 141–47.

12 Nolan, *Short History*, 8.

13 "Report on the Rise and Progress of the Acad. of Nat. Sciences of Phil by Dr. John Barnes 1816." Coll. 502(A), ANSP Meeting Miscellaneous (1812–1836), file 7.

14 Nolan, *Short History*, 9; and Patsy A. Gerstner, "The Academy of Natural Sciences of Philadelphia: 1812–1850," in *The Pursuit of Knowledge in the Early American Republic: American Scientific and Learned Societies from Colonial Times to the Civil War*, ed. Alexandre Oleson and Sanborn C. Brown (Baltimore: Johns Hopkins University Press, 1976), 176. The Seybert collection in its original cabinet is still at the Academy.

15 Gerstner, *Pursuit of Knowledge*, 177.

16 Ibid.

17 ANSP Archives, coll. 502, "Contribution for Consideration of the Academy of Nat. Sciences of Philad."

18 *Minutes of the Academy of Natural Sciences of Philadelphia*, 17 March 1812, ANSP.

19 Gerstner, *Pursuit of Knowledge*, 175.

20 ANSP Archives, coll. 502, folder 5, "Conversations in the Society." The founders were misinformed. Galileo was not burned alive but forced to recant his views on the Copernican system and condemned to house arrest. They were thinking of Giordano Bruno, whom the Inquisition burned alive for believing that the earth revolves around the sun. *Complete Dictionary of Scientific Biography*, vol. 2 (Detroit: Charles Scribner's Sons, 2008), 539–44.

21 *Poulson's American Daily Advertiser*, 7 April 1812.

22 John C. Greene, *American Science in the Age of Jefferson* (Ames: Iowa State University Press, 1984), 4.

23 Alexander Wilson, *American Ornithology, or the Natural History of Birds of the United States* (Philadelphia: Bradford and Inskeep, 1808), vol. 3, viii, preface.

24 Gerstner, *Pursuit of Knowledge*, 176.

25 Thomas Say's mother, Ann Bonsall Say, was a granddaughter of John Bartram's, thus a niece of William Bartram's.

26 ANSP *Minutes*, 12 June 1812, vol. 1, coll. 502, f. 5.

27 For more on William Maclure, see *William Maclure, The European Journals of William Maclure*, ed. John S. Doskey (Philadelphia: American Philosophical Society, 1988); and Leonard Warren, *Maclure of New Harmony: Scientist, Progressive Educator, Radical Philanthropist* (Bloomington: Indiana University Press, 2009).

28 ANSP *Minutes*, 18 April 1812, vol. 1, coll. 502, f. 5, 58.

29 Samuel George Morton, *Memoir of William Maclure*, read before the ANSP, 1 July 1841, ANSP (Philadelphia: Merrihew and Thompson, 1841).

30 Stroud, *Thomas Say*, 34–36. ANSP *Minutes*, 21 December 1813, vol. 2, 117.

31 ANSP Archives, coll. 502, f. 7, 89A.

32 According to Edward Nolan, Samuel Jackson of the University of Pennsylvania, who had suggested the Academy's title to its members, refused to join the society because he thought its lack of respect for religion would be detrimental to his budding medical career. Nolan, *Short History*, 6.

33 ANSP *Minutes*, 23 July 1816, vol. 2. Founded in 1814, the Linnaean Society of New England was the forerunner of the Boston Society of Natural History.

34 ANSP *Minutes*, 16 August 1814 and 9 September 1817, vol. 2, 158, 380.

35 Ibid., 18 April and 1 August 1815, vol. 2, 226.

36 Ibid., 4 August 1818, vol. 2, 240.

37 The expedition was under the command of Nicholas Baudin (1750–1803).

38 These men included Anselme Desmarest (1784–1838), French zoologist and author; Pierre-André Latreille (1762–1833), French priest and naturalist known for his studies in insect classification; and Henri de Blainville (1777–1850), French zoologist and anatomist who was appointed in 1830 to succeed Jean-Baptiste Lamarck in the chair of natural history at the French Academy of Sciences.

39 Stroud, *Thomas Say*, 42.

40 "A Brief History of Academy Publications," ed. Maurice Phillips, *ANSP Proceedings*, vol. 100 (1948), xiii.

41 Phillips, "Brief History," 2.

42 Thomas Jefferson to Charles Thomson, 20 September 1787, quoted in Silvio A. Bedini, *Thomas Jefferson: Statesman of Science* (New York: Macmillan, 1990), 180.

43 Jefferson to Reuben Haines, Monticello, 10 May 1818, Quaker Collection, Haverford College.

44 George Ord, "Account of a North American Quadruped, Supposed to Belong to the Genus Ovid," *Journal of the ANSP* 1, pt. 1 (1817): 8–13.

45 Thomas Say, "An Account of the Crustacea of the United States. Read Aug. 5, 1817," *Journal of the ANSP* 1, pt. 1 (1817): 57.

46 Joseph Jackson, "Bass Otis, America's First Lithographer," *Pennsylvania Magazine of History and Biography* 37, no. 4 (1913): 385–94. For more on early American lithography, see the forthcoming publication *Philadelphia on Stone: The First Fifty Years of Commercial Lithography in Philadelphia, 1828–1878*, ed. Erika Piola (University Park: Penn State University Press for the Library Company of Philadelphia, 2012).

47 C. A. Lesueur, "Descriptions of the Five New Species of the Genus *Cichla* of Cuvier," *Journal of the ANSP* 2, pt. 2 (1822): 214–21.

48 Jackson, "Bass Otis," 388.

49 Joseph Kastner, *A Species of Eternity* (New York: E. P. Dutton, 1978), 254.

50 Thomas Nuttall, "Observations on the Genus Eriogonum, and the Natural Order Polygoneae of Jussieu," *Journal of the ANSP* 1, pt. 1 (1817): 33.

51 Thomas Say, "Some Account of the Insect Known by the Name of Hessian Fly, and of a Parasitic Insect That Feeds on It," *Journal of the ANSP* 1, pt. 1 (1817): 45.

52 Annette Gordon-Reed, *The Hemingses of Monticello: An American Family* (New York: W. W. Norton, 2008), 463.

53 Correspondence of Reuben Haines, Quaker Collection, Haverford College.

54 George Ord to Reuben Haines, Paris, 23 March 1829, Wyck Papers, American Philosophical Society.

55 Thomas Nuttall, "Observations on the Geological Structure of the Valley of the Mississippi," *Journal of the ANSP* 2, pt. 1 (1821): 34.

56 The original paper, "Description of Three New Genera of Fluvatile Fish," appeared in the *Journal of the ANSP* 1, pt. 2 (1818): 417–22. The retraction appeared in December, in vol. 1, pt. 2, 485–86.

57 Henry Savage Jr., *Discovering America: 1700–1875* (New York: Harper and Row, 1979), 182.

58 A. Hunter Dupree, *Asa Gray* (Cambridge, MA: Belknap Press of Harvard University, 1959), 100.

59 Stroud, *Thomas Say,* 128.

60 Florida passed to the United States beginning in 1819 with the Adams-Onis Treaty of 1819 (also called the Transcontinental Treaty of 1819), and passage concluded in 1821. The United States paid residents' claims against Spain as part of the treaty. These claims included those of not only Florida residents but residents in other areas through the Rockies to the Pacific. (Information given to the author by Dr. Thomas Peter Bennett, former president of the Academy of Natural Sciences.)

61 Say to Jacob Gilliams, St. Mary's, Georgia, 30 January 1818, Say Papers, Manuscript Collection, vol. 7, Historical Society of Pennsylvania.

62 William Bartram, *Travels through North & South Carolina, Georgia, East & West Florida* (Philadelphia, 1791).

63 Say to John Melsheimer, Philadelphia, 10 June 1818, Thomas Say Papers, ANSP.

64 Lesueur's book on fish was never published, but his manuscript is in the Museum of Natural History in his native city of Le Havre, France.

65 Dr. Thomas Peter Bennett, "The 1817 Florida Expedition of The Academy of Natural Sciences," *Proceedings of the Academy of Natural Sciences of Philadelphia*, vol. 152 (October 2002): 14.

TOP Agricultural seed samples collected by Charles F. Kuenne, 1948. ANSP Botany Department. Purcell photograph.

BOTTOM Volcanic ash specimens from the Adam Seybert Mineral Collection, 1812. ANSP Mineralogy Department. Purcell photograph.

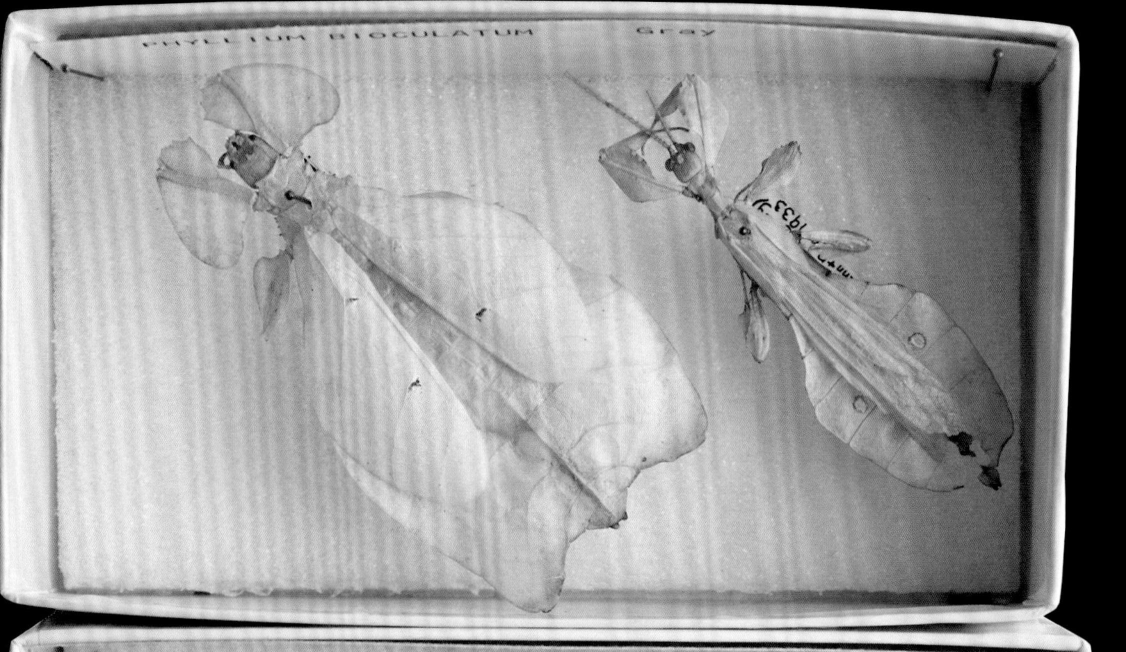

Leaf insects" (a lineage of tropical walking sticks). These remarkable Phasmida are found in rainforest canopies of tropical Asia. Included n this group are some newly described specimens from the Philippines. The others are from New Guinea and the Seychelles. ANSP Entomology Department. Purcell photograph.

Chapter 2

The Lure of the West: Exploration and Exodus

We cannot but hope . . . that the time will arrive, when we shall no longer be indebted to the men of foreign countries for a knowledge of any of the products of our own soil, or for our opinions in science.
—Thomas Say in *Account of an Expedition*, 1823

Although only seven years old and much smaller in size, the Academy of Natural Sciences had by 1819 established a reputation on a par with the American Philosophical Society, the prestigious institution considered the American equivalent of the Royal Society of London. Thus, when the U.S. government planned to organize a small team of scientists to accompany a one-thousand-man military expedition to the West, Secretary of War John C. Calhoun turned for advice not only to the Philosophical Society but also to the Academy. The primary purpose of the expedition was to establish a U.S. presence in and thus secure the upper Mississippi valley frontier against British and French fur traders who were making profitable alliances with the native populations.

Although the military component of the expedition was canceled because of a financial crisis brought on by the Panic of 1819, Congress agreed that the expedition's scientific contingent should proceed as planned. Secretary Calhoun selected Major Stephen Harriman Long (1784–1864) of the Corps of Topographical Engineers to lead the party. Long was a skilled explorer and cartographer who had taught mathematics at West Point and had led a previous expedition to the West. A corresponding member of the Academy, Major Long turned to his colleagues in Philadelphia to serve as the expedition's personnel. He enlisted Thomas Say as the principal zoologist for the expedition because at the time Say was considered the most brilliant zoologist in the country.[1] Say's assignment was to document the flora and fauna of the area and to study the culture and habits of the Indians whom the exploring party would encounter, in addition to compiling Indian vocabularies. Titian Peale, an Academy curator, was employed as Say's assistant, to help collect specimens and prepare them. He was also to create a visual record of the expedition, a task for which he had been well trained by his artist father, Charles Willson Peale. Samuel Seymour, a British watercolorist and associate of Thomas Birch, Thomas Sully, and other well-known artists in Philadelphia, although not himself an Academy member, was recruited to paint geological features, "landskips," important meetings with the Indians, and Indian portraits (figure 2.2). William Baldwin, a young Quaker doctor, was to act as surgeon and botanist, and Augustus Jessup as geologist.

The expedition's assignment, to explore the trans-Mississippi West, would give these Academy members an opportunity to forge new ground in American natural science. By focusing entirely on natural history, the party hoped to accomplish even more than Lewis and Clark had on their expedition sixteen years before.

The party was to explore the expanse of land from the Missouri River to the sources of the South Platte, Arkansas, Red, and Canadian Rivers in present-day Colorado and New Mexico. The

2.1 Thomas Say was a pioneer in the field of malacology at the Academy, where mollusks remain a major focus of research today. These cone shells were collected for the museum by A. J. Ostheimer III in Tanzania, Dutch New Guinea, and the Palau Islands during the 1950s. With more than ten million specimens, the Academy's shell collection is one of the three largest in the world. ANSP Malacology Department 212603, 207187, and 200387. Purcell photograph.

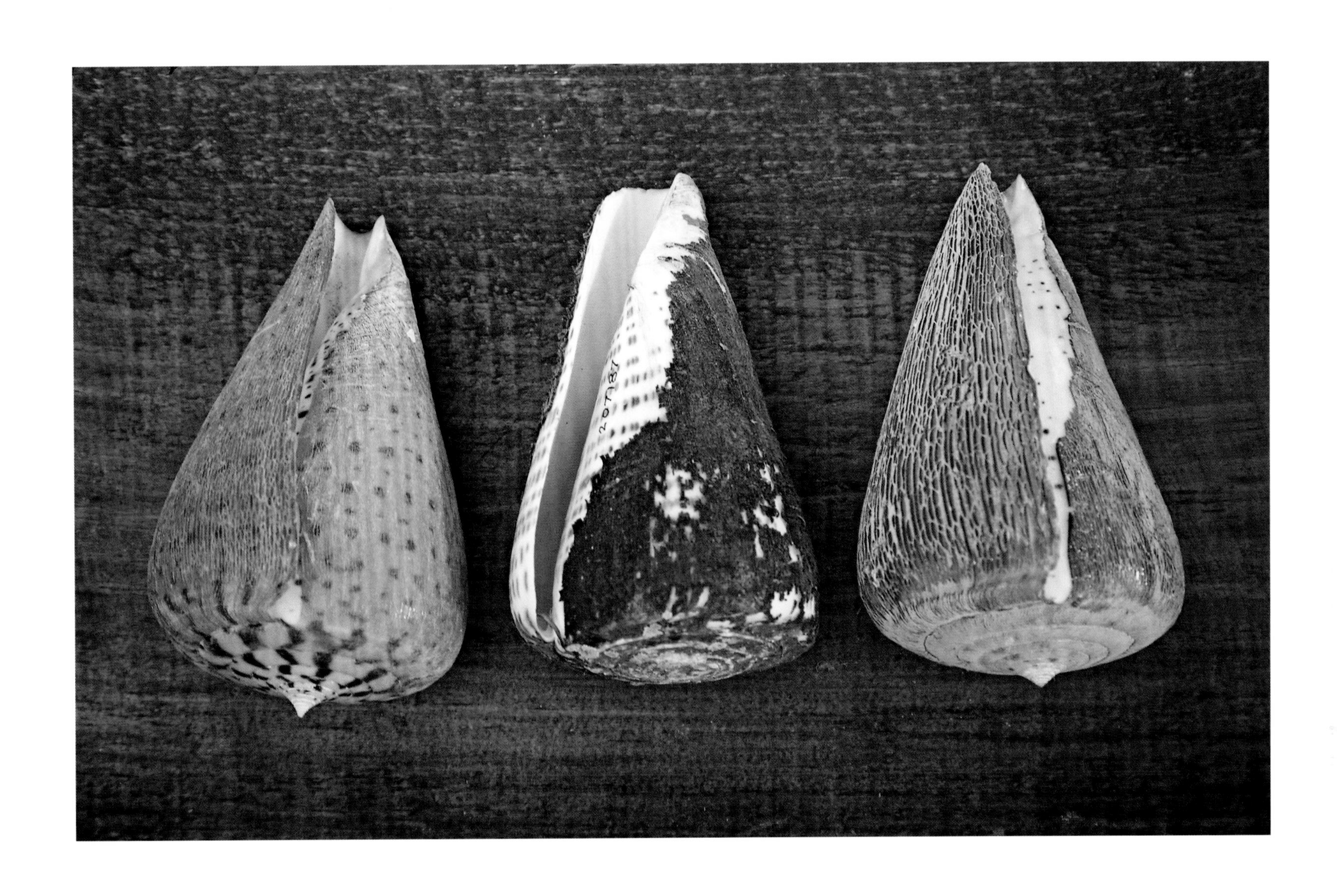

Red River was especially important because it marked part of the disputed boundary of the Louisiana Purchase. Undeniably, the journey would be dangerous and difficult. "The continent was covered by penumbras," according to the historian Daniel J. Boorstin, "between the known and the unknown, between fact and myth, between present and future, between native and alien, between good and evil."[2]

Unfortunately, Baldwin, who suffered from a pulmonary disease, probably tuberculosis, died in Franklin, Missouri, early in the trip, and Jessup resigned for undisclosed reasons during the winter encampment at Council Bluffs on the Missouri. Edwin James, an energetic young Vermont native, took both their places, acting as botanist, geologist, and surgeon. James, who had studied medicine after graduating from Middlebury College, proved worthy of his many assignments and was elected to Academy membership in 1823. He must also be credited as the first white man to reach the summit of Pike's Peak after four arduous days of climbing, including a night spent just below the snow line without food or blankets. Though Zebulon Pike recorded seeing the mountain twelve years earlier, he had not climbed it, so Long named it after James. However, since it was closely associated with Pike's discovery, it has always been known as Pike's Peak.[3] James's name may not be attached to a mountain, but his written account of the Long Expedition to the Rocky Mountains has become a minor classic of American expeditionary literature.[4]

During the winter before the explorers left Philadelphia, Charles Willson Peale painted portraits of Long, Say, and Peale's son Titian, commenting dryly that "if they did honour to themselves in that hazardly expedition that they might have the honour of being placed in the Museum and if they lost their skalps, their friends would be glad to have their portraits"(figures 2.3, 2.4).[5] Say's portrait hangs today in the Academy's library, Long's is at Independence National Historical Park in Philadelphia, and Peale's is in the Philadelphia Museum of Art.

The party left Pittsburgh on 4 May 1819 aboard the *Western Engineer,* a colorful early steamboat constructed to draw only a small amount of water in order to navigate the difficult waters of the Ohio, Mississippi, and Missouri Rivers, where sunken trees and branches were hazardous to all forms of navigation (figure 2.5). The design included a "bulletproof" house for the helmsman, extensive lockers for storing food, guns, preserving apparatus, and Indian presents, and an extensive library of

2.2 *Oto Council* by Samuel Seymour, 1819. Watercolor. ANSP Archives coll. 79. Thomas Say, with his profile accented by heavy dark sideburns, sits erect to the left of Major Long.

exploration and natural science literature. Peale noted in his journal that "on the [boat's] right hand wheel is [painted] James Monroe in capitals, and on the left J.C. Calhoun, they being the two propelling powers of the Expedition."[6] With its figurehead a huge serpent spewing steam from enormous nostrils and its stern wheels lashing the waves, the boat was designed to impress the Indians with the power of the U.S. government. But the impression appears to have had a negative impact that was not the designer's intent. A local resident along the Missouri was quoted as saying, "As the steam escapes from [the boat's] mouth, it runs out a long tongue to the perfect horror of all Indians that see her. They say, 'White man bad man, keep a great spirit chained and build a fire under it to make it work a boat.' "[7]

As they left the Pittsburgh waterfront, the "gentlemen of science," as other expedition personnel referred to the Academy members on board, were resplendent in special brass-buttoned uniforms designed exclusively for them. At Cincinnati, they visited Dr. Daniel Drake, one of the earliest corresponding members of the Academy, who had been instrumental in establishing a science museum in that Ohio town. Here the party also met John James Audubon, hired by Drake to prepare animal specimens for display. Audubon showed the explorers his portfolio and later recalled "how [Major Long], Messrs. T. Peale, Thomas Say and others stared at my drawings of birds at the time."[8] A few years later, these three would have a chance to see Audubon's elegant pictures again, this time in Philadelphia, when the artist came to the Academy of Natural Sciences in an attempt to find backing for his proposed book on the birds of North America.

At the end of 1820, the Long Expedition returned to Philadelphia, where the explorers regaled their Academy friends with tales of their western adventures. They also described the numerous natural history specimens they had collected, detailed accounts of which were published in the Academy's *Journal.* Many of the insects were described and illustrated in Say's pioneering book, *American Entomology, or Descriptions of the Insects of North America* (1824–28). Most of the zoological and ethnographic specimens brought back by the expedition were deposited in Peale's museum by order of the government, just as Lewis and Clark's specimens had been before them. Peale had repeatedly petitioned the federal government to take over and

2.3 *Thomas Say* (1787–1834) by Charles Willson Peale, 1819. Oil on canvas. ANSP Library coll. 286. Say is shown in the uniform specially made for the scientists on the first Long Expedition to the Rocky Mountains (1819–20).

2.4 *Titian Ramsay Peale* (1799–1885) by Charles Willson Peale, 1819. Oil on canvas. Philadelphia Museum of Art, bequest of Robert L. McNeil, Jr. Peale was an artist, hunter, and assistant naturalist for the Long Expedition (1819–20).

2.5 *Western Engineer* by Titian Ramsay Peale, 1819. Pencil sketch. American Philosophical Society. The boat was designed for the Long Expedition to the Rocky Mountains (1819–20).

run his museum as a national institution, but he had been as consistently turned down. Nevertheless, his museum was the obvious repository for discoveries made on a government-sponsored expedition. Although the Academy had been incorporated on 24 March 1817 so that it was by then a verifiable institution, it did not have the longtime standing of Peale's Philadelphia Museum nor the space needed for large collections or public viewing.

Although the Academy was not permitted to take charge of the collections, Say made scientific history by technically describing forty-three new animals from the expedition. His "discoveries" included the coyote, the swift fox, and the plains gray wolf, as well as several species of birds, reptiles, rats, squirrels, shrews, bats, insects, and land snails previously unknown to naturalists (figure 2.6). The historian Bernard Jaffe has written that Say "helped not only to illuminate many of the hidden biological corners of the continent, but also to pile up information which helped in the creation of a new and fundamental biological synthesis."[9]

The work of preparing a record documenting the zoology of the country for the War Department fell to Edwin James, assisted by Long, but especially to Say, who wrote much of the report. Aware that he was now in the vanguard of establishing America's own natural science, Say was devoted to proving its credibility. As had Thomas Jefferson before him, he strongly believed that it was time for American naturalists to name and describe their endemic flora and fauna.

By the time the Long Expedition's explorers returned in the fall of 1820, the Academy had been in new quarters for three years, a rented hall newly built in the vacant lot behind a house belonging to Jacob Gilliams's father (figure 2.7). Presumably, the architect William Strickland had designed the building, since he had been on the committee that oversaw its construction. Not only was this building used for the Academy's meetings and to house its collections, but the members also utilized it to reach out to the community. In the spring of 1820, Thomas Nuttall was granted permission to deliver his course of botanical lectures in the Academy's hall. These were not the first public lectures offered by the Academy. As early as 1814, Dr. Gerard Troost had spoken on mineralogy and crystallography, while Drs. Daniel Barnes and Benjamin Waterhouse had given courses in botany.

These last were so successful that they were repeated the following year and attended by "upwards of two hundred ladies" and "a considerable number of gentlemen."[10] So popular were these programs, it had been necessary to rent a separate building to accommodate such large gatherings.

Just as the various scientific institutions around the world were connected by correspondence and the exchange of specimens, the naturalists associated with them were also linked in interesting ways. The Abbé José Francisco Corréa da Serra (1751–1823),[11] a learned and distinguished Portuguese historian, statesman, and botanist with an international reputation who served for many years as his country's ambassador to the United States, presented the Academy with a slice of the trunk of a *Maclura aurantiaca*, the Osage orange (figure 2.8). Thomas Nuttall, who had first seen the tree in 1810 in the St. Louis garden where the *aurantiaca* had been raised from seeds from Osage Indian territory given to Meriwether Lewis, described it for science and named it in honor of William Maclure, a patron of the botanist's western journey and president of the Academy.[12] Corréa da Serra had also subsidized Nuttall's expedition. In 1821, as a result of this trip, Nuttall published his influential *Journey of Travels into the Arkansas Territory*.

Because of Maclure's largesse to the Academy, its library included many important books: Aldrovandi's *Monstrorum*

2.6 *Swift Fox* from John James Audubon, *The Viviparous Quadrupeds of North America* (1846–54). ANSP Library.

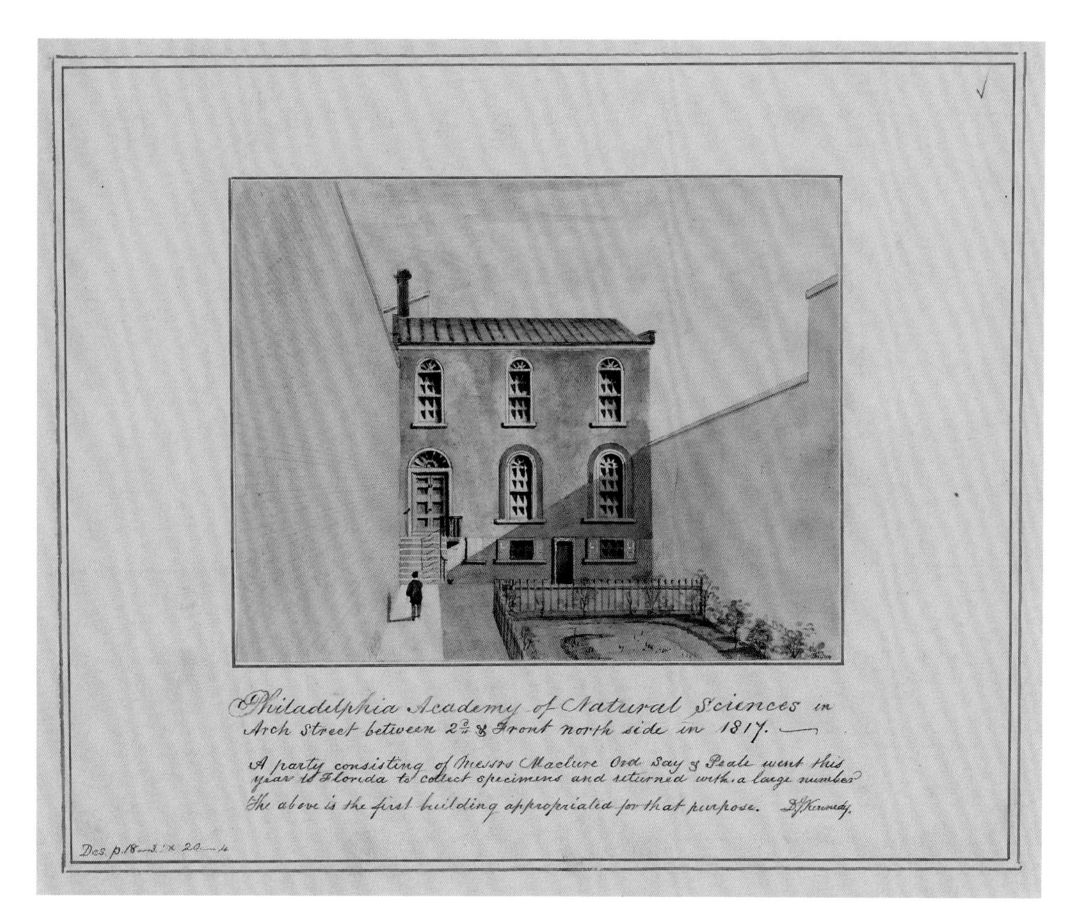

2.7 *The ANSP in 1817* (Gilliams Hall) by David Kennedy. Watercolor. Historical Society of Pennsylvania.

historia (1642) with its pictures of fantastic monsters (figure 2.9), the thirty-nine volumes of Buffon's *Histoire naturelle et particulière avec la description du cabinet du roy* (1749), and ten volumes of Buffon's *Histoire naturelle des oiseaux* (1770). Among many others were nine different publications by Linnaeus; André Michaux's *Flora boreali-americana* (1803); Redouté's *Liliaces* (1802), the artist's paintings of the flowers in the Empress Josephine's garden at Malmaison outside Paris; and on minerals, medicine, antiquities, agriculture, and taxidermy.[13] Maclure was not only president of the institution but also its principal financial supporter. He would continue in both capacities until his death in 1840 in Mexico, where he had gone to live because of his health.

At this time, many eminent scientists from around the world were being elected corresponding members at every Academy meeting: Gaspard, Comte de Sternberg, from Prague, Bohemia; Charles Joseph van Hoorebeke of Ghent, Belgium; "two gentlemen from Peru"; a Dr. James Leighton Jr. from St. Petersburg, Russia; N. W. Almroth, professor of chemistry in Stockholm, Sweden; and most important, the eminent Dr. William Henry Hooker, professor of botany from Glasgow, Scotland.[14]

As for Philadelphia, Dr. John Davidson Godman, a dedicated naturalist, married to Rembrandt Peale's daughter Angelica, was elected a resident member that year (figure 2.10).[15] Godman's *American Natural History*, published in 1828, was based almost entirely on specimens in Charles Willson Peale's Philadelphia Museum, many of which Say had brought back from the West. Say's *American Entomology* was also based on Peale's insects, because microscopic insect predators had destroyed his own collection while he was away.

Partly compensating for the loss of Say's specimens was the continuing arrival of other collections from at home and abroad. The explorer Henry Schoolcraft donated "a handsome collection of shells from Lake Michigan"; the banker Nicholas Biddle, compiler of the Lewis and Clark journals, presented examples of silver and copper from mines in Mexico; and Baron von Struve of Hamburg sent thirty-five mineral specimens from Norway, Sweden, and Germany. This fruitful exchange worked the other way as well. In 1822, Academy members sent a collection of minerals to the Academy of Sciences in Turin, Italy, including samples of anthracite from the Lehigh, the Schuylkill, and the Susquehanna Rivers in Pennsylvania.[16]

FACING PAGE 2.8 *Maclura aurantiaca* (Osage Orange) from Thomas Nuttall, *The North American Sylva*, 1859. ANSP Library.

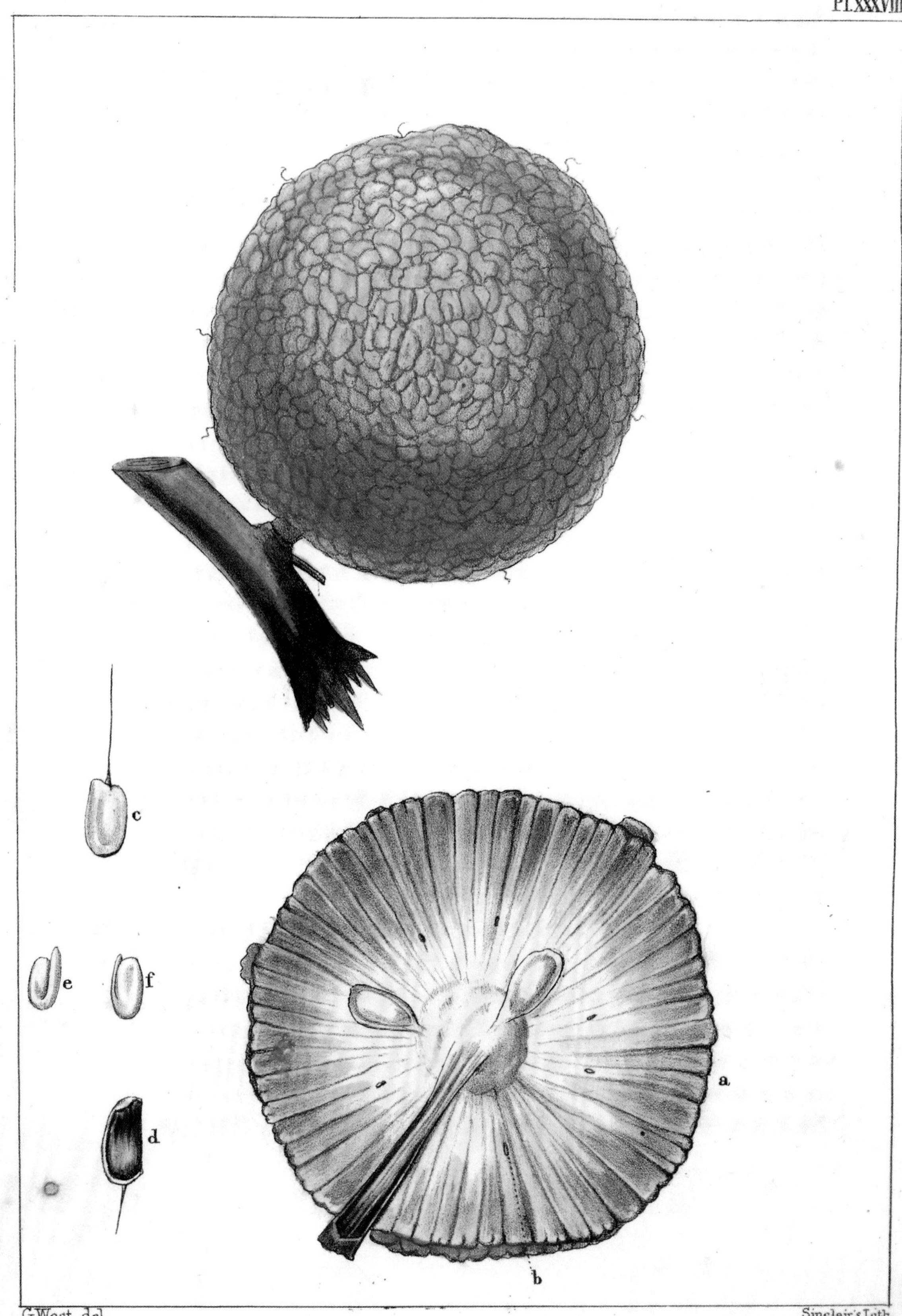

G. West del.

Sinclair's Lith.

Maclura Aurantiaca.

Osage Orange. *Bois d'Arc.*

2.9 *Monstrorum historia* by Ulisse Aldrovandi. Bononiae, Typis Nicolai Tebaldini, 1642. ANSP Library.

In the spring of 1823,[17] Say was again engaged as zoologist and antiquary (paleontologist) for an expedition with Major Long, this time to the northern bend of the Missouri to establish the boundary of the United States on the 49th parallel. The expedition's principal purpose was to stop hostilities between American and British traders, much the same object as the aborted military expedition of 1819.

Secretary Calhoun had ordered Long to travel from Philadelphia to Lake Superior, during which trip he would survey the intervening territory, describe the plant and animal life, and study Indian customs. Because Edwin James missed connecting with the expedition and had to turn back, Say added botanical collecting to his other duties. William H. Keating (1799–1840), the twenty-four-year-old recording secretary of the Academy, signed on as mineralogist and geologist. This talented young man had studied at the School of Mines in Paris and held a master's degree from the University of Pennsylvania, where he had been teaching. Two years earlier, he had published a book on mining that was said to be the first scientific work written on the subject by an American. Keating would also act as the expedition's journalist, compiling the account after the men returned in late October.[18]

When the official narrative was published the following year, the *North American Review* in Boston praised the expedition's accomplishments while deriding the government for not providing sufficient funds to achieve all its goals. The *Review* mentioned Say's 150-page appendix on natural history, "the merits of which we do not pretend to speak, being fully convinced that no better pledge of its value can be desired by the public, than the name of the author."[19]

Although Say collected mainly insects on this expedition, he also brought back a group of plants that Long gave Thomas Nuttall for identification and description. Nuttall completed descriptions of only five species before he set sail for Liverpool, saying he would continue his work in the spring. But when nothing was heard from him by the first of July, Keating turned the herbarium specimens over to the noted botanist Lewis David von Schweinitz, a Moravian minister from Bethlehem, Pennsylvania, and a corresponding member of the Academy, who published them in the appendix to the expedition report.[20]

The birds that Say had brought back from the first Long Expedition were still undescribed for science, as there was no one at the time to take on the task. But in the winter of 1824, the collection was turned over to Prince Charles-Lucien Bonaparte for study (figure 2.11). Bonaparte had come to America to visit his uncle and father-in-law, Joseph Bonaparte, Napoleon's older brother, who was living in exile on his eighteen-hundred-acre estate near Bordentown, New Jersey.[21] Bonaparte's scientific knowledge was made known to the Academy's members when Say read at a meeting the prince's learned essay on the storm petrels he had observed during his "protracted voyage across the Atlantic." This article was subsequently published in the *Journal*,[22] and Bonaparte was elected an Academy member in February 1824.

The charismatic young prince was a brilliant systematist but entirely a "cabinet" naturalist. He did not go on expeditions as had his mentor Say but instead devoted his entire time to classification and nomenclature from the comfort of home. His first ambitious project was to correct the nomenclature in Alexander Wilson's *American Ornithology; or, the Natural History of Birds*

Inhabiting the United States (1808–13), the first book on indigenous birds published in America. In a long article for the *Journal,* spread out over many issues, Bonaparte sorted out females and young birds that had been identified as separate species and clarified the confusion over the summer and winter "dress" of certain birds.[23] His study led directly to the book he published in four volumes (1825–33), continuing Wilson's *Ornithology*. In it, he incorporated descriptions of some of the birds brought back from the west by Say, such as the burrowing owl (*Speotyto cunicularia*), which lives in the ground in prairie dog holes; the band-tailed pigeon (*Columba fasciata*); and the western phoebe (*Sayornis saya*), which he named "Say's Phoebe" in honor of his friend.[24]

In his preface, Bonaparte praised Wilson for his book—his "vast enterprise"—and said that his own "undertaking was not precipitately decided on, nor until the author had well ascertained that no one else was willing to engage in the work." He said that he was "aware of his inability to portray the history and habits of birds in a style equal to that of his distinguished predecessor, principally because he does not write in his own language." Further, he thanked Say and John Godman "for the care they have bestowed in preventing the introduction of foreign expressions, or phrases not idiomatic, into my composition."[25] The publication was quite an achievement for a twenty-two-year-old Frenchman, only two years in the United States, who had spent the early part of his life in Italy and for whom English was a third language.

Bonaparte was treading on dangerous ground correcting Wilson at all, because he was doing so under the critical gaze of Wilson's closest friend and champion, George Ord, then a vice president of the Academy. Ord had finished the ninth and last volume of Wilson's *Ornithology,* was the executor of his estate, and was fiercely defensive of his friend's reputation. That June, after Bonaparte began his series of essays continuing and correcting Wilson's work, Ord wrote to tell him that "there was present [at an Academy meeting] an Englishman recently *come over* [possibly his friend Charles Waterton], who cannot see the propriety of a Frenchman's attempting, or presuming, to meddle

2.10 *John Davidson Godman* (1794–1830) by Rembrandt Peale. Oil on canvas. Private collection.

2.11 *Charles-Lucien Bonaparte* (1803–1857) by T. H. Maguire, 1849. Stipple engraving. ANSP Archives coll. 457. "He was a very handsome man in the Bonaparte style," recalled Malvina Lawson, who hand-colored much of Bonaparte's book on American birds.

Lewis David von Schweinitz

Father of American Mycology and More

Lewis David von Schweinitz (1780–1834), Moravian clergyman and botanist, is generally considered the father of North American mycology. In the early nineteenth century, he described more than a thousand new species of fungi.

Born in Bethlehem, Pennsylvania, Schweinitz was the great-grandson of Nicholas Lewis Count Zinzendorf, the founder of the Moravian Brethren. He was educated at Nazareth Hall, in the Moravian community near Bethlehem, where he learned botany and drawing.

In 1798 he entered the Moravian Theological Seminary in Niesky, near the Polish border in what is now Germany. While there, he drew and painted thousands of fungi from the surrounding countryside. Many of these original watercolors are preserved in five volumes of his unpublished "Fungorum Nieskiensium Icones," four of which are housed in the Academy's archives. Schweinitz and his seminarian mentor, Johannes Baptista von Albertini, published *Conspectus fungorum in Lusatiae superioris agro Niskiensi crescentium* in 1805. This classic work on fungi earned Schweinitz an honorary doctorate from the University of Kiel.

Upon his return to the United States in 1812, Schweinitz continued working on fungi while holding church positions in Salem, North Carolina, and Bethlehem, Pennsylvania. He published *Synopsis fungorum in Carolinae superioris* in 1822 and *Synopsis fungorum in America Boreali* in 1832—major works that formed the foundation of North American mycology. He also worked with flowering plants and published papers describing new species of violets and sedges. He was elected a corresponding member of the Academy of Natural Sciences in 1822.

Schweinitz assembled thousands of his own mycological and botanical collections, along with those of correspondents from diverse parts of the world, into what was then the largest herbarium in the New World. Estimated to contain twenty-three thousand species, it became the foundation of the Academy's herbarium after his death in 1834.

—*Alfred E. Schuyler*
Curator Emeritus
Department of Botany

254.

255.

256.

with the nomenclature of Wilson's Ornithology." He added, with a sarcasm not unusual for Ord, "I should suppose, that so long as that *Frenchman* does not presume to review Wilson's poetry [Wilson was also a poet], our fastidious critic (who by the by represents himself as an ornithologist) would be satisfied."[26] Bonaparte would not be the only one criticized by Ord.

In 1825, a major schism occurred at the institution that would disrupt and alter the Academy for many years. Robert Owen, a wealthy Welsh reformer who had just bought an entire town on the frontier in Indiana and sought to establish a "new moral world" there, sowed the seeds of change. It was his involvement with William Maclure that drew in the Academy of Natural Sciences. All that winter and into the summer, Academy members, along with the Philadelphia intelligentsia in general, hotly debated the ideas of this charismatic man and the utopian experiment he was launching in the west.

Owen, early in life, had made a fortune in the cotton-spinning industry. An innovator, he had been the first industrialist in Great Britain to import American Sea Island cotton grown on the islands off South Carolina and Georgia. In 1800, he took over a large mill in New Lanark, Scotland, which he bought with several partners. It was the beginning of the industrial revolution, and factory conditions all over Britain were appalling. At New Lanark, Owen pioneered reforms to ease his workers' long hours and squalid living conditions. Although he had benefited from his country's economic growth, he nevertheless saw the means as too exacting to justify the results. Over the next twenty-five years, Owen developed his ideas of a utopian community where people would share everything they had and the institutions of religion and marriage would be abolished—a proposition put forward earlier by the influential English reformer William Godwin. Owen's plan was decidedly unpopular in Great Britain, so he turned to the New World. An opportunity for the practical application of his concepts occurred when the German evangelist George Rapp offered for sale the entire town of Harmonie (later changed to New Harmony), Indiana, which he had recently vacated with his followers to set up another establishment he called "Economie," outside Pittsburgh.[27] Owen bought the town and surrounding twenty thousand acres outright.[28]

He then spent two winter months in 1825 promoting his plan in Washington, even delivering two speeches to a large crowd in the Capitol's Hall of Representatives. President Monroe and the newly elected John Quincy Adams were in the audience for Owen's second speech, along with many members of the cabinet, Supreme Court, and Congress.[29]

On arriving in Philadelphia in late March, Owen wrote to Maclure, who was still overseas: "I am surrounded by your friends here & we have had much conversation respecting you & your charitable objects. The result of which is a great desire on the part of all of them to see you here & to have your direction in various important matters which they have before them."[30] These important matters included whether to join Owen's enterprise and especially to ascertain if Maclure would be willing to become Owen's partner and provide financial backing for those who considered going to New Harmony (figure 2.12).

By the summer, Owen's ideas and plans were all the rage. The Quaker Reuben Haines, corresponding secretary of the Academy, wrote to his vacationing cousin Ann: "As I know thee likes a little excitement, I may perhaps as well let thee know how much thee has lost by not accepting my invitation to return home last 7th day. Our breakfast company this morning were Mr. Owen, Dr. [William] Price,[31] Mr. Phiquepal[32] . . . T. Say." Had Ann been present, he said, she could have taken part with Owen in a discussion of almost three hours. He added that he and Say planned to accompany Owen on a steamboat to Point Breeze,[33] the estate of Joseph Bonaparte, where Joseph's nephew, the ornithologist Charles-Lucien, was living. A member of Owen's entourage recorded in his journal that Joseph Bonaparte "had been most anxious to make [Owen's] acquaintance, & sent his carriage down to the landing place on the banks of the Delaware to receive him."[34]

Throughout the first quarter of the nineteenth century, the life of William Maclure paralleled that of Owen. Maclure was also a successful businessman grown wealthy by middle age. Imbued with ideas of social change fostered by his sympathy with the French Revolution, Maclure believed that only through education could life be improved for the masses. Convinced that the old "monkish system" of classical education had been useless

2.12 *View of New Harmony* by Karl Bodmer, 1832. Watercolor. Joslyn Art Museum, Omaha, Nebraska.

in his mercantile life, he saw no reason to provide it for the common man; scientific knowledge would be more beneficial. Combining this theory with the advanced educational reforms of the Swiss educator Johann Heinrich Pestalozzi, who incorporated physical labor in his regimen, Maclure had already put his concepts into practice in a small school he started in Philadelphia. At first he was dubious about Owen and his New Harmony plans, but he was encouraged to change his mind by the enthusiasm of Mme. Marie Louise Duclos Fretageot, a Pestalozzian teacher he had brought to Philadelphia. Three Academy founders—Gerard Troost, Say, and John Speakman—were also interested in the experiment.

While he was still mulling over the pros and cons of Owen's enterprise, Maclure continued to pursue his love of geological exploration. In late August 1825, he hired a coach and gig and arranged an extensive excursion to examine important mineralogical sites in Pennsylvania and New Jersey. Coal, particularly the recently discovered anthracite (hard coal), was of special interest because of its increasing importance and value as a replacement for wood in heating American buildings. Accompanying him were Academy members Say, Lesueur, Troost, Reuben Haines, James Carmalt, and Dr. William Price. Haines recounted their adventures in letters to his wife, while Lesueur made nearly one hundred sketches of the towns and sites they visited. Among these were Valley Forge, Pottstown, Lebanon, Mauch Chunk (now Jim Thorpe), the Lehigh and Delaware Water Gaps, Nazareth, and Easton in Pennsylvania and Franklin Furnace and Schooley's Mountain in New Jersey.[35]

"We travel leisurely," Haines wrote his wife. "Maclure is willing to stop just where we please—a part [of the group] walk on in advance, Lesueur with his gun and Say with flycatchers [nets]. Lesueur sketches all the interesting views . . . Maclure and Troost examine the rocks, I attend to the Taxedermy and botanys."[36] He said that Maclure thought that no coal deposits anywhere else in the world were comparable to those at Mauch Chunk, and Troost declared he never expected to see such a mass of coal and such an extensive mine as long as he lived.[37] Seven weeks later, the scientists returned to the Academy, loaded with specimens for further study.

Owen returned to Philadelphia in early November and on his first evening back conferred with Maclure, Say, Haines, and other members of the Academy. Maclure apparently made the decision then to join Owen and his enterprise at New Harmony. Haines was immensely enthusiastic, but his wife seems not to have been interested in such a venture. A family friend, Deborah Norris Logan, wrote in her diary in mid-November: "In the

2.13 *The Founders* by John Chappellsmith, 1827. Indiana State Museum and Historic Sites. Silhouette of the principals involved in the New Harmony experiment. From left: Robert Owen, William Maclure, Thomas Say, Charles-Alexandre Lesueur, and Madame Marie Louise Duclos Fretageot. The banner expresses one of Owen's precepts, "Man Makes His Own Existence."

afternoon I went . . . to drink tea with Jane Haines, and paid a very agreeable visit as I always do there. . . . Reuben Haines is altogether absorbed at present in a contemplation of Robert Owen and his schemes, and would, I doubt not, if it were not for the Ballast of the female part of his family, embark his all in the furthurance of 'Friend Owen's' plans."[38] It was a different matter for Say and Lesueur, neither of whom was married.

Maclure wanted to establish a "school of industry" in New Harmony where science and the "useful arts" would replace "latin, greek, and lilliputian," as Say put it to Charles-Lucien Bonaparte.[39] Maclure, who financially supported both Say and Charles-Alexandre Lesueur, wanted them to teach science and art in his proposed school and invited them to "visit" the experimental town. Say believed that by accompanying Maclure to New Harmony, he was only to have a look at the community and then go on to Mexico with his benefactor. He evidently did not anticipate a permanent move. He wrote to his colleague Thaddeus Harris in Boston, "I expect to set out in the course of 2 or 3 days on a journey for the winter in company with Mr. Maclure; we shall visit Mr. Owen's establishment on the Wabash & thence proceed down the Mississippi."[40] Like many of his scientific colleagues at the time, Say had no means of support for his chosen profession except to rely on patronage, and Maclure filled that necessary role for him. Lesueur may have already committed to a New Harmony move, as had Troost and Speakman, but Say—who was not particularly interested in social reform but entirely devoted to science—certainly had not expected to leave Philadelphia for good and take up life on the frontier.

In the early afternoon of 8 December 1825, a keelboat, named *The Philanthropist* in honor of Owen and Maclure and later dubbed "The Boatload of Knowledge" by one of the passengers because of all the eminent scientists aboard, pulled away from its Pittsburgh mooring (figure 2.14). With the boat's departure, the Academy of Natural Sciences lost three of its most important active members: Maclure, Say, and Lesueur. Two of its seven founders, Troost and Speakman, had already gone ahead to New Harmony (figure 2.13).

Shortly after the party left Philadelphia, things began to fall apart at the Academy, as arguments erupted between the members. Factions developed, which had not been there before. It was Say who had held the society together; his even temper and constant presence created an atmosphere of order and amiability. Chaos followed his departure. He had taken almost sole charge of the *Journal,* and there was no one to replace him. A prolific writer, he had presented a record number of five papers on different subjects at the meetings preceding his departure, all of which were approved for publication in the *Journal.*[41]

At the very next meeting after he left Philadelphia, two of Say's papers were rejected, something that had never happened before.[42] Politics or jealousy appears to have been the cause, rather than the accuracy of Say's science, and Richard Harlan (1796–1843) may have been behind these rejections. One of Say's articles dealt with reptiles, an area of study that Harlan had made his own. According to historian Patsy A. Gerstner:

> Most of Harlan's colleagues found him unpleasant and argumentative, and many questioned his abilities as a scientist. As an American pioneer in the study of fossils, especially the reptiles, Harlan made some astute observations but committed some serious errors as well. But his poor relations with his peers seem to have been generated as much by his irascible nature and their dislike of him as by his errors.[43]

After the New Harmony contingent left Philadelphia, discord at the Academy reached such a pitch that members appointed to the *Journal*'s committee quit one after another in exasperation, and nothing was published in 1826. Charles-Lucien Bonaparte wrote to William Cooper in New York: "We shall see into whose hands is the Journal now to fall." He thought it possible that Zacheus Collins ("one of the most disgusted"), Robert Vaux, Godman, Peale, Gilliams, and "of course the American Linnaeus [Say] etc etc will never again appear in the Ac[ademy]. (I do not speak of my humble self)." As for George Ord, Bonaparte

2.14 *Aboard the* Philanthropist *descending the Ohio River to New Harmony* by Charles-Alexandre Lesueur, 1826. Modified pencil sketch. Muséum d'histoire naturelle, Le Havre, France.

2.15 Hall of the Academy of Natural Sciences, a former Swedenborgian church occupied by the Academy in 1826, by C. Burton. Hand-colored lithograph. ANSP Archives coll. 49.

observed that he had "united with the rabble . . . but after all he is with all his faults probably an honest man."[44]

The following July, the Academy *Minutes* note that the entire publication committee resigned en masse.[45] Cooper suggested to Bonaparte that he hoped Academy members would "be content with shedding a few gallons of ink, instead of a drop of blood; and sharpen their pens rather than draw their swords."[46]

Pens had indeed been sharpened, and they proved as deadly as swords. After Richard Harlan published his *Fauna Americana, Being a Description of the Mammiferous Animals Inhabiting North America* in 1825, John Godman, in a scathing review, accused Harlan of plagiarizing information already published by a French naturalist in 1820.[47] A furious Harlan issued a small pamphlet refuting these charges by asserting that, as he had stated in his book, he had intended to correct and add to the Frenchman's classifications.[48]

In this publication, Harlan turned the knife on Godman, his fellow physician, by claiming that Godman's book *Anatomical Investigation, Comprising Descriptions of Various Fasciae of the Human Body* (1824) was merely a rehash of old ideas.[49] "Unfortunately for the parent his immature productions were not 'dead born' from the press," Harlan wrote acidly, "but will descend to posterity, like flies in amber, preserved by the medium that surrounds them."[50]

In a sense, Harlan was signaling the transition in natural science studies taking place in the later 1820s. Dealing with American living and fossil species, Harlan's book was basically a work of nomenclature and classification. Godman thought such systems were arbitrary and failed to realize that a universally adopted system of nomenclature and classification could benefit science. Earlier naturalists, like Alexander Wilson and William Bartram, had written pages of lively, delightful anecdotes about American fauna with little specific scientific description. Both men were "field naturalists," having made extensive forays into the American wilderness to witness nature at first hand. Men like Harlan represented the emerging trend toward exacting descriptions with relatively few or no anecdotal "histories." The era of the self-taught amateur naturalist was ending, and the inevitable sequence to trained professionals was beginning to occur.

By early 1826, despite the disruptive exodus of some of its more prominent members to New Harmony, the Academy was on the brink of a real estate purchase. The pending acquisition was not universally popular. From New York, Bonaparte wrote excitedly to Isaac Hays, head of the publishing committee: "The public voice frightens me! They say the Academy has ruined themselves by purchasing a lot, and announce the death of our Journal of which they say another number will never be

published! Is it possible! . . . and if so would it not be better to help by funds the Journal than the building?" (figure 2.15).[51]

The building, offered for $4,300, was a former Swedenborgian church at 12th and George (now Sansom) Streets, only a few blocks away from the hall the Academy was renting from Jacob Gilliams's father. More space was desperately needed for a larger meeting room and to house the Academy's specimen collections and books. The *Minutes* for 25 October 1825 state: "Resolved that it is expedient for the Academy to procure a new Hall possessing more extensive accommodations and in a more central situation than the present."[52]

William Strickland had designed the church only eight years earlier and probably suggested it as a possible Academy headquarters. He may have thought the interior light would be particularly effective for illuminating the collections. The edifice supported a large dome and a glass lantern, or skylight, ten feet in diameter and thirty feet from the floor that, according to Strickland, admitted "a great body of light into the dome, which is thence reflected throughout the church."[53] In spite of some protests, the purchase went ahead, and the members, to save money, moved the collections themselves in the spring of 1826. By the end of February, the Academy had received donations amounting to $1,400, and a loan was taken out at 6 percent interest for the remainder.[54] Maclure assisted the Academy yet again by contributing $500 toward the purchase price.[55]

Maclure's continued benevolence since his first involvement with the institution had not gone unrecognized. Only three weeks after he left Philadelphia for an extended period, the *Minutes* for 13 December 1825 records: "Dr. [Richard] Harlan on behalf of forty-two member subscribers presented to the Academy a portrait of the President William Maclure Esquire, executed by Mr. [Thomas] Sully at their request."[56] This portrait hangs today in the reading room of the Academy's library (figure 2.16).

In spite of the acquisition of a new building and the unity over Maclure's portrait, there was more to come in the way of schism. Disagreements prompted a group of disgruntled members, including Bonaparte, Gilliams, Dr. John Sharpless, and others, to set up a rival society they called the Maclurean Lyceum of the Arts and Sciences. The first number of their publication states diplomatically that "in consequence of an increasing taste for scientific pursuits, it is thought advisable to form another institution in this city, which should afford additional facilities for the acquisition of knowledge. With this view the society was established in May 1826."[57] The members named their new

FACING PAGE 2.16 *William Maclure* by Thomas Sully (1783–1872), 1825. Oil on canvas. ANSP Library coll. 286.

2.17 *Scallop shells* by Lucy Way Say, drawn for Thomas Say's pioneering book on American shells, *American Conchology* (1830–36), plate 56. Original hand-colored drawings. ANSP Archives coll. 433B. Lucy Say was the first woman elected to membership in the Academy.

organization after Maclure because of all he had done to promote American natural science and no doubt also in the hope that he would aid them financially. Maclure was still president of the Academy, but this apparent conflict did not bother the Lyceum founders.

The Maclurean Lyceum lasted only a few years. Say, who had been nominated president of the society, lived hundreds of miles away in Indiana; there were no funds to support the institution because Maclure was apparently not interested; and Bonaparte, the driving force behind the Lyceum, left in the fall of 1827 for an extended trip to Europe and then moved permanently to Italy in the winter of 1828. Many of the Maclurean Lyceum members rejoined the Academy, where several friends, including Jacob Gilliams and Isaac Hays, remained and attempted to champion the interests of the defectors.

Like the defunct Lyceum in Philadelphia, Owen's "new moral world" of communism in New Harmony also failed. Too many inhabitants depended on too few to support them, and the idea of doing away with religion and marriage brought damaging criticism to the community. Say and Lesueur stayed on nevertheless, involved with their scientific pursuits. Maclure funded a printing press that was used to publish his own manuscript, which he sent from Mexico (*Opinions on Various Subjects Dedicated to the Industrious Producers*), and most of Say's pioneering *American Conchology, or Descriptions of the Shells of North America Illustrated from Coloured Figures From Original Drawings*. Say was forced to remain in the frontier community at Maclure's pleasure because he had no other means of support, and while in New Harmony, he married, which increased his financial responsibilities. His wife, Lucy Way Sistaire (1801–1886), a young teacher who had been aboard "the Boatload of Knowledge," was a gifted artist who would illustrate in exquisite detail the shells Say published in his *Conchology* (figure 2.17).[58]

By the closing years of the 1820s, after surviving the loss of several key players, the Academy began to regain its cohesion with new men at the helm.

PTS

Notes

1. William H. Goetzmann, *Exploration and Empire: The Role of the Explorer and Scientist in the Exploration and Development of the American West, 1800–1900* (New York: Alfred Knopf, 1967), 183.
2. Daniel J. Boorstin, *The Americans: The National Experiment* (New York: Random House, 1965), 222.
3. Patricia Tyson Stroud, *Thomas Say: New World Naturalist* (Philadelphia: University of Pennsylvania Press, 1992), 114–15.
4. Edwin James, ed., *Account of an Expedition from Pittsburgh to the Rocky Mountains in the Years 1819, 1820 by Order of the Hon. J. C. Calhoun, Secretary of War, under the Command of Maj. S. H. Long of the U.S. Top. Engineers, Compiled from the Notes of Major Long, Mr. T. Say, and Other Gentlemen of the Party*, 3 vols. (Philadelphia, 1823).
5. Brooke Hindle, "Charles Willson Peale's Science and Technology," in Edgar P. Richardson, Brooke Hindle, and Lillian B. Miller, *Charles Willson Peale and His World* (New York: Harry N. Abrams, 1983), 166.
6. Titian R. Peale, "Journal from the Long Expedition," Library of Congress. Lillian B. Miller, ed., *The Collected Papers of Charles Willson Peale and His Family*, microfiche (Millwood, NY, 1980).
7. Edwin James, ed., *Account of an Expedition from Pittsburgh to the Rocky Mountains*, vol. 15 of *Early Western Travels, 1748–1846*, ed. Reuben Gold Thwaites (Cleveland, OH: Arthur H. Clark, 1905), 178n.
8. Maria R. Audubon, *Audubon and His Journals* (New York: Dover, 1994, reprint of Charles Scribner's Sons, 1897), vol. 1, 37.
9. Bernard Jaffe, *Men of Science in America* (New York: Simon and Schuster, 1944), 153.
10. *Journal of the ANSP* 1, no. 2: 32.
11. See J. E. Agan, "Corréa da Serra," *Pennsylvania Magazine of History and Biography*, vol. 49, no. 1 (1925): 1–43.
12. Thomas Nuttall, *The North American Sylva: or, A Description of the Forest Trees of the United States, Canada and Nova Scotia not Described in the Work of F. Andrew Michaux* (Philadelphia, 1857), 1: 141–42.
13. *Journal of the ANSP* 2, pt. 1 (1821): i–xv.
14. Original handwritten *Minutes of the ANSP*, vol. 2, for 1821.
15. Original handwritten *Minutes of the ANSP*, vol. 2: 24 July, 566; 28 August, 569; 16 October, 25, 573; December 1821, 579.
16. Original handwritten *Minutes of the ANSP*: 5 March 1822; 8 June 1824, 550; 10 August 1824, 72; 5 March 1822, 631.
17. Ibid., 29 April 1823, 791.
18. William H. Keating, ed., *Narrative of an Expedition to the Source of St. Peter's River, Lake Winnepeek, Lake of the Woods, etc. Performed in the Year 1823, by Order of the Hon. J.C. Calhoun, Secretary of War, under the Command of Stephen H. Long, U.S.T.E. Compiled from the Notes of Major Long, Messrs. Say, Keating & Colhoun.* 2 vols. (London, 1825).
19. *North American Review* 21 (July 1825): 128ff.
20. Jeannette E. Graustein, *Thomas Nuttall, Naturalist: Explorations in America, 1808–1841* (Cambridge, MA: Harvard University Press, 1967), 195.
21. See Patricia Tyson Stroud, *The Emperor of Nature: Charles-Lucien Bonaparte and His World* (Philadelphia: University of Pennsylvania Press, 2000).
22. Charles-Lucien Bonaparte, "An Account of Four Species of Stormy Petrels, Read January 13th, 1824," *Journal of the ANSP* 3, pt. 2: 227.
23. Charles-Lucien Bonaparte, "Observations on the Nomenclature of Wilson's Ornithology," *Journal of the ANSP* 3, pt. 2: 340–71 (read 9 March 1824); 4, pt. 1: 24–66 (read 23 March 1824); 4, pt. 1: 163–200 (read 9 November 1824); 4, pt. 2: 251–277 (read 23 November 1824); 5, pt. 1: 57–106 (read 31 May 1825). The last installment was marked "to be cont'd," but it never was.
24. Charles Lucian [sic] Bonaparte, *American Ornithology; or, the Natural History of Birds Inhabiting the United States Not Given by Wilson*, 4 vols. (Philadelphia: Carey, Lea, & Carey, 1825, 1826, 1827, 1833).
25. Preface to Bonaparte's *American Ornithology*, 4–5.
26. George Ord to C.L. Bonaparte, Philadelphia, 2 June 1824 (MS 2608, Muséum national d'histoire naturelle, Paris).
27. The historic core of Rapp's settlement is preserved at Old Economie Village, in Ambridge, Pennsylvania.
28. For more on New Harmony, see Arthur Bestor, *Backwoods Utopias, the Sectarian Origins and the Owenite Phase of Communitarian Socialism in America, 1663–1829* (Philadelphia: University of Pennsylvania Press, 1970). Also, Robert Dale Owen, *Threading My Way, An Autobiography* (New York: Augustus M. Kelley, 1967), first published by G. W. Carleton, New York, 1874.
29. Stroud, *Thomas Say*, 172.
30. Robert Owen to Maclure, Philadelphia, 27 March 1825, in Arthur E. Bestor Jr., ed., *Education and Reform at New Harmony: Correspondence of William Maclure and Mme. Fretageot, 1820–1833* (Indianapolis: Indiana Historical Society, 1948; reprint, New York: Augustus M. Kelley, 1973), 304–5.
31. William Price (1788–1860) was the third child and first son of Philip and Rachel Price, superintendent and matron of the Quaker Westtown School outside Philadelphia. One of the earliest pupils enrolled there, he must have been in the same class as Thomas Say (1787–1834), who was among the first students at the school, as was Reuben Haines. Price attended the University of Pennsylvania, where he studied medicine with Benjamin Rush, Philip Syng Physick, and Caspar Wistar. Josephine Mirabella Elliott, ed., *Partnership for Posterity: The Correspondence of William Maclure and Marie Duclos Fretageot, 1820–1833* (Indianapolis: Indiana Historical Society, 1994), Appendix E, 1067.
32. William S. Phiquepal (Guillaume Sylvan Casimer Phiquepal d'Arusmont) was a Pestalozzian teacher brought from Paris by Maclure to instruct pupils in a school set up by Maclure in Philadelphia.
33. Reuben Haines to Ann Haines, Germantown, 6 July 1825, Wyck Papers, American Philosophical Society.
34. Donald MacDonald, *The Diaries of Donald MacDonald, 1824–1826, Indiana Historical Society Publications* 14, no. 2 (1942): 209.
35. Elliott, *Partnership for Posterity*, 284. These drawings are now in the natural history museum in Le Havre.
36. Haines to Jane Haines, Reading, 23 August 1825, Wyck Papers, American Philosophical Society.
37. Troost quoted by Haines to Jane Haines, 1, 3, 6 September 1825, Wyck Papers, American Philosophical Society.
38. Deborah Norris Logan's *Diary*, vol. 9 (1825): 6 (Winterthur Museum Library).
39. Say to Charles Bonaparte, Columbus, Ohio, 13 July 1826 (Muséum national d'Histoire naturelle, Paris; film 542, American Philosophical Society).
40. Say to Thaddeus Harris, Philadelphia, 21 November 1825, Museum of Comparative Zoology Archives, Harvard University, Cambridge, MA.
41. Original handwritten *Minutes of the ANSP*, vol. 4 (29 November 1825): 373.
42. Original handwritten *Minutes of the ANSP*, vol. 4 (6 December 1825): 377.
43. Patsy A. Gerstner, Richard Harlan entry in *American National Biography* (New York: Oxford University Press, 1999), vol. 10, 102–4.
44. Charles Bonaparte to William Cooper, Point Breeze (Bordentown, NJ), 2 April 1826 (film 1514, American Philosophical Society).
45. Original handwritten *Minutes of the ANSP*, vol. 4 (31 July 1827): 864.

46 William Cooper to Charles Bonaparte, 5 April 1826, Charles-Lucien Bonaparte Correspondence with American Scientists, MNHN, microfilm at American Philosophical Society.

47 John Godman, "Remarks on an Article in the North American Review," *Journal of the Franklin Institute*, no. 1 (1826): 19–20. See Patsy A. Gerstner, "The Academy of Natural Sciences of Philadelphia, 1812–1850," in *The Pursuit of Knowledge in the Early American Republic: American Scientific and Learned Societies from Colonial Times to the Civil War*, ed. Alexandra Oleson and Sanborn C. Brown (Baltimore: Johns Hopkins University Press, 1976), 184–85.

48 Richard Harlan, *Refutation of Certain Misrepresentations Issued Against the Author of "Fauna Americana," in the Philadelphia Franklin Journal No. 1, 1826, and in the North American Review No. 50* (Philadelphia: W. Stavely Printer, 1826), 25.

49 John Godman, *Anatomical Investigation, Comprising Descriptions of Various Fascias of the Human Body* (Philadelphia: H. C. Carey & I. Lea, 1824).

50 Harlan, *Refutation*.

51 Bonaparte to Isaac Hays, New York, 14 January 1826 (Isaac Hays Papers, American Philosophical Society).

52 Original handwritten *Minutes of the ANSP*, vol. 4 (25 October 1825): 361.

53 "First New Jerusalem Temple in Philadelphia" (1817), posted by Kirsten Gyllenhaal, *New Church History Fun Fact*, 24 January 2007.

54 Original handwritten *Minutes of the ANSP*, vol. 4 (28 February 1826): 599.

55 Original handwritten *Minutes of the ANSP*, vol. 4 (15 August 1826): 644.

56 Original handwritten *Minutes of the ANSP*, vol. 4 (13 December 1825): 379.

57 *Contributions of the Maclurean Lyceum to the Arts and Sciences* 1, no. 1 (Philadelphia, January 1827).

58 Thomas Say, *American Conchology, or Descriptions of the Shells of North America Illustrated From Coloured Figures From Original Drawings* (parts 1–6, New Harmony, Indiana, 1830, 34; part 7, Philadelphia, 1830).

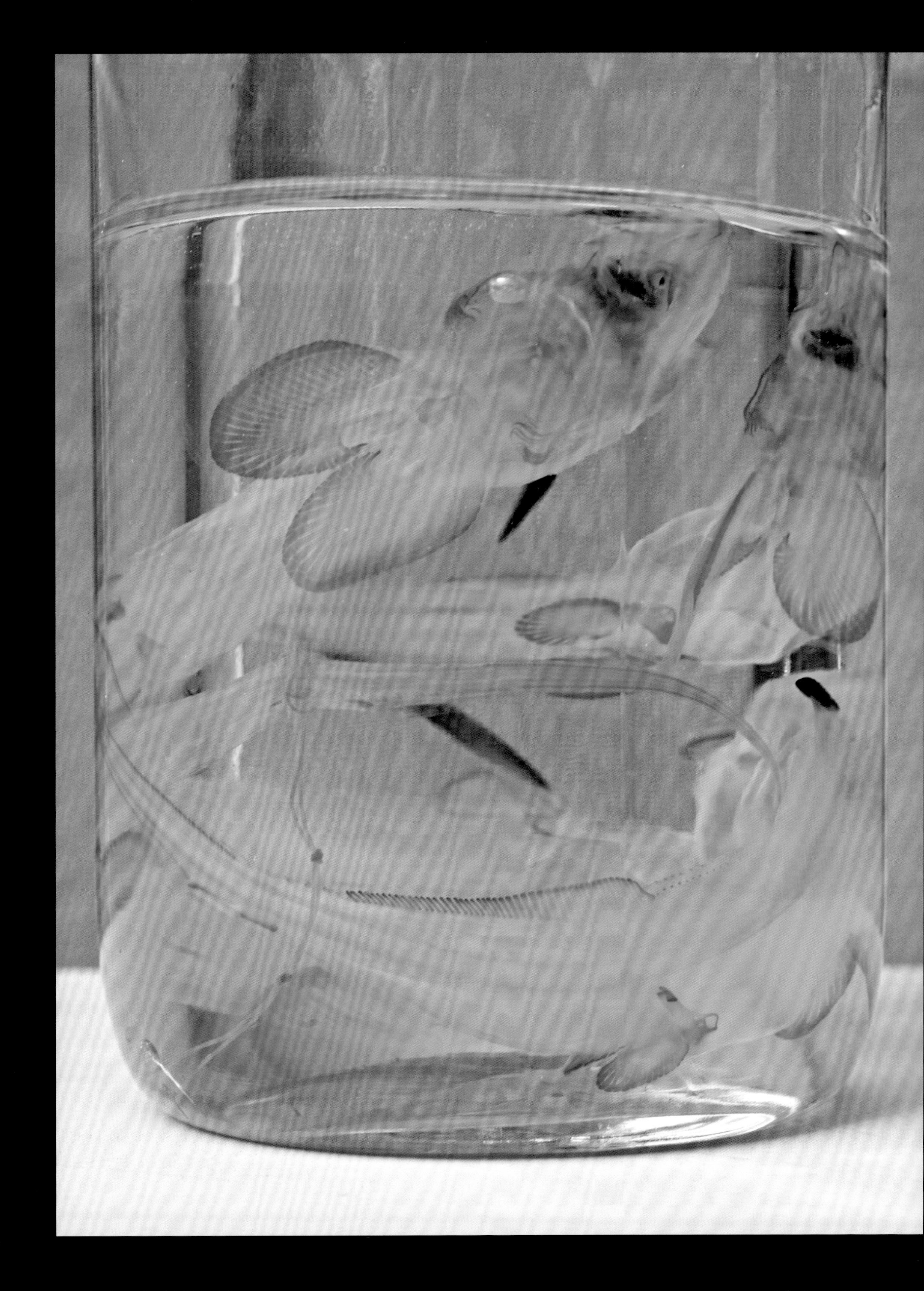

Cleared and stained ratfish (*Callorhinchus milii*) collected in New Zealand by D. Didier in 1999. ANSP Ichthyology Department

FOLLOWING PAGES Collection of sphinx or hawk moths (*Hemaris thysbee*) from eastern North America. ANSP Entomology Department.

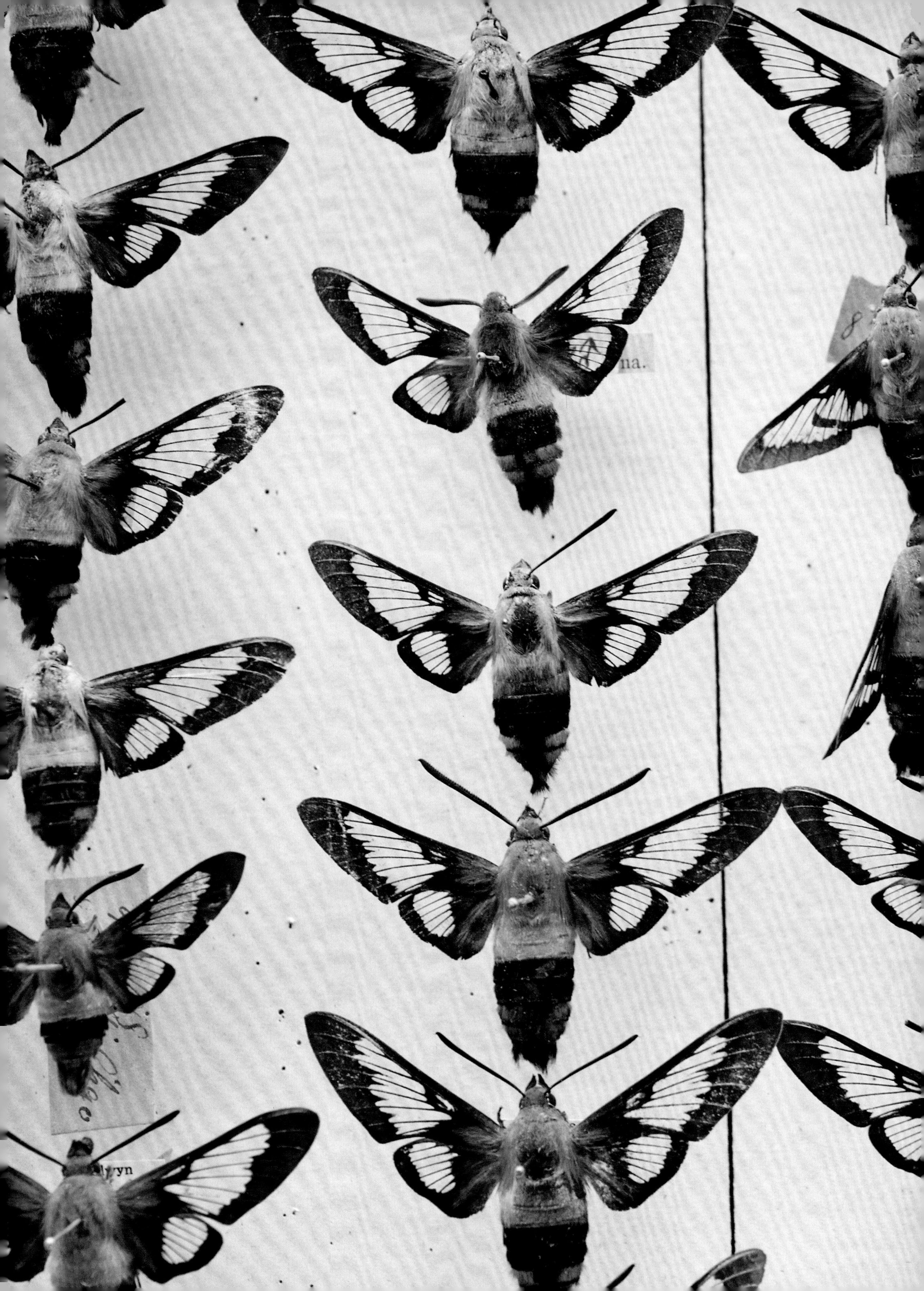

Chapter 3

A Widening Sphere

The opportunities for a Naturalist have been vastly greater than I anticipated, at the same time there have been more dangers than I bargained for.
—Charles Pickering to Samuel G. Morton
"U. S. ship *Vincennes*, Feegie Islands," 7 August 1840

In 1828, after fourteen years of sponsoring popular public lectures, the Academy extended its reach into the community by opening its doors to the public on Tuesday and Saturday afternoons. New visitors could see, without charge, the thousands of shells, minerals, birds, fish, insects, mammals, fossils, and plants the Academy's members had been acquiring since 1812. William Maclure, though living in Mexico, undoubtedly encouraged this civic endeavor because of his abiding belief in the value of public education. However, not everyone was as pleased with the easy access championed by the Academy's absentee president.[1]

The *Minutes* of 15 March 1831 note a reaction against the decision to allow public access to the museum. To control admissions, the members resolved that "no person can be admitted to the hall of the Academy unless they present to the doorkeeper a note signed by a member."[2] Many felt that there was a need for more supervision, as children had broken some of the cabinets and had otherwise been destructive. Later in the year, Samuel Morton proposed an increase in the annual insurance on the Academy's property. He also moved that the curators be authorized to wall up the window under the stairs "as a measure of precaution."[3] Both proposals reflect the members' unease about offering unlimited access to the Academy's fragile collections.

While the 1830s saw a gradual professionalization, the Academy remained dominated by amateurs. Dr. W. S. W. Ruschenberger, surgeon in the U.S. Navy, Academy member, and later its president, pointed out in his 1852 history of the institution that nearly all its members, of necessity, were occupied daily in their professional practices of physician, lawyer, cleric, merchant, or manufacturer, while the study of natural science was "an agreeable occupation for their leisure hours." "Theoretically," he observed, "this Society does not consist of men already learned; but of men who desire to become acquainted with Nature and her laws, and who are fully sensible of the great benefits which increased knowledge of the creation must confer upon mankind. In a word, the institution is an academy, a school of learners rather than an association of learned men."[4]

This was not entirely true, for men like Thomas Say (entomology), Charles-Lucien Bonaparte (ornithology), and Thomas Nuttall (botany) devoted their lives to natural history and had been pioneers in the establishment of their respective disciplines. Many Academy members were associated with and gave lectures on nature to other organizations in Philadelphia, including the American Philosophical Society (1743), Peale's Museum (1784), the Athenaeum of Philadelphia (1817), and the Franklin Institute (1824), as well as the Lyceum of Natural History in New York (1817). Many members of these institutions were in turn associated with the Academy.

As the financial difficulties of the early years subsided and the budget of the young society stabilized, publication of the

3.1 Globe used by John Cleves Symmes Jr. (1742–1814) in the 1820s and 1830s to promote his theory that the earth was hollow with openings at the poles. ANSP Library.

NORTH
AMERICA
Canada
UNITED STATES
Bermudas
Bahama
Portorico
Jamaica
Hayti
SSW
SWBS
WSW
WBS

J. T. WETHERILL.
Mastodon Giganteum.
BIGBONE LICK.

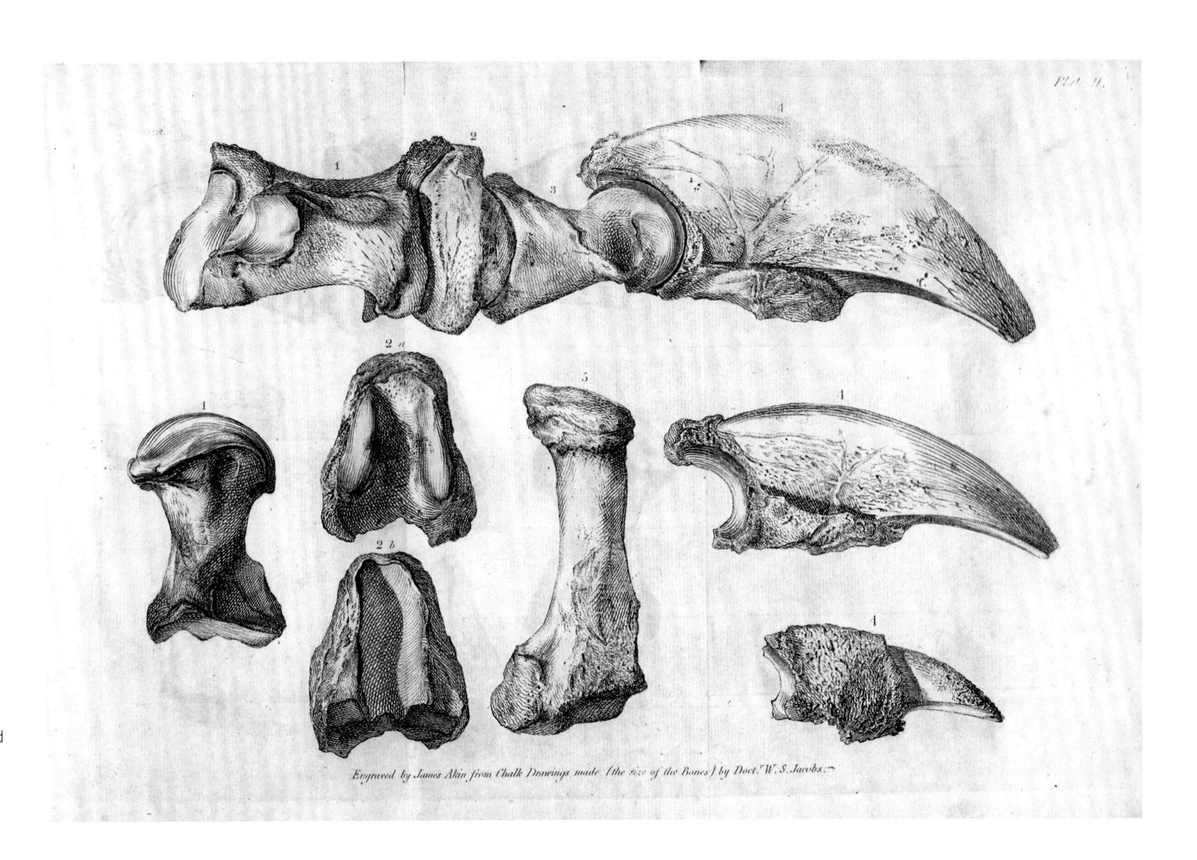

3.3 *Megalonyx* bones, illustration from Caspar Wistar, "A description of the Bones deposited by the President [Jefferson], in the Museum of the Society, and represented in the annexed plates." *Transactions of the American Philosophical Society* 4, no. 71 (1799): 526–31. ANSP Library.

Academy's *Journal* resumed in 1829. The first issue after the two-year hiatus contained a long article on dipterous insects (a large order including flies, gnats, and mosquitoes) by Thomas Say, which must have reconciled some of the older members who had been infuriated when several of Say's essays were rejected after his departure for New Harmony.[5] The volume also contained six papers on reptiles by Richard Harlan and six on fossils by Samuel George Morton (1799–1851). These two doctors were by then among the most active members of the Academy.

Harlan published more than thirty papers on natural history throughout his career, and his interests in nature and comparative anatomy were combined in a lifelong fascination with fossil animals, especially reptiles. Although he was unpopular with his peers and was often criticized for his errors, his pioneering work focused attention on a rich fossil fauna. He added a great deal to the Academy's fossil collections and helped to enlighten the European community on the work of American scientists.[6]

Morton had a decidedly different personality from Harlan, and he was much liked and admired by his fellow Academy members. His most influential work was on human craniology, a pursuit that evolved from his interests and anatomical teaching at the Pennsylvania Medical College. Well respected as a scientist, Morton served as an officer of the Academy in many capacities, including president (1849–51) near the end of his life.[7] (A more thorough discussion of Morton's activities can be found in chapter 5.)

Because Maclure was in Mexico, George Ord, vice president, acted as his "zealous and efficient substitute"[8] until John Price Wetherill succeeded him in 1834, for reasons discussed later. Wetherill, like Maclure, was a generous benefactor of the institution. In the summer of 1829, when Joseph Dorfeuille, proprietor of the Cincinnati Natural History Museum, offered a large collection of fossil bones for sale, Richard Harlan persuaded Wetherill to buy it for the Academy (figure 3.2). Among these relics were special treasures from a saltpeter cave in what is now West Virginia that Harlan wrote about for the *Journal*. These were the bones of the *Megalonyx*, the extinct clawed beast that Jefferson had thought was an enormous lion when he deposited similar bones at the American Philosophical Society some three decades earlier. The society published Jefferson's paper on the animal in its *Transactions* of 1799. It was a marvelous discovery, which Jefferson used to refute claims by the famous French naturalist Georges-Louis Leclerc Count de Buffon that animals in the New World were smaller and therefore inferior to those of the Old World. Jefferson measured the largest claw at 7.5 inches, huge compared to a lion's claw of 1.41 inches.[9] The Philadelphia anatomist Caspar Wistar gently discouraged his friend Jefferson's lion identification and described the animal as an extinct giant

3.2 Tooth from an American mastodon (*Mammut americanum*). John Price Wetherill Fossil Collection from Big Bone Lick, Kentucky, 1829. ANSP Paleontology Department #13155. Purcell photograph.

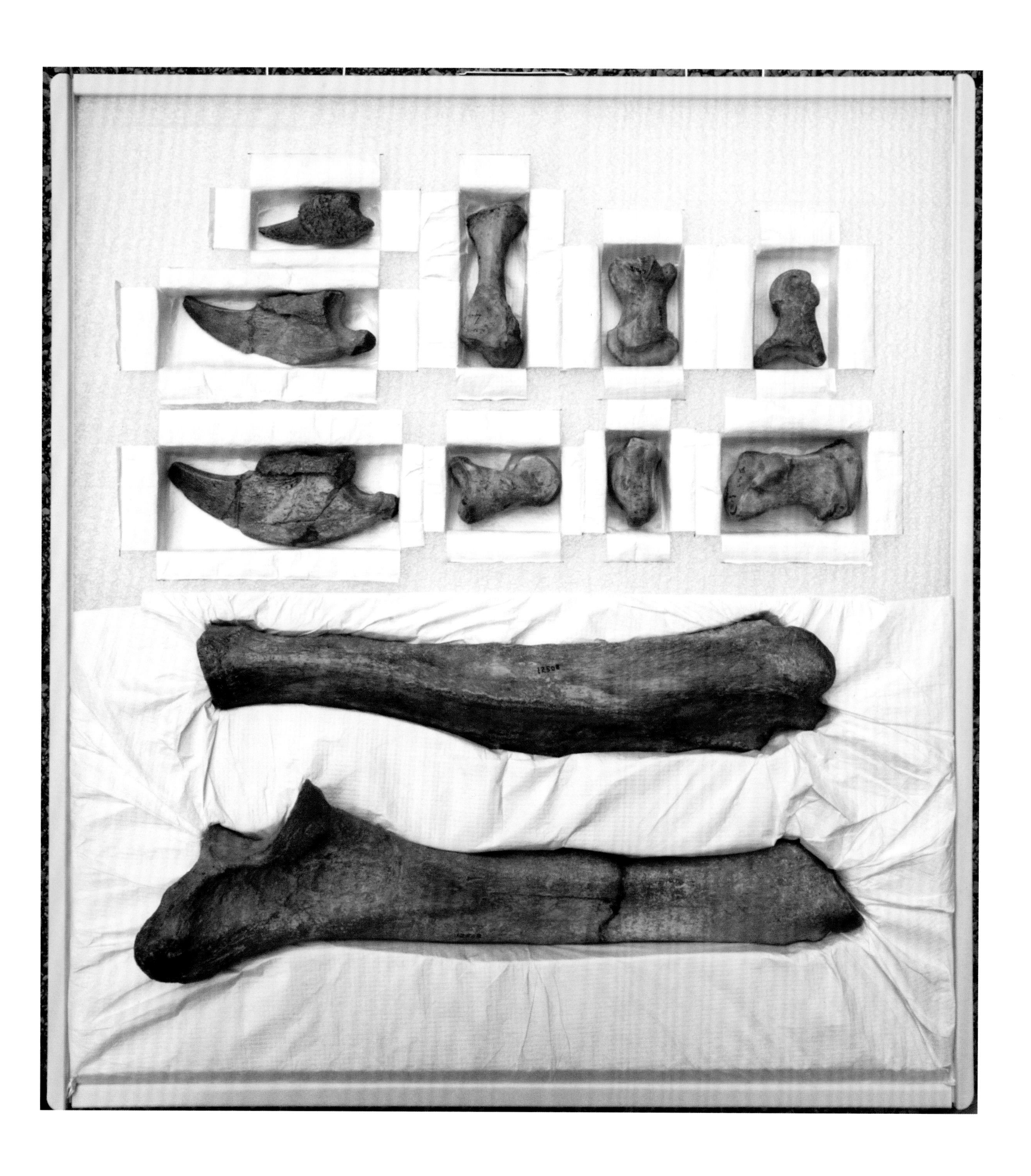

ground sloth. In 1822, the French naturalist Anselm Desmarest, an Academy corresponding member (elected 1817), named the creature *Megalonyx jeffersonii* in Jefferson's honor (figure 3.3).[10] The fossil bones used by Jefferson in his initial publication are now at the Academy, having been transferred by the American Philosophical Society in 1849 and given to the Academy in 1987 (figure 3.4).[11]

In 1834, the British naturalist Thomas Nuttall gave the Academy an immense and important collection of 4,000 species of plants from the Cape of Good Hope, New Holland (Australia), New Zealand, Antarctica, Tierra del Fuego, the Strait of Magellan, the East and West Indies, and Siberia, as well as "some of the rarest European plants." Among the collection were twenty-four botanical specimens collected by Johann and Georg Forster, father-and-son naturalists, who sailed on Captain Cook's second voyage to the Pacific (1772–75) (figure 3.5).[12] The same *Minutes* that mention Nuttall's donation state that Thomas Nuttall and John Kirk Townsend "left the city [Philadelphia] on Thursday 13th instit for the mouth of the Columbia River."[13] Nuttall had resigned from his position as director of the botanical garden at Harvard University to join Nathaniel Jarvis Wyeth's expedition to the northwest coast. Townsend, funded by the Academy and the American Philosophical Society, was to collect bird specimens, while Nuttall would focus on plants. Both men would make important collections of previously undescribed species on this grueling journey.

In addition to new collections, many important members were added to the Academy's roster during its third decade, including several of special distinction. In 1829, Maclure proposed Joel Roberts Poinsett, the first United States minister to Mexico (1825–1829), whose name was to become well known through the bright red flower that honors him (*Poinsettia pulcherrima*) (figure 3.6). Poinsett was an avid horticulturist who exchanged seeds and plants with collectors worldwide. Maclure, living in Mexico City at the time, encountered Poinsett there, although it is possible that he had known him previously in

FACING PAGE 3.4 *Megalonyx* claw bones, Jefferson Fossil Collection, ANSP Paleontology Department, #12507. Purcell photograph.

ABOVE LEFT 3.5 *Casuarina nodiflora*, botanical specimen collected by Georg Forster on Captain James Cook's second voyage to the South Pacific (1772–75). ANSP Botany Herbarium.

ABOVE RIGHT 3.6 *Joel Roberts Poinsett* (1779–1851) by Thomas Sully, 1827. Oil on canvas. American Philosophical Society.

France, where Maclure had lived at various times. Poinsett, a cosmopolitan diplomat and statesman, had met many influential people of the day during his years abroad. These included the emperor Napoleon I, the French financier Jacques Necker and his brilliant daughter Madame de Staël, and Prince Klemens von Metternich, the Austrian statesman and arbiter of European power politics. Named Secretary of War in 1837 by President Martin Van Buren, Poinsett oversaw several of the nation's important exploratory expeditions, including the Wilkes Expedition to the South Seas and North Pacific coast (1838–42). Poinsett would become a founder of the Smithsonian Institution.[14]

In the fall of 1831, John James Audubon was elected an Academy corresponding member. The following summer, the artist gave the first of several donations to the institution by presenting twenty bird skins he had collected in Florida.[15] (For more on Audubon, see chapter 4.) Another important addition to the membership rolls came in 1832, when the Academy elected as correspondent the eminent Scottish geologist Charles Lyell (1797–1875), who was then in the midst of publishing his great work, *The Principles of Geology* (three vols., 1830–33). Lyell, along with several others before him, would develop revolutionary ideas about the earth's history: that it is millions of years old rather than thousands, as had been believed, based on the Bible. (Today, the earth is believed to be 4.5 billion years old.) Lyell and his geological theories would have a strong influence on Charles Darwin, who was just then embarking on his *Beagle* voyage with volume 1 of Lyell's work in hand. Several months later, the Academy acquired a benefactor second only to Maclure when Dr. Thomas B. Wilson joined the society.[16]

Sadly, the decade of the 1830s also saw the death of some important young colleagues. In 1830, John Godman, the author of *American Natural History* (1826–28), died of tuberculosis. A year later, Reuben Haines, the Germantown Quaker farmer who for years had been the Academy's faithful corresponding secretary, took his own life after a long struggle with depression.[17]

Perhaps the greatest loss to the Academy and to American science in general came in the fall of 1834, when Academy founder Thomas Say died in New Harmony, Indiana, from complications of typhoid fever. Dr. Benjamin Coates, in his memoir of Say, delivered to Academy members in mid-December, used a vivid metaphor to describe Say's contributions to natural science: "It is not that there is no more gold in the mine; but in raising his own ore, Mr. Say has opened the shafts and galleries, pointed out the veins, and indicated, by his example, the best manner of working them."[18] Coates concluded his remarks by saying: "Thomas Say was the greatest American naturalist of his day, with a field of knowledge in vertebrates, insects, mollusks, crustacea and fossils not equaled by an American of his time."[19]

Coates's memoir of Say was particularly welcomed by Academy members because it had been preceded a week earlier by an unfortunate address given at the American Philosophical Society by George Ord. The Academy's vice president had characterized the transition of the institution's early days, from amateur to more professional, with a self-serving dismissal of his predecessors' work. Particularly repugnant was his description of the founders as a "club of humorists" who held weekly meetings "merely for the purpose of amusement" and whose specimen collections could only "excite merriment." He spoke of Say as being deficient in "elementary learning" and "indifferent to polite literature."[20] This was the Say who so often quoted Shakespeare and the Bible in his letters. Ord's remarks were considered a cruel characterization of the Academy, a libel of Say, and in the end a mere eulogy of Ord himself. His remarks so infuriated Academy members that they refused to publish them in the *Journal*.

The contentious Ord was surprised by and resentful of the negative reaction to his remarks by those Academy members who had been present at his talk. He wrote to his friend Charles Waterton in England:

> This Discourse [his memoir of Say] was read to the Philo. Society, and received with approbation; but there were three or four persons, also members of the Academy of Natural Sciences, present, who thought fit to take exception to it; and who, communicating to the other members of the Academy an erroneous idea of my sentiments and intention, were the cause of a cabal, which deprived me of the chair of Vice President, a situation which I had held for many years. At the head of this conspiracy was Audubon's dear friend, Rich. Harlan, M.D., a fellow whose moral character is so infamous in Phila., that he is excluded from the society of gentlemen.[21]

Ord's description of this incident gives some insight into the warring factions that had existed at the Academy for some time, seriously disrupting its business. To further demonstrate the Academy's high regard for Say, despite Ord's controversial comments, a committee was assigned "to take measures to procure for the Academy a portrait in oil of their late fellow

3.7 *Slender-billed Guillemot,* illustration from John James Audubon, *The Birds of America* (1827–38), plate 430. ANSP Library. The specimens illustrated were collected by John Kirk Townsend in the Pacific Northwest and sold to Audubon by the Academy in 1836.

member."[22] A few months, later it was reported that a committee had been appointed "to wait upon the Trustees of the Philadelphia Museum to request the loan of Mr. Say's portrait [by Charles Willson Peale, painted in 1819] from that institution for the use of Mr. Rembrandt Peale."[23] Rembrandt Peale, who had known Say personally, depicted him as slightly older and more serious and sedate than in his father's rendering of the naturalist. To make the portrait look more like that of a typical Academy member, he put Say in contemporary civilian clothes rather than in the uniform of the Long Expedition.

The geologist Benjamin Silliman of Yale wrote in his *American Journal of Science and Arts,* for which Say, in 1818, had contributed a pioneering paper on the uses of the fossil record in geology as a guide for dating rock strata: "It is no exaggeration to assert, that [Say] has done more to make known the zoology of this country, than any other man."[24]

An American context of pragmatism had shaped the character of Say's scientific endeavors, which early pointed the way toward professionalism in the natural sciences. By applying his knowledge of insects to medicine and agriculture, he contributed significantly to fields that were vitally important to the American economy.

Say's death marked a transition in the Academy's history from an era dominated by the institution's founders—the pioneers of natural history study in America, few of whom had been full-time scientists—to that of the next generation, represented by men like Richard Harlan, Samuel George Morton, and Timothy Abbot Conrad, who continued Say's work in conchology, illustrated by Say's wife, Lucy. Although both Harlan and Morton were practicing physicians, along with Conrad they developed as well an increasingly professional expertise in their fields of natural history. By the beginning of 1834, the sciences at the Academy were departmentalized. Members were now grouped into one of five standing committees, mineralogy, geology, zoology, ornithology, and botany.[25]

The Academy library received a posthumous gift from Say when Lucy donated her husband's entomological collection and the books "relating to his favorite pursuit." Maclure also gave the Academy a large collection of books that he had accumulated and had left in New Harmony principally for Say's use. Academy curator and former librarian Dr. Charles Pickering (1805–1878) traveled to that pioneer town in the far southwestern corner of Indiana to select and bring back Maclure's rare and beautiful volumes. In 1841, Lucy Say gave the Academy all of the sixty-eight copper plates from her husband's pioneering work on shells, *American Conchology*[26]—she had painted most of the original watercolors. Six months later, by unanimous vote, she was elected the first woman member of the Academy.[27]

Once again, there was a protracted interim between editions of the *Journal*; in fact, three years lapsed between the publication of parts 1 and 2 of volume 7 (1834–37). The first article of part two (1837) describes twelve birds collected by John Kirk Townsend on the Columbia River during the Wyeth Expedition, which Townsend and Nuttall had joined three years earlier. Nuttall had brought back the specimens, traveling by ship via Hawaii around the tip of South America to Philadelphia, while Townsend remained at Fort Vancouver on the northwest coast and then traveled on to Hawaii and the South Pacific (figure 3.7). The article's editor notes that Audubon, on his recent visit to Philadelphia, had "kindly offered to give figures of these species in the coming number of his splendid work on the 'Birds of America.'"[28] "Kindly offered" is a euphemism for Audubon's eager attempts to get hold of these unknown birds to paint for his book. Apparently, Townsend's friends had given Audubon permission to include them in his own work, "as it secured to [Townsend] these important discoveries in science, which owing to his protracted absence, were liable to be anticipated by others."[29] Perhaps because the Academy had partially funded Townsend, the members felt free to dispose of his specimens without his explicit permission. In return for the favor, Audubon named one of the new western species Townsend's Warbler (*Dendroica townsendi*).[30]

American science made an important step into the international arena when the United States Exploring Expedition to the South Seas (often called the Wilkes Expedition after its commander, Lt. Charles Wilkes) left Hampton Roads, Virginia, on 18 August 1838. The idea for this major scientific endeavor, which would circumnavigate the world for the next four years, charting and mapping coastlines and collecting natural history and anthropological materials, had been born many years earlier, when a veteran of the War of 1812, John Cleves Symmes Jr. (1742–1814), attempted to promote an unusual theory that the earth was hollow and entrances to the interior could be found by sailing to the North or South Pole (see figure 3.1).

There were many who supported his idea, not so much for its scientific import but because they saw that a United States–sponsored expedition could advance their respective interests. Chief among these enthusiasts were merchants looking to find undiscovered whaling grounds in northern and southern oceans, U.S. Navy personnel anxious to burnish their careers by exploring unknown lands, and scientists keen for new discoveries in natural history.

To find expedition personnel for this last category, Secretary of the Navy Mahlon Dickerson sent a letter to the Academy at the end of August 1836, requesting suggestions and nominations for "scientific gentlemen, of suitable age to be employed as members of this Corps, who may be well acquainted with Geology and Mineralogy, with Botany, with Meteorology, magnetism, electricity and other subjects connected with natural history . . . I would also respectively ask of your Society a series of inquiries as to subjects of natural history."[31] In response, twenty-five members signed a petition supporting the nomination of Academy curator Charles Pickering as naturalist.[32]

Two years elapsed before the expedition set sail. The government was inexperienced in outfitting such an undertaking, and Secretary Dickerson proved indecisive, if not incompetent and hostile.[33] When Martin Van Buren became president in 1838, he put the expedition into the hands of his efficient Secretary of War, Joel Poinsett, thus more solidly connecting the expedition to the Academy and its mission. Poinsett, an Academy member for nearly ten years, stressed the scientific nature of the enterprise.

Nine civilian scientists were selected from a projected corps of twenty-five. Among these were two active members of the Academy: Charles Pickering, a medical doctor as well as a naturalist, and Titian Peale, who joined the expedition as a naturalist and also served as an artist, along with Joseph Drayton. Other participants included Academy corresponding member James Dwight Dana Jr., a geologist who anticipated modern global tectonic theory and became one of America's scientific leaders; Joseph P. Couthouy, a conchologist (elected an Academy corresponding member in 1837); and William D. Brackenridge, a botanist.

Scientific expeditions were the graduate schools of the early nineteenth century for many of the best European scientists, according to the historian A. Hunter Dupree; men such as Alexander von Humboldt, Darwin, T. H. Huxley, Alfred Russel Wallace, and Joseph Hooker founded their careers on journeys to far places in the world.[34] The same would be true for Americans, as this was the first time in American history that civilian and naval personnel joined in a peacetime expedition focused on science.[35]

It was a difficult assignment, with conditions on shipboard cramped and unpleasant. Peale wrote to his young daughters from the *Peacock* on the night before leaving: "The little stateroom in which I live is just about as large as your mother's bedstead; in it I have a little bed over and under which is packed clothes, furs, guns, books and boxes without number, all of which have to be tied to keep them from rolling and tumbling about, and kept off the floor as it is sometimes covered with water." He added that "the water we drink is kept in barrels and Iron tanks, it is very warm and now smells very bad, but as we do not come on board ship to be comfortable we content ourselves with anything we can get."[36]

The expedition's flotilla of six naval vessels, the *Vincennes* (Wilkes's flagship), the *Peacock*, the *Porpoise*, the *Relief*, the *Flying Fish*, and the *Sea Gull*, set sail in 1838, the same year the first steamboat crossed the Atlantic (figure 3.8). This expedition, both the first and the last government-sponsored expedition under sail,

3.8 *U.S.S. Vincennes in "Disappointment Bay," January 1840.* The Peabody Essex Museum, Salem, Massachusetts. The *Vincennes* was Wilkes's flagship for the United States Exploring Expedition (1838–42).

The Titian R. Peale Insect Collection

Peale used different techniques for mounting his moths and butterflies. Some were grouped together and arranged by family or geographical location. Others were encapsulated in pairs of curved glass covers. In most boxes the specimens were anchored to the glass by a system of numbered corks and were visible from both the top and bottom without opening the glass covers, a method of storage and display that protected them from insect damage and the wear of physical contact. ANSP Entomology Department. Purcell photograph.

Titian Ramsay Peale (1799–1885) was born in Philadelphia, the youngest son of prominent artist Charles Willson Peale. The elder Peale founded the nation's first natural history museum in 1786, and his interactions with most of the prominent scientists of his day led to many opportunities for his precocious son. Titian Peale became a protégé of Thomas Say's (1787–1834), a cofounder of the Academy of Natural Sciences and the father of American entomology. While still in his teens, Peale began preparing the scientific illustrations for Say's seminal publication, *American Entomology* (1824–28). Peale accompanied Say on expeditions to collect plant and animal life in the western and southeastern states, and he began to amass a personal collection of insects, consisting primarily of Lepidoptera (moths and butterflies). Eventually Peale joined a team of explorers who circumnavigated the globe between 1838 and 1842 as part of the Wilkes Expedition, a trip sponsored by the U.S. government.

His insect collection (the oldest insect collection in the Americas) survives to this day and forms part of the Academy's entomology collection. Most of the original Peale collection is housed in 101 boxes constructed by Peale, which have been restored, repaired, and rehoused in cabinets as part of a recent preservation project. The survival of this nineteenth-century collection may be due in part to the remarkably well-designed storage boxes that are now known as Peale boxes. Specimens were pinned to corks in a double-glazed, hermetically sealed box lined with tinfoil and then baked to kill any eggs or larvae of carpet beetles or other museum pests that might attack the dried specimens within. The inner box was then bound like a book, with each specimen's data handwritten on the inside back cover of the "book box." The surviving Peale collection is worldwide in scope, although the majority of specimens originated in the mid-Atlantic region. Some of the insects collected by Peale on the Wilkes Expedition are present (primarily those collected along the Atlantic coast of Brazil in 1838), but most of the specimens from the latter part of this expedition were lost when Peale's ship sank at the mouth of the Columbia River in 1841. This remarkable collection was donated to the Academy by Peale's family after his death in 1885.

—*Jason D. Weintraub*
Collection Manager
Department of Entomology

Limenitis
Thecla Niphon ?
Papilio
No 12
Now place
Hub.

would crisscross the Atlantic and sail around Cape Horn—with several ships visiting Antarctica, where the *Sea Gull* was lost with all hands—then up the western coast of South America, across to the Pacific Islands, and on to Australia. The second year, the squadron visited New Zealand and cruised the Antarctic, then went on to the Fiji Islands and the Sandwich Islands (Hawaii).

From Fiji, Charles Pickering wrote a long letter to Samuel George Morton at the Academy:

> We have now got through the worst part of the Cruise, and I am doing very well . . . however we are not yet out of the woods. The opportunities for a Naturalist have been vastly greater than I anticipated, at the same time there have been more dangers than I bargained for. In South America I made two trips to the crest of the Andes (near Valparaiso, & again almost 2000 miles further north, near Lima); either of them might well form an era in the life of a Naturalist, or indeed of any intelligent person.[37]

The botanist Asa Gray (1810–1888) gave a vivid description of one of Pickering's experiences in the southern mountains in the Academy's *Proceedings*: "Perhaps the most singular peril . . . was that in which this light-framed man once found himself on the Peruvian Andes, when he was swooped upon by a condor, evidently minded to carry off the naturalist who was contemplating the magnificent ornithological specimen."[38]

In his letter to Morton, Pickering summed up the vast portion of the earth he had seen on this expedition: "I am also in a fair way of seeing every part of the Southern Hemisphere, having touched at 3 points in Austral America; at Rio Negro in N. Patagonia, Terra del Fuego; & Valparaiso; at N. Zealand, at Sydney, and we expect to put in at the Cape on our way home."[39]

From the Sandwich Islands, several months later, Pickering wrote again: "We are all here now [the squadron of ships] & are beginning to look toward home, as we have the promised termination of the cruise in May, 1842."[40] They had reached the halfway point, with their return to the United States still nineteen months away. The expedition went on to make important explorations along the northwest coast of North America. There the *Peacock* sank at the boiling mouth of the Columbia River. Although its crew, including Titian Peale, was saved, many valuable notes and specimens were lost in the accident. The reduced squadron then headed west to Asia, returning to North America by way of Singapore, the Cape of Good Hope, and St. Helena Island, eventually arriving in New York harbor.[41]

The explorers' reception after four long years away was decidedly disappointing. An unfriendly Congress, a public that had lost interest, and a series of courts-martial involving the naval officers tarnished the expedition's reputation. Fortunately, in spite of the loss of many specimens from the wreck of the *Peacock*, the collections of fauna and flora were prodigious, and a great number were new to science. Some of them came to the Academy as the personal collections of the Academy members who participated in the trip, but most were considered U.S. government property. Because the government was not equipped to receive the massive number of specimens, much was lost before the collections came under systematic control. Nearly fifteen years went by before the ultimate destination of the specimens was determined. By that time, the Smithsonian Institution had been established as the nation's natural history museum, and what was left of the expedition's findings went there, where much of it remains today.[42]

The nineteen volumes of reports and atlases that eventually resulted from the journey were a landmark in the emergence of the United States into international science.[43] However, the meager number of copies of the expedition narrative that were printed angered officials at the Academy of Natural Sciences. At a meeting in February 1846, the members resolved that "the present number of 100 copies only, printed on public account, is utterly inadequate to supply the demand for this work, especially as about one half of that number is understood to be distributed in donations to Foreign Governments, while none are allowed to Scientific Societies at home." The printing of one hundred copies "scarcely deserves the name of a publication," the *Minutes* added indignantly.

Such a small print run was decried as unjust to the nation that sponsored the expedition as well as to "the meritorious individuals who have performed the scientific duties." Having been consulted by the Navy Department for advice in organizing and planning the expedition, the Academy "deems itself justified in complaining of the treatment which, in common with all the other scientific bodies then consulted [including the American Philosophical Society], it receives by the existing arrangement."[44]

Titian Peale was even angrier about the results of the expedition than the institutions that helped to plan it. Since space aboard the ships had been limited, he had sent back specimens from ports of call whenever possible. These eventually arrived either at the Peale Museum in Philadelphia or at the National

3.9 Musk Parrots, fig. 1 (*Aprosmictus splendens*); fig. 2 (*Aprosmictus personatus*) collected by Titian Ramsay Peale and illustrated by Peale in the official report of the *United States Exploring Expedition, during the years 1838, 1839, 1840, 1841, 1842 under the command of Charles Wilkes, U.S.N.* vol. 8: Atlas, *Mammalogy and Ornithology* (Philadelphia: J. B. Lippincott, 1858). ANSP Library.

1. Aprosmictus splendens. (Peale.) 2. Aprosmictus personatus. (G.R.Gray.)

T.R.Peale del. T.House sc.

3.10 Red Shining-Parrot (*Prosopei tabuensis*—formerly *Plactycercus artrogularis*) collected by Titian Ramsay Peale in 1840 in the "Cannibal Islands" (Fiji) during the U.S. Exploring Expedition (1838–42). ANSP Ornithology Department #22762. Purcell photograph.

FACING PAGE 3.11 Mounted Ruby-Cheeked Sunbird. Duc de Rivoli collection. Given to the Academy by T. B. Wilson in 1846. ANSP Exhibits Department. Purcell photograph.

ACADEMY OF NATURAL SCIENCES OF PHILADELPHIA, No. 18167
RIVOLI COLLECTION
Anthothreptes phoēnicotis
(Temm.)
Ruby-cheeked Sun-bird
Male
GIFT OF THOS. B. WILSON, M. D.
Borneo

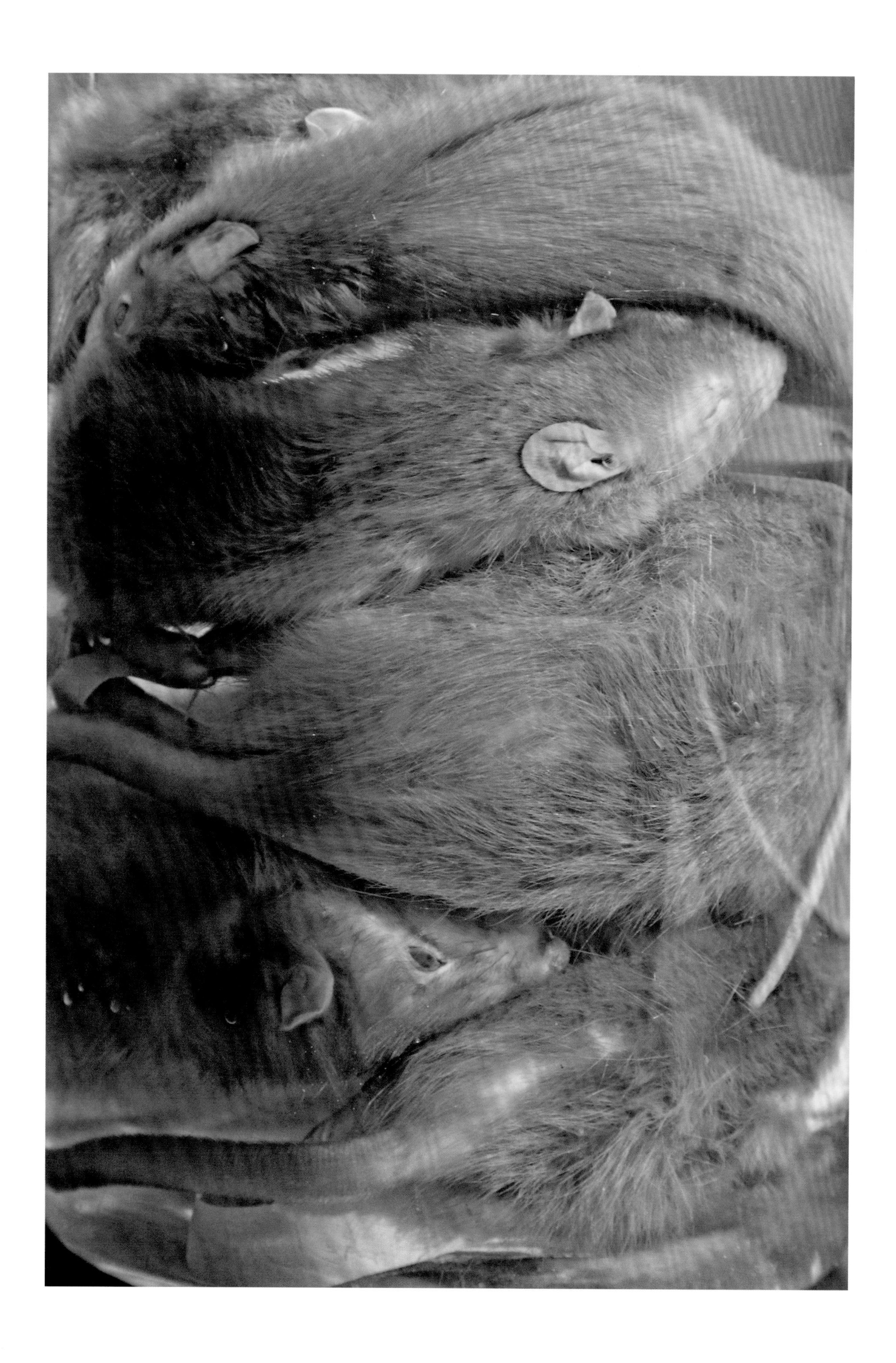

Institute, the forerunner of the Smithsonian, in Washington. Often they were unpacked carelessly and labeled poorly, and valuable data was misplaced or lost. Peale wrote to a friend that "one hundred and eighty specimens of birds which I collected are missing, including some new species."[45]

Disappointing as this was, after Peale had collected and preserved thousands of specimens, made notes with descriptions of species, habitat, and other pertinent information, and executed numerous drawings of birds and landscapes, he was devastated when his manuscript describing the discoveries was rejected. In 1852, John Cassin, Peale's colleague at the Academy, was hired to revise and publish the descriptions of the birds and mammals that Peale and his expedition teammates had discovered.

Peale was helpless to change Wilkes's decision. "The materials forming our scientific stores, were peddled to [other] authors," he wrote with bitterness many years later, "not so much for aid, but to supersede us; our positions held during the voyage were misrepresented, and higher salaries than ours offered, besides greater facilities to consult libraries, and other collections."[46] He was by then in Washington, D.C., working in the U.S. Patent Office, and sorely missed his access to Academy resources.

At the time, Cassin was considered the most competent systematic ornithologist in America. As curator at the Academy, he had the advantage of unlimited access to the most complete natural history library in the country and the most comprehensive bird collection in the world.[47] This collection was one that had once belonged to Victor Masséna, duc de Rivoli, a rich French nobleman, and it had been acquired through the largesse of Academy member Dr. Thomas B. Wilson. The Academy *Proceedings* record on 30 June 1846 that the magnificent collection, then still in Paris, contained ten thousand specimens, mounted and named (figure 3.11).[48] A committee was formed to create the space needed to accommodate this great gift, and Wilson paid for an enlargement of the building at Broad and Sansom Streets to house it.

Four years later, in 1850, Wilson added another superb collection, that of Charles-Lucien Bonaparte (figure 3.12). At the time, Bonaparte was living in Holland, in exile from Rome for political reasons, and was badly in need of funds. After his offer to sell his

FACING PAGE 3.12 Norway Rat specimens (*Rattus norvegicus*) collected in Italy by Charles-Lucien Bonaparte. ANSP Mammalogy Department. Purcell photograph.

3.13 Freshwater fish specimen (*Leuciscus scardafa)* collected in Italy by Charles-Lucien Bonaparte in 1837. ANSP Ichthyology Department #17002. Purcell photograph.

specimens to the British Museum was refused, his friend John Edward Gray, keeper of the natural history collections, suggested that he contact Edward Wilson, Thomas Wilson's brother, then living in Paris, who acted as his agent. According to George Ord, Wilson paid $3,500 for Bonaparte's collection.[49]

Aside from providing the large group of European fish specimens that eventually came to the Academy (figure 3.13), Napoleon's nephew played an indirect role in another development in American natural science. For some time he had known Louis Agassiz (1807–1873), the Swiss geologist, whom he first met at a scientific conference in Freiburg, Germany, in 1838. This friendship between colleagues had begun with their correspondence in 1834, through their mutual interest in fishes, and caused them, at Bonaparte's suggestion, to plan a trip together to America.[50]

In the end, personal reasons prevented Bonaparte from accompanying Agassiz, and the two friends had a falling out over Bonaparte's refusal to loan Agassiz money for the trip, but Agassiz determined to go anyway.[51] In the meantime, Agassiz had reached out to the Academy. The *Proceedings* record the extract of a letter of April 1846, from "M. Agassiz of Neuchâtel, Switzerland," acknowledging his election as a corresponding member and "announcing his intention of visiting this country for scientific purposes during the present year."[52]

Agassiz sailed to America in 1846. By the following year, he was deeply involved in his work in the New World and requested through a friend the loan of a collection of fossil fish from the Academy for a description in a "memoir" he was preparing.[53] In 1848, he secured a professorship at Harvard, where he would spend the rest of his life. In spite of their disagreements, Agassiz's distinguished career at Harvard may have originated with Bonaparte's encouragement that he should try his luck in America.

In 1841, the Academy began publishing the *Proceedings* in place of the *Journal*, as "a medium of communicating to the scientific public discoveries and observations at short intervals of time, and thus often enabling the claim to priority to be securely established."[54] By 1843, nearly 150 copies of the *Proceedings* were being sent throughout the United States and all over the world. Private individuals as well as societies, institutions, and academies in London, Paris, Brussels, Stockholm, Madrid, Turin, Berlin, Munich, Moscow, and St. Petersburg and the Asiatic Society of Bengal and the Egyptian Society at Cairo received the Academy's *Proceedings*. In the United States, distribution of the *Proceedings* included the American Philosophical Society, the Franklin Institute, and the Athenaeum in Philadelphia and the Albany Institute, the New York Lyceum of Natural History, the Natural History Society of Boston, the National Institute in Washington, the Franklin Society of Providence, and the U.S. Naval Lyceum at Brooklyn.[55]

In the mid-1840s, donations of collections poured in from all parts of the world: fossils from the southern foot of the Himalayas sent by corresponding member Captain P. T. Cauley and exotic shells from the Indian Ocean and the Cape of Good Hope from Robert Evans Peterson, part of a Philadelphia family of publishers who had served as an Academy auditor.[56] Mr. Hodge of Philadelphia deposited a perfect Egyptian mummy, enclosed in its original sarcophagus, from the Catacombs of Thebes,[57] and Elisha Kent Kane gave "a very fine specimen of *Ursus maritimus*, or polar bear, killed by him during a recent voyage to the Arctic regions in search of Sir John Franklin,"[58] the English explorer who lost his life attempting to find the fabled Northwest Passage. And most spectacular of all, in September 1846, the *Proceedings* record that "the Rivoli Collection has arrived and is deposited in the Hall of the Academy" (figure 3.14).[59]

As knowledgeable as Academy members were becoming in the ways of "Nature and Her Laws," there was as yet little concern with the conservation of species. At a meeting in early March 1844, Academy ornithologist John Cassin reported:

> During the past winter, the Snowy Owl (*Strix nyctea*) had been observed in the vicinity of Philadelphia, and in the adjoining states of New Jersey and Delaware, in numbers unusually great for this latitude. At a moderate calculation, he supposed that not less than 100 specimens had been shot and exposed for sale in the markets, and at a very trifling price. Several [owls] had been seen for many days successively upon the roofs of houses, chimney tops, and other elevated places, in the city itself, during the very severe weather of January. As this is a bird confined almost exclusively to extreme northern latitudes, its appearance here, in such larger numbers, is to be solely attributed to the great severity of the season to the North.[60]

There was no thought of condemnation for one hundred rare Snowy Owls being shot and sold in the markets, unthinkable in the twenty-first century (figure 3.15).

PTS

3.14 Mounted bird skeleton. Duc de Rivoli collection. Given by T. B. Wilson in 1846. ANSP Exhibits Department. Purcell photograph.

3.15 *Snowy Owl (Nyctea scandiaca)*. John James Audubon, *The Birds of America* (1827–38), plate 121. ANSP Library.

Notes

1 As the historian Nathan Reingold has written, "The development of the cabinets [of specimens], the study of their contents, the use of the contents in classrooms and in popular lectures was the life of science for many Americans for well into the [nineteenth] century. Practicing scientists were quite frequently annoyed, if not outraged, that the cabinets so often provided materials for display, not research." Nathan Reingold, ed. *Science in Nineteenth-Century America: A Documentary History* (New York: Octagon Books, 1979), 30.

2 ANSP *Minutes*, vol. 5 (15 March 1831): 275.

3 Ibid., vol. 5 (22 November 1831 and 13 December 1831): 296, 298.

4 W. S. W. Ruschenberger, M.D., *A Notice of the Origin, Progress, and Present Condition of the Academy of Natural Sciences of Philadelphia* (Philadelphia: T. K. and P. G. Collins, 1852), 14.

5 Thomas Say, "Description of New Dipterous Insects of the United States," *Journal of the ANSP* 6, pt. 1 (1829): 149–78.

6 Patsy A. Gerstner, "Richard Harlan," in *American National Biography* (New York: Oxford University Press, 1999), vol. 10, 102–4.

7 Michael M. Sokal, "Samuel George Morton," in *American National Biography*, vol. 15, 959–61. Morton served as recording secretary of the Academy (1825–29), corresponding secretary (1831), vice president (1840), and president (1849).

8 Edward J. Nolan, *A Short History of the Academy of Natural Sciences of Philadelphia* (Philadelphia: Academy of Natural Sciences, 1909), 10.

9 Keith Thompson, *The Legacy of the Mastodon: The Golden Age of Fossils in America* (New Haven, CT: Yale University Press, 2007), 35.

10 Silvio A. Bedini, *Thomas Jefferson: Statesman of Science* (New York: Macmillan, 1990), 272.

11 Earle E. Spamer, Edward Daeschler, and L. Gay Vostreys-Shapiro, *A Study of Fossil Vertebrate Types in the Academy of Natural Sciences of Philadelphia: Taxonomic, Systematic, and Historical Perspectives*, Special Publication #16 (Philadelphia: Academy of Natural Sciences, 1995), 19–20.

12 Joan Apfelbaum, "Collections of J. R. and J. G. A. Forster in the Herbarium of the Academy of Natural Sciences of Philadelphia," *Notulae Naturae*, no. 437 (April 1971), and ANSP *Frontiers* 34, no. 5 (1970): 30.

13 Original handwritten ANSP *Minutes*, vol. 5: 10 May 1831, 278; 7 August 1832, 352; vol. 6: 18 March 1834, 418.

14 Andrew Roller, "Joel Roberts Poinsett," in *American National Biography*, vol. 17, 615–17.

15 See ANSP *Minutes*, 17 July 1832. Among them was the *Columba zenaida*, a pigeon named by Charles Bonaparte after his wife, Zenaïde.

16 Original handwritten ANSP *Minutes*, vol. 5: 27 March, 338, and 25 June 1832, 347.

17 For a letter relating to Haines's death, see Marie Fretageot to William Maclure, "Bettwin Louisville and Cincinnaty," 8–26 November 1831, Maclure-Fretageot Correspondence, Workingman's Institute Library, New Harmony, Indiana, quoted in Patricia Tyson Stroud, *Thomas Say: New World Naturalist* (Philadelphia: University of Pennsylvania Press, 1992), 235–36.

18 Benjamin Coates, M.D., *A Biographical Sketch of the Late Thomas Say, Esquire, read before the Academy of Natural Sciences of Philadelphia, December 16, 1834* (Philadelphia: W. P. Gibbons, 1835), 19.

19 Coates, *A Biographical Sketch*, 20.

20 George Ord, "A Memoir of Thomas Say, Foreign Member of L. S. and Z. S. London, Read before the American Philosophical Society on the 19th of December 1834." Rejected for APS publication but printed in Thomas Say, *The Complete Writings of Thomas Say on the Entomology of North America*, ed. John L. Le Conte (New York: Bailliere Brothers, 1859; reprint, New York: Arno Press, 1978), vol. 1, x.

21 George Ord to Charles Waterton, Philadelphia, 17 April 1835, George Ord Papers, APS.

22 Original handwritten ANSP *Minutes*, vol. 6 (23 December 1834): 439.

23 ANSP *Minutes*, vol. 6 (7 April 1835): 456.

24 Benjamin Silliman, *American Journal of Science and Arts* 27 (January 1835): 394.

25 Original handwritten ANSP *Minutes*, vol. 6 (28 January 1834): 413.

26 ANSP *Proceedings* 1, no. 5 (3 August 1841): 63. Today, the original copperplates for Say's *American Conchology* are in the collection 441 in the Academy archives. Charles Pickering was the grandson of the Federalist politician Timothy Pickering, who had so vehemently opposed Jefferson on the Embargo Act.

27 ANSP *Proceedings* 1, no. 10 (25 January 1842): 145.

28 John Kirk Townsend, "Description of Twelve New Species of Birds, Chiefly from the Vicinity of the Columbia River," *Journal of the ANSP* 7, pt. 2 (1837): 187–93.

29 Ibid.

30 For more on Townsend and Audubon, see Barbara Mearns and Richard Mearns, *John Kirk Townsend, Collector of Audubon's Western Birds and Mammals* (Dumfries, Scotland: privately printed, 2007).

31 Mahlon Dickerson to Samuel George Morton, U.S. Navy Department, 31 August 1836. Samuel George Morton Papers, APS. For a history of the U.S. Exploring Expedition, see Nathaniel Philbrick, *Sea of Glory: America's Voyage of Discovery: The US Exploring Expedition, 1838–1842* (New York: Viking, 2003).

32 Patsy A. Gerstner, "The Academy of Natural Sciences of Philadelphia, 1820–1850," in *The Pursuit of Knowledge in the Early American Republic: American Scientific and Learned Societies from Colonial Times to the Civil War*, ed. Alexandra Oleson and Sanborn C. Brown (Baltimore: Johns Hopkins University Press, 1976), 187.

33 Reingold, ed. *Science in Nineteenth-Century America*, 108.

34 A. Hunter Dupree, *Asa Gray: 1810–1888* (Cambridge, MA: Belknap Press of Harvard University Press, 1959), 68.

35 Herman J. Viola and Carolyn Margolis, eds., *Magnificent Voyagers: The US Exploring Expedition, 1838–1842* (Washington, DC: Smithsonian Institution Press, 1985), 10.

36 Jessie Poesch, *Titian Ramsay Peale, 1799–1885, And His Journals of the Wilkes Expedition* (Philadelphia: American Philosophical Society, 1979), 67.

37 Charles Pickering to Samuel George Morton, "U.S. ship *Vincennes*, Feegie Islands," 7 August 1840, Samuel George Morton Papers, American Philosophical Society, B/M43.

38 Asa Gray, "Charles Pickering," ANSP *Proceedings* 13 (1878): 441–44.

39 Pickering to Morton, 7 August 1840, APS, B/M843.

40 Pickering to Morton, Sandwich Islands, 12 October 1840, Samuel George Morton Papers, APS, B/M843.

41 Poesch, *Titian Peale*, 66.

42 Viola, *Magnificent Voyagers*, 21–22.

43 Ibid., 22.

44 Original handwritten ANSP *Minutes*, vol. 7 (17 February 1846): 461.

45 Poesch, *Titian Peale*, 96.

46 Peale's remarks quoted in ibid., 102.

47 Viola, *Magnificent Voyagers*, 50.

48 ANSP *Proceedings* 3, no. 3 (30 June 1846): 75.

49 Patricia Tyson Stroud, *The Emperor of Nature: Charles-Lucien Bonaparte and His World* (Philadelphia: University of Pennsylvania Press, 2000), 268–69. George Ord to Charles Waterton, Philadelphia, 6 October 1850, George Ord papers, APS.

50 Stroud, *Emperor of Nature*, 186.

51 Ibid., 209–10.

52 ANSP *Proceedings* 3, no. 2 (7 April 1846): 35.

53 ANSP *Proceedings* 3, no. 11 (26 October 1847): 299.

54 ANSP *Proceedings* 1, no. 33 (26 December 1841): 323.

55 ANSP *Proceedings* 1, no. 28 (25 July 1841): 283.

56 ANSP *Proceedings* 2, no. 2 (9 March 1944), 21; and vol. 2, no. 3 (21 May 1844), 21. Peterson's publishing house, Childs and Peterson, published Eliza Kent Kane's important *Arctic Explorations* in 1857.

57 ANSP *Proceedings* 3, no. 3 (9 June 1846): 21.

58 Ruschenberger, *A Notice*, 19.

59 ANSP *Proceedings* 3, no. 2 (1 September 1846): 97.

60 ANSP *Proceedings* 2, no. 2 (5 March 1844): 19.

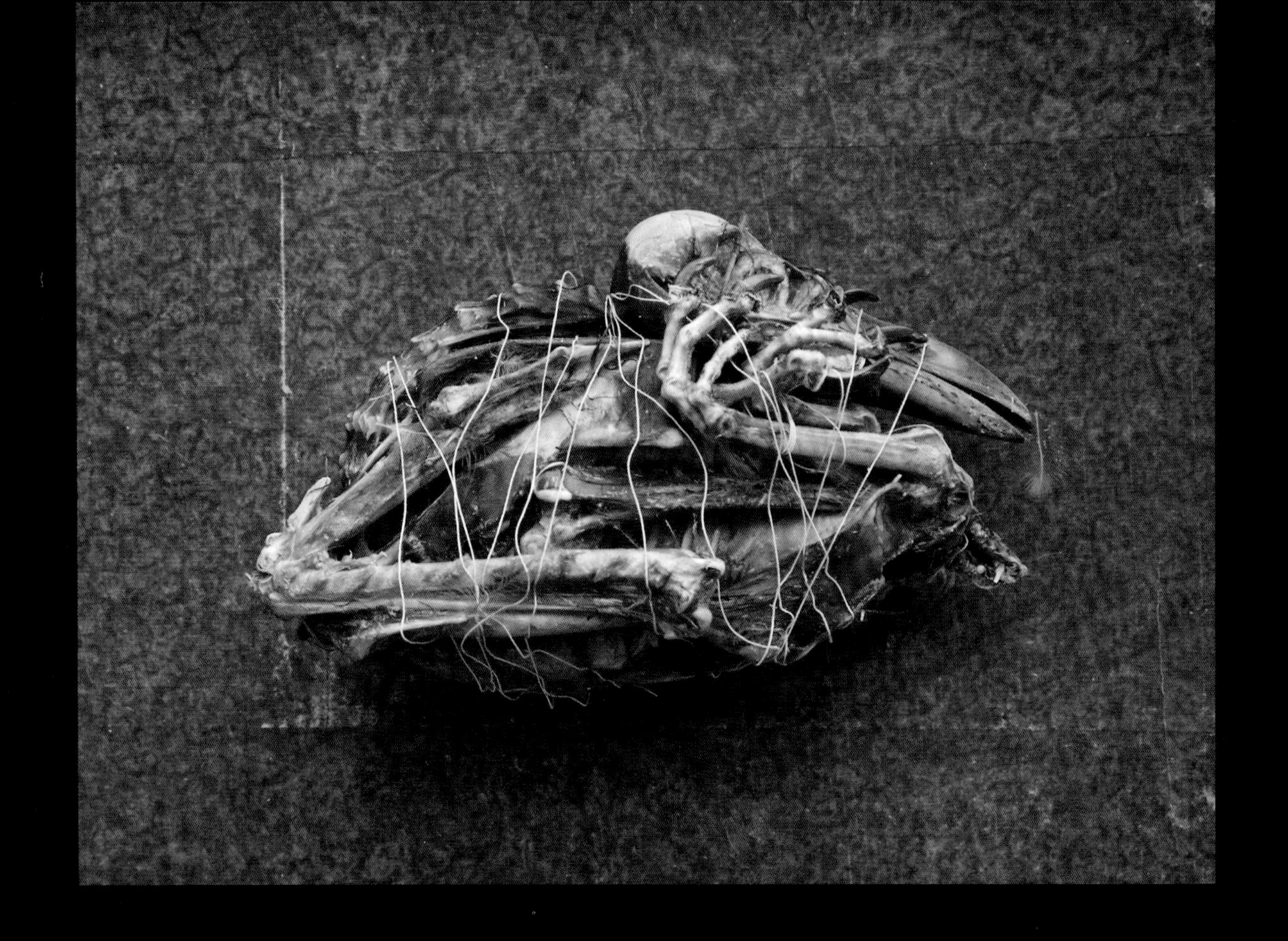

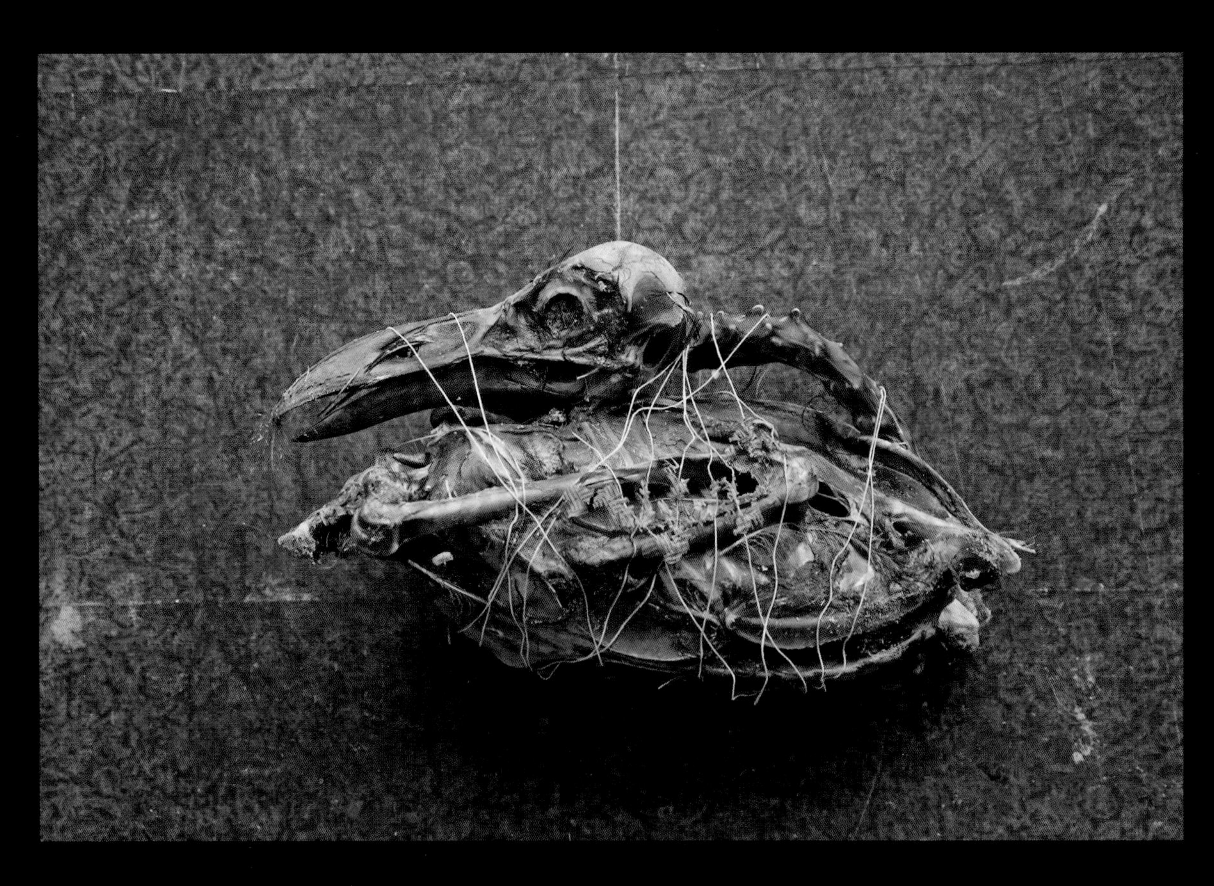

NOT WANTE

2ND

S.

NAME

FROM

Chapter 4

The American Woodsman Comes Calling: John James Audubon and the Academy

I think your work extraordinary for one self-taught, but we in Philadelphia are used to seeing very correct drawing.
—Alexander Lawson to John James Audubon, April 1824

The Society of Natural Sciences of Philadelphia has *at last* elected me one of their members.
—John James Audubon to Lucy Audubon, 7 November 1831

As John James Audubon stood on the brink of international fame through the publication of his great book *The Birds of America* in December 1826, his thoughts turned from the ink-spattered studio of the Scottish engraver William Lizars to the Academy of Natural Sciences, some five thousand miles and a four-week boat journey away. The self-described "American Woodsman" had become the toast of the town in Edinburgh, where he had been embraced by the leading naturalists of the city, wined and dined by the most prominent members of society, and declared an exceptional model of humanity by Scotland's leading phrenologist.[1] His paintings had been exhibited to great acclaim, and the first parts of the book he had been working on for almost two decades were about to be published. But despite these achievements, he could not shake from his mind the frosty reception he had received at the Academy of Natural Sciences a little more than two and a half years before. Taking pen in hand, he drafted a letter to his friend and fellow painter Thomas Sully: "As soon as my first number will be finished, say one month hence—I will forward you a copy," he began:

> I have to beg of you that you will take the trouble of presenting it in my name to that institution who thought me unworthy [of] being one of their members. There is no malice in my heart and I wish no return from them. . . . Merely let them know (if you please) that humble talents ought to be fostered first in one's own country.[2]

The rejection that had so deeply affected Audubon occurred in the spring of 1824, when the young artist made his first formal visit to the Academy. He arrived at the museum building in Gilliams Court in a state of high expectation, carrying a portfolio of drawings that he believed to be the best ever made of North American birds. He was accompanied by Charles-Lucien Bonaparte (1803–1857), the nephew of Napoleon and a prominent member of Philadelphia's scientific community, who had offered to introduce the newcomer to the members of the Academy in hope of winning their support for Audubon's ambitious publishing scheme.[3]

Aside from their common language and interest in birds, the two men could not have been more different. Bonaparte, whose family connections and genial personality made him a popular figure in Philadelphia's stratified society, was highly regarded as a knowledgeable and accomplished ornithologist. His election to membership in the Academy two months earlier had formalized his acceptance into the exclusive inner circle of American natural history. Bonaparte was twenty-one at the time.

His taller, older, and more flamboyant companion was neither wealthy nor socially prominent. His credentials in ornithology

4.1 Reddish Egret egg collected by John James Audubon in Florida. ANSP Ornithology Department #190717. Purcell photograph.

Florida
ANSP #190717
Ardea rufa Bodd
1041
Florida
J. J. Audubon

were yet to be established. John James Audubon was born in Santo Domingo in 1785, the bastard son of a French sea captain. At the age of eight, he was formally adopted by his father's forgiving wife, who had cared for the boy and his illegitimate half sister at the family home in France since 1789. In 1803, Audubon emigrated to the United States. He settled at Mill Grove, a Pennsylvania estate owned by his father. Except for a few years there and a trip to France in 1805, Audubon spent the next twenty years traveling throughout the United States, trying to make ends meet through a series of ill-fated business ventures (figure 4.2). Although his first two decades in America were financially unsuccessful (his business failures culminated in bankruptcy and a stint in debtors' prison in 1819), they afforded Audubon an opportunity to indulge his lifelong passion for birds.[4] After years of travel, observation, and painting, he had a collection he was proud to display. He was thrilled at the thought of showing his work to the most knowledgeable—and so, presumably, most appreciative—audience in America. He would soon learn, however, that his audience's knowledge did not translate into the appreciation he expected.

Ironically, the greatest obstacle to his acceptance in Philadelphia was the legacy of the very person who had inspired his mission there. Fourteen years earlier, in 1810, while operating a small general store in Louisville, Kentucky, Audubon had met an enterprising Scottish immigrant named Alexander Wilson (1766–1813). Wilson was writing and illustrating an ambitious treatise on North American birds (figure 4.3). His book, *American Ornithology,* published between 1808 and 1814, eventually included nine volumes. It was the first book devoted exclusively to the subject of North American birds.

At the time Audubon met him, Wilson was en route from Philadelphia to New Orleans to solicit subscribers for his publication (figure 4.4). Audubon was impressed by the sample volumes Wilson displayed, and he was on the verge of subscribing when his business partner, Ferdinand Rozier, discouraged him, pointing out that Audubon could hardly afford the $100 outlay and, more tactfully, that Audubon's own bird paintings were as good or better than Wilson's. Audubon realized that Rozier was right and refrained from buying Wilson's book, an action that later caused the moody Scot to complain in his journal that in Louisville, he "neither received one act of civility . . . , one subscriber, nor one new bird. . . . Science or literature has not one friend in this place."[5] Audubon recalled their meeting somewhat differently, explaining in his *Ornithological Biography* that he had shown Wilson every hospitality and even lent him some of his own paintings during the course of Wilson's stay.

What actually transpired between the two men in Louisville may never be known, but the meeting undoubtedly gave Audubon the idea that a book of his own paintings could far exceed Wilson's in drama, importance, and scope. This idea became an obsession, and demoralizing financial reverses only strengthened Audubon's resolve to make a success of his bird painting. For five years, he devoted himself entirely to that end. He had studied the birds and made the paintings; now what he needed was an engraver and financial backing to launch his book.

On his 1824 debut at the Academy, Audubon found himself surrounded by Wilson's friends and colleagues. Wilson had died in 1813, more than a decade before (and just one year after being elected to membership in the Academy), but the members still revered his memory and considered his work on American ornithology definitive.

4.2 *John James Audubon* (1785–1851) by John Syme, 1826. Oil on canvas. White House Historical Association.

FACING PAGE 4.3 *Red Owl (Screech Owl)* by Alexander Wilson, hand-colored engraving by Alexander Lawson for Wilson's *American Ornithology* (1808–13). ANSP Library.

1. Red Owl. 2. Warbling Flycatcher. 3. Purple Finch. 4. Brown Lark.

12

Attending (and possibly chairing) the meeting at which Audubon displayed his paintings was the Academy's vice president, George Ord (1781–1866), Wilson's closest friend, biographer, editor, and executor and the ghostwriter of the posthumously published ninth volume of *American Ornithology* (figure 4.5). There is no official record of what took place at the Academy on that April day,[6] but it is evident from subsequent events that Audubon's presentation was little short of disastrous. If he had let the quality of his work speak for itself, he might have carried the day, but one can imagine that Audubon, unnerved by the quiet reserve of the assembled group, began to praise his own creations as he pulled them, one by one, from their protective case. The reactions of some who were there suggest that he also criticized Wilson's work, describing the illustrations as lifeless and dismissing much of the ornithological information contained in the accompanying text as incomplete or faulty. This was more than George Ord could tolerate. Short-tempered and vitriolic by nature, Ord was more than a match for Audubon in open debate. Incensed by the newcomer's brash and tactless remarks, he rose to Wilson's defense, challenging Audubon's scientific credentials and integrity. By the end of the meeting, it was clear that any possibility of the Academy supporting Audubon's project had vanished (figure 4.6).[7]

Despite the unfortunate tenor of the meeting, many Academy members recognized the importance of Audubon's work and befriended the artist. Bonaparte demonstrated his continuing support by offering to buy Audubon's paintings if Alexander Lawson (1772–1846), Philadelphia's most accomplished engraver, could be induced to engrave them. But here again, Audubon encountered the enduring influence of Alexander Wilson, for Lawson counted among his proudest credits the engraved plates for Wilson's book. When Audubon visited Lawson's studio to explore the possibility of having his own paintings engraved, his reception was cool. Lawson had heard about Audubon's unflattering opinion of Wilson's work and relished the opportunity to return the favor. "I think your work extraordinary for one self-taught," he told Audubon after a patronizing survey of his paintings, "but we in Philadelphia are used to

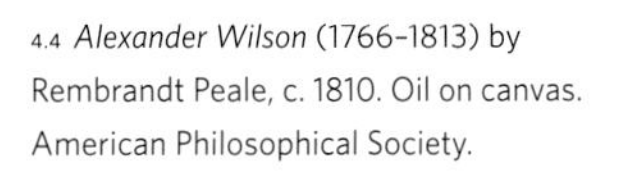

4.4 *Alexander Wilson* (1766–1813) by Rembrandt Peale, c. 1810. Oil on canvas. American Philosophical Society.

4.5 *George Ord* (1781–1866) by John Neagle, 1829. Oil on canvas. ANSP Library coll. 286.

FACING PAGE 4.6 Audubon was blackballed for membership in the Academy following his controversial visit in 1824. He was subsequently elected a corresponding member in 1831. Academy ballot box with balls (detail). ANSP Archives coll. 115. Purcell photograph.

seeing very correct drawing."[8] When Bonaparte explained his interest in buying Audubon's paintings, Lawson closed the discussion: "You may buy them," he snapped, "but I will not engrave them. . . . Ornithology requires truth in forms and correctness in the lines—here are neither!"[9] Audubon was devastated by the double rejection he received in Philadelphia. "The world owes to me the adoption of the plan of drawing from animated nature," he wrote.[10]

In many ways, Lawson's petty refusal to engrave Audubon's paintings and the Academy's criticism of his sometimes excessive artistic license served him well in the end. The first forced him to look elsewhere for an engraver (resulting in prints far superior to any Lawson could have created); the second induced caution and increased care in his work.

Audubon's *The Birds of America*, published in parts, would eventually contain 435 hand-colored plates with life-size depictions of 1,065 individual birds (figure 4.7). It was printed in "double-elephant" format (approximately 39½ by 26½ inches), the largest that nineteenth-century technology would permit, but its conceptual scale exceeded even its physical dimensions. Never before had such an undertaking been attempted. Never again would a project of its kind be carried off with such brilliant success.

Ultimately published in Edinburgh and London from 1827 to 1839, *The Birds of America* took Audubon twelve years and the assistance of two engraving firms to complete.[11] During that time, the artist traveled extensively in Europe and twice journeyed back to the United States to seek subscribers and to improve and augment his collection of American bird paintings. While no record exists of Thomas Sully ever presenting the first number of Audubon's great work to the Academy, as Audubon had proposed in December 1826, Charles-Lucien Bonaparte brought his first number (a packet containing the first five plates of the book) to Philadelphia within a few months of its publication in Great Britain.[12] The institution's minutes record that "M. Bonaparte exhibited the first number of Audubon's Ornithology" to the assembled body on the sixteenth of October 1827. It was one of the earliest showings of Audubon prints in North America, but it apparently left his Philadelphia critics unimpressed. After Bonaparte displayed the life-size plates and some other business was transacted, the meeting was adjourned with no official comment on Audubon's magnificent work (figure 4.8).[13]

A year and a half later, Audubon's friend Richard Harlan (1796–1843) made an effort to have the Academy purchase its own copy of the naturalist's book. He presented the members with a prospectus for *The Birds of America*, along with a glowing review of the book by the eminent French naturalist and Academy member Baron Georges Cuvier. "I had commenced raising a subscription for the price of the work among my friends at the Acad. of Nat. Sciences of Phila and could have succeeded," wrote Harlan to a friend in October 1829, "but Mr. A. declined it on such terms. He has a great deal of pride on this subject."[14]

Perhaps Audubon's unwillingness to take Harlan's semi-private subscription on behalf of the Academy was due to his continuing resentment of the Academy's refusal to support his publication or elect him to membership in 1824. George Ord had not helped to reduce Audubon's animosity in the intervening years, for, with the assistance and encouragement of his English friend Charles Waterton, Ord had gone out of his way to identify errors in Audubon's published plates and had publicly accused him of plagiarism.[15]

While some of his criticisms may have had merit, Ord's relentless rants about Audubon and his work seemed increasingly petty and found less and less sympathy among the Academy's membership. As *The Birds of America* took shape in print after magnificent print, volume after impressive volume, the artist's stature steadily increased. By 1831, Audubon's contribution to ornithology was so secure that his friends in Philadelphia were able to overcome Ord's objections and elect him a corresponding member of the Academy.[16]

In the years following his initial rejection, Audubon resentfully declared he would never accept an Academy membership, even if offered it, but this was no longer his attitude in 1831.[17] On hearing from the Academy, he wrote to his wife to share the good news: "The Society of Natural Sciences of Philadelphia has *at last* elected me one of their members, the papers say Unanimously."[18] On the title page of all of his books published after that date, Audubon proudly listed his Academy membership among his many honorific associations.

Eager to please the savants in Philadelphia, Audubon had already presented the Academy with volume 1 of his *Ornithological Biography*.[19] In the summer after his election, he sent twenty bird skins from a recent journey to Florida.[20]

At about the same time as Audubon's election, the Academy took up a subscription among members to secure a copy of *The Birds of America* for the Academy's library.[21] This time, there was no objection from Audubon, who was grateful for both the money and the acceptance that such a subscription represented. The magnificent publication, bound in four volumes, would become one of the Academy's greatest treasures.

Audubon's relationship with the Academy waxed and waned in the years following his election, becoming quite intense in 1836 when the artist was eager to obtain access to a number of

4.7 *American Flamingo* by John James Audubon, hand-colored engraving by Robert Havell Jr. for *The Birds of America* (1827–38). ANSP Library.

Drawn from Nature by J. J. Audubon, F.R.S. F.L.S.
Engraved, Printed and Coloured by Robt Havell. 1838
American Flamingo.
PHŒNICOPTERUS RUBER, Linn.
Old Male.
1._Profile view of Bill at its greatest extension.
2._Superior, front view, of upper Mandible.
3._Interior, front view of upper Mandible.
4._Inferior front view of lower Mandible.
5._Interior, front view of lower Mandible with the Tongue in.
6._Profile view of Tongue.
7._Superior front view of Tongue.
8._Inferior front view of Tongue.
9._Perpendicular front view of the foot fully expanded.

4.8 *Carolina Parrot* [Parakeet] by John James Audubon, hand-colored engraving by Robert Havell Jr. for *The Birds of America* (1827–38). ANSP Library.

FACING PAGE 4.9 Carolina Parakeet (detail) collected by John James Audubon, Fort Leavenworth Kansas, 1843. ANSP Ornithology Department #136786. Purcell photograph.

Above Fort Leavenworth
May 4. 1843.

new bird specimens collected at the mouth of the Columbia River by Thomas Nuttall (1786–1859) and John Kirk Townsend (1809–1851) (figure 4.10). While these as yet unpublished discoveries were of enormous interest to everyone at the Academy (which had sponsored the two explorers on their overland trip), they were of even greater interest to Audubon, who was eager to include pictures of them in *The Birds of America* before that publication came to a close.

On 12 September 1836, he wrote to his friend and fellow member Edward Harris to make his case:

> You well know how anxious I am to make my work on the birds of our country as complete as possible within my power. . . . But I am becoming old, and though very willing, doubt whether I could support the fatigues connected with a journey of several years and separated from my dear family. Well, the desiderata has come to Philadelphia at least in part, and if I could be allowed to portray the new species now [that] there is an appendix to *The Birds of America*, I should be proud and happy to do so, but do you think that the Academy is likely to indulge me in this wish?[22]

Unfortunately for Audubon, Townsend had not come back to Philadelphia with the specimens, for he had extended his journey to explore the biologically rich islands of Hawaii. Townsend's friends in Philadelphia were understandably reluctant to let another ornithologist take his hard-won treasures and rush them into print, thus denying him and the Academy the credit for their discovery. This prompted a flurry of correspondence between Audubon and his friends in Philadelphia, as Audubon lobbied his supporters to gain access to the precious specimens. He even met with Nuttall in Boston to seek his help in negotiating a sale of the skins. In the end, a compromise was struck. A list of the birds would be published in the Academy's *Journal* with credit going to Townsend. Duplicate skins would be sold to Audubon for inclusion in his book, but Audubon had to agree to Nuttall's approval of all new names. Formal credit for the discovery of new species among the western bird skins would go to Townsend.

Even after Audubon completed *The Birds of America*, his dependence on Academy resources continued. Inspired by the success of his ornithological publications and driven by a need to create "something new rather than tread in old shoes upon people's heels,"[23] Audubon launched an ambitious plan to publish a definitive book on what he called the "viviparous quadrupeds" (literally, four-footed animals bearing living offspring) of North America (figure 4.12). Initially, he envisioned it as a single-volume work, "rather less than half the size of *The Birds of America* in about one hundred plates giving all that can be given in such fauna of the size of life accompanied by a one vol[ume] of letter press, the whole to be finished (God granting me life and health) in two years."[24] The publication eventually grew to a three-volume work, with an additional three volumes of text, and took fifteen years to complete.

Audubon's coauthor for the mammal book was John Bachman (1790–1874), a man widely recognized as "the most distinguished student of Mammalogy" in America[25] and a corresponding member of the Academy since 1832. A passionate and meticulous naturalist, Bachman was also the much-loved pastor of St. John's Lutheran church in Charleston, South Carolina. He had befriended Audubon on the artist's first visit to Charleston in October 1831.[26] As they prepared *The Viviparous Quadrupeds*

4.10 *John Kirk Townsend* (1809–1851), portrait miniature on ivory, possibly by Thomas E. Barratt, ca. 1833. ANSP Archives coll. 457.

FACING PAGE 4.11 White-tailed jackrabbit (*Lepus townsendii*) collected by Wharton Huber, Bear River Expedition (Utah), 1927. ANSP Mammalogy Department #14492. Purcell photograph.

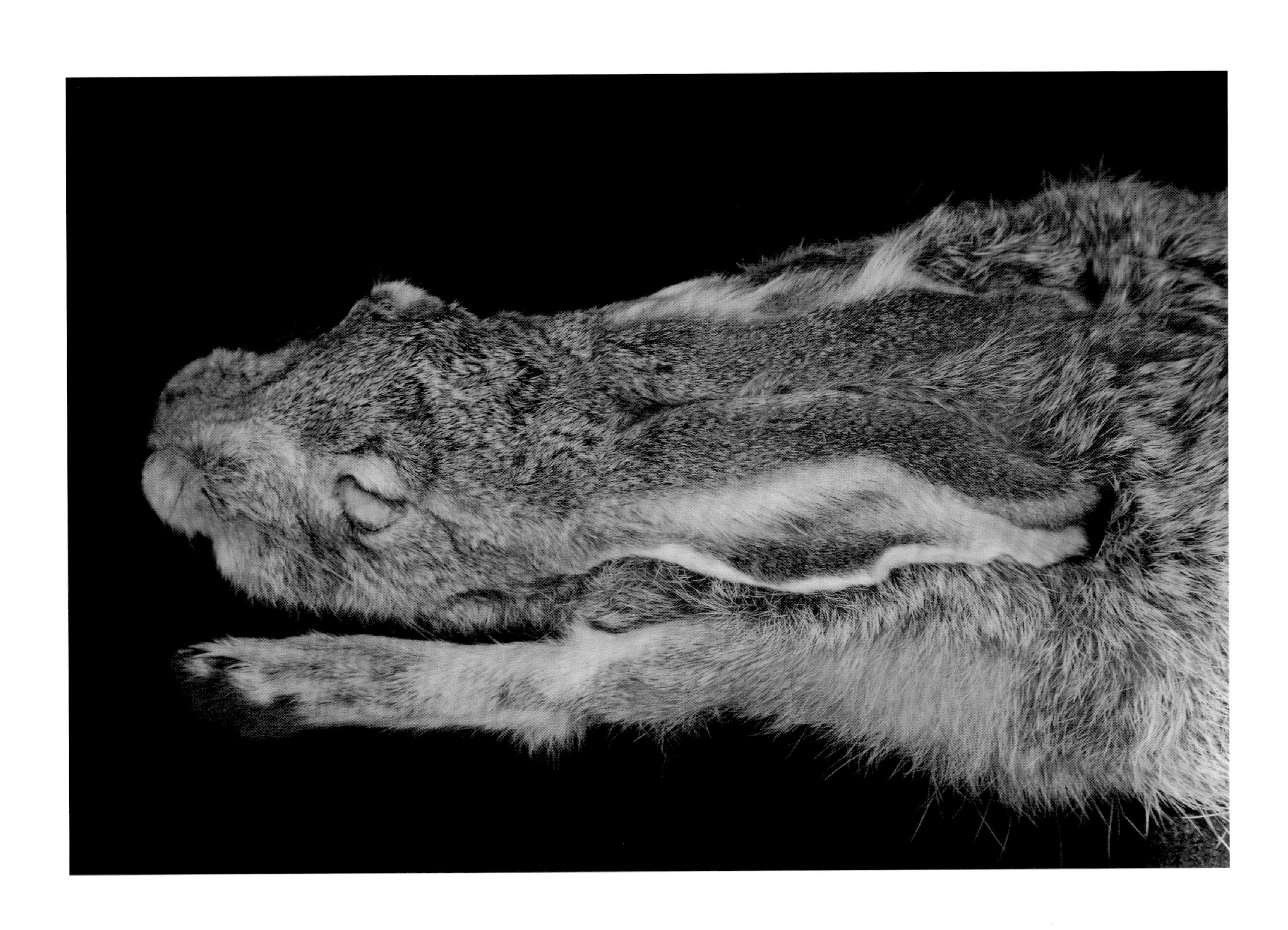

of North America for publication a decade later, both Bachman and Audubon corresponded extensively with other Academy members and gleaned information from the Academy's library, collections, and publications.

The Academy's response to Audubon's new book was enthusiastic and in marked contrast to the cold reception some of its members had given *The Birds of America* fourteen years earlier. The *Minutes* of the March 1843 meeting record that the chairman "called the attention of the Society to the first number of Mr. Audubon's 'Quadrupeds,' which is placed in the Hall for inspection, and for the subscription of such members and others as may desire to possess this truly splendid and invaluable work."[27] The Academy's own subscription copy still resides in the library, along with *The Birds of America*, while more than one hundred skins of the birds and mammals Audubon, Harris, and their party collected on the Missouri River expedition are housed elsewhere in the museum (see figure 4.9).

Another interesting bit of ephemera that attests to the artist's association with the institution survives in the form of three simply bound copies of the Academy's *Journal*.[28] Each volume bears Audubon's bold signature proclaiming his ownership of the journals and celebrating the coveted Academy membership they represent. As if to chide the Academy for his belated recognition, Audubon added the letters *FRS* to his name to remind anyone else who might see the volumes that more than a year before his acceptance by the Academy, he had been elected a Fellow of the Royal Society in England.[29]

At the time of Audubon's death in 1851, the institution that had once been so reluctant to embrace him passed a unanimous resolution praising his achievements and expressing "profound regret" over the death of "their esteemed and venerable Colleague."[30]

Audubon's nemesis, George Ord, outlived the artist by fifteen years, but he did not sustain his influence. The minutes indicate that the man one colleague called "the Nestor of American Naturalists"[31]—but that others now dismissed as a vitriolic curmudgeon—did not attend the meeting at which the resolution praising Audubon was passed.[32]

RMP

Notes

1 John Chalmers, *Audubon in Edinburgh* (Edinburgh: NMS Publishing, 2003), 50–52.

2 This draft appears in Audubon's hand in his 1826 diary, now owned by the Field Museum in Chicago. It is not known whether the letter was ever sent. There is no record in the Academy minutes of Thomas Sully ever presenting anything from Audubon.

3 For more information about Bonaparte, see Patricia Tyson Stroud, *The Emperor of Nature: Charles-Lucien Bonaparte and His World* (Philadelphia: University of Pennsylvania Press, 2000).

4 For examples of his earliest ornithological illustrations, see Scott V. Edwards et al., *Audubon: Early Drawings* (Cambridge, MA: Harvard University Press, 2008).

5 Alexander Wilson, *American Ornithology* (Philadelphia, 1808–25), vol. 9, 39.

6 Although minutes for all Academy meetings exist for this period, there is no mention of Audubon's visit. This suggests that Audubon made his presentation during an informal meeting of the Academy or that George Ord was so upset by what transpired that he excised all references to Audubon from the official records of the Academy.

7 Audubon was nominated for corresponding membership in the Academy a short time after his visit (27 July 1824), but his nomination was rejected. See Academy nomination records, archives collection #115.

8 Alexander Lawson, quoted in Alice Ford, *John James Audubon* (Norman: University of Oklahoma Press, 1964), 145.

9 Alexander Lawson quoted in William Dunlap, *A History of the Rise and Progress of the Arts of Design in the United States* (New York: Dover, 1969 [1834]), vol. 2, 403–4; quoted in Richard Rhodes, *John James Audubon: The Making of an American* (New York: Alfred A. Knopf, 2004), 222.

10 Alexander B. Adams, *John James Audubon* (New York, 1966), 267.

11 For a detailed account of Audubon's struggle to publish *The Birds of America*, see Duff Hart-Davis, *Audubon's Elephant* (London: Weidenfeld & Nicolson, 2003).

12 The prints shown would have included the famous turkey cock and the yellow-billed cuckoo, among others.

13 ANSP *Minutes*, 16 October 1827, coll. 502, box 2, folder 1, vol. 3a, 875.

14 Letter from Harlan to William Swainson, 29 October 1829, Linnean Society of London, quoted in Waldemar H. Fries, *The Double Elephant Folio: The Story of Audubon's Birds of America* (Chicago: American Library Association, 1973), 225.

15 Ord and Waterton specifically accused Audubon of copying several of Wilson's compositions. They also criticized him for errors in scientific accuracy and of using, without attribution, the artistic and scientific work of others.

16 The election occurred at the meeting on 25 October 1831. Audubon was informed of it in a letter from the Academy's corresponding secretary (ad interim), Thomas McEuen. See *Minutes*, ANSP coll. #502, and Beinecke Rare Book and Manuscript Library, Yale University, Audubon Papers, gen. mss. 85, box 3, folder 118.

17 This sentiment was expressed in the draft of his letter to Thomas Sully, 20 December 1826, contained in his European diary now at the Field Museum in Chicago.

18 Letter from Audubon to Lucy Audubon, 7 November 1831, in *Letters of John James Audubon, 1826–1840* (Boston: Club of Odd Volumes, 1930), 148.

19 Volume 1 of the Academy's five-volume set of *Ornithological Biography* (1831–39) bears the following inscription: "For the Acad. Nat. Sc. Phila., From the Author, Oct. 1831." The minutes of the 4 October meeting indicate that that is when the presentation of that volume took place. Volume 5 bears the inscription "Acad. Nat. Sciences from the Author, July 27 1839." Volumes 3 and 4 have no inscriptions but bear bookplates indicating that these were gifts of the author. Inexplicably, volume 2 appears to have been purchased by subscription and presented to the Academy's library by several members. Perhaps the original copy was borrowed and lost and so had to be replaced. Possibly it was never received. Interestingly, all five volumes are heavily (and rather critically) annotated by Audubon's nemesis, George Ord.

20 *ANSP Minutes*, 17 July 1832.

21 The subscription appears to have begun in 1831, perhaps at the time of, or slightly before, Audubon's election to membership. Payments were still being made as late as 1847. See Fries, *The Double Elephant Folio*, 225–27.

22 Letter from Audubon to Harris, 12 September 1836, quoted in Rhodes, *John James Audubon*, 394.

23 Audubon to Benjamin Philips, 23 June 1839, location of original letter unknown, from typescript, Audubon papers, Princeton University Library, quoted in Robert McCracken Peck, "Audubon and Bachman: A Collaboration in Science," in *John James Audubon in the West: The Last Expedition*, ed. Sarah E. Boehme (New York: Harry N. Abrams, in association with the Buffalo Bill Historical Center, 2000), 72.

24 Ibid.

25 *Literary World: A Gazette for Authors, Readers and Publishers* 1, no. 6 (13 March 1847): 128. The same description of Bachman was repeated in *Southern Quarterly Review* 22 (October 1847): 290.

26 Their meeting occurred on 16 October 1831, just two weeks after Audubon's 4 October visit to the Academy and nine days before his election as a corresponding member. For an account of the Bachman-Audubon friendship, see Jay Shuler, *Had I the Wings: The Friendship of Bachman and Audubon* (Athens: University of Georgia Press, 1995).

27 Ibid.

28 They are volumes 1, 2, and 4 (published in 1817, 1821, and 1824, respectively). Volume 1 was given to the Academy by Dr. Gordon Sauer in 1992. Volumes 2 and 4 were given to the Academy in 2003 by Mr. and Mrs. William L. McLean III, Sandra McLean, Dr. and Mrs. James Baker, Mr. and Mrs. James Bodine, and Mr. and Mrs. Robert Wolf. The resolution to present the copies of the journal to Audubon appears in the minutes of the Academy of Natural Sciences, 1 November 1836, 520. For more on these journals, see Robert McCracken Peck, "Audubon Makes His Mark in Philadelphia," in *Other People's Books: Association Copies and the Stories They Tell* (Chicago: Caxton Club, 2011), 71–76.

29 Audubon was elected a member of the Royal Society in the spring of 1830. It was a membership of which he was justifiably proud, as few other Americans had received this honor up to that time. See Rhodes, *John James Audubon*, 341.

30 The original resolution, dated 4 February 1851, is in the Audubon papers, Beinecke Rare Book and Manuscript Library, Yale University. It was published in the *Proceedings of the ANSP* 5 (1851): 146.

31 John Cassin in *Proceedings of the ANSP* 5 (February 1851): 154–55.

32 Parts of this essay have appeared in other essays by the author, including an article on Audubon's watercolors in *FMR*, January 2009, and on Audubon's rabbit drawings in *Archives of Natural History* 31, no. 2 (October 2004).

FACING PAGE 4.12 White-tailed jackrabbit (*Lepus townsendii*) by John James Audubon, hand-colored lithograph by Bowen & Co. from *The Viviparous Quadrupeds of North America* (1846–54). ANSP Library.

Coyote skulls (*Canis latrans Say*) collected by John Kirk Townsend (1809–1851) in "Indian country" west of the Missouri River in 1834. ANSP Mammalogy Department. #2267

C. LUPUS
Germany
Prince de Wied
2258

Chapter 5

The American Golgotha: Defining Race in the Early Republic

The craniological treasures which you have been so fortunate as to unite in your collection, have in you found a worthy interpreter.
—Alexander von Humboldt to Samuel G. Morton, 17 January 1844

[Morton was] one of the brightest ornaments of Philadelphia and of the country.
—*American and Gazette*, obituary, 16 May 1851

A bank of unobtrusive gray metal cabinets in the basement of the University of Pennsylvania's Museum of Archaeology and Anthropology contains one of the most expensive and unsettling acquisitions ever made by the Academy of Natural Sciences. On shelf after wooden shelf, the cases hold hundreds of human heads that peer back at the occasional researcher in disquieting silence. Assembled over a period of twenty years by Academy curator and president Samuel George Morton (1799–1851), the collection was purchased for the Academy's museum from Morton's estate in 1853 for the then enormous sum of $4,000.[1] It contains well over one thousand skulls, all carefully numbered, labeled, and catalogued as part of Morton's pioneering—and controversial—study of human diversity.[2] There are tattooed Maori heads from New Zealand; Thugg skulls from India; mummified heads from Egypt and Peru; skulls from the Sandwich Islands, Mexico, and Malaysia; and hundreds of others gathered by Morton's network of correspondents and collectors from Asia, Africa, Europe, and the Americas. Almost one-third of the collection is made up of the skulls of American Indians. Except for some handwritten inscriptions and the yellowed classification label attached to the forehead of each skull, most of the individuals remain anonymous. They were assembled for scientific, not personal reasons—a massive set of data points to be compared and analyzed without sentiment. Nevertheless, a few are accompanied by poignant, personal stories.

Such is the case of skull #14, "the celebrated Mrs. Fortesque," an "Anglo-American" who, according to Morton, "was in succession seduced, thrown upon the town & abandoned" before going insane and dying in an almshouse at the age of thirty. "She was of respectable family & had been remarkably beautiful," noted the observant doctor, who was himself only one year older than his ill-fated study specimen. He may well have known her as a patient, for he began his medical career at the Philadelphia Almshouse, where she expired in 1830.[3]

Morton almost certainly did not know the occupants of the other heads, but his records still note many of their personalities, gleaned, we assume, from information sent by the thirty or so friends, colleagues, and fellow collectors who procured them on his behalf. According to Morton's catalogue, specimen #434, one of the largest and most striking skulls in his collection, was from a Dutchman "of noble family" and high military rank who was "handsome, not deficient in talent, and of an amiable disposition, but devoted to conviviality and dissipation, which finally destroyed him" while he was still "under thirty years of age" (figure 5.2).[4]

Some of the other craniums are from individuals who sound less sympathetic than the abandoned Mrs. Fortesque or the convivial, if dissipated, noble Dutchman. Morton noted that skull #42 was from a "Celtic Irishman, age 21, imprisoned for larceny, and in all respects a vicious and refractory character,"[5] and that

5.1 "A Native of South Australia." Human skull, Morton Collection #1328, University of Pennsylvania Museum. Purcell photograph.

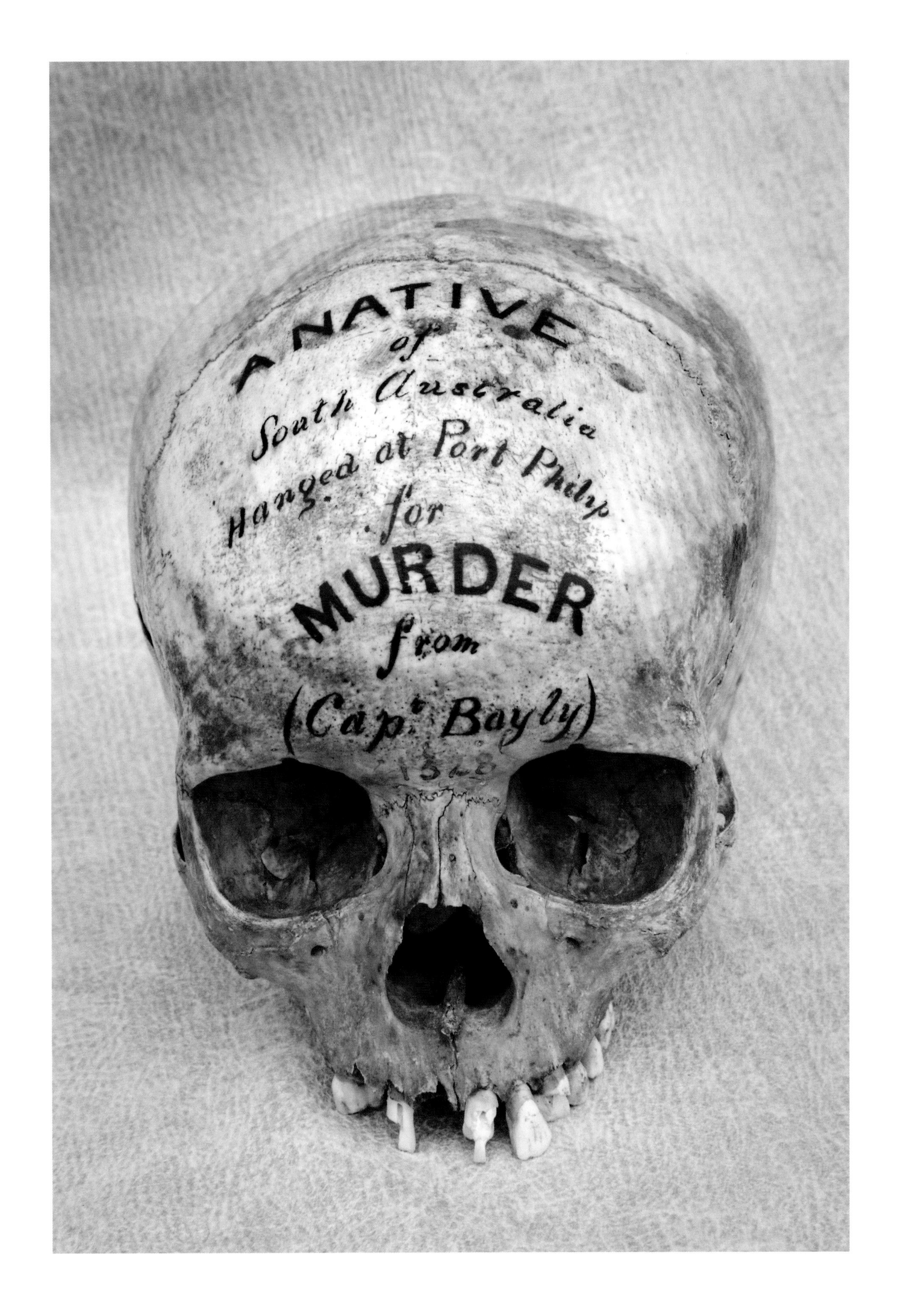
A NATIVE
of
South Australia
Hanged at Port Philip
for
MURDER
from
(Capt. Bayly)

skull #539 was from a man "who was executed for piracy and murder, at Philadelphia, May 19, 1837."[6] The inked inscription on a third skull proclaims that it was from "a NATIVE OF SOUTH AUSTRALIA, Hanged at Port Philip for MURDER" (figure 5.1).

The most notorious and famous specimen in Morton's collection is a British-Australian prisoner named Alexander Pierce (specimen #59; see figure 5.12), whose grisly story of devouring his fellow convicts was provided by the skull's donor.[7]

With or without such personal biographies, all of Morton's specimens served to support the systematic study of comparative anatomy that earned the doctor a worldwide following in his lifetime and an even larger number of critics in more recent years.[8] Many contemporaries considered this founder of modern physical anthropology a model of scientific integrity whose efforts to gather, quantify, and publish information about the branches of the human family tree made him "one of the brightest ornaments of Philadelphia and of the country."[9] At the time of his death, the *New York Tribune* wrote that "probably no scientific man in America enjoyed a higher reputation among scholars throughout the world."[10] In the 1970s and 1980s, a number of prominent scientific and social historians criticized him as a stealthy racist who applied a misleading veneer of scientific objectivity to a biased study of humans that helped to justify the horrors of slavery.[11] A more recent appraisal of his work has helped to restore his reputation as a careful scientist whose findings were misused for social causes in which he did not believe.[12] Regardless of its interpretation (and misinterpretation), Morton's skull collection still stands at the foundation of modern physical anthropology.

According to Morton, it was the need for representative teaching examples of the "five races of man" for a class in comparative anatomy in 1830 that first stimulated his interest in collecting skulls.[13] That initial interest ultimately led Morton to amass the largest collection of human crania in the world.[14] His collection became known as the "American Golgotha."[15] The collection was on view in the Academy's museum until the early twentieth century, at which time the skulls were put into storage. The collection was deposited at the University of Pennsylvania in the 1960s and formally given to the university in the 1990s. In recent years, the collection has received renewed attention. Between 2002 and 2007, with support from the National Science Foundation, all of Morton's skulls were digitally photographed in three dimensions. They are now available online, where they can be consulted by anthropologists and ethnographers from around the globe.[16]

His own head was captured on canvas for the Academy by the Hudson River School painter Paul Weber (1823–1916) (figure 5.3). Morton was described by his friend and fellow Academy member Dr. George B. Wood as being "of large frame, though somewhat stooping," with "bluish grey eyes, light hair, and a very fair complexion," and by another contemporary as being "mild and courteous in his demeanor."[17] According to those who knew him, Morton balanced his urbane sophistication with a quiet and unassuming manner. Like his near contemporary, Joseph Leidy, he commanded the respect and the loyalty of almost everyone who knew him. A seemingly gentle soul, he was, in the words of historian William Stanton, "an altogether improbable person to foment revolution in American science."[18]

In light of the unsettling human collections he brought to the Academy and the books he later published about them, it seems surprising that Morton was raised as a member of the Society of Friends, which promoted abolition of slavery and other liberal causes and held at its core a deep respect for the rights of individuals. His Irish-born father, George Morton, died just six months after his son's birth, leaving the boy's rearing to his Quaker mother, Jane Cummings, and her second husband, Thomas Rogers, an amateur mineralogist and also a Quaker. As a young man, Morton attended a number of different schools, including the Westtown School, a highly regarded Quaker boarding school outside Philadelphia, where he received spiritual guidance and a strong grounding in natural history. There he came to know many of the people with whom he would later associate at the Academy of Natural Sciences.[19] Morton went on to receive degrees in medicine from the Pennsylvania Medical School (in 1820) and the University of Edinburgh (in 1823). After two years of study and travel abroad, he returned to Philadelphia to establish a medical practice in his native city. There the Academy of Natural Sciences would become his institutional home.[20] His involvement with the Academy came, at least in part, through the influence of Richard Harlan, whom he met in medical school.[21] Harlan helped to reinforce Morton's early interest in natural history, but the two men later had a serious falling out over different approaches to medicine and their competing medical practices.[22]

Plagued by a lifetime of ill health, variously described as either a liver ailment or heart disease, Morton nevertheless was able to take time from his large medical practice and from his duties as professor of anatomy at Pennsylvania Medical College to serve as the Academy's recording secretary (1825–29), curator (1830–32), corresponding secretary (1831–40), vice president

5.2 "Dutchman." Human skull, Morton Collection #434, University of Pennsylvania Museum. Purcell photograph.

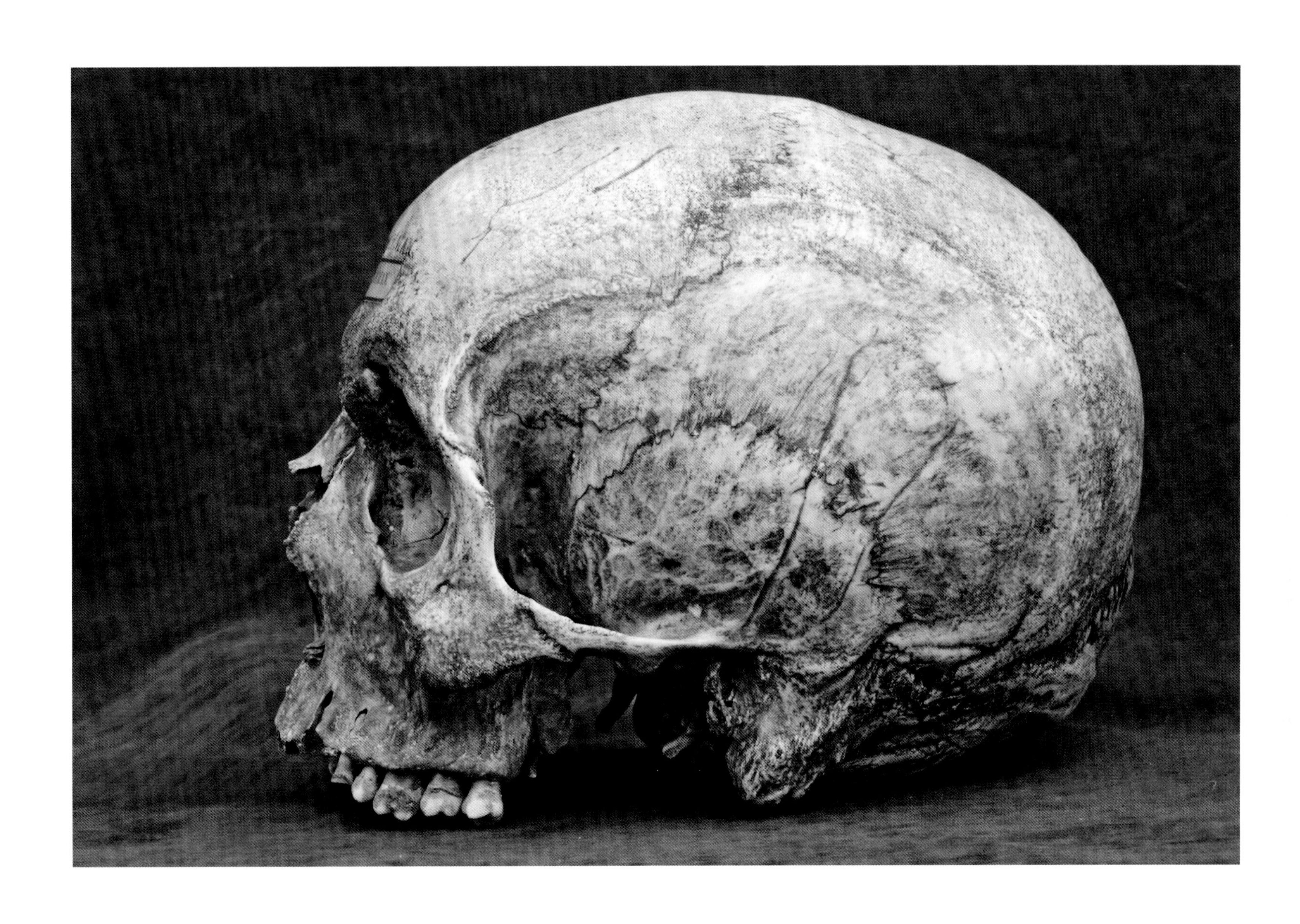

(1840–49), and president (1849–51). He also made time to write and publish articles on invertebrate paleontology, anatomy, mineralogy, geology, and ethnography in the Academy's *Proceedings* and *Journal*.[23]

Morton's earliest scientific recognition in a nonmedical field came with his work on invertebrate fossils, which he published in a number of articles accompanied by his own lithographic illustrations in the *Journal of the Academy of Natural Sciences* beginning in 1829.[24] He summarized and expanded this research in a book titled *Synopsis of the Organic Remains of the Cretaceous Group of the United States* (1834), which helped to establish Morton as the founder of invertebrate paleontology in the United States.[25] The geologist Benjamin Silliman of Yale, editor of the *American Journal of Science*, thought so highly of Morton's ability as a geologist and paleontologist that he asked Morton to keep him "informed of the progress of this branch of knowledge of which I . . . regard you as the particular guardian & supervisor."[26]

Despite the wealth derived from his lucrative medical practice and a sizable inheritance from an uncle in Ireland, Morton, his wife, and eight children lived in relative simplicity in keeping with their Quaker traditions and beliefs. In a letter inviting John James Audubon to stay with him in Philadelphia, Morton apologized in advance for his "plain way of living."[27]

Morton's poor health, growing family, and busy professional life discouraged his own travel to distant areas, so he found vicarious pleasure and great stimulation in exchanging letters and scientific specimens with friends from around the world. As corresponding secretary of the Academy, he was often the first to receive news of new discoveries and the adventures that accompanied them. In the 1830s, his fellow Westtown School alumnus John Kirk Townsend sent him bird, mammal, and insect specimens from the West Coast and Hawaii (and Morton arranged for some of these to be sold to Audubon). William S. W. Ruschenberger sent specimens of shells, plants, birds, and mammals from Peru and the South Pacific, where he was serving as a surgeon with the U.S. Navy. The American consuls from Lima, Cairo, and Constantinople each sent contributions, and Charles Pickering wrote him long and detailed letters from exotic ports of call during his five years as a naturalist on the U.S. Exploring Expedition (1838–42) and then sought his support when hoping to become secretary of the Smithsonian in 1846.[28] Some of his specimens

FACING PAGE 5.3 *Samuel George Morton* (1799–1851) by Paul Weber, 1851. Oil on canvas. ANSP Library coll. 286.

5.4 Black bear skull (detail), 1840. Morton Collection. ANSP Mammalogy Department #2209. Purcell photograph.

came from closer to home. In 1840, he procured the skull of a famous menagerie bear named Jack Downing that had killed a child in Camden, New Jersey. He carefully noted the bear's name and offense on its cranial crest and added it to the collection (figure 5.4).

As natural history specimens arrived from near and far, the lines of ownership and control became blurry. Morton separated vertebrate material from the rest, organizing it into a single collection that he kept in the Academy's museum. Because he spent his own money to augment the donated specimens—between $10,000 and $15,000, according to one friend—he began to think of the entire collection as his own, thus justifying his executors' posthumous sale of the collection to the Academy, even though many of the donors believed that their gifts had been to the museum and not to the individual who served as its corresponding secretary.[29]

When it came to crania, an impressive list of donors it was! Academy founder and first president Gerard Troost sent Indian skulls from Tennessee; William Maclure, the Academy's second president, sent skulls and other items of interest from Mexico, where he had moved for reasons of health. John James Audubon sent Morton at least one Indian skull and the remains of four Mexican soldiers killed at the battle of San Jacinto (1836) during the U.S. war with Mexico.[30] The explorer and U.S. presidential aspirant John Frémont sent Morton several Indian skulls from California,[31] while the American journalist, explorer, and archaeologist John Lloyd Stephens—whose classic book *Incidents of Travel in Central America, Chiapas, and Yucatan* was described by Edgar Allan Poe as "perhaps the most interesting travel book ever published"—provided Morton with Mayan skulls from Tikal.[32]

In the 1830s and 1840s, interest in human heads and willingness to plunder both recent and ancient tombs were more common than one might suppose (figure 5.5). Troost wrote that he had "opened a great number of graves,"[33] and William Byrd Powell of Tuscumbia, Alabama, proudly reported that his own collection of Indian heads numbered "about two hundred crania," including "Choctaw, Cree, Alabama, Uchee, Hitchetee, Cherokee, Chickasans, Steackapaw, Monumental and Natchez Indians," most of which he had obtained from ancient burial sites and mounds made more accessible by the Indian Removal Act of 1830.[34] With seemingly no qualms about disturbing the deceased, many collectors were willing to provide Morton with the cranial evidence he was seeking, some for money, and others for the honor of doing so.

By 1837, Morton had gathered enough cranial specimens to satisfy his needs for teaching comparative anatomy and to form an opinion on one of the important issues of early nineteenth-century ethnography—whether man had evolved from one or

5.5 Excavating Indian burial mounds, detail from Mississippi Valley panorama by John J. Egan, c. 1850. Tempera on lightweight fabric. St. Louis Art Museum, Eliza McMillan Trust.

from many sources. Those who adhered to the teachings of the Bible cited Genesis as confirmation of a single source. This was the "unity of man" theory. Other theories, for which Morton became the leading proponent, said that the physical diversity of humans suggested multiple origins.[35] He articulated these ideas about what became known as the "American School" of ethnography in a series of books and scholarly papers over the rest of his professional career.

Morton's first book on ethnography was a sumptuously illustrated volume on American Indians titled *Crania Americana or a comparative View of the Skulls of Various Aboriginal Nations of North and South America* (1839).[36] In it, Morton explained that a careful analysis of all the skulls in his collection had convinced him that the German anatomist Johann Friedrich Blumenbach (1752–1840) was correct in suggesting that there were five races of humans. Among these, Morton identified twenty distinct "families."

In preparing the text and plates for *Crania Americana*, and in his subsequent study of skulls "from tribes inhabiting almost every region of both Americas,"[37] Morton was surprised to discover that each possessed nearly identical osteological structures (figure 5.7). Based on this observation, he concluded that "the American Indian, from the southern extremity of the continent to the northern limit of his range, is the same exterior man."[38] Morton thus used the physical evidence contained in his collection to determine that despite their many different languages, tribal affiliations, and external appearances, "the American nations, excepting the Polar tribes, are of one Race and species" that stood "isolated from the rest of mankind."[39] It was a stunning observation that caused much excitement in the still small but growing field of ethnography.

Although Morton accepted "the fundamental principles of Phrenology, that the brain is the organ of the mind, and that its different parts perform different functions,"[40] he rejected the more popular applications of this pseudoscience, in which the shapes of individual skulls were considered accurate indicators of personality.[41] He nevertheless bowed to the public's fascination with such thinking by including in his book a

5.6 Funeral scaffold of a Sioux chief by Carl Bodmer, engraving from a portfolio of plates created to illustrate Prince Maximilian of Wied-Neuwied's *Travels in the Interior of North America, 1832–34*, given to the Academy by the prince in 1844. ANSP Library.

1420
L-606
Peruvian from Pisco
685
L-606
685. Peruvian from Pachacamac

5.8 "Interior of the Temple of Appolinopolis at Elfou," engraving from Vivent Denon, *Voyage dans la Bas et la Haute Egypte Pendant les Campagnes de Bonaparte 1798 et 1799* (1807). ANSP Library.

phrenological diagram and an essay on the subject by George Combe (1788–1858), a British lawyer and educator whose work he had encountered during his time in England.[42] While Combe was not the first to promote the theory, he had become the world's best-known proponent of phrenology, which was then viewed not as a parlor trick or a fraudulent scam for charlatans but as a potentially promising field of research.[43]

Because of Combe's popularity—his lectures attracted huge audiences in England, Europe, and the United States—the inclusion of his essay in Morton's book gave the book, and Morton, far more visibility than they might have had otherwise, shifting a highly technical publication from the realm of pure science to that of popular discourse. Although he went to some effort to distance himself from Combe and any analysis of "organs which are beyond our sight and reach,"[44] Morton's reputation as the dean of American anatomists was undoubtedly boosted (and later tarnished) by his association with Combe.

Among the extensive network of interested amateurs who helped Morton acquire the specimens needed to pursue his ethnographic studies was a colorful Englishman named George Gliddon (1809–1857), who served as the U.S. vice-consul to Egypt from 1832 to 1840 and introduced Egyptology to America. Gliddon was a complex and, at times, contradictory character, described by one historian as "A name-dropper, a sponger, a swinger on the shirttails of the great, a braggart, pretender, and scatologist, Gliddon was also courageous, generous, warm-hearted and loyal, and a friend worth having."[45]

After living in Egypt for almost two decades, Gliddon traveled extensively in the United States in the 1840s with a moving panorama of the Nile River and a live performance in which he presented captivating stories of life in ancient Egypt. Standing onstage between two large tables, "one piled high with copies of the chief works on Egyptology, the other with relics of Egypt," he lectured to rapt audiences "while majestic scenes of Ancient Egypt moved slowly along the walls and soft strains of appropriate oriental music filled the hall".[46] The effect was mesmerizing. "Once placed within a hall thus adorned," wrote one enthusiastic viewer, "the visitor found himself in a new and magic region; the present vanished, and the men and the events of thirty and forty centuries back arose before his gaze. In such a scene, the most dull could not fail to be impressed, the coldest could not resist the contagion of enthusiasm."[47] Gliddon's audiences ranged from a few hundred to several thousand, averaging five hundred per performance. By 1849, it was estimated that more than one hundred thousand people had heard him speak.[48] Through Gliddon's popular publications, still more became aware of his

FACING PAGE 5.7 Two "Flat Head Indian" skulls from Pisco, Peru, Morton Collection #1420 and #1374, University of Pennsylvania Museum. Purcell photograph.

colorful accounts of Egypt's ancient past. His books sold in large numbers, and one of his articles went through twelve editions and sold twenty-four thousand copies.[49]

Morton first came into contact with Gliddon in 1837, when the vice-consul visited the United States to purchase machinery for cotton, rice, and oil mills on behalf of Mohammed Ali, viceroy of Egypt. What began as an informal request by Morton for Gliddon to keep an eye out for Egyptian crania, turned into a growing obsession for the acquisitive diplomat, who wrote Morton from Egypt in March 1839 that he had "gone on collecting, and putting by, without thinking of the number, until I have 93 skulls in my house . . . 93 relics of humanity, 'grinning horribly their ghostly smiles,' out of Cupboards . . . and shelves."[50]

Like many of the people who obtained skulls for Morton, Gliddon seemed to lack all scruples in the methods by which he obtained the desired relics, even confessing "a sort of rascally pleasure" in obtaining cranial specimens from recent burial sites and ancient tombs that today would be protected by law (figure 5.9).[51] In letter after letter to his new acquaintance in Philadelphia, he gloated over his escapades as a tomb robber:

> Many a chuckle I have had with my Snake Hunter [his assistant tomb robber] . . . at the success of our ruses [in] spite of the vigilance of Priests, Skieckas [?], Guards, and Families, to extract skulls from Convents, Tombs, Sanctuaries, and Mummy Pits, especially when we [were] required to keep our operations secret, lest poor "Mousa" [another assistant] should get the sack, which is our Egyptian mode of "Lynching."[52]

5.9 "Tomb of Sarbout el Cadem," engraving from Leon Laborde, *Voyage de l'Arabie Pétrée* (1830). ANSP Library.

Gliddon asked his "Snake Hunter," Haggi Moresen, to exhume skulls of contemporary Egyptians "from the modern cemeteries around Cairo" and billed Morton $25 to compensate Moresen "for the risks encountered if he had been discovered by the Police in procuring 37 specimens."[53] He took a direct hand in collecting more aged skulls, in sites where there were fewer chances of meeting government officials in need of "inducement" or a cadaver's angry next of kin. "You have been chiefly benefitted by a vagabond spirit that has kept me traveling since last summer," he wrote Morton in May 1840:

> [I have been] wandering in the Eastern and Western Deserts, visiting all the localities interesting to the Antiquary or exciting to the Hunter... I have been able to drag from the inmost recesses of tombs and Sanctuaries myself the best portion of the collection now sent. Some of these Skulls are above 4,500 years old affording vast fields of conjecture....[54]

In thanks for Gliddon's many efforts on his behalf, and to encourage his further collecting activities abroad, Morton nominated Gliddon for corresponding membership in the Academy in 1841, and Gliddon was duly elected.

The cranial specimens that Gliddon provided from Egypt became the basis for Morton's next book, *Crania Aegyptiaca* (1844). This publication was less elaborate and less beautiful than *Crania Americana*, but it was no less scholarly. It contained more of the detailed skull measurements for which he had become well known in his earlier book, but this time Morton focused on the skulls of ancient and contemporary Egypt. Central to both his American and Egyptian studies were Morton's calculations of relative intelligence based on brain size, a figure he calculated by filling the skull with lead shot and then weighing it.[55] The conclusion he drew from his measurements and articulated in both books confirmed earlier findings by Johann Blumenbach that there was a consistent pattern of different cranial capacities among the races, and that this somehow suggested a corresponding difference in intelligence.[56] In his 1981 book *The Mismeasure of Man*, the science historian and paleontologist Stephen Jay Gould suggests that Morton's measurements were skewed to match "every good Yankee's prejudice—whites on top, Indians in the middle, and blacks on the bottom; and among whites, Teutons and Anglo-Saxons on top, Jews in the middle, and Hindus on the bottom."[57] The racial bias that may or may not have underlain Morton's work has been a subject of controversy ever since.

At its most basic level, Morton's book challenged the orthodox views of the religious establishment, which supported both the unity of man and the acceptance of Archbishop Usher's claim that humans had been created just six thousand years before, in 4004 B.C.[58] If the physical aspects of man had remained unchanged for most of that period, as Morton's Egyptian specimens seemed to demonstrate, then how was it possible that so many human varieties had emerged in such a short time following their creation and then abruptly stopped changing? Or as one nineteenth-century writer observed in a review of a later book on the subject, "Either there were separate creations of different types of mankind, or men must have existed on earth for Chiliads of years."[59]

Morton's increasingly influential publications brought both criticism and praise from his fellow Academy members. The Reverend John Bachman, a longtime correspondent of Morton's, engaged in a series of rebuttals of Morton's theories in the *Charleston Medical Journal* and *Charleston Courier*. By contrast, the eminent German explorer and naturalist Alexander Von Humboldt (1769–1859), an Academy corresponding member since 1843, wrote Morton that

> The craniological treasures which you have been so fortunate as to unite in your collection, have in you found a worthy interpreter. Your work is equally remarkable for the profundity of its anatomical views, the numerical detail of the relations of organic conformation, and the absence of those poetical reveries which are the myths of modern physiology.[60]

Humboldt marveled at the "evidence of artistic perfection" in *Crania Americana*, and in an awkward attempt to compliment Morton, he patronizingly expressed his surprise that "you could produce a work that is a fitting rival of whatever most beautiful has been produced either in France or England" (figure 5.10).[61]

Nineteenth-century interpretations of Morton's analysis seem dreadfully misguided by the standards of the twenty-first century, but in his day Morton was almost universally lauded for his scrupulous objectivity. He made no claims he did not believe he could substantiate with facts derived from the tangible evidence of his collection. Even his harshest twentieth-century critics claimed "no sign of fraud or conscious manipulation" in his

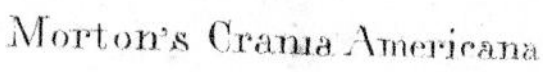

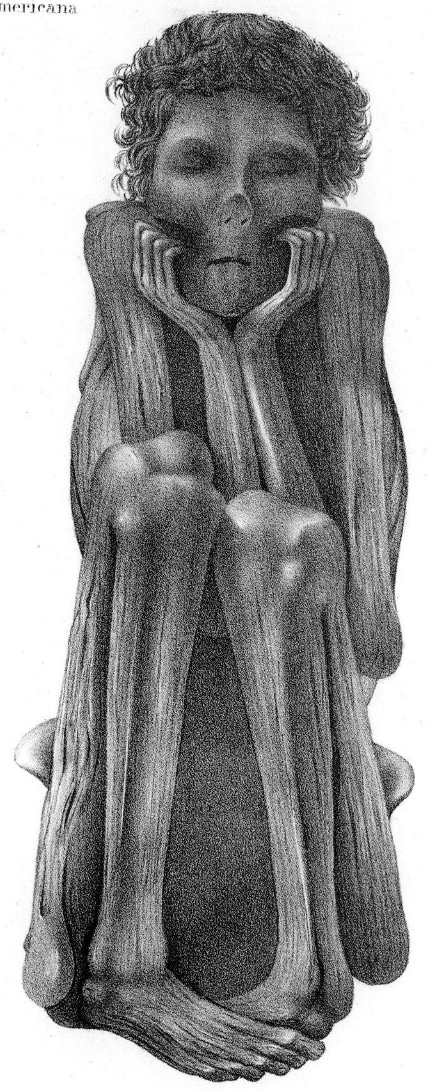

5.10 "Natural Mummy, Muysca Indian," lithograph from Samuel G. Morton, *Crania Americana* (1839). ANSP Library.

findings.[62] Subsequent analysis seems to confirm that his measurements were accurate.[63] Still, the use and misuse of Morton's findings on race has left a taint on his reputation.

After Morton's premature death from pleurisy in 1851, his associates George Gliddon and Josiah Nott carried on his work in ethnography. The two men often invoked Morton's name to support their own overtly biased views on matters of race and religion. Nott, who was elected a corresponding member of the Academy in 1845, is today remembered not just as a cofounder (with Morton) of the "American School" of anthropology but also, more positively, as the doctor who promoted the germ theory of disease and who discovered that yellow fever could be spread only by an intermediate host.[64] With additional contributions from Louis Agassiz, Gliddon and Nott published a massive, eight-hundred-page volume on ethnography in 1855. Titled *Types of Mankind*, it was dedicated "to the memory of Morton" and included a long essay on their mentor's life and scholarly achievements. The book was a compendium of existing ideas, including a few unpublished Morton manuscripts, brought together to support the authors' belief in the specific diversity of humankind. According to its opinionated author-editors, the book was intended to vindicate "scientific truth" over "fake theology."[65] Focusing as it did on subjects that were of increasing national and international concern in the years leading up to the American Civil War, *Types of Mankind* proved to be extremely popular. With a subscription list of 992 patrons, the first printing sold out immediately. It reappeared in at least nine editions before the end of the century, further solidifying Morton's reputation as the nineteenth-century dean of physical anthropology.[66]

At the Academy, it was hoped that Joseph Leidy might continue Morton's work, but Leidy declined this opportunity, "not from a want of interest in ethnographic science, but because other studies occupied my time."[67] Four years earlier, Morton and Leidy had had a falling out when Morton mounted what Leidy called "a very violent opposition" to Leidy's successful candidacy as chairman of the curators at the Academy, in favor of William Gambel (1823–1849), who had recently returned from a overland trip to California and wished to take a more active role in Academy affairs. Leidy may still have harbored hurt feelings about the incident and so had little inclination to advance Morton's interests within the Academy following Morton's death.[68] Instead, it was Morton's student and protégé, Dr. James Aitken Meigs (1829–1880), who continued to build his mentor's collection and contribute to the study of ethnology, which he described as "this youngest, most intricate, and most important of the sciences."[69] In addition to serving as the Academy's librarian from 1856 to 1859, Meigs chaired a standing committee on ethnology for all but three years between 1854 and 1876, when the committee was dissolved.

A less well-known Academy member with a parallel interest in the classification of race—and with a collection that was almost as unusual as Morton's—was a Philadelphia lawyer and naturalist named Peter Arrell Browne (1782–1860). In the turbulent decades of the 1830s, 1840s, and 1850s, as America struggled to stabilize its young democracy and establish its place in the world, Browne tried to understand the diverse population of this nation of immigrants by looking at its members' hair. His many donors—some willing, some not—ranged from well-known artists, writers, and scientists to lunatics and the criminally insane. They also included mayors, governors, senators, congressmen, and the first fourteen presidents of the United States (figure 5.11).

Browne's remarkable assemblage was made not for sentiment's sake or as a gathering of celebrity souvenirs but, like Morton's skulls, to supply keys to a number of scientific puzzles that he hoped to solve. It was assembled for posterity and with patriotic zeal by a man who devoted much of his life to trying to decipher mysterious patterns in the natural world.[70]

Browne's death in 1860, at the age of seventy-eight, brought to a close his unusual collection, but not before he published several papers and a book outlining his classification system.[71] He died convinced that much more research was wanting in this promising field. The individuals whose locks he had collected included Maori chiefs, Inuit seal hunters, Bavarian clock makers, and Corsica's most famous native son, Napoleon Bonaparte. But Browne's pride and joy was his presidential collection, which extended from George Washington to Zachary Taylor.

Unlike Morton's many skulls, which were given to the University of Pennsylvania in 1997, Browne's hair collection still resides in the archives of the Academy of Natural Sciences. As yet unexamined by forensic historians, the samples' physical properties and the DNA they contain may hold the key to questions Browne himself might never have thought to ask.

RMP

5.11 Hair sample from Samuel G. Morton, collected by Peter A. Browne, ANSP Archives coll. 756. Purcell photograph.

Notes

1 William Stanton, *The Leopard's Spots: Scientific Attitudes Toward Race in America, 1815–1859* (Chicago: University of Chicago Press, 1960), 28. The $4,000 purchase price was contributed by forty-two "gentlemen" whose names read like a who's who of Philadelphia society and Academy history (Biddle, Meigs, Lea, Mercer, Vaux, Farnum, Pepper, Cooke, Morris, Wood, Merrick, Wetherill, etc.). See J. Aitken Meigs, *Catalogue of Human Crania in the Collection of the Academy of Natural Sciences* (Philadelphia: J. B. Lippincott, 1857), 3; see also *Proceedings of the Academy of Natural Sciences* 6 (1853): 321, 324.

2 At the time of Morton's death, the collection contained 867 human crania, as well as a large number of nonhuman vertebrate skulls. Following Morton's death, his friend and fellow member of the Academy Dr. James Aitken Meigs continued to build the collection and published Morton's catalogue until it numbered 1,225 specimens. The collection was deposited at the University Museum in 1966 and given to that institution in 1997. For more on the collection and its current use at the university, see Emily S. Renschler and Janet Monge, "The Samuel George Morton Cranial Collection," *Expedition* 50, no. 3 (Winter 2008): 30–38.

3 Quoted in Stanton, *The Leopard's Spots*, 29.

4 Samuel George Morton, *Catalogue of Skulls of Man and the Inferior Animals in the Collection of Samuel George Morton* (Philadelphia: Turner & Fisher, 1840), 13.

5 Samuel George Morton, *Catalogue of Skulls of Man and the Inferior Animals in the Collection of Samuel George Morton* (Philadelphia: Merrihew and Thompson, 1849).

6 Ibid.

7 This history was published in Morton, *Catalogue of Skulls* (1840), 8. The same skull is also described in the third edition of Morton's catalog, published by Merrihew & Thompson in 1849, specimen 59. The information was provided by a Mr. William Cobb Hurry of Calcutta, who evidently acquired the skull for Morton. The original letter to Morton from which he gleaned this information, dated 5 April 1839, is contained in the Library Company's Morton Papers housed at the Historical Society of Pennsylvania (Yi2 7388 F9). A book about Pierce, published by a descendant, claims that he was not guilty of the terrible crimes of which he was accused and for which he was executed, but that he was, in fact, a political prisoner fighting for penal reform who was illegally killed by his British captors to silence him. Dan Sprod, *Alexander Pearce of Macquarie Harbour: Convict, Bushranger, Cannibal* (privately printed: Cat and Fiddle Press, 1977).

8 Among Morton's most outspoken critics is Stephen Jay Gould, whose article "Morton's Ranking of Races by Cranial Capacity," *Science* 200 (5 May 1978): 503–9, was later put in more popular form to serve as the title essay in his book *The Mismeasure of Man* (New York: W. W. Norton, 1981). See also Ann Fabian, *The Skull Collectors—Race, Science and America's Unburied Dead* (Chicago: University of Chicago Press, 2010), and Nicholas Wade, "Scientists Measure the Accuracy of a Racial Claim," *New York Times*, 14 June 2011.

9 Anon., "Death of Dr. Ord," *American and Gazette*, Philadelphia, 16 May 1851.

10 *New York Tribune*, quoted in Gelenn White, "American Golgotha," *SK&F Psychiatric Reporter*, no. 18 (Jan.–Feb. 1965): 9.

11 See especially Stephen Jay Gould, *The Mismeasure of Man*.

12 See Renschler and Monge, "Samuel George Morton Cranial Collection," 30–38. See also Jason E. Lewis et al., "The Mismeasure of Science: Stephen Jay Gould versus Samuel George Morton on Skulls and Bias," in PLoS Biology, June 2011, http://www.plosbiology.org. In this paper, the authors concluded that contrary to Gould's claims, there was "no clear evidence that Morton manipulated his data or analysis to support a priori racial beliefs. Instead, we discovered errors by Gould which support his own a priori beliefs" (from the abstract).

13 Morton, introduction to *Catalogue of Skulls* (1849), iii.

14 Stanton, *The Leopard's Spots*, 28.

15 "Golgotha" is an archaic term for graveyard that was popularized by Shakespeare in *Richard II* (4,1—Bishop Carlisle). The German craniologist Johann Friedrich Blumenbach (1752–1840) amassed a similar, if smaller, collection of animal and human skulls in the late eighteenth century. It was known as "Dr. B's Golgotha," which probably prompted the slightly different title for Morton's collection. See Richard Holmes, *The Age of Wonder* (New York: Pantheon Books, 2008), 310.

16 The skulls were scanned using CT (computed tomography) technology. The scanning project grew from a 2002 grant from the University of Pennsylvania's Research Council, which helped to establish the Open Research Scan Archive (ORSA), a collection of high-resolution CT scans of human and nonhuman cranial and postcranial remains. A multiyear National Science Foundation grant (no. 0447271) to Tom Schoenemann (James Madison University) and Janet Monge (University of Pennsylvania Museum) created a database that provides worldwide access to scholars interested in comparative CT scan data. For more information on this project, see Janet Monge, "ORSA: The Open Research Scan Archive," *Expedition* 50, no. 2 (Winter 2008): 35. The skulls are today in the University of Pennsylvania's Museum of Archaeology and Anthropology.

17 G. B. Wood, *Biographical Memoir of Samuel G. Morton* (Philadelphia: T. K. and P. G. Collins, 1853), 19; and Anon., "Death of Dr. Morton," *American Gazette*, 16 May 1851.

18 Stanton, *The Leopard's Spots*, 27. The same could be said of Charles Darwin, whose book *The Descent of Man*, published twenty years after Morton's death, made much of what Morton was trying to discover about human origins irrelevant.

19 According to Helen Hole, "The study of science [at Westtown School was] emphasized to an extent unusual in other schools of the time." See Helen G. Hole, *Westtown Through the Years, 1799–1942* (Westtown, PA: Westtown Alumni Assoc., 1942), 91. Among the other students at Westtown School who would later play an important role at the Academy of Natural Sciences were Thomas Say, Reuben Haines, John Kirk Townsend, Edward Drinker Cope, Isaac Hays, John Cassin, Isaac Sharpless, William L. Bailey, Howard Pennell, Samuel Rhoads, Isaac Wistar (Academy president, 1891–95), Francis Pennell, and T. Chalkey Palmer (Academy president, 1926–28). For a roster of Westtown alumni, see *Catalog of Westtown Through the Years, 1799–1942*, comp. Susanna Smedley, 1945. We are indebted to Mary Brooks and Kevin Gallagher at the Westtown School for the information they have provided about the school and its graduates.

20 Morton was elected a member of the Academy in 1820.

21 Harlan was an assistant instructor for Joseph Parrish, Morton's principal instructor and mentor at the medical school of the University of Pennsylvania. For Harlan's influence on Morton, see Charles D. Meigs, *A Memoir of Samuel George Morton, M.D.* (Philadelphia: T. K. and P. G. Collins, 1851), 15.

22 Whatever affection or mutual respect may have existed between the two physicians had evaporated by the end of 1831. A letter to Morton from Harlan dated 9 December of that year reveals that whatever their differences on scientific matters, at the core of their falling out was a dispute over medical procedures and competition for business in the medical field. In his brutally curt message, Harlan states that he had "long since lost all confidence in your probity as a man and talent as a Physician." He goes on to threaten Morton with legal action and the destruction of his reputation as a physician if he does not abstain from interfering with Harlan and his patients. See letter from Harlan to Morton, 9 December 1831, Morton Papers, American Philosophical Society.

23 For a complete list of Morton's contributions to Academy publications, see Edward J. Nolan, ed., *An Index to the Scientific Contents of the Journal and Proceedings of the Academy of Natural Sciences of Philadelphia, 1812–1912* (Philadelphia: Academy of Natural Sciences, 1913), 139–41.

24 In the copy of his *Synopsis of the Organic Remains of the Cretaceous Group of the United States* that he presented to the Academy (QE734 M891), Morton bound in plates 4 and 5 from the Academy's *Journal* (vol. 6), adding to one the following note: "This is among the earliest Lithograph executed in Philadelphia, and was drawn by me in 1827–28. S.G.M."

25 Stanton, *The Leopard's Spots*, 26.

26 Silliman to Morton, quoted in Stanton, *The Leopard's Spots*, 26.

27 Letter from Samuel Morton to J. J. Audubon, 10 September 1839, Audubon collection, Beinecke Rare Book Collection, Yale University, gen. mss. 85, box 5, folder 256.

28 The letters written by Pickering to Morton while he was on the U.S. Exploring Expedition are in the Morton Papers at the American Philosophical Society. The letter from Charles Pickering to Samuel Morton requesting support for his application to the Smithsonian is dated 9 September 1846, and can be found in the Morton Papers owned by the Library Company of Philadelphia and on deposit at the Historical Society of Pennsylvania (Ti2 7389.F28).

29 George B. Wood, *Biographical Memoir*, 13.

30 In the 1849 edition of Morton's catalog, the Indian skull (#1229) is dated 1845, but if it was collected by Audubon on his 1843 expedition, that must be the year of its presentation. The Mexican skulls are numbered 555–558, inclusive. Audubon tried unsuccessfully to obtain skulls for Morton during his time in London. See Ann Fabian, "The Curious Cabinet of Dr. Morton," in *Acts of Possession: Collecting in America*, ed. L. Dilworth (New Brunswick, NJ: Rutgers University Press, 2003), 116.

31 In the 1849 catalog, these are listed as #1446–1449, the last being sent through Edward M. Kern.

32 Victor Wolfgang von Hagen, *Maya Explorer: John Lloyd Stephens and the Lost Cities of Central America and Yucatan* (San Francisco: Chronicle Books, 1990), 198.

33 Letter from Gerard Troost to Samuel Morton, 6 December 1838, Morton Papers, American Philosophical Society.

34 Letter from William Byrd Powell to Samuel Morton, 6 August 1838, Morton Papers, American Philosophical Society. The 1830s saw an increased national focus on Native Americans as a number of legislative and military actions forced the "removal" of many Indian tribes from their traditional areas of residence in the American South and relocated them in the "West." Many skulls were illegally exhumed from Indian burial grounds, and many more were obtained as a result of the Indian fatalities that occurred during this period of conflict. The Indian Removal Act of 1830 made Indian skulls more readily available to collectors because the descendants of those buried were no longer present to defend their ancestral graves. For the names of other collectors of the period, see Ann Fabian, "The Curious Cabinet of Dr. Morton," 133n2. See also Fabian, *The Skull Collectors*.

35 The term often used for this is "polygenesis."

36 Samuel Morton, *Crania Americana, or a Comparative View of the Skulls of Various Aboriginal Nations of North and South* America (Philadelphia: J. Dobson, 1839). The seventy-two lithographic plates and two hundred "minor illustrations" in Morton's book were by John Collins (a few based on original drawings by other artists). The publication was financed by Morton with the assistance of his uncle, James Morton, of Clonmel, Ireland; the Academy's president, William Maclure; and fifteen subscribers.

37 Samuel George Morton, *An Inquiry into the Distinctive Characteristics of the Aboriginal Race of America* (Philadelphia: John Pennington, 1844), 5.

38 Ibid., 7.

39 Ibid., 39, 7.

40 Morton, *Crania Americana*, i.

41 Morton actually lectured against the concept of phrenology at the Pennsylvania Medical College; see Morton, *Brief Remarks on Diversities of the Human* [illegible] (Philadelphia: Merrihew & Thompson, 1842), 16.

42 Morton may have been introduced to Combe by his friend John James Audubon, whose distinctive phrenological features Combe examined at a meeting of the Phrenology Society at Combe's home in Edinburgh in November 1826; see John Chalmers, *Audubon in Edinburgh* (Edinburgh: National Museums of Scotland, 2003), 50–51.

43 In her essay "The Curious Cabinet of Dr. Morton," Ann Fabian says that Morton "sometimes relied on phrenological principles to interpret the heads he collected," but his own writing seems to contradict this. See Fabian, *Acts of Possession*, 115.

44 Morton, *Crania Americana*, i; Morton repeated his warning with this same phrase in his *Brief Remarks* (1842), 16.

45 Stanton, *The Leopard's Spots*, 46.

46 Ibid., 49.

47 Luke Burke, introduction to George Gliddon, *Otia Aegyptiac: Discourses on Egyptian Archaeology and Hiroglyphical Discoveries* (London: J. Madden; New York: Bartlett and Welford, 1849), 5–6.

48 Stanton, *The Leopard's Spots*, 49.

49 The article was distributed through the magazine *New World*; see Stanton, *The Leopard's Spots*, 49.

50 George Gliddon to Samuel Morton, 31 March 1839, Morton Papers, American Philosophical Society.

51 Ibid.

52 Ibid.

53 Gliddon to Morton, 15 October 1838, Morton Papers, American Philosophical Society.

54 Gliddon to Morton, 21–24 May 1840, Morton Papers, American Philosophical Society.

55 In his earlier experiments, Morton had used mustard seeds to calculate the volume, but he found these too variable and less accurate than the uniform volume of BBs, which provided replicable results.

56 Blumenbach had published his findings in his influential book *Comparative Anatomy*, 1807.

57 Gould, *The Mismeasure of Man*, 53–54.

58 Stanton, *The Leopard's Spots*, 213.

59 A. L., "On the Unity of the Human Race," *Quarterly Review* X (1854): 273–304, quoted in Stanton, *The Leopard's Spots*, 169.

60 Letter from Alexander von Humboldt to Samuel Morton, 17 January 1844, quoted in Meigs, *A Memoir*, 48.

61 Letter from Alexander von Humboldt to Samuel Morton, quoted in Henry S. Patterson, "Memoir of Morton," in Josiah C. Nott and George R. Gliddon, *Types of Mankind; or, Ethnological Researches* (Philadelphia: Lippincott, Grambo, 1854), xxxiv–v, quoted in Fabian, *The Skull Collectors*, 90. Humboldt had long had an interest in anthropology; see Holms, *The Age of Wonder*, and Peter J. Kritson et al., "Exploration, Headhunting, and Race Theory" in *Literature, Science, and Exploration in the Romantic Era* (Cambridge: Cambridge University Press, 2004).

62 Gould, *The Mismeasure of Man*, 69. This same conclusion was reached by a team of anthropologists who remeasured Morton's skulls in 2009–10. See Jason E. Lewis et al., "The Mismeasure of Science."

63 See Lewis et al., "The Mismeasure of Science."

64 Stanton, *The Leopard's Spots*, 65.

65 Quoted in ibid., 163.

66 Ibid., 163.

67 Joseph Leidy, quoted in Ales Hrdlika, "Physical Anthropology in America: An Historical Sketch," *American Anthropologist* 16:508-554 (1914): 516.

68 Letter from Joseph Leidy to S. S. Haldeman, 1 February 1847. ANSP coll. 1. In this letter, Leidy acknowledges that Morton "honestly owned" his opposition "and gave for the reason that he was compromised [biased] in the matter toward Mr. [William] Gambel."

69 J. Aitken Meigs, *Catalogue of Human Crania in the Collection of the Academy of Natural Sciences of Philadelphia* (Philadelphia: Merrihew & Thompson, 1857), 11.

70 Browne also made important contributions to the field of geology and was a founding member of the Geological Society of Pennsylvania.

71 Peter A. Browne, *Trichologia mammalium: or, a Treatise on the Organization, Properties and Uses of Hair and Wool: Together with an Essay upon the Raising and Breeding of Sheep* (Philadelphia: Commonwealth of Pennsylvania, 1853).

5.12 Skull of "Pierce" [Peirce], the cannibalistic Englishman from Australia. Morton Collection #1328. University of Pennsylvania Museum. Purcell photograph.

Bottled specimens from the Academy's herpetology collection. Purcell photograph.

Chapter 6

Gorillas Grab the Limelight: Paul Du Chaillu, John Cassin, and the Professionalization of Science

Nearly six feet high, with immense body, huge chest and great muscular arms, with fiercely glaring large deep grey eyes, and a hellish expression of face which seemed to me like some nightmare vision: thus stood before us the king of the African forest.
—Paul Du Chaillu in *Explorations and Adventures in Equatorial Africa*, 1861

At a country auction in the 1960s, an antiques collector named Elmer Payne noticed a Victorian pocket watch with an intriguing inscription inside. On a silver plate separating the works from the outside cover, a shield of engraved letters almost too small to see bore a message that read, in part:

> . . . This watch I Carried in my Explorations in AFRICA it has Often been used by me in Astronomical Observations. While wearing it I Shot Gorillas and many other Wild Beasts, it has been a good Companion to me, and may it prove so to Friend Arthur [Tallmadge].
> P. B. Du Chaillu
> Feb. 26th 1868[1]

Even though the names in the inscription were not familiar to Payne, he thought that the odd history of the timepiece was sufficiently interesting to justify a modest bid. He had little competition, and so within minutes of his having seen it, the watch was his. His satisfaction over its bargain price turned to disappointment when he discovered that the key required to wind the watch was missing. Berating himself for his hasty and impulsive purchase, he put it away and soon forgot it. Then, in 1978, he was reminded of his auction treasure when he saw the Academy's gorilla diorama during a family visit to the museum. A few days later, he contacted the Academy to offer the watch as a gift. Richard Estes, an expert on African wildlife who was then curator of mammals at the museum, received the call and happily accepted the offer on behalf of the Academy.

When it was deposited in the archives a short time later, the unusual timepiece joined a collection of letters revealing a long and, at times, contentious relationship between the Academy and its original owner, the French American explorer Paul Belloni Du Chaillu (1831–1903), who burst on the world stage in the late 1850s by introducing the gorilla to a wide-eyed public following an Academy-sponsored expedition to West Africa. Although the species had been described in a scientific paper a few years before, Du Chaillu was the first Westerner to see the gorilla in the wild and report his experience. His description of his first encounter with a gorilla was as dramatic and theatrical as anything anyone had ever offered (figure 6.2):

> He stood about a dozen yards from us and was a sight I think never to forget. Nearly six feet high, with immense body, huge chest and great muscular arms, with fiercely glaring large deep grey eyes, and a hellish expression of face which seemed to me like some nightmare vision: thus stood before us the king of the African forest. He was not afraid of us. He stood there, and beat his breast with

6.1 Paul Du Chaillu's pocket watch with a plate from *Explorations and Adventures in Equatorial Africa* (1861). ANSP Library. Purcell photograph.

his huge fists till it resounded like an immense bass-drum ... meanwhile giving roar after roar. ... His eyes began to flash fiercer fire as we stood motionless on the defensive, and the crest of short hair which stands on his forehead began to twitch rapidly up and down, while his powerful fangs were shown as he again sent forth a thunderous roar. And now truly he reminded me of nothing but some hellish dream creature—a being of that hideous order, half man half beast, which we find pictured by artists in some representations of the infernal regions.[2]

So sensational were Du Chaillu's accounts of the gorillas, and so effective was he at associating himself with them, that the two became inseparable in the public's mind. Together, Du Chaillu and his charismatic find became a sensation the likes of which the world had never seen before. That Charles Darwin would publish his theories on evolution just months after Du Chaillu's emergence from Africa no doubt added enormously to the impact of Du Chaillu's discovery, for it put the explorer at the center of a philosophical and religious debate that continues today.

In an era dominated by figures whose names and bigger-than-life personalities will be forever linked with the map of Africa—men like David Livingstone, Henry Morton Stanley, Richard Burton, and John Hanning Speke—Paul Du Chaillu is little remembered today, partly because his fame was achieved almost entirely through the discovery of a single species of animal rather than the geographic questions he was able to resolve or any national influence he was able to exert through his travels. Nevertheless, Du Chaillu was among the best-known explorers of the Victorian era. The name Mr. Payne found at the end of the long watch-case inscription, though unfamiliar to him, was once so famous as to be virtually synonymous with African exploration.

Although Du Chaillu was such a prominent public figure (the author of some thirteen books and a lecturer who could draw an audience of thousands), his personal history is surprisingly vague. Various biographers have given his place of birth as

6.2 "Fierce Attack of a Gorilla," engraving from Paul Du Chaillu's *Wild Life Under the Equator* (1868). ANSP Library.

New Orleans, the West Indies, West Africa, or France, and his birth date as anywhere from 1830 to 1840. It seems most likely that he was born in Paris about 1831.[3] His father's status as a trader and supply master for French naval and colonial stations in the Gabon estuary in West Africa is well documented. Unfortunately, we know nothing about his mother. Du Chaillu appears to have had two sisters, but since there is no official record of his birth, it is quite possible that the boy was illegitimate. In any case, after ten or thirteen years in France, he joined his father in Gabon, where he received some informal education from American missionaries. As a teen, he may have been given some minor clerical jobs by the French colonial government. After four or five years of this, at the urging of his American missionary friends, Du Chaillu decided to start a new life in the United States. In 1852, at the age of about twenty, he arrived in New York City with a few letters of introduction, some scientific specimens, and a valuable cargo of ebony wood. He soon found a job teaching French at a school for girls in Carmel, New York, where the students made fun of him for his diminutive stature and his poor command of English.

Du Chaillu soon found that what set him apart from others and earned him some measure of respect from his students, fellow teachers, and Americans in general were his experiences in Africa. Everyone he met wanted to hear about this colorful land and its exotic inhabitants, both wild and human. Du Chaillu was charming, resourceful, and a good storyteller who soon developed a reputation as an engaging raconteur (figure 6.3).

With the encouragement of his friends, he approached the editors of the *New York Tribune* and offered to write some articles about Gabon. He also traveled to Boston, Washington, and Philadelphia, where he gained the attention of natural historians with his firsthand accounts of the flora and fauna of West Africa.

In Philadelphia, Du Chaillu met with the members of the Academy of Natural Sciences, to whom he gave some of the bird and mammal specimens he had brought with him from West Africa. Excited by obtaining such novelties, several of the leading members of the Academy, especially the curator of birds, John Cassin, expressed interest in having Du Chaillu return to Africa to collect more examples of its little-known wildlife for additional study. Their interest may have been further enhanced by Du Chaillu's stories of gorillas, the first evidence of which had been recently presented to the Boston Society of Natural History by an American medical missionary, Thomas Savage, and described by Academy member Jeffries Wyman.[4] Savage had acquired skulls and other skeletal material in Gabon from another missionary, J. Leighton Wilson. Du Chaillu's friendship with Wilson, with whom he had lived for a while during his time in Gabon, and with Savage, whom he had met in West Africa during Savage's visit there, further enhanced his reputation as an authority on the flora and fauna of the region and a collector with the potential of bringing back some highly desirable specimens.

So, with both the academic imprimatur and the financial backing of the Academy, Du Chaillu embarked aboard an American sailing ship for Africa in October 1855.[5] For the next four years, the young explorer traveled up and down the West African coast and inland to areas with which Westerners had had very little, if any, interaction before. Du Chaillu's earlier experiences

6.3 Paul Belloni Du Chaillu (1831–1903), photographed shortly after his return from his first collecting trip to Africa in 1859. Private collection.

Academy of Nat. Sciences, Philadelphia.
DuChaillu's Coll.
Academy of Nat. Sciences, Philadelphia.
DuChaillu's Coll.
Academy of Nat. Sciences, Philadelphia.

in the region gave him several advantages in this risky venture. He could speak many of the local languages, and he had friends and contacts among the missionaries, traders, and tribal leaders in the area. These proved invaluable in enabling him to penetrate the unexplored interior (where he hoped to discover the source of the Congo River).

Du Chaillu's collections of birds and mammals, sent back to Philadelphia as the opportunities arose, pleased John Cassin and others at the Academy (figure 6.4). Cassin even named some of the new discoveries after their discoverer in a series of scientific papers published in the Academy's *Proceedings* between 1856 and 1859.[6] For a while, the relationship was a mutually beneficial and amicable one, but when it was discovered that the prolific collector was selling some of his most unusual specimens to other institutions and private collectors, his relationship with the Academy began to sour.

Of course the biggest prize—and the biggest bone of contention—was his gorilla specimens, which Du Chaillu quickly recognized would provide his opportunity to achieve both fame and fortune. When he made his triumphant return to America from Africa in the summer of 1859, after forty-two months away and with some twenty gorilla specimens in tow, Du Chaillu found a public eager to meet him and hear his tales of adventure. They especially wanted to know more about the new species of primate he had seen and collected in his travels. Charles Darwin's publication of *On the Origin of Species* in the very year of Du Chaillu's return helped stimulate an even greater public response than Du Chaillu could have imagined.

Seizing his opportunity, the young explorer honed his lecturing skills and toured the major cities of the United States with a series of hugely popular public talks. He also signed contracts with American, English, and (later) French publishers for a book in which he would spell out his adventures in riveting detail.

When Du Chaillu traveled to England in the spring of 1861, news of his discoveries and his reputation as a public speaker preceded him. On 11 May, he was invited to present highlights of his African adventures at a special meeting of the Royal Institution, and a few days later, he made a similar presentation to the Royal Geographical Society. With the renowned geologist Sir Roderick Murchison in the chair and such prominent

FACING PAGE 6.4 Bird skins collected in West Africa by Paul Du Chaillu. ANSP Ornithology Department. Purcell photograph.

6.5 "My First Gorilla," engraving from Paul Du Chaillu's *Explorations and Adventures in Equatorial Africa* (1861). ANSP Library.

personalities as the chancellor of the exchequer and future prime minister William Gladstone in the audience, Du Chaillu addressed standing-room-only crowds of more than eight hundred per lecture.

The consummate showman, he appeared onstage flanked by some twenty gorilla skeletons and mounted specimens and with a set of maps and large illustrations of African scenery and wildlife. In his heavy French accent, he presented enough hair-raising accounts of his close encounters with each animal to simultaneously amaze, delight, and horrify his overflow audiences.

When published by John Murray a few weeks later, Du Chaillu's book, *Explorations and Adventures in Equatorial Africa*, was an instant success. Though unusually large and expensive for its genre (it contained 480 pages and cost one guinea), the book sold more than ten thousand copies in Great Britain alone in its first two years in print.[7] London's infatuation with Du Chaillu and his gorillas was so widespread that the satirical magazine *Punch* seized on it in its 1861 edition. One of several gorilla cartoons in the issue shows a footman introducing a well-dressed gorilla to British society (figure 6.6).

In a long, illustrated review in that year, the *Illustrated London News* declared Du Chaillu's book a "very remarkable work" for which "ordinary terms of eulogy" were deemed inadequate.[8] Many other papers and magazines gave the book and its author equally enthusiastic attention. Unfortunately for Du Chaillu, not everyone was so positive about his achievements. Some questioned the geographic accuracy of his report. Others challenged the veracity of his scientific discoveries.

Within weeks of Du Chaillu's first lecture in England, almost every leading scientific figure with an interest in Africa or an opinion on evolution was involved in the debate—from Sir Richard Burton, who defended some of Du Chaillu's ethnographic and geographic claims, to Thomas Huxley, Darwin's champion and another Academy correspondent, who was severely critical of where Du Chaillu and his champion, Richard Owen, had tried to put the gorilla on the evolutionary ladder.

As others increasingly found fault with the geographic and biological details of Du Chaillu's narrative, the Academy's leadership began to distance itself from their onetime collector, even refusing to reimburse Du Chaillu for some of the expenses he had incurred during his African travels.[9] George Ord led the attack, encouraging his English friend, Charles Waterton, to mount an increasingly unpleasant and personal assault on Du Chaillu in the British press.[10] In his correspondence about Du Chaillu, Ord frequently invoked the name of his Academy colleague and friend John Cassin, who he said had suspected and rejected some of Du Chaillu's fraudulent claims of geographic discovery even before they were detected by others.[11]

Although Cassin never made any public statements about the affair, it must have been a particularly awkward situation for him, for it was he who had brought Du Chaillu into the Academy's orbit. (He had first met him in Paris in the taxidermy studio of Édouard and Jules Verreaux when negotiating the purchase of the Duc de Rivoli bird collection for the Academy in 1846.)[12] And it was Cassin who had benefited the most from Du Chaillu's efforts by publishing many of his important discoveries in the Academy's *Proceedings*. Now he feared being tainted by association with a suspected charlatan.

In a comment to George Ord, which Ord relayed to Waterton in July 1861, Cassin allegedly declared Du Chaillu "an entire humbug—no naturalist nor anything else."[13] Cassin's "humbug" criticism grew from Du Chaillu's disputed claims of having seen and killed gorillas in the wild (as opposed to buying them from local traders, as Cassin and others suspected). The "no naturalist" dismissal carried a more specific meaning, which would have been clearly understood by his contemporaries. By this, Cassin meant that Du Chaillu had no formal training in natural history and therefore lacked the skills to identify and scientifically classify the organisms he was collecting. Such skills—and the knowledge base that supported them—were jealously guarded by Cassin and the other so-called closet naturalists of the period.[14] Du Chaillu was by no means alone in being negatively categorized and dismissed in this way. Cassin had used exactly the same "no naturalist" phrase to describe John James Audubon sixteen years earlier.[15]

While explorer-collectors like Audubon and Du Chaillu cited their firsthand experience with wild nature as their most important credential as naturalists, Cassin rejected this notion. He believed a true naturalist could only be someone thoroughly conversant with scientific nomenclature and steeped in the professional literature of the day. He bridled at the suggestion that field naturalists could ever achieve a status of importance that was superior or even equal to his own:

> There is an indescribably pitiful display of ignorance and meanness of the idea in arrogating, as some writers have done, a superior position for the "field naturalist" over the "closet naturalist." As well might he who navigates a ship presume on being the greatest of astronomers, or the practical gauger pretend to be the only mathematician.

6.6 "The Lion of the Season," wood engraving, *Punch* (1861). Private collection.

THE LION OF THE SEASON.

Alarmed Flunkey. "MR. G-G-G-O-O-O-RILLA!"

> Great is life in the woods, say we, and the greatest of all sports is bird-collecting; but, to become a scientific ornithologist, is quite another business, and a very much more considerable consummation.[16]

As Cassin himself acknowledged, the superior position he so proudly assumed was made possible, in large measure, by his association with the Academy of Natural Sciences. In the preface to his book *Illustrations of the Birds of California, Texas, and Oregon, British and Russian America* (1856), he referenced the unrivaled resources of the institution at which he had been an unpaid curator since 1842. Although he was the sole author of the book, he invoked the editorial "we" in addressing his readers:

> Our advantages for study have been much superior to those possessed by former writers in America. There never was in the United States, until within the last ten years, a library of Natural History, approximating in any considerable degree to completeness nor affording the necessary facilities for the study of Ornithology. Nor until within that period was there any collection sufficiently comprehensive to answer the purpose of comparison and general research. . . . These most important and desirable objects have been fully accomplished in the formation of the Library and Museum of the Academy of Natural Sciences of Philadelphia.[17]

Without access to the specimens contained in the Academy's museum and the scientific literature contained in its library, amateurs like Du Chaillu (and even Audubon) had a very difficult time gaining the acceptance of the scientific elite, thus dooming them to perpetual outsider status.

Unable to compete in the academic arena, Du Chaillu believed he could regain legitimacy only by returning to the field to vindicate his earlier activities. Enriched by the sale of his specimens and the commercial success of his book, Du Chaillu decided to make a second expedition to Africa in 1863. This time

6.7 Senegal Bushbaby (*Galago senegalensis*), collected in West Africa by Paul Du Chaillu. ANSP Mammalogy Department #1044. Purcell photograph.

he would take the necessary scientific instruments to verify his geographic discoveries and a camera to prove beyond doubt the veracity of his anthropological and zoological claims. He now knew how aggressive his critics could be, and he was determined to answer their charges with proof positive that what he had described in his book was true.

Du Chaillu's second expedition, while far better planned and financed than his first, was plagued with difficulty from the start. While off-loading his tons of equipment on the African coast, one of the tenders capsized, and the critical scientific instruments he had been so careful to take with him were lost. He requested more from England, but these would take many months to arrive. While he waited, he practiced his newly learned skills as a photographer, working with some live gorillas he was able to procure with the help of native tribes.

Du Chaillu already spoke a number of local languages and taught himself more. This proved invaluable in his investigations, which were remarkably thorough and sympathetic for their time. He was among the very first Westerners to encounter a warlike tribe known as the Fang or Fan. His colorful descriptions of their alleged cannibalism added enormously to the public's interest in his first expedition. On his second trip, Du Chaillu made an equally sensational ethnographic discovery when he became the first European to encounter and describe Pygmies in their jungle fastness (figure 6.8). His claims about both groups, while contested by some, were ultimately confirmed by subsequent explorers.

While most of Du Chaillu's encounters with native people were peaceful, some were not. His second expedition came to an abrupt and violent end when, in an unfortunate mishap, one of his party killed two members of the Ashango tribe. The Ashangos, understandably, retaliated. Du Chaillu and his men were forced to retreat in disarray, abandoning the explorer's precious scientific instruments and photographs. The only piece of equipment to survive this disastrous episode was the pocket watch that Du Chaillu later presented to his friend Arthur Tallmadge, and that Elmer Payne gave to the Academy (figure 6.1).

Fortunately, Du Chaillu had taken the precaution of creating three copies of his field notes, one of which survived. These provided lots of interesting details for his next round of public lectures in England and America and served as the basis for his next book, *A Journey to Ashango Land* (1867).

By the mid-1860s, most of his most controversial claims from the first expedition had been confirmed as truthful and his

6.8 "Dinner with the Dwarfs," engraving from Paul Du Chaillu's *The Country of the Dwarfs* (1872). ANSP Library.

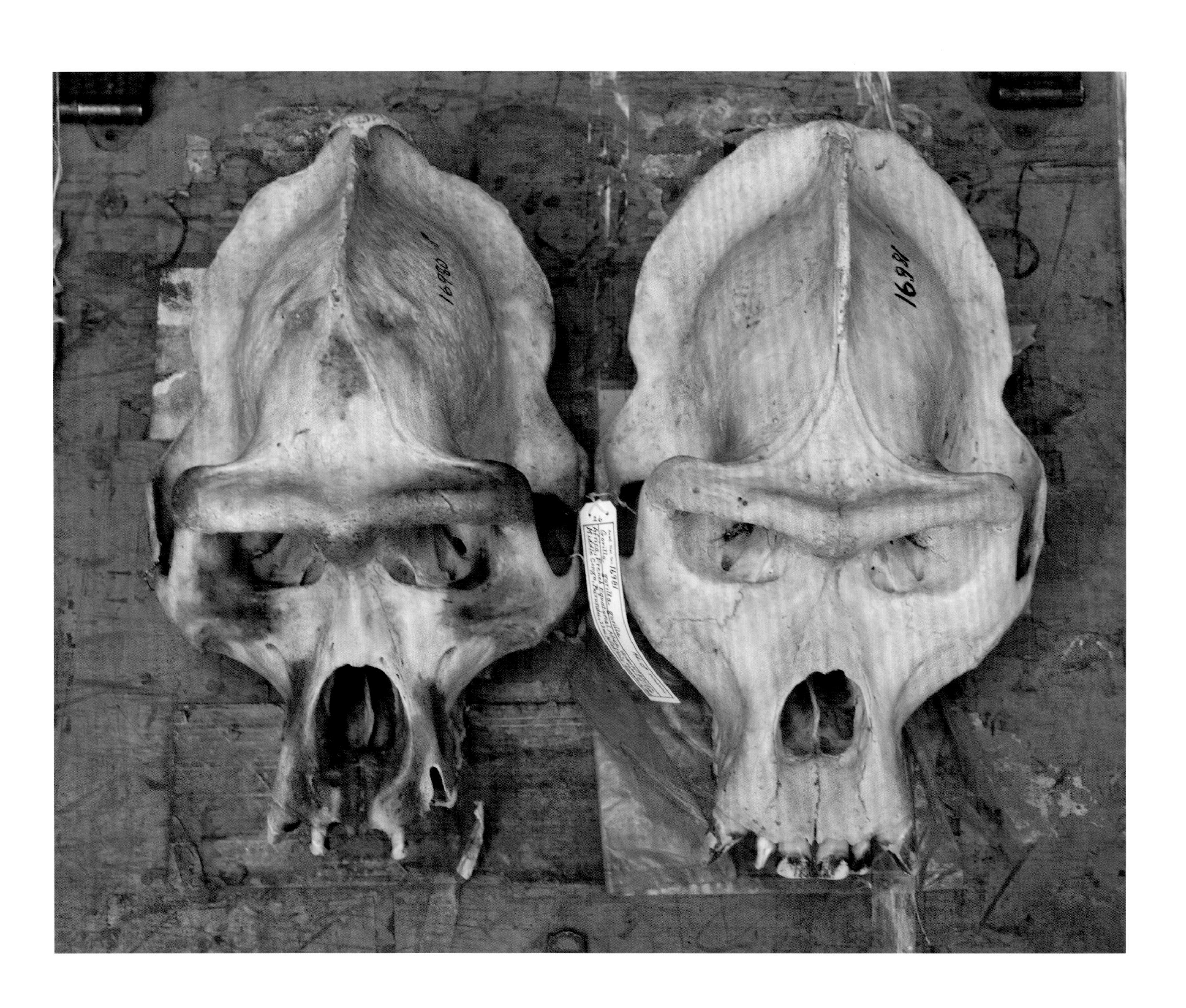
16980
16981

newest accounts of adventure were accepted without serious dissent. Du Chaillu's reputation as an explorer was now secure. A born storyteller, Du Chaillu spent the rest of his life telling and retelling the stories of his travels in Africa. He published a series of highly successful books on the subject, including several aimed at adolescents, thus creating a special niche market for young readers.[18]

It is hard to overstate the impact of Du Chaillu's writing on the public's perception of Africa. Only David Livingstone and Richard Burton match him in terms of numbers of volumes published and read. His books went into multiple editions and were available in public and private libraries around the world.

In the twentieth century, as parts of West Africa became more easily accessible, more and more people with an interest in gorillas traveled there to see what they could see and what they could bring home. Some did it for the adventure. Some did it with serious scientific intent. Somewhere in between were the collectors who represented the natural history museums of the world. Their mission, like Du Chaillu's, was usually to secure specimens for public display.

The Academy's Lowland Gorilla diorama, which helped trigger Elmer Payne's gift of Du Chaillu's watch in 1978, was created in 1937 from specimens collected on George Vanderbilt's African expedition two years before (figure 6.9). Even though the central male gorilla in that diorama was mounted by Louis Paul Jonas almost three-quarters of a century after Du Chaillu's first gorilla specimens were displayed in Philadelphia, New York, and London, its posture was based on the frontispiece of Du Chaillu's first book, *Explorations and Adventures in Equatorial Africa,* for Du Chaillu was still recognized as the authority on the subject of gorilla behavior.[19]

Du Chaillu tried to move on to other topics—he spent years researching and writing about Scandinavia—but he would be forever linked in the public's mind to Africa and the gorilla. When he died in St. Petersburg in 1903 while researching a book on Russia, obituaries around the world focused first and foremost on his two African expeditions and his "discovery" of the gorilla.

Although John Cassin's contributions to science were far greater than Du Chaillu's, he never achieved anything like the public notoriety of his onetime collector. Overshadowed by Audubon, whose credentials as an ornithologist he challenged, and by Du Chaillu, whose principal "discovery" was far more sensational than any of his own, Cassin spent his life in the relative obscurity of academia, producing solid technical studies of birds from his native country and from around the world.

6.9 Gorilla skulls collected on the Vanderbilt Expedition to West Africa in 1934–35. ANSP Mammalogy Department. Purcell photograph.

From the time of his first involvement with ornithology, Cassin was a prolific writer, authoring scientific papers on dozens of new birds from Africa, Asia, South America, and western North America, specimens of which were purchased abroad or collected for him by friends at the Academy and an ever-expanding circle of fellow researchers. His first full-length book, *Illustrations of the Birds of California, Texas, Oregon, British and Russian America,* revealed the full extent of his ornithological knowledge and established him as the leading American ornithologist of his day.

The book, issued in parts from 1852 to 1855 and as a whole in 1856, 1862, and 1865, contained color plates and detailed information on fifty species of birds "not given by former American authors."[20] It also offered a "general synopsis of North American Ornithology" in which Cassin compared American bird species with others from around the world. No American had ever offered such a comprehensive analysis. Before Cassin, none had possessed the knowledge to attempt it (figure 6.10).

Ironically, the very specimens that made the Academy's ornithological collections the most complete in the world[21]—and enabled Cassin to achieve preeminence in this still-nascent field of study—were also the source of Cassin's greatest misery. They may even have been responsible for his death.

John Cassin was born near Media, Pennsylvania (about twenty-five miles west of Philadelphia), on 6 September 1813. He moved to Philadelphia in his early twenties to pursue a career in trade, but he appears to have retained at least part ownership of the modest forty-acre farm that had served as the home of his grandfather Luke and father, Thomas, before him.

Records of his early life are sketchy, but we know that he was educated at the Westtown School, the same Quaker boarding school responsible for the education of so many of America's pioneering naturalists and fellow Academy members, including Thomas Say (1787–1834), Samuel Morton (1799–1851), John Kirk Townsend (1809–1851), and Edward Drinker Cope (1840–1897).[22]

Cassin's interest in birds appears to have been well developed even before he arrived at the Westtown School. In a letter of 1842, he cited ornithology, botany, and mineralogy as three subjects to which he had "been devoted since childhood."[23]

Included among his other natural history interests were the study of shells (of which he had amassed a large collection by 1846) and insects.[24] It was, at least in part, his original contributions in the latter field that helped establish his reputation as a serious naturalist among the scientific cognoscenti in Philadelphia. Although he described his knowledge of insects as "limited," Cassin was the first person to recognize that the so-called

seventeen-year locust (*Cicada septendecim*) consisted of several different species of cicada.[25] The distinctions he noticed during the cicada outbreak of 1834 were confirmed by others seventeen years later. One of three seventeen-year locust species was eventually named *Cicada cassini* in his honor.[26]

Following his election to the unpaid position of curator in 1842, Cassin began working weekends and weekday evenings at the museum, inventorying, labeling, and expanding the Academy's substantial assemblage of bird specimens. He used the professional contacts he had developed in the import-export business to acquire new specimens from around the world. Birds he could not acquire as gifts he obtained in trade for duplicates or purchased outright with the financial backing of Thomas B. Wilson (1807–1865), a wealthy Philadelphia physician who had joined the Academy a decade earlier and went on to become the Academy's president in 1863. In little more than a decade, Cassin and Wilson managed to expand the Academy's bird collection from the largest in America to the largest in the world.[27]

Cassin's interest in birds focused at first on African and European species, but it soon shifted to those of the American West. Just as Alexander Wilson examined new species of birds collected by Lewis and Clark, and Audubon studied western species collected by Thomas Nuttall and John Kirk Townsend before taking his own trip west in 1843, so Cassin was stimulated by exposure to new discoveries from Texas, California, Oregon, and the greater Northwest then coming from a variety of new sources.

In a letter of August 1845, Cassin revealed the joy he felt in seeing some of these new species for the first time:

> Eureka! Gambel is here [at the Academy] with his California birds & others—not very many, but some of the most magnificent specimens I ever saw—he has four new species (in addition to those already described). . . . He has also most beautiful specimens of well known birds & others not so well known . . . decidedly the gem of the collection is a most superb specimen of *Leptostoma longicauda* [Roadrunner], a beautiful cuckoo-like bird which walks on the ground. . . . He and I have done little else for two afternoons & evenings—last evening 'till 12 o'clock & I am now going to meet him again.[28]

Cassin was describing the collections of William Gambel (1821–1849), a young doctor and Academy member who died of typhoid fever on a second expedition to California just four years

6.10 John Cassin (1813–1869), ca. 1860. ANSP Archives coll. 457.

6.11 Gambel's Partridge (Gambel's Quail), hand-colored lithograph by George White from John Cassin's *Illustrations of the Birds of California, Texas, Oregon, British and Russian America* (1856). ANSP Library.

later. During his brief career, Gambel made significant marks on several fields of natural history. He is remembered today through the names of two of his California discoveries: Gambel's Quail (*Callipepla gambelii*) (figure 6.11) and Gambel's Oak (*Quercus gambelii*).

The Texas war for independence in 1836 and the subsequent Mexican-American War (1846–48) increased military activity in the West. This provided new opportunities for bird collecting for ornithologically inclined soldiers and thus brought many new birds to Cassin's attention at the Academy. Excited by the opportunity for a major publication on these discoveries, Cassin put aside his work on African and European species to focus on the novelties of his own country. It was in part a collection made by John James Audubon many years before that stimulated Cassin to undertake the work that eventually led to his first book.

Despite Cassin's skepticism about Audubon's merits as an ornithologist, he had enough business sense to know that because Audubon had set the standard for American bird books, anything he might do to augment or improve Audubon's work would have to be associated in some way with *The Birds of America*. It was for this reason that he first described his own book as a supplement to Audubon's popular "octavo edition" (a smaller, inexpensive version of *The Birds of America* that had been available since 1844).[29]

On the initial prospectus and two of the part-title sheets for Cassin's *Illustrations*, Audubon's name is larger than Cassin's. Other titles refer to his work as containing "Descriptions and Figures of all North American Birds not given by Former American Authors."

At the time Cassin began work on *Illustrations*, he already had considerable experience in publishing, primarily through the Academy's internationally distributed *Journal* and *Proceedings*. He soon discovered, however, that producing and selling a book of his own without institutional support and financial backing were considerably more difficult than contributing papers to an already established journal.

His original plan was to publish *Illustrations* in thirty parts. Each part would contain five color plates and appear about every other month. The price was to be $1 per part, payable to Cassin on delivery.[30] This meant that, in addition to his time, Cassin would have to advance the costs for artists, paper, printing, lithography, hand-coloring, and distribution.

Cassin calculated that he would need at least 250 subscribers to defray his expenses and break even on the project, but he also knew that it would be difficult to enlist subscribers without a sample section of the book to show.[31] So in April 1852, with only forty subscribers committed to the enterprise, Cassin published and distributed the first part of his great work.[32] A disappointing

public response caused him to abort the project after only ten of the proposed thirty parts. "With my book, it is now a struggle to save myself from losing money," he wrote in June 1854. "I am bound to quit with the 10th No. if not before . . . Lippincott is willing to go on to the end of 3 vols. but I am not—not at all, no how."[33]

Although Lippincott eventually published three editions of Cassin's book (1856, 1862, and 1865), the print runs were small. It is quite likely that never more than a few hundred copies of *Illustrations* were printed, and subsequent losses have added to its scarcity. No contemporary reviews can be found to explain the poor sales of the book, but it may be surmised that inadequate marketing was only part of the problem. Unlike Audubon's *Birds of America*, which depicted birds commonly seen by Americans east of the Mississippi, Cassin's *Illustrations* showed birds few Americans had seen or were ever likely to encounter. The text, while interesting in parts, was far drier than Audubon's, which, like his paintings, contained an artistic dimension never achieved by Cassin. Finally, the ornithological synopses that did so much to establish Cassin's reputation in the scientific community were far more technical than most laymen were willing to read. It was, in short, a book without a sufficient market to justify Cassin's tremendous financial and intellectual investment.

Time has not been kind to Cassin's *Illustrations*. Increasing knowledge of western birds soon overtook Cassin's preliminary information, thereby permitting more comprehensive publications to supplant *Illustrations* soon after it appeared.[34] Even his friend Spencer Baird at the Smithsonian, who had helped obtain some of the specimens for Cassin to describe, acknowledged the paucity of information available to Cassin in a letter to a friend: "His plates are excellent but his biographies are nothing at all."[35] The book's limited distribution—a significant factor in its value to today's collectors—severely reduced its potential to make an impact on public perceptions of natural history. Its greatest value today is not as an ornithological treatise but as a historical document—a time capsule of ornithological exploration in the American West. The sources whose firsthand accounts Cassin quoted include some of the most colorful and adventurous naturalists of the nineteenth century.[36] Many risked illness, frostbite, starvation, Indian attack, and various other privations to gather the information cited by Cassin. At least five—William Gambel, Adolphus L. Heermann, Richard H. Kern, Caleb B. R. Kennerly, and Robert Kennicott—subsequently lost their lives in the name of science.

Cassin clearly enjoyed the pleasures of nature and often complained to Baird of the press of business that kept him from the field, but he also realized that it was his indoor activities at the Academy that set him apart from the growing number of field naturalists then exploring the American West.

In the end, Cassin's life as a closet naturalist proved more dangerous than confronting rattlesnakes, desert heat, or Indians, for in his academic zeal, this champion of civilized living was slowly poisoning himself. Arsenic, then a primary ingredient in the preservation of bird and mammal skins, appears to have been the culprit. He began to show symptoms of poisoning soon after the Academy received an immense bird collection—some twenty-five thousand specimens—purchased from the French nobleman Victor Massena, duke of Rivoli (for whom Cassin later named the Massena Partridge [Montezuma Quail]).[37] "The constant exposure to the Arsenical dust and odours . . . from the Rivoli specimens has brought on me a soreness of the throat," he wrote in December 1847, "attended, I fear at this time, by chronic inflammation of the Larynx and parts adjacent which gives me much uneasiness and has totally suspended my progress in the further arrangement of the collection."[38]

His letters for the next twenty years chronicle the increasingly debilitating effects of arsenic poisoning:

> I have broke down again, and more seriously I fear—Dr. Wilson and I undertook an examination of the collection of owls—it is a very Arsenicy job, but I thought I could stand it, as it would not take more than a few weeks—I could not however,—I labeled about half the collection after reexamining (after Wilson had examined the species) and was taken with *congestion of the lungs* and most violent head ache and fever,—I was cupped and physicked, and I have now just returned from a trip of 4 days in the country—I am much improved though not well by any means—pain in my breast and some headache. [11 November 1848]
>
> My health continues rather bad—pain in my breast—I do not go amongst the Arsenic. [12 December 1848]
>
> My health this winter has been good except a most abominable ulceration of my tongue, which has been exceedingly bad since October last. I did not eat a single meal for three months—had to live upon mush and milk, rice milk, soup, etc. [14 February 1850][39]

Cassin often referred to his work with the Academy's birds as selling himself to the devil or "mortgaging myself by perpetual lease to Arsenic and Liver complaint" in return for "solacing pleasures" or "trifling amusement."[40] But the "amusement" of

6.12 Kirtland's owl (Saw-whet owl), hand-colored lithograph by George White from John Cassin's *Illustrations of the Birds of California, Texas, Oregon, British and Russian America* (1856). ANSP Library.

ornithology was so compelling for John Cassin that he could not stay away from it, no matter what the consequences.

When he briefly suspended work on the Academy's bird collection in 1849, he compared the experience to death. "I have done nothing for a month and am wretched," he wrote. "I am now so entirely and thoroughly habituated to constant study and thought on Natural History that, without it, I am at a loss, as it were, for my accustomed sustenance."[41]

Cassin's temporary withdrawal from ornithology, designed to convince T. B. Wilson and the Academy that he should be paid for his services, had the additional and seemingly unexpected advantage of giving him more time with his family, which seems to have gotten short shrift because of his scientific activities.[42] Of the many hundreds of letters of Cassin's personal and professional correspondence that survive, in only a handful does he discuss his daughter, Rachel, and son, Will. He never mentions his wife, Hannah, at all.

Cassin's closest friendship through the years was with Spencer Fullerton Baird (figure 6.13). Fortunately for both men, the friendship had as many practical as personal applications, including a steady exchange of books and specimens. With Baird's help, Cassin also was given the opportunity to describe the many new birds then being collected by the U.S. government's expeditions at home and abroad. His most important publications included the volume on ornithology and mammalogy for the U.S. Exploring Expedition of 1838–42 (1858) and the ornithological sections for Matthew C. Perry's report on the American Squadron expedition to Japan and the South China Sea of 1852–54 (1856), James M. Gilliss's U.S. Naval Expedition to the Southern Hemisphere of 1849–52 (1856), and the so-called Rogers and Ringgold North Pacific Exploring Expedition of 1853–56 (1862). With Baird and Lawrence, he also coauthored the ornithological volumes associated with the Pacific Railroad Surveys (1858).

During the publication of *Illustrations*, Cassin had become intrigued by the process of lithography and the physical production of the color plates required for any bird book to succeed. He decided to become more involved with publishing. Cassin's business acumen and political contacts proved extremely valuable to William T. Bowen, the lithographer with whom he worked on the publication of his book, but the benefits of the relationship were mutual. Cassin's dominant position in the field of American ornithology, which began with his authorship of *Illustrations*, was greatly enhanced by his association with Bowen, who was considered one of the finest lithographers in North America. Cassin formalized the relationship in the spring of 1858 when he joined Bowen's widow as half owner and president of the lithographic company. "I am now the solemnly constituted head of a large establishment," Cassin wrote Baird, "[with] printers, colorists, and draughtsmen, having been so *de facto* for the last 18 months [since Bowen's death in late 1856]. I intend to hunt up the best draughtsmen in the U.S.!—and want all the work I can get, perhaps more!"[43]

For the first time in his career, Cassin's vocation and avocation were completely intertwined. Now even the few American natural history texts that were not written by Cassin were nevertheless under his control, for Bowen and Company was the acknowledged leader in natural history publication in the United States.[44] Fortunately for Cassin, Audubon's books, *The Birds of America* and *The Viviparous Quadrupeds of North America*, were now Cassin's books as well. They remained the firm's bread-and-butter products while Cassin lobbied to secure new natural history commissions from Baird and his friends in Washington.

John Cassin was no stranger to politics. He was actively involved with Philadelphia's Democratic Party and served as superintendent of public stores for the U.S. Customs House from 1854 to 1857.[45] "Even in Science," he observed, "things go so much by influence and a mutual assistance understanding, that

6.13 Spencer Fullerton Baird (1823–1887), Cassin's closest friend, served as Assistant Secretary of the Smithsonian Institution from 1850 to 1878 and then as Secretary until his death. Photograph ca. 1855. ANSP Archives coll. 457.

FACING PAGE 6.14 Cassin's Kingbird (*Tyrannus vociferans*), one of four North American bird species to bear Cassin's name. Lithograph from *Report on the United States and Mexican Boundary Survey* by William H. Emory (1857–59). ANSP Library.

TYRANNUS VOCIFERANS.

absolute merit is frequently smothered and often overlooked."[46] Though Cassin preferred to stress merit over influence, he was well aware of the importance of both.

Politics played a part in every aspect of his life, even when he was describing new species of birds. Since the describer and not the discoverer of a new species is given the privilege of naming it for posterity, Cassin, Baird, and a handful of their contemporaries had a tremendous impact on the scientific nomenclature of North and South America, Africa, and Asia, from which new specimens were appearing on a regular basis. Because of their influence, many North American bird names reflect the small circle of ornithologists they chose to honor—Harris's Hawk, Le Conte's Sparrow, Bell's Vireo, Baird's Sandpiper, Kirtland's Warbler, Harlan's Hawk, Heermann's Gull, Xantus's Murrelet, Sprague's Pipit, MacGillivray's Warbler, Bullock's Oriole, Lawrence's Goldfinch, Brewer's Sparrow.

Since Cassin described more birds than any of his American contemporaries, he usually had the opportunity to name them whatever he chose. On the occasions when someone else was doing the naming, or when someone suggested a name he should use, he did not hesitate to voice his opinion on the subject. In 1855, Cassin urged Baird to drop his plan to honor Lt. John W. Gunnison (1812–1853), who had been killed by Indians while leading a railroad survey, and to name the bird instead for his friend and fellow Academy member T. Charlton Henry (1825–1877). "Cannot *Chordealis Gunnison* be called *C. Henryi*?" requested Cassin. "Name some other bird after Lt. Gunnison. Henry ought to have one named after him."[47]

Cassin, who dispensed bird names like military honors in his own publications, was himself so honored more than any other American ornithologist. Four North American birds—an auklet, a sparrow, a kingbird, and a finch—and a number of other animal species still bear his name (figure 6.14).

John Cassin died on 10 January 1869, after a brief illness contracted during a trip to Washington the month before. "Coming up from Washington in the night, I took cold so severely that I have scarcely been able to do anything ever since," he wrote Baird. Focused on birds to the very end, Cassin promised to return some specimens "if I die by it!"[48] He was fifty-five years old.

His passing left a large void in the still small profession. "With Cassin's death we lost our only Old World ornithologist," wrote Baird to the American ornithologist Thomas M. Brewer (1814–1880) on receiving the news.[49] To Cassin's widow he was more personal: "John was the nearest and dearest friend I had in the world, and in losing him it seems as if all interest in science had gone with him. As a friend, councillor and in all matters of mutual interest, I always found him ready with his help."[50]

Next to Baird and Cassin's immediate family, the person most affected by Cassin's death was Elliott Coues (1842–1899), a man who by the end of the century would inherit Cassin's place as the most prominent figure in American ornithology. In a letter to James G. Cooper (1830–1902) some six weeks after Cassin's death, Coues eulogized Cassin in heartfelt prose:

> Of course you heard the sad, sad news that John Cassin's labors are ended. The loss to science none of us can measure; nor can those privileged to call him friend adequately express the depth of that bereavement. And many as are our American ornithologists—high as some stand in American Ornithology—there is none left in all our land who can lift up the mantle that has fallen from his shoulders. . . . Since Audubon passed away from the scene of his usefulness, death has struck no such cruel blow to our beloved science. As Dr. Brewer has said to me, "Which one of our younger ornithologists will undertake to stand, after thirty-five years of training, where Cassin stood at his death?"[51]

In the years since Cassin's death, his important contributions to the professionalization of American ornithology have been largely forgotten. Several bird names and the ornithological journal *Cassinia* (published by the Delaware Valley Ornithological Club) comprise the most conspicuous and lasting tributes to the man who described close to two hundred new species of birds. Even his book *Illustrations of the Birds of California, Texas, Oregon, British and Russian America* has failed to provide the sort of literary and ornithological immortality for which Cassin longed. Until it was reproduced in facsimile by the Texas State Historical Association in 1991, the book's scarcity had made it all but impossible for anyone to see. Fortunately, the bird collection that he helped to secure and organize, including some 130 specimens collected by Paul Du Chaillu, survives as his most lasting contribution to the field of science he so loved.[52]

RMP

Notes

1. Inscription on gold pocket watch once owned by Paul Du Chaillu, ANSP coll. 237.
2. Paul Du Chaillu, *Explorations and Adventures in Equatorial Africa* (New York: Harper & Bros., 1861), 98–101.
3. For a discussion of Du Chaillu's birth, see Lysle E. Meyer, *The Farthest Frontier: Six Case Studies of Americans and Africa, 1848–1936* (Selinsgrove, PA: Susquehanna University Press, 1992), 34–35. See also Henry H. Bucher, Jr., "Canonization by Repetition: Paul Du Chaillu in Histiography," *Revue Française d'Histoire d'Outre-Mer* 66 (1979): 15–32.
4. Thomas S. Savage and Jeffries Wyman, "Notice on the External Characters and Habits of the *Troglodytes gorilla*, a new species of Orang from the Gabon River . . . Osteology of the Same," *Boston Journal of Natural History* 5 (1847): 417–41.
5. In a later dispute over the Academy's responsibility for the expenses incurred by Du Chaillu during his first expedition, the Academy claimed that individual members had agreed to sponsor his trip and not the Academy itself. Such a distinction was difficult for Du Chaillu to understand and created ill will on both sides.
6. See John Cassin, *Proceedings of the Academy of Natural Sciences of Philadelphia*, vol. 7 (April 1855): 324–28; vol. 8 (Dec. 1856): 1–7; vol. 9 (April 1857): 1–7; vol. 11 (June 1859): 133–44. See also Michel Vaucaire, *Paul Du Chaillu, Gorilla Hunter* (New York: Harper & Brothers, 1930), 82; and Jean-Marie Hombert and Louis Perrois, *Coeur d'Afrique: Gorilles, cannibals et Pygmees dans le Gabon de Paul Du Chaillu* (Paris: CNRS Editions, 2007), 121–23.
7. Vaucaire, *Gorilla Hunter*, 135.
8. *Illustrated London News*, 11 May 1861, 443, and 18 May 1861, 472.
9. See Du Chaillu correspondence in ANSP coll. 567 and summaries of the internal committee reports regarding his claims in coll. 464.
10. Richard Conniff has suggested that some of the attacks against Du Chaillu may have been racially motivated. See Richard Conniff, "Race, Sex and the Trials of a Young Explorer," *New York Times*, 13 February 2011; and Conniff, *The Species Seekers* (New York: Norton, 2011), 285–303.
11. See Ord-Waterton correspondence, American Philosophical Society.
12. For more on the Verreaux brothers and their worldwide trade in mounted birds and mammals, see Erwin Stresemann, *Ornithology from Aristotle to the Present* (Cambridge, MA: Harvard University Press, 1975), 162–63. For Cassin's early introduction to Du Chaillu at the Verreaux shop in Paris, see letter from George Ord to Charles Waterton, 21 July 1861, Ord-Waterton papers, American Philosophical Society.
13. Letter from George Ord to Charles Waterton, 21 July 1861, Ord-Waterton papers, American Philosophical Society. A calling card with Cassin's comments (sent to Waterton by Ord and copied in his letter of 21 July) is contained in the Waterton Papers (Moore Collection), Stonyhurst College, United Kingdom.
14. After his return from Africa in 1858, Du Chaillu published descriptions of some of his discoveries in the *Proceedings of the Natural History Society of Boston*. Although Du Chaillu was assisted by Dr. Kneeland, a professor of zoology at the Massachusetts Institute of Technology (MIT), Cassin may well have taken umbrage at what he considered a transgression by Du Chaillu from his role as a collector to that of a bona fide naturalist. See *Proceedings of the Natural History Society of Boston*, 1860, vol. 5, 417, and vol. 7, 296–304 and 358–67.
15. Letter from John Cassin to Spencer F. Baird, 23 June 1845, Baird Correspondence, Record Unit 7002, Box 17, Smithsonian Institution.
16. John Cassin, *Illustrations of the Birds of California, Texas, Oregon, British and Russian America* (Philadelphia: J. B. Lippincott, 1856), 280.
17. Ibid., iv.
18. For a complete bibliography of Du Chaillu's writings, see Hombert and Perrois, *Coeur d'Afrique*, 215.
19. The image of the gorilla with one arm supporting itself from the branch of a tree was popularized by Du Chaillu, but it first appeared as an illustration (by Marie Firmin Bocourt) that accompanied Geoffrey Saint-Hilaire's "Description des mammiferes nouveaux . . . Quatrieme Memoire, Famille des singes, second supplement," in *Archives du Muséum d'Histoire Naturelle*, 1858–61, vol. 10. Because it referred back to these two scientific sources, and because it made a good structural design, this posture was adopted by taxidermists for many years. The male gorilla in a mounted group of Du Chaillu gorillas in the natural history museum in Melbourne, Australia, is posed in a similar way, as is one shown in an illustration in "Our Taxidermists at Work" by Frederick Lucas, in *Ward's Natural Science Bulletin*, 1 January 1883.
20. John Cassin, *Illustrations*, title page.
21. In 1857, Dr. P. L. Sclater described the Academy's collection as "superior to that of any museum in Europe and therefore the most perfect [i.e., complete] in existence" (*Proceedings of the Zoological Society of London*, 1857, 1). More recently, Erwin Stresemann has written: "The bird department of the Philadelphia Museum [i.e., the Academy of Natural Sciences] became in 1856 the largest and richest in the world, with 29,000 specimens (23,000 in glass cases), far surpassing even the Leiden Rijksmuseum." Stresemann, *Ornithology*, 243.
22. I am grateful to Alice Long, archivist at the Westtown School, Chester County, Pennsylvania, for her assistance in providing records for the period of John Cassin's attendance at the school. For further information on the Westtown School, see Watson W. Dewees and Sarah B. Dewees, *The History of Westtown Boarding School, 1799–1899* (Philadelphia: Sherman & Co., 1899), and Susan Smedley, *Westtown Through the Years* (Philadelphia: Lyon and Armor, 1945).
23. Cassin to S. Haldeman, 26 July 1842, ANSP coll. 221C.
24. Cassin to Charles M. Wheatly, 5 June 1846, coll. B-W558, American Philosophical Society Archives.
25. Cassin to S. Haldeman, 26 July 1842, ANSP coll. 211C.
26. Other closely related cicadas were found to emerge every thirteen years. One of this group, *Magicicada tredecassini*, was also named for Cassin. For a summary of this family and the history of its nomenclature, see Richard D. Alexander and Thomas E. Moore, "The Evolutionary Relationships of 17-Year and 13-Year Cicadas and Three New Species," *Miscellaneous Publications*, Museum of Zoology, University of Michigan, no. 121 (24 July 1962).
27. *Proceedings of the Zoological Society of London*, 1857, 1; see also Stresemann, *Ornithology*.
28. Cassin to Spencer F. Baird, 14 August 1845, Baird Collection, Smithsonian Institution.
29. See Cassin to Baird, 10 January 1852, Baird Collection, Smithsonian Institution.
30. This was exactly the same price per part as Audubon's one-hundred-part octavo edition of *The Birds of America*, published from 1839 to 1844. See Nicholas B. Wainwright, *Philadelphia in the Romantic Age of Lithography* (Philadelphia: Historical Society of Pennsylvania, 1958), 54.
31. In a letter to Baird dated 9 February 1852, Cassin writes: "I must have 100 subscribers to begin with—and the book will not pay expenses under 250," Baird Collection, Smithsonian Institution.
32. For a list of his early subscribers, see the Texas State Historical Association reprint (1991), I-39.
33. Letter to Baird, 4 June 1854, Baird Collection, Smithsonian Institution.

34 To Cassin's credit, he encouraged and even played a central role in assembling these publications. The transcontinental railroad surveys that followed the American Civil War resulted in a series of publications, the ornithological sections of which Cassin helped to write. Other books that reduced the usefulness of *Illustrations* were Baird's *The Birds of North America*, coauthored by Cassin and George N. Lawrence (Philadelphia, 1860), Robert Ridgeway's *A History of North American Birds*, coauthored with Baird and Thomas M. Brewer (New York, 1874 and 1884), and Charles E. Bendire's *Life Histories of North American Birds* (Washington, D.C., 1895).

35 Baird to Andrew Jackson Grayson, 15 December 1856, Baird Collection, Smithsonian Institution, quoted in Lois Chambers Stone, *Andrew Jackson Grayson, Birds of the Pacific Slope* (San Francisco: Arion Press, 1986), 62.

36 For biographical information about many of the collectors quoted in Cassin's text, see Edgar Erskine Hume, *Ornithologists of the United States Army Medical Corps* (Baltimore: John Hopkins University Press, 1942).

37 The collection was purchased for the Academy by Thomas B. Wilson in 1846.

38 Cassin to Baird, 3 December 1847, Baird Collection, Smithsonian Institution.

39 All letters are from Cassin to Spencer F. Baird, Baird Collection, Smithsonian Institution.

40 Cassin to Baird, 6 June 1849, ibid.

41 Cassin to Baird, 5 April 1849, ibid.

42 Cassin was not paid for his services to the Academy until 1865. On 5 April 1866, he wrote to Spencer Baird at the Smithsonian that he was celebrating "my first year as a paid curator." Baird Papers, Smithsonian Institution.

43 Cassin to Baird, 5 April 1858, Baird Collection, Smithsonian Institution.

44 For more on the Bowen Lithography Company and the history of lithography in Philadelphia, see Erika Piola, ed., *Philadelphia on Stone: The First Fifty Years of Commercial Lithography in Philadelphia, 1828–1878* (University Park: Penn State University Press for the Library Company of Philadelphia, forthcoming).

45 Cassin's customs-house post was almost certainly arranged as a reward for his work on behalf of the Democratic Party and the then-mayor of Philadelphia, Richard Vaux. See letters to Baird, 16 May 1856 and 29 June 1857, both in Record Unit 7002, Box 17, Baird Collection, Smithsonian Institution. Cassin's earlier political post, that of superintendent of public stores, is more difficult to document. A letter dated 31 May 1854, in the archives of the Historical Society of Pennsylvania addresses Cassin by that title, but the Philadelphia Business Directory lists him as proprietor of John Cassin & Co. at that time. It was not uncommon for government officials to have full-time businesses in addition to their part-time positions of public service, and this may have been the case with Cassin. Judging from his correspondence, he often worked at several jobs simultaneously. For futher information, see *McElroy's Philadelphia Directory* (Philadelphia: Edward C. and John Biddle, 1854–57).

46 Cassin to Baird, 16 February 1847, Baird Collection, Smithsonian Institution.

47 Cassin to Baird, 26 December 1854, ibid.

48 Ibid.

49 Baird to Thomas Brewer, 11 January 1869, Baird outgoing correspondence, Smithsonian Institution.

50 Baird to Mrs. Cassin, 11 January 1869, ibid.

51 Elliott Coues to James G. Cooper, 21 February 1869, present location unknown. The letter was reprinted in the *Condor* and *Cassinia*.

52 The database of the bird collection (as of 2010) shows 131 bird skins collected or presented to the Academy by Paul Du Chaillu. This number includes eight types. The Mammal Department has forty-one mammals and nine reptiles and amphibians from West Africa also collected by Du Chaillu. Our thanks to Nate Rice and Ned Gilmore for providing this information.

A bushman model, created in 1937 by Eda Kassel for the Hall of Earth History, peers from a museum window. ANSP Exhibits Collection. Purcell photograph.

FOLLOWING PAGES A collection of slime molds (*Myxomycetes*) assembled by George A. Rex in the late nineteenth century. ANSP Botany Department. Purcell photograph.

MYXOMYCETES.
Lindbladia Effusa (Ehr.) Rost
Var. Simplex Rex
GEO. A. REX, Philad'a, Pa.
MYXOMYCETES.
Lamproderma Arcyrionema Rost
MYXOMYCETES.
Dictidium Cernuum (Pers.)
Pt. Pleasant N.J.
22 July 1891 A.P.B.
37 AMOS P. BROWN,
NICETOWN, PHILADELPHIA, PA.
GEO. A. REX, Philad'a, Pa.
MYXOMYCETES.
Lamproderma + W
88
MYXOMYCETES.
Lindbladia effusa (Ehrb.) Anglesea N.J.
24 July 1892 A.P.B.
8 AMOS P. BROWN,
NICETOWN, PHILADELPHIA, PA.
MYXOMYCETES.
Comatricha typhina Ro
var pumila.
Wingate Col.
45 AMOS P. BROWN,
NICETOWN, PHILADELPHIA, PA.
MYXOMYCETES.
Lindbladia
effusa (Ehrb.)
1891 Rex.
86 AMOS P. BROWN,
NICETOWN, PHILADELPHIA, PA.
MYXOMYCETES.
Stemonitis
Germantown Pa
July 1891 A.P.B.
60 AMOS P. BROWN,
NICETOWN, PHILADELPHIA, PA.
MYXOMYCETES.
Lindbladia Effusa (Ehr.) Rost
Var. Simplex
GEO. A. REX, Philad'a, Pa.
GEO. A. REX, Philad'a, Pa.
Comatricha Longa Pk
MYXOMYCETES.
Orthotrichia microcephala Wing.
Point Pleasant N.J.
10 Aug 1891
AMOS P. BROWN,
NICETOWN, PHILADELPHIA, PA.

MYXOMYCETES.
Lindbladia effusa (Ehrb.)
Anglesea N.J.
24 July 1892 A.P.B.
9
AMOS P. BROWN,
NICETOWN, PHILADELPHIA, PA.
MYXOMYCETES.
Dictidium cernuum (Pers.)
Hamburg, Berks Co. Pa.
10 July 1892 A.P.B.
39 AMOS P. BROWN,
NICETOWN, PHILADELPHIA, PA.
GEO. A. REX, PHILADELPHIA, PA.
GEO. A. REX, Philad'a, Pa.

Chapter 7

The Marvelous Bipedal Masterpiece: Religion, Politics, and Public Display

Then the institution, with its noble past and with its world-famous record of great things accomplished in the field of natural science, begins another era in its mission of usefulness, an era in which its scope and powers ... will be well in the van of modern ideas of method and progress.
—Editorial about an addition to the Academy's building at 19th and Race Streets, *Telegraph*, September 1896

In an effort to avoid distractions from their scientific pursuits and reduce the opportunities for destructive personal animosity, the Academy's founders explicitly prohibited the discussion of either politics or religion at the society's weekly meetings. As the nineteenth century progressed, however, these wise prohibitions were sometimes overlooked, sparking conflict among the members.

Because of its broad reach into almost every American family, the Civil War was a subject on which every member held deep feelings. Discussions of any aspect of this topic inevitably risked bringing simmering tensions to the fore. In his centennial history of the Academy, Edward Nolan attributed the notably short tenure of Thomas B. Wilson as president of the Academy (December 1863 to June 1864) in part to his "distinctly Southern" sympathies, which were "not in harmony with the rather aggressive patriotism of nearly all his associates in the Academy."[1] The year after Wilson's resignation, in an unprecedented gesture of patriotism and without any apparent reference to the recipients' interest in natural history, the members elected Generals Ulysses S. Grant and George G. Meade (the Union commander at the Battle of Gettysburg) to membership in the Academy. According to Nolan, who knew the participants, a proposal to commit $210 of Academy funds to pay the cost of expanding this honor to life memberships without the requirement of any dues prompted heated arguments, or what Nolan diplomatically described as "lively debate," among the members, some of whom must have opposed spending scarce Academy resources on what were clearly political (nonscientific) appointments.[2]

Opinions on matters of religion were equally difficult to banish from discussion. In the aftermath of Darwin's publication of *On the Origin of Species* in 1859, some turned to institutions like the Academy for reassurance that the world's flora and fauna were not evolving on their own but were all part of a grand design whose order only needed further scrutiny to make sense. "The great objects of our museums," said the influential Swiss naturalist (and anti-Darwinian) Louis Agassiz, "should be to exhibit the whole animal kingdom as a manifestation of the Supreme Intellect."[3] Many Academy members agreed. As if to reinforce Agassiz's view, in their short history of the Academy written in 1876, W. S. W. Ruschenberger and George W. Tryon Jr. distanced themselves from the antireligious rhetoric of earlier publications and claimed that the Academy had always been "devoted to the acquisition and diffusion of knowledge of the works and laws of the Creator."[4]

If the Academy's collections were to be viewed as tangible evidence of divine creation, then their care became a sacred trust, and those who studied them were increasingly perceived as divinely anointed intermediaries and interpreters. "Original

7.1 Leg bones from the Moa of New Zealand, an enormous, flightless bird that was driven to extinction before European contact in the eighteenth century. The specimens shown were purchased for the Academy by T. B. Wilson in 1848. W. Mantell Collection, ANSP Paleontology Department #12163. Purcell photograph.

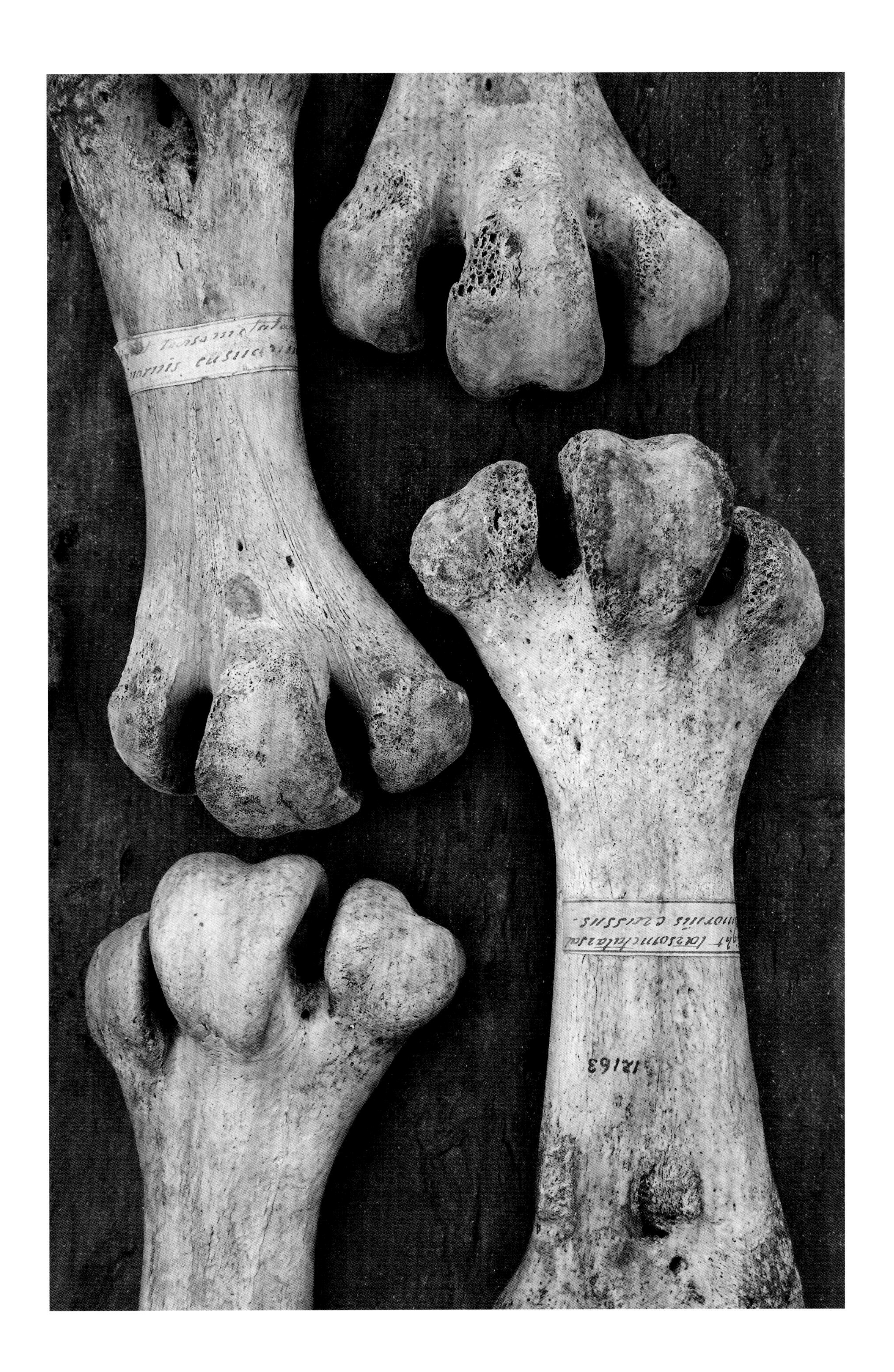

scientific research rarely brings pecuniary profit," wrote Ruschenberger and Tryon,

> but . . . the reward of the discoverer in natural science is . . . great. To stand, as it were, between God and man in the laboratory, the mine, the study, anywhere, and feel . . . there has stolen into his mind what has hitherto been known to God alone . . . is a privilege so high, and a pleasure so overwhelming as to sink in utter insignificance, not only the toils of research, but all the emanations of jealousy and prejudice which so often attend the first coming of truths before the world.[5]

Joseph Leidy was the very embodiment of this privileged intermediary. With his handsome (some thought Christlike) looks and brilliant command of so many subjects, he had become, in the words of one of his many admirers, "a high priest of Nature standing at the threshold of her mysteries" (figure 7.3).[6] Obsessively hardworking ("I get up at day-light and continue my scientific researches until midnight," he wrote a friend in 1847),[7] Leidy was an example of a new breed of professional who could embrace with equal enthusiasm both pure research and public education. Leidy's work in the field of paleontology helped put American science on a par with European science and in so doing, according to one eulogist, "turned the eyes of the savants of the Old World to their younger brethren in the West."[8]

An early supporter of Charles Darwin, with whom he exchanged cordial correspondence and whose election to membership he advanced as early as 1859, Leidy was the first scientist to describe a dinosaur from North America (in 1856) and the first to discover that the cause of trichinosis was microscopic parasites in uncooked pork. But it was Leidy's description of a large Cretaceous dinosaur from New Jersey and his subsequent collaboration with a colorful English artist to put that creature on public display that thrust the studious doctor into the public eye. Their collaboration changed the way the Academy—and

7.2 The main exhibit hall in the Academy of Natural Sciences' new 1876 building shortly after opening. ANSP Archives coll. 49.

every other natural history museum—would interpret natural history in the years to come.

Leidy's description of the dinosaur he called *Hadrosaurus foulkii* (Foulke's bulky lizard) was based on the discovery of a nearly complete collection of fossil bones in a marl pit near Haddonfield, New Jersey, by Academy member William Parker Foulke (1816–1854) in the 1840s. When Leidy published his description of the creature in the Academy's *Journal* in 1859, and suggested that it probably stood on its hind legs like a kangaroo, it attracted a good deal of attention in the scientific world. What changed the *Hadrosaurus* from an important scientific discovery (one of 375 extinct North American animals Leidy described) to a popular sensation was the way in which the dinosaur was ultimately displayed. This came about not through Leidy's intentional decision to go public with his find, but rather through the serendipitous arrival of Benjamin Waterhouse Hawkins (1807–1894), who turned Leidy's discovery into a spectacular exhibit and made the Academy an irresistible destination for Philadelphia's citizens and out-of-town visitors alike.

Hawkins was a multitalented English artist who had begun his career by illustrating the scientific publications of such notable figures as Richard Owen, Charles Darwin, and Thomas Henry Huxley (all corresponding members of the Academy). He went on to create the first life-size sculptures of dinosaurs and other prehistoric creatures at the Crystal Palace exposition in Sydenham, South London, in 1854. After a hiatus of several years, during which he continued to create illustrations for books and periodicals, Hawkins came to the United States in 1868, ostensibly to deliver a series of lectures on evolution.[9] His lecture tour quickly turned into something more significant.

Within a few months of his arrival in the United States, Hawkins was employed by the commissioners of Central Park in New York City to create a grand "Palaeozoic Museum" in which he would "undertake the reconstruction of a group of animals of the former periods of the American continent" (figure 7.4).[10] New York had already used Central Park to create its own "Crystal Palace" in which to house an Exhibition of the Industry of All Nations in 1853. Now the city's leaders envisioned creating a paleontological installation comparable to the one at the Crystal Palace in Sydenham.

With the demonstrated popularity of his London display and official endorsements from influential friends in Britain (including Charles Darwin), Hawkins had had little difficulty convincing New York's city officials that he was the right man to lead such a project.[11] Just what to put in this new Palaeozoic Museum was another matter. Scientific understanding of ancient life had evolved dramatically since 1853, when Hawkins had developed designs for the dinosaurs in Sydenham under the scientific direction of Richard Owen. The intervening discovery of *Hadrosaurus foulkii* by Joseph Leidy revolutionized concepts of dinosaur anatomy, posture, and locomotion. Additional finds in England and further analysis of existing fossil specimens had confirmed that Hawkins's heavy, mammal-like reconstructions of *Iguanodon* and *Megalosaurus* at the Crystal Palace had been incorrect. Hawkins's new challenge was to create a dynamic display of life-size extinct animals from America, all, as he phrased it, "clothed in the forms which science now ventures to define."[12]

Because neither the specimens nor the scientific expertise he required were available in New York (the American Museum of Natural History was not founded until 1869 and did not have any significant paleontological collections until the end of the nineteenth century), Hawkins traveled to Philadelphia, to seek the advice of Joseph Leidy.[13] Despite a heavy teaching commitment

7.3 Joseph Leidy (1823–1891), artist unknown. Charcoal on paper. Private collection.

at the University of Pennsylvania and administrative responsibilities at the Academy, where he had been a curator since 1846, Leidy was generous in sharing with Hawkins his deep knowledge and innovative thinking about dinosaurs.[14] He also was willing to give Hawkins access to the important fossil specimens in his care at the Academy (figure 7.5).

Up until that time, dinosaur bones, if they were exhibited at all, were usually shown as isolated paleontological phenomena, without context or meaning to any but a very few specialists.[15] Under Leidy's direction, and with additional advice from Leidy's younger colleague, Edward Drinker Cope, Hawkins took a new approach to the ancient bones.[16] In September 1868, with Leidy's encouragement and the approval of the Academy's curators, Hawkins began the painstaking process of molding the fossil bones of *Hadrosaurus foulkii.* He carefully suspended the plaster casts, bone by bone, from a metal armature, filling in the missing bones with plaster duplicates and topping his model with an invented skull, based loosely on the skull of a modern-day iguana (figure 7.6). In little more than two months of feverish activity, he created the first fully articulated dinosaur skeleton display in the world.[17]

The *Hadrosaurus* was not the only dinosaur whose bones Hawkins was allowed to copy while he was at the Academy. Another recent dinosaur discovery about which Hawkins would have heard a great deal from Leidy and Cope was *Laelaps aquilunguis* (later renamed *Dryptosaurus*), which had been discovered by Cope in a New Jersey marl pit in 1866. In his published description of the creature, Cope declared it "nothing more or less than a totally new gigantic carnivorous Dinosaurian, probably of Buckland's genus *Megalosaurus*."[18] Coming from the same time and location as *Hadrosaurus foulkii,* Cope's *Laelaps* made another ideal subject for Hawkins's Palaeozoic Museum. In a printed concept drawing and in one of his few surviving sketches of the planned museum (now in the Academy's archives), Hawkins shows a sinuous *Laelaps* attacking its larger rival, *Hadrosaurus.*[19] A photograph of Hawkins's New York studio

7.4 Benjamin Waterhouse Hawkins's rendering of a proposed "Palaeozoic Museum" for New York City, ca. 1869. Pencil and chalk on paper. ANSP Archives coll. 803.

(on the second floor of the Arsenal in Central Park) from about the same period (1869–70) shows a partially completed armature supporting his cast of the *Laelaps* skeleton.[20]

In November 1868, Hawkins presented the Academy with a completed *Hadrosaurus* skeleton mount, made entirely at his own expense. He was happy to do so, he wrote Leidy, "as a public acknowledgement of my indebtedness to the liberality of the Academy of Natural Sciences for the facilities they have afforded me in obtaining so large a portion of the unique materials necessary for the great work I have undertaken for the Commissioners for the Central Park in New York."[21] Without much discussion or appreciation of the impact such a display might have, Hawkins's mount was put on exhibit in the Academy's Broad Street museum. The public's response to the new exhibit was overwhelming. Even though the museum was open to outsiders (with sponsorship from Academy members) only two afternoons a week and closed for the month of August, nearly one hundred thousand people came to see the *Hadrosaurus foulkii* in 1869—almost twice as many as had visited the museum in the previous year.[22]

Most of the Academy's members were highly appreciative of Hawkins's gift, but not everyone was pleased with its unexpected consequences. "The increased attractiveness of the Museum of the Academy," complained one Academy official "[has] led to such an increase in the numbers of visitors as greatly to endanger the condition of the more perishable portions of the collections." In a year-end report to members, the secretary described the mayhem the new *Hadrosaurus* display was causing in the formerly staid Academy:

> In consequence of the very small amount of unoccupied room, the visitors move in nearly continuous streams through the narrow intervals of the cabinets, affording little opportunity for the examination of specimens. Besides this disadvantage to those who would really wish to examine the collections, the crowds lead to many accidents, the sum total of which amounts to a considerable destruction of property, in the way of broken glass, light wood work, &c. Further, the excessive clouds of dust produced by the moving crowds, rest upon the horizontal cases, obscuring from view their contents.[23]

The curators concluded that the only way to reduce the "excessive number of visitors" was, for the first time in the Academy's history, to charge an admission fee.[24]

A photograph of Hawkins standing under his novel reconstruction suggests a ghostly quiet in the Academy's Broad Street museum, but clearly the atmosphere was quite the opposite (figure 7.7).[25] With his very first construction on the continent, Hawkins had introduced "dinosauromania" to America.

Sadly, Hawkins's Palaeozoic Museum was never completed. On 3 May 1871, vandals broke into his New York studio and

7.5 Joseph Leidy with the bones of *Hadrosaurus foulkii* in the Academy's museum. The photograph was taken shortly after their discovery in a New Jersey marl pit in 1859. ANSP Archives coll. 9.

7.6 Having found no skull with the otherwise fairly complete remains of *Hadrosaurus foulkii*, Leidy and Hawkins were forced to speculate on its appearance. To top off the articulated skeleton, Hawkins created this plaster skull, basing it on a modern-day iguana. It is all that remains of his original 1868 mount. ANSP Exhibits Collection.

completely destroyed his work of three years.[26] Some reports attributed its destruction to the religious beliefs of Peter Sweeny, one of William Marcy "Boss" Tweed's cohorts, who purportedly considered Hawkins's Palaeozoic Museum a blasphemous display of evolutionary thinking.[27] More likely, the corrupt Tweed ring, then running New York, simply wasn't getting enough graft from the project, and Hawkins's public criticisms of their blatant mismanagement of the city evoked retaliation of a particularly brutal kind.

Although most of the molds for Hawkins's American dinosaurs were destroyed, those he had created for *Hadrosaurus foulkii* were not—they may have remained at the Academy or somewhere nearby in Philadelphia—so Hawkins was able to fulfill requests for several additional copies of the now iconic reptile. These came from Princeton University (in 1874), the Smithsonian Institution (in 1875), and the organizers of the Centennial Exposition in Philadelphia (in 1876). Because of changing scientific thinking and the physical deterioration of Hawkins's mounts, none of these mounts survived past World War II, but their impact was more widespread and long lasting than any other public display originating in the Academy's museum. The bipedal marvel represented the powerful convergence of contemporary science with modern display, which changed forever the way in which the public viewed "deep time."[28]

In a rare expression of his opinion on evolution and extinction, and on the influence of museum displays, Joseph Leidy wrote to the Smithsonian's Secretary, Joseph Henry, in 1875 to say that he supported the idea of the Smithsonian mounting a cast of the Academy's *Hadrosaurus* skeleton. "The great service these restorations render," he wrote, "is that they lead people to reflect;—those who are not students. They break up old and rather fixed views about the world being created just as we now see it. Nothing tends so much to lead people to believe in the existence of former races of animals, as such restorations." Leidy went on to observe that individual bones, skulls, and teeth were "not sufficiently familiar objects with people generally to impress them with the idea that they are evidence of the former existence of animals which are extinct," while a complete skeletal mount of an extinct creature "at first startles, and then convinces them of the wonderful changes our earth passed through before man appeared."[29]

While it did not foster as deep a political divide as had discussion of the Civil War, the overwhelming public response to Hawkins's display of Leidy's *Hadrosaurus* brought into focus an increasingly contentious discussion about just how public the Academy should become. The exponential growth of the society's library and scientific collections during the nineteenth century provided a tangible indication of the institution's global reach and intellectual dominance. Leidy's scientific work had contributed significantly to this growth, but so had the benevolent generosity of former presidents William Maclure and Thomas Wilson and countless other members whose gifts, large and small, had been reverently and appreciatively recorded in the Academy's *Journal* and *Proceedings* through the years (see figure 7.1 and other Purcell photographs).

For the Academy's founders and many of its later members, the amassing of collections lay at the base of all understanding of the natural world. If properly organized, they believed, the ever-growing number of specimens in their care would help to reveal the structure of the universe. The more specimens the Academy owned, the more good work it could do in unraveling the mysteries of nature, but all of these required care and organization. Without the proper organization, the collections could quickly

7.7 Benjamin Waterhouse Hawkins standing under the newly articulated skeleton of *Hadrosaurus foulkii* in the Academy's museum, 1869. ANSP Archives coll. 803.

7.8 The Academy's growing library was given a prime location on the first floor of the 1876 building. Later, it was moved into an even larger area on the second floor. ANSP Archives coll. 49.

become a chaos of meaningless curiosities. In its belief in the need for order, the Academy was not alone. According to the historian Steven Conn,

> Classification, and the systems through which objects were arranged and displayed, became the keys to unlocking the knowledge contained in museum collections. As metonyms, these objects only assumed their full significance through their proper arrangement. It was the way museums organized and displayed their objects that would turn them from "cemeteries," which killed ideas, into the "nurseries" which would help them grow.[30]

The Academy's curator and obsessive collector Edward Drinker Cope expressed the need for ever more specimens to study in gastronomical terms: "An institution without new collections," he wrote, "is a stomach without food."[31] But ever-expanding collections—the base of Cope's metaphorical diet and of the Academy's authority over scientific knowledge—required ever-expanding facilities to house them. Between 1815, when the Academy acquired its first building in Gilliams Court at 35 Arch Street, and 1876, when it moved to its present location, the institution grew through four buildings, each significantly larger than the one before. Erecting these new buildings and caring for the collections that filled them put an ever-increasing financial burden on the Academy and its members. It also challenged the viability of the Academy's organizational structure, which was built on the premise that volunteer curators could adequately run the museum and care for its collections.

In 1813, when the voluntary curator structure was established, the Academy's library contained just a few books and journals and its scientific collections consisted of the odd mineral, shell, and stuffed bird. By the early 1860s, the library held more than twenty thousand volumes and the collections had grown to two hundred thousand specimens, sixty thousand of which were on display in glass-fronted cabinets in the museum.[32] Through all that period, with the exception of occasional contract labor, the Academy's only regularly paid employees were a custodian, varyingly referred to as either a doorkeeper or a janitor, the librarian (one half day a week), and the chairman of the curators, who was granted a modest annual stipend beginning in 1847.[33]

One of the first people to suggest altering the Academy's organization to reflect the changing reality of its size and complexity was Joseph Leidy.[34] In 1848, when he was elected chairman of the Academy's four curators, Leidy recommended creating up to fifteen committees in specific scientific fields to help supervise and care for the collections. In an effort to protect the Academy's fragile specimens from damage and increase their usefulness to researchers both within and outside the Academy, he proposed

The American Entomological Society

The American Entomological Society is the oldest continuously operating scientific organization devoted to the study of insects in the United States. It began in 1859, when its founders gathered at the Philadelphia home of Academy member Ezra T. Cresson (1838–1926) to establish "a Society whose object shall be to ascertain the name, locality, habits, time etc. of insects taken within the United States of America."

From the outset, the society emphasized the importance of establishing and maintaining a research collection and a reference library, producing scholarly and popular publications, and educating the public about insects. Many of the Academy's leading naturalists, including Joseph Leidy and Edward Drinker Cope, were counted among its members. The influential coleopterist and Academy vice president John L. LeConte was the society's first president.

In 1876, when the Academy of Natural Sciences moved to its present location, the society applied to form the Academy's Entomological Section. This close relationship with the Academy continues to this day. The society's extensive library and insect collection are housed at the Academy, and its Philadelphia meetings are held there.

—*G. W. Cowper*
Corresponding Secretary
The American Entomological Society

FACING PAGE Members of the American Entomological Society on a collecting trip circa 1900. ANSP Archives coll. 922.

increasing the number of curators and restricting the people to whom curators could lend keys.[35]

Seventeen years later, in 1865, he reported that at least three departments (birds, insects, and plants) were in serious jeopardy because of the amateurish supervision they had received. "It is no longer possible for volunteer or amateur curators to give them that degree of attention which is required for the preservation of destructible collections," he reported.[36] To avoid further damage and loss, Leidy recommended that a committee of five be appointed to raise a public subscription for "a fund to be especially devoted to the employment of suitable persons to take charge of various collections, to provide for their safety and to insure as far as practicable their permanency and increase."[37] Although an expanded committee structure was eventually approved, Leidy's proposed fund was never realized, and four volunteer curators remained nominally in charge of the collections until 1876, when a modification of the bylaws permitted a restructuring of the organization and minimal payment to the curators.[38]

The tension between those who wished to make the collections accessible to the public and those who felt they should be restricted for purely research purposes had existed within the Academy almost from the first year of its founding.[39] For the first decade or so, the space available for Academy activities was so limited as to preclude use by anyone other than the members. The purchase by the Academy of an old Swedenborgian church at 12th and George (Sansom) Streets in 1826 made the public display of its collections possible for the first time, but even then, visitation was strictly limited to members and their guests (figure 7.9).

In 1827, the threat of legislative action to deny tax exemption to the Academy prompted the formation of an internal committee to explore whether to loosen the society's restrictive rules about access.[40] The committee ultimately recommended that the museum should be opened "two half days in every week," not just for members but for anyone interested, provided an Academy member was willing to vouch for his or her good character by supplying the visitor with a signed ticket of admission.[41] So, in September 1828, the Academy began a tradition of public display that has continued to the present day.

One of the principal supporters of this open-door policy, and of public education in general, was William Maclure, who served as the institution's president and primary benefactor from 1817 until his death in 1840. It was Maclure who paid for most of the cost of purchasing the former Swedenborgian church to serve as the Academy's headquarters, and it was he who helped to underwrite the dissemination of knowledge through the publication of the Academy's *Journal*. Even when he was not present for Academy meetings, he continued to promote his wish to have the Academy make its facilities and collections open to public view. In 1839, while living in Mexico, Maclure addressed a letter to the Academy in which he strongly suggested "the necessity of some legal instrument specifying the liberality of the Academy in gratuitously giving access to all classes." He went on to suggest a series of objectives that he hoped the Academy would meet. Included on his wish list were the following desiderata:

> That the library, museum, etc. shall be open to the public gratis,
> That the library shall be furnished with tables, chairs, etc. heated and lighted,
> That the library shall be accessible 4 or 5 days in the week,
> That the library shall be open to the public the whole of Sunday or Sunday night,
> That there shall be a librarian who shall attend at least 6 hours on the days of the admission and the nights until 9 o'clock,
> That the lecture room shall be open for the use of any one who wishes to lecture, particularly on Sundays on any species of science, etc. etc. admission for males 5 cents and females gratis.[42]

7.9 The old Swedenborgian church at 12th and George (Sansom) Streets, purchased by the Academy in 1826, made the public display of its collections possible for the first time. ANSP Archives coll. 49.

7.10 The Wagner Free Institute of Science, established by William Wagner in 1855. Wood engraving, ca. 1865. Wagner Free Institute of Science.

At a meeting later that summer, the Academy's bylaws were amended to satisfy several of Maclure's requests.[43]

Despite the philanthropic impulse that Maclure championed, support for the "diffusion of useful, liberal knowledge" was not universally accepted as the Academy's responsibility by its later members. During Maclure's lifetime, and increasingly after his death in 1840, a small but influential group of members viewed the institution as a place better suited to pure research and scientific training than public education. They believed that the museum's collections should be strictly regulated for their own protection and that the energies of the members should be directed to creating new knowledge rather than disseminating information that already existed.[44]

Philadelphia's surge in population and wealth in the years surrounding the Civil War added to the growing tension between those who advocated public access and the protective academics who felt that educational activities should be relegated to colleges, universities, and other institutions. In part to fill this void, the Wagner Free Institute of Science had been established in northeastern Philadelphia in 1855 (figure 7.10). The city's third museum of natural history (following Peale's American Museum and the Academy's),[45] the institute was founded by William Wagner (1796–1885), a successful businessman, protégé of Philadelphia merchant, banker, and philanthropist Stephen Girard, and a member of the Academy since 1815. It was Wagner's wish to create a "public institution" (as opposed to a private academy) that could offer "gratuitous instruction in the Natural Sciences" in such fields as "Geology, Mineralogy, Metallurgy, Mining, Botany, Chemical Agriculture, with their application to the arts and other kindred sciences."[46] Staffed by a paid faculty, Wagner's institute was empowered by the Commonwealth of Pennsylvania to grant degrees at the time of its founding. To the present day, the institute maintains both a collection and a faculty with a primary focus on education.

If it was Peale's goal to showcase the "natural curiosities" of the young nation in a "romantic and amusing manner" for the purpose of "instruction and amusement,"[47] and Wagner's to provide a free education in science to the working classes of Philadelphia, some believed that it fell to the Academy to advance research on the mysteries of the natural world without the accompanying encumbrances and expense of a public education program. Massive additions to its collections in the 1840s, 1850s, and 1860s had established the Academy as the preeminent natural history museum in the country. It now had to grapple with that responsibility by reconfirming—or redefining—its mission. The erection of *Hadrosaurus foulkii* in 1868, a little more than fifty years after the institution's founding, brought into focus its increasingly disparate activities and stimulated discussion about how the Academy might define itself for the next half century and beyond.

Had the Academy's officers thought of the institution in the same way that Charles Willson Peale thought of his museum, or William Wagner thought of his, they might have been proud of the record number of visitors generated by *Hadrosaurus foulkii* and thought of ways to turn such a spectacle into a welcome source of revenue. But in reading the meeting minutes of the period, one senses that they found *Hadrosaurus*'s unanticipated celebrity not only a nuisance but a bit of an unseemly embarrassment. In their 1876 *Guide to the Museum*, W. S. W. Ruschenberger (figure 7.11) and George W. Tryon Jr. report with some satisfaction that the 10 cent admission fee that was imposed soon after the *Hadrosaurus* went on display reduced the number of visitors "from thousands to hundreds, and even less" per day.

During his tenure as president (1869–81), Ruschenberger stated repeatedly that he believed the Academy's first priority should be original research,[48] and yet, as an amateur naturalist who had been sending specimens to the museum since the early 1830s,[49] he was reluctant to see the Academy give up its role as a gathering place for lovers of natural history (literally, amateurs) seeking "rational amusement" in one another's company. Resisting Joseph Leidy's push toward professionalization in the 1860s, Ruschenberger was even more resistant to a more aggressive push in the same direction by Edward Cope a decade later. The

conflicting visions for the Academy held by these two men soon led to a personality clash that precluded any hope of compromise or cooperation.

In an editorial in *The American Naturalist* in January 1880, in which he lobbied to put "the direction of the museum and the scientific work dependent on it in the hands of thirteen professors" (a slight variation on the plan Leidy had promoted several years before but which Ruschenberger had effectively blocked), Cope accused Ruschenberger of incompetence, libel, and injuring the interests of the institution.[50] "Is the 'Academy' to be an academy of original research in the sciences, or, shall we say a trustee school, which will tolerate original research provided it be not too extensive or important?" chided the irascible Cope (figure 7.12). Describing Ruschenberger's administration of the Academy as a "return to the days when learned men were the property of priests or the mere ornaments of the governing classes of society," Cope pledged to do whatever was necessary to wrest control of the institution from Ruschenberger and his friends, whom Cope considered "a collection of generally worthy gentlemen who know more of everything else than of science and its needs."[51]

While much of Cope's political maneuvering took place behind the scenes, Ruschenberger recorded his impression of one particularly unpleasant face-to-face confrontation with the angry academic in a memo to the files:

> About 5 o'clock PM July 11, 1883 I met Mr. E.D. Cope on Broad Street in front of the St. George Hotel. He admitted my charge that he had packed the meeting of the Academy last night with his friends . . . [and] that no matter how long it might take, he wanted to succeed sooner or later in upsetting my policy [of keeping the curatorial positions at the Academy honorific and nonsalaried] which he believed obstructed the progress of young naturalists. . . . My conviction is that Mr. Cope is ambitious, selfish, unscrupulous and wholly unreliable and unfaithful.[52]

Relying on a narrow reading of the Academy's founding documents and a reservoir of goodwill from members grateful for his central role in building the new museum, Ruschenberger managed to block repeated efforts to change the Academy's

7.11 William S. W. Ruschenberger (1807–1895) by William K. Hewitt. Oil on canvas. ANSP Library coll. 286.

7.12 Edward Drinker Cope (1840–1897), ca. 1876. ANSP Archives coll. 457.

administrative structure. Even after he surrendered the presidency to Joseph Leidy in 1881, he and a like-minded group of friends and associates worked to keep the Academy focused on its role as a gathering place for knowledgeable amateurs and an institution more focused on research than public education.

In 1889, eight years after Joseph Leidy succeeded Ruschenberger as president, the University of Pennsylvania gave the institution another opportunity to review its goals and objectives when it invited the Academy to move its massive collections into a new, purpose-built facility near the university's new campus in West Philadelphia (figure 7.13).[53] William Pepper, a longtime member of the Academy and provost of the university, was behind the proposal, which he hoped would ultimately involve all of Philadelphia's cultural institutions.[54] Pepper recognized that if the Academy agreed to move to West Philadelphia, it would give immediate authority to the university's nascent school of biology and help it to solidify its emerging role as the city's academic center.

In his proposal, dated April 10, 1889, Pepper argued that combining the scientific and educational activities of the two institutions would create a mutually beneficial synergy that would give Philadelphia "advantages not enjoyed by any city in the world."[55] Although the press referred to the university's plan as an "amalgamation" of the two institutions,[56] Pepper was sensitive to the delicate issue of the Academy's independence and tried to reassure its members that their autonomy would be protected and even enhanced by the proposed move. "The juxtaposition of several great independent institutions administered in harmony and for the common purpose of the increase and diffusion of knowledge," Pepper argued, "would . . . vastly increase the dignity of each institution, and render its work more economical and effective." If the Academy failed to grasp this opportunity, he warned, the university would be "forced to push ahead in lines of teaching and of collecting altogether parallel to those of the Academy." He further predicted that if the Academy refused to move and the two institutions continued "to work at a distance, the future discloses nothing but an incalculable waste in competition, duplication of teachers, laboratories, and museums."[57]

Pepper's proposal skillfully contrasted the rosy opportunities occasioned by the move with the sinister repercussions of its rejection. It described in detail the new location on offer: a tract of ten acres that the university had recently purchased from the city "in immediate proximity to the University of Pennsylvania, extending from Thirty-fourth Street to the lines of the Pennsylvania R.R. at South Street and from Locust to Spruce Street." Anticipating objections to its distance from Center City, the provost touted "its central and highly accessible position, adjoining South Street Station, five minutes from Broad St. Station; twenty minutes from Broad and Walnut Streets by the Walnut St. bridge, ten minutes from Powelton Avenue Station."

Out of "a sense of public duty," and on behalf of the university, Pepper proposed to sell the site at or below its cost on a fee simple basis, "subject to no condition or covenant whatsoever." He offered to help the Academy pay for the move by creating

7.13 The University of Pennsylvania's new campus in West Philadelphia at about the time Provost William Pepper wanted the Academy to move there (ca. 1889). University of Pennsylvania Archives.

"a syndicate . . . formed by friends of the two institutions" that would "take over from the Academy at Nineteenth and Race Streets, the purchase money being advanced to enable the Academy to erect its new building upon its new site in West Philadelphia."[58]

Joseph Leidy, who had long served on the faculty of the University of Pennsylvania, saw many advantages to such a move (most notably the opportunity for unlimited expansion), but as president of the Academy, he had to be sensitive to an apparent conflict of interest. When he referred the subject to the Academy's council of curators for discussion, he discovered that the majority of his fellow officers were strongly opposed to the idea. When council members were probed by reporters for their reaction to the proposal, one dismissed it as "preposterous." Another called it "brazen." A third was more expansive:

> We have the finest and most complete library of natural history in the world, and we have a rare collection of natural history specimens in the museum. We have a desirable and convenient location at Nineteenth and Race streets, in contrast to what would be a very inconvenient and out-of-the-way neighborhood in West Philadelphia.

"It is all one-sided," he concluded. "The Academy has all to lose—its reputation and individuality as a scientific institution; and the University has everything to gain—the museum of collections, the library, and everything else."[59]

Strongly in favor of the move was Edward Drinker Cope, who, like Leidy, served on the university's faculty. Although he failed to attend any of the members' meetings at which the matter was discussed, he took his views directly to the press, as became his habit in future years, and through it, tried to embarrass and bully his opponents into agreement or silence. In a particularly vicious personal attack in the *Telegraph*, he singled out the Academy's longtime librarian and recording secretary Edward Nolan and the geologist and curator Angelo Heilprin (whom he dismissed as being Irish and Russian, respectively). "Who are these men [who oppose the university's proposal] and why do they take such a vital interest in this question of moving the site of the Academy?" he asked rhetorically. "They are both salaried *attaches* of the Academy and . . . would have to do some work for their salaries if the Academy were taken over to the University, and the chances are that they would be removed altogether from their soft places."[60]

Following Cope's lead, several of the city's newspapers jumped into the debate by decrying the Academy, "an old fogy institution," for resisting the proposed move. "The people who would suffer," opined one editorial, "would be the mere potterers and hobby-riders who always hang on to such an institution in the hope of gaining an importance from its importance."[61]

Despite the heated rhetoric in the press, the proposal appears to have been dead on arrival at the Academy. The council recommended a vote against the move, and a committee, chaired by former president Ruschenberger, was appointed to draft a statement explaining why.[62] Leidy, who had been only one of two members of the council to support Provost Pepper's proposal, was diplomatically absent when the full Academy met on 28 May 1889 to vote on the issue.[63] In its coverage of the meeting, the *Times* reported that the vote against the move was an overwhelming sixty-eight to three.[64] After outlining the reasons for the decision, the newspaper went on to report the depth of feeling behind it:

> After the meeting the proposition of Dr. Pepper was the only subject discussed, and nearly every member had some decided opinion to express against it. Mr. Binder characterized it as a dish of "Pepper sauce," while Professor Houston said it was a proposition something like one [that] would be coming from a growing town in the West to one of the ancient cities in the Old World to come and bring their city and superior learning or else the Western town would be compelled to build a city. . . . Such remarks as "They want to get our library" were common among the members as they were preparing to leave.[65]

William Pepper's plan to bring the cultural institutions of Philadelphia together in one place was not an entirely original one. Twenty years before, the Academy had joined with the American Philosophical Society, the Franklin Institute, the Library Company of Philadelphia, and the Pennsylvania Academy of the Fine Arts in petitioning the state legislature for permission to construct new buildings on Center Square (or Penn Square, as it was then called).[66] It was the hope of each of these institutions that all would relocate around the still-open lot at the intersection of Broad and Market Streets. Newspapers from Philadelphia to London praised the plan as a model of civic improvement, extolling the virtues of cooperative and proximate cultural activities in the very heart of the city. The state senate

approved the bill, but special interests, seeing more money to be made in erecting a city hall on the site with poorly supervised government funds, waged an effective campaign against it. The plan's opponents claimed that it would be undemocratic to allow "elite" institutions to take over such a public space. When the proposal failed to pass the House of Representatives, the Academy and the other institutions behind the initiative were forced to look elsewhere for expansion.

The Academy's hardworking building committee, led by W. S. W. Ruschenberger, eventually secured a lot on Logan Square (in April 1868) for $65,298. Many members considered this location too remote from city life, so other sites were actively explored. In 1875, John H. Towne, a trustee of the University of Pennsylvania, bequeathed $10,000 to the Academy with the stipulation that it "remove to the vicinity of the University of Pennsylvania in West Philadelphia," anticipating William Pepper's proposal by fourteen years.[67] At about the same time, another supporter offered to give all of the members of the Academy free passes on the railroad for a year "and an acre of ground, and to use his influence to add $30,000 to the building fund" if the Academy would move to Germantown.[68] Both ideas were rejected. With no other prospects in sight, the committee moved ahead with plans to build a museum, library, and meeting hall at the corner of 19th and Race Streets. The cornerstone was laid on 30 October 1872, and the work began (figure 7.14). Three years later, the Broad Street building, which had been the home of the Academy since 1840, was sold for $60,000. When the collections and library were moved to the still unfinished museum on Logan Square, it took ten weeks and "312 furniture car loads" to accomplish the task.[69]

The collegiate gothic building, designed by James H. Windrim (1840–1919), was planned with room for growth. When the museum opened in the winter of 1875–76, at a cost of $239,160, parts were left intentionally "exposed, like a beggar's deformity, appealing to the generous and intelligent to contribute means to cover it."[70] The original building covered 27,275 square feet, or more than three-fifths of an acre, all of which was "fully occupied by the collections" almost from the day it opened.[71] This may be why, when the University of Pennsylvania made its formal offer of land thirteen years later in 1889, the proposal was greeted by some with interest.

Although it had been soundly rejected in 1889, the idea was floated again in 1890, when the Academy's plans for expansion at its Logan Square site ran into difficulties. Once again, the Academy rejected the university's invitation.[72]

Despite its failed attempt to get legislative support for its move to Center Square in 1867, the Academy continued to petition government officials for financial backing. Finally, in 1889, after many rejections, Angelo Heilprin and a number of well-connected Academy members were able to convince the state

7.14 As depicted in this newspaper engraving of 1872, the original design for the Academy's building on Logan Square included impressive towers that were never built. A somewhat less flamboyant building was opened in time for the nation's centennial in 1876. The original serpentine stone façade and collegiate gothic features specified by architect James Windrim (1840–1919) were replaced with a brick surface and a more federal appearance in 1910. ANSP Archives coll. 49.

The Academy's Logan Square Building (1876)

> On the day of the opening of the new edifice . . . the galleries were thronged with visitors from ten o'clock A.M. until after the hour for closing, ten P.M., and the vast building presented an animated and brilliant spectacle.

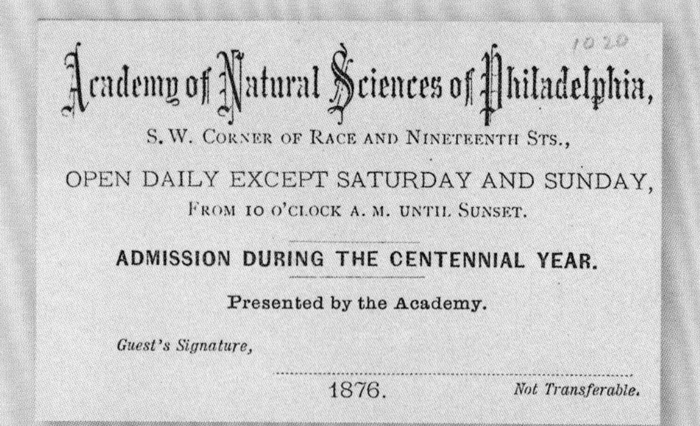
Academy of Natural Sciences of Philadelphia,
S. W. Corner of Race and Nineteenth Sts.,
OPEN DAILY EXCEPT SATURDAY AND SUNDAY,
From 10 o'clock A. M. until Sunset.
ADMISSION DURING THE CENTENNIAL YEAR.
Presented by the Academy.
Guest's Signature,
1876. *Not Transferable.*

An admission ticket from 1876. ANSP Archives coll. 417B.

FACING PAGE Top: cutaway view of the main exhibition hall before extra floors were inserted in 1936; bottom: the Logan Square building's 19th Street façade circa 1876. ANSP Archives coll. 49.

So begins an account of the grand opening of the Academy's new museum building that appeared in the 13 May 1876 edition of the *Friends' Intelligencer,* a Quaker newspaper. The writer, identified only as "L.J.R.," went on to describe in great detail not just the displays, which were "excellent," but also many of the architectural features of the new building. He likened the museum, with its abundant windows, glass roof, and "slender but powerful" iron balusters, to a house of worship, filled with the Creator's marvelous works.

The perceived resemblance was no accident. Like many natural history museums of its time, the Academy's new building closely followed the basic plan of a Gothic cathedral. Designed by Philadelphia architect James H. Windrim, the exterior was faced with green-hued serpentine stone and featured arched doorways and windows, soaring spires, and even gargoyles. Visitors entered through a vestibule that led into a soaring central hall flanked on either side by aisles. In place of pews were row upon row of cabinets, some filled with specimens, others with books.

In the original interior arrangement of the Academy, the library occupied the first floor, ringed by a single balcony. The museum was located directly above it, ringed by two balconies. These balconies offered extra display space while still allowing light from the windows and an eighty-foot-long skylight to penetrate into the main hall.

Despite its grand proportions, the Academy's new building was designed from the beginning to be merely the first wing of a much larger structure. To demonstrate this fact, the trustees instructed the builders to leave a wide brick patch on the south wall "exposed, like a beggar's deformity, appealing to the generous and intelligent to contribute means to cover it." It would remain exposed for the next sixteen years until sufficient funds were raised to construct what would be the first of many additions to the Academy.

—Barbara Ceiga

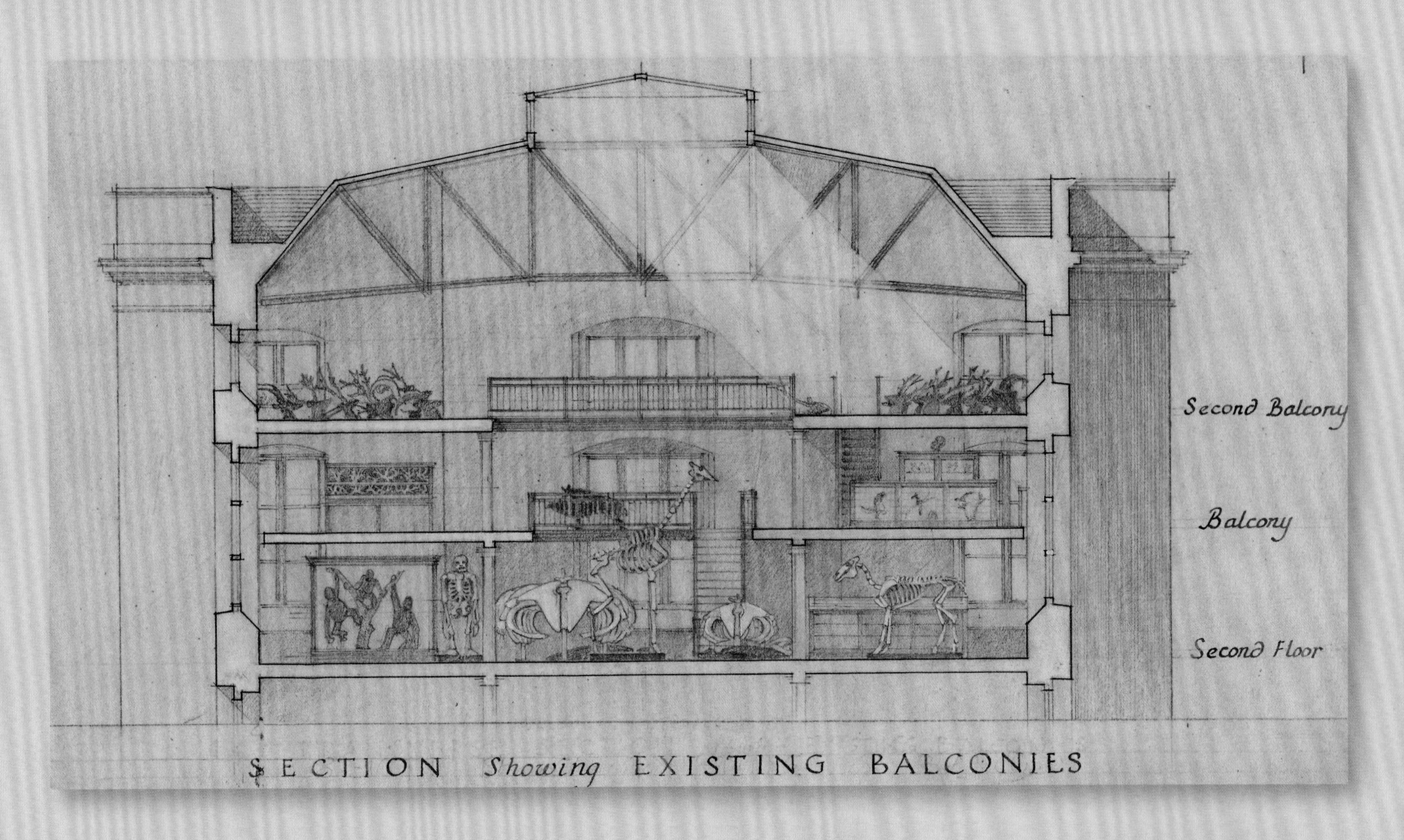
Second Balcony
Balcony
Second Floor
SECTION Showing EXISTING BALCONIES

legislature to appropriate $50,000 to assist the Academy with the construction of a badly needed addition that would be "an ornament to Philadelphia and a monument to science."[73] Another $50,000 was appropriated in 1890–91. And so, in early 1892, a two-story "connecting museum," consisting of public exhibition space and a lecture hall with a balcony large enough to hold 350 people, was opened. Four years later, in October 1896, a new four-story museum wing completed the expansion. The additions were heralded by the press, as were the housecleaning efforts inspired by the new space. At last, cheered the *Telegraph*, the Academy would have ample space for displaying "treasures of natural history which were gathered years ago and lay forgotten in the basement of the building at Nineteenth and Race streets. . . . Box after box and crate after crate, being dragged from their hiding places and cleaned amid an epidemic of coryza [coughing and sneezing], disclosed treasures that must infallibly make all the academies of natural history in the world turn green with envy." The newspaper continued:

> The Herculean task to which the officers and curators of the Academy of Natural Sciences set themselves years ago is at last nearing its end. About October 15 [1896], and not earlier, the new museum, and the wealth of specimens in the old museum, all rearranged, are to be opened to the public. Then the institution, with its noble past and with its world-famous record of great things accomplished in the field of natural science, begins another era in its mission of usefulness, an era in which its scope and powers, while nothing is sacrificed of the old dignity and the old force, will be well in the van of modern ideas of method and progress.[74]

RMP

7.15 A group of young naturalists, including ornithologists Witmer Stone and Samuel Rhoads, frolic with mounted specimens in the Academy's museum, about 1894. Historical Society of Haddonfield.

Notes

1 Edward Nolan, unpublished history of the Academy of Natural Sciences, coll. 463, 347. See also Nolan, *A Short History of the Academy of Natural Sciences of Philadelphia* (Philadelphia: Academy of Natural Sciences, 1909), 20.

2 Nolan, unpublished history, coll. 463, 352.

3 Agassiz, quoted in Paul Boller, *American Thought in Transition: The Impact of Evolutionary Naturalism 1865–1900* (Chicago: Rand McNally, 1969), 13. These thoughts parallel those expressed by Charles Willson Peale a century earlier. See Coleman Sellers, *Mr. Peale's Museum: Charles Willson Peale and the First Popular Museum of Natural Science and Art* (New York: W. W. Norton), 1980.

4 W. S. W. Ruschenberger and George W. Tryon Jr., "Summary History of the Academy of Natural Sciences of Philadelphia," in *Guide to the Museum of the Academy of Natural Sciences of Philadelphia* (Philadelphia: Academy of Natural Sciences, 1876), 99. The founders had intentionally excluded reference to God in their charter, stating instead that the society was formed because its members were "desirous of becoming more fully acquainted with the natural productions of our country," a patriotic and intellectual purpose without religious intent. ANSP coll. 109, 188.

5 Ibid., 116–17.

6 Nolan, unpublished history, coll. 463, folder 9 (1891), 423.

7 Joseph Leidy to Samuel Haldeman, 6 March 1847, ANSP, coll. 1.

8 Newspaper clipping, 12 May 1891, ANSP coll. 417, quoted in Stephen Conn, *Museums and American Intellectual Life, 1876–1926* (Chicago: University of Chicago Press, 1998), 48.

9 Despite his close working relationship with Charles Darwin and many of his supporters, Hawkins himself did not believe in evolution and lectured widely on what he believed to be the theory's flaws. See Valerie Bramwell and Robert Peck, *All in the Bones: A Biography of Benjamin Waterhouse Hawkins* (Philadelphia: Academy of Natural Sciences, 2008), 82–86.

10 This directive, which appeared in the twelfth annual report of the Board of Commissioners of the Central Park (1868), is quoted in Edwin H. Colbert and Katherine Beneker, "The Paleozoic Museum in Central Park, or the Museum that Never Was," *Curator* 2 (1959): 140; and Alan A. Debus and Steve McCarthy, "A Scene from American Deep Time: New York's Paleozoic Museum—Revisited," *The Mosasaur, Philadelphia: Journal of the Delaware Valley Paleontological Society* 6 (May 1999): 107. For a further discussion of this subject, see Edward Colbert, *Men and Dinosaurs* (New York: E.P. Dutton, 1968).

11 For a list of the well-known scientists and public figures who provided letters of support for Hawkins, see Bramwell and Peck, *All in the Bones,* Appendix 2, 106–7.

12 Hawkins's letter to Andrew Green, May 1868, reprinted in the Twelfth Annual Report of the Commissioners of the Central Park, 1868, 132–34, quoted in Colbert, "The Palaeozoic Museum in Central Park," 49–50.

13 Hawkins also visited the Smithsonian Institution, Yale's Peabody Museum, and other American institutions on the East Coast.

14 For more on Leidy's influence in the history of vertebrate paleontology, see Ronald Rainger, "The Rise and Decline of a Science: Vertebrate Paleontology at Philadelphia's Academy of Natural Sciences, 1820–1900," *Proceedings of the American Philosophical Society* 136, no. 1 (1992): 1–32.

15 The American artist and museum impresario Charles Willson Peale (1741–1827) excavated, articulated, and exhibited a mastodon skeleton in his Philadelphia Museum in 1801, so there was precedent for such a display, but few, if any, other fossil creatures had been fully articulated for public exhibition. Hawkins's was the first dinosaur skeleton to be treated in this way. For more on Peale's mastodon, see Sellers, *Mr. Peale's Museum.*

16 For more on Cope, see Jane Pierce Davidson, *The Bone Sharp: The Life of Edward Drinker Cope* (Philadelphia: Academy of Natural Sciences, 1997).

17 For a further discussion of Hawkins's *Hadrosaurus foulkii* display, see Richard C. Ryder, "Hawkins' Hadrosaurus: The Stereographic Record," *The Mosasaur, Philadelphia: The Journal of the Delaware Valley Paleontological Society* 3 (Nov. 1986): 169–79; and Earle E. Spamer, "The Great Extinct Lizard: *Hadrosaurus foulkii,*" *The Mosasaur* 7 (May 2004): 109–25.

18 Edward Drinker Cope, "Remarks on the Remains of a Gigantic Extinct Dinosaur from the Cretaceous Greensand of New Jersey," *Proceedings of the Academy of Natural Sciences* 18 (1866): 275–79.

19 Thirteenth Annual Report of the Board of Commissioners of the Central Park (1870). For a reproduction of the plate in this report, see Simon Baatz, *Knowledge, Culture, and Science in the Metropolis: The New York Academy of Sciences, 1817–1970, Annals of the New York Academy of Sciences* 584 (1990): 100. The original drawing is in ANSP coll. 803.

20 ANSP coll. 803.

21 B. W. Hawkins to J. Leidy, 14 September 1868, coll. 567.

22 According to the annual reports in the Academy's *Proceedings,* 51,520 persons visited the museum in 1867 and 65,769 visited in 1868.

23 *Proceedings of the Academy of Natural Sciences* (1869): 235.

24 The admission fee proposed by the curators was ten cents, which, they believed, "will probably be sufficient to moderate the crowds of visitors, and at the same time will be no obstruction to those who are desirous of seeing the Museum," *Proceedings* (1869): 235.

25 The picture was reproduced in the Twelfth Annual Report of the Board of Commissioners of Central Park (1869). An original, heavily retouched print of the photograph was retained by Hawkins and included in his personal scrapbook, which is now in the archives of the Academy of Natural Sciences (coll. 803). It is proudly labeled (by Hawkins) "Restored Skeleton of *Hadrosaurus Foulkii* from 76 fossil fragments by B. Waterhouse Hawkins, 1868."

26 For an account of the proposed Palaeozoic Museum and its destruction, see Colbert and Beneker, "The Palaeozoic Museum in Central Park," 137–50; and Adrian J. Desmond, "Central Park's Fragile Dinosaurs," *Natural History* (October 1974): 64–71.

27 *New York Times,* 17 June 1871, cited in Debus and McCarthy, "A Scene from American Deep Time," 110.

28 For more on the visualization of deep time, see Martin J. S. Rudwick, *Scenes from Deep Time: Early Pictorial Representations of the Prehistoric World* (Chicago: University of Chicago Press, 1992).

29 Joseph Leidy to Joseph Henry, 18 December 1875, Smithsonian Institution Archives, Joseph Henry Papers, Record Unit 26, vol. 154, #146.

30 Conn, *Museums,* 21.

31 Edward D. Cope, "The Academy of Natural Sciences," *Penn Monthly* 7 (March 1876): 174–76, quoted in ibid., 61.

32 Conn, *Museums,* 43. The sixty thousand figure comes from Ranger, "The Rise and Decline," 10. The library statistic comes from J. H. Slack, *Handbook to the Museum of the Academy of Natural Sciences of Philadelphia,* 1866, 12.

33 Nolan's unpublished history, coll. 463, 343. The bylaws were adjusted in October 1846 to provide the part-time salary of the librarian and the chairman of the curators (proposed 27 October, adopted 24 November 1846). Some curators made a modest income from lecture fees charged for their public lectures.

34 Leidy was elected to membership in 1845, elected a curator in 1846, and became chairman of the curators in 1848. He served as president of the Academy from December 1881 until his death in April 1891.

35 Edward Nolan, unpublished history, coll. 463, 271–72. Other changes proposed by Leidy can be found in the minutes of the Academy for 21 September 1852 and 5 October 1852. See also Ranger, "The Rise and Decline," 10.

36 Nolan's unpublished history, coll. 463, 354. His proposals are also cited in *Guide to the Museum of the Academy of Natural Sciences of Philadelphia* (1876), 105, and quoted in Rainger, "The Rise and Decline," 23.

37 Nolan's unpublished history, coll. 463, 354.

38 Nolan, *A Short History*, 24.

39 After two years of internal lectures by Thomas Say and others, the Academy's first series of public lectures began in 1814, allowing the treasurer, Richard Randolph, to report in 1815 that they had caused the Academy to "rise into public notice," attract "the attention of strangers," and establish "a character and reputation far exceeding our most sanguine expectations." Treasurers Report, Academy minutes, 1815, quoted in Patsy A. Gerstner, "The Academy of Natural Sciences of Philadelphia 1812–1850" in *The Pursuit of Knowledge in the Early American Republic* (Baltimore: Johns Hopkins University Press, 1976), 177.

40 For the full report of the committee, see collection 940.

41 Nolan's unpublished history, coll. 463, 155; quoted in Gerstner, "The Academy of Natural Sciences," 180. The members approved a resolution opening the museum on Tuesdays and Saturdays from one o'clock until sundown. See meeting minutes 12 August 1828, coll. 940.

42 Letter from William Maclure, 23 June 1839, incorporated into the minutes of 13 August 1839; quoted in Gerstner, "The Academy of Natural Sciences," 182. Also see letter from Maclure to S. G. Morton, 15 June 1839, American Philosophical Society.

43 The changes were reported in the 27 August 1839 minutes. See also Gerstner, "The Academy of Natural Sciences," 182.

44 In response to the University of Pennsylvania's 1889 proposal that the Academy more closely ally itself with the university, the Academy stated, "The aims and requirements of the two institutions are largely distinct, the legitimate work of the Academy being where that of the University ceases. The one endeavors to increase knowledge, while the other imparts it." Abstract from the 21 May 1889 minutes, coll. 435.

45 Pierre Eugène Du Simitière (1737–1784) also owned and operated an eclectic museum that contained some natural history and ethnographic artifacts. The museum operated in Philadelphia from 1782 until it was dispersed by sale in 1785; see *Pierre Eugene DuSimitiere: His American Museum 200 Years After* (Philadelphia: Library Co.), 1985.

46 Wagner Free Institute of Science, Articles of Incorporation, 1855.

47 C. W. Peale, quoted in David Brigham, *Public Culture in the Early Republic: Peale's Museum and Its Audience* (Washington, DC: Smithsonian Institution Press, 1995), 19, 20, 28, 36, 38.

48 W. S. W. Ruschenberger, *Report on the Condition of the Academy of Natural Sciences of Philadelphia on Moving into Its New Edifice* (Philadelphia: Collins, 1876), 40, and Ruschenberger, "On the Value of Original Scientific Research," *Penn Monthly* 4 (November 1873): 18–19.

49 In a letter to the Academy from Verpariso dated March 1832, Ruschenberger lists birds, insects, shells, minerals, fish, and ethnographic items among his presentations, reflecting his wide range of interests. Coll. 567.

50 [A.S. Packard Jr. and] E. D. Cope, "Editor's Table," *The American Naturalist* 14, no. 1 (January 1880): 38–42.

51 [A. S. Packard and] E. D. Cope, *The American Naturalist*, 40.

52 Undated manuscript note from W. S. W. Ruschenberger, ANSP coll. 567, quoted in Conn, *Museums*, 54.

53 The University of Pennsylvania established its West Philadelphia campus in 1872.

54 William Pepper was elected to membership in the Academy in August 1867 with support from Joseph Leidy, John Cassin, and Harrison Allen.

55 The university's proposal is given in its entirety in an abstract of the minutes of the meeting of the Academy of 7 May 1889, coll. 435.

56 Unidentified newspaper, 29 May 1889, coll. 417A, vol. 2, 17.

57 Letter from William Pepper to the Academy of Natural Sciences, 10 April 1889, transcribed in the minutes of 7 May 1889, QH71 A16 A14 v.14, pp. 462–64. A typescript is contained in coll. 435.

58 Ibid.

59 *Telegraph*, 8 May 1889, from the publicity clipping book in the Academy archives, coll. 417A, vol. 2, 15.

60 Anonymous newspaper clipping titled "An Old Fogy Institution," May 1889, coll. 417A, vol. 2, 15. Cope's quotation was also included, in expanded form, in "Voting Against It," *Telegraph*, 27 May 1889, coll. 417A, vol. 2, 16.

61 Anonymous newspaper clipping titled "A Diffusion of Energy," 15 May 1889, coll. 417A, vol. 2, 15.

62 The entire committee consisted of Angelo Heilprin, Charles Norris, Edward C. Nolan, John W. Redfield, and chairman W. S. W. Ruschenberger. See collection 435 for documents relating to the University of Pennsylvania proposal and the Academy's response.

63 In his unpublished history of the Academy, Edward Nolan described Leidy as being "possessed of an almost morbid dislike of disagreement and contention," coll. 463, folder 9 (1891), 424. "To him controversy and conflict were always repugnant," concurred W. S. W. Ruschenberger; "A Sketch of the Life of Joseph Leidy," *Proceedings of the American Philosophical Society* 30 (25 April 1892).

64 Voting against the motion to reject the university's proposal were Ryder, Norris, and Dougharty. Leidy and Cope were not present for the vote. For a complete list of those voting in favor of the rejection, see the ANSP minutes of 28 May 1889, coll. 502, vol. 11, 354–59.

65 Anonymous article titled "The Academy of Natural Sciences: Rejection of Provost Pepper's Proposal to Remove," *Times*, 29 May 1889, in coll. 417A, vol. 2, 17.

66 The initial meeting to discuss this plan was held at the American Philosophical Society on 29 April 1867. Documents relating to this initiative can be found in ANSP collection 502.

67 Quoted in Ruschenberger, *Report on the Condition of the Academy*, 19.

68 Ibid., 19.

69 According to Ruschenberger, the move of the collections began on 2 November 1875, and was completed on 11 January 1876. See ibid., 20.

70 Ibid., 23–24.

71 Ibid., 23.

72 The idea would surface again in 1970 when the Academy was considering a move to the Delaware waterfront. See Chapter 16.

73 Anon., "The Academy of Natural Sciences," *The American*, no. 501 (15 March 1890): 436, from coll. 417A, vol. 2.

74 Anon., "Nature's Wonders," *Telegraph*, 19 September 1896, newspaper clipping in coll. 417A, 100.

Fossilized teeth of a giant ground sloth (*Megalonyx wheatleyi*) from Port Kennedy Cave, Pennsylvania, collected by Henry C. Mercer (1856–1930) circa 1894. ANSP Vertebrate Paleontology Department #218. Purcell photograph.

FOLLOWING PAGES Ammonite (detail). ANSP Invertebrate Paleontology Department. Purcell photograph.

Chapter 8

Fossils, Finders, and Feuds: Leidy, Hayden, Cope, and Marsh

Night into day . . . I felt as though I had groped about in darkness, and that all of a sudden, a meteor flashed across the sky.
—Joseph Leidy to Charles Darwin, 1859

Fossil bones have been collected in various parts of the North American continent for centuries. In modern times, archaeologists have found evidence to suggest that pre-Columbian Indians gathered them to use for "medicine," providing a basis for their legends of colossal beasts that roamed the earth.[1] Thomas Jefferson, who thought it possible that mastodons and other extinct creatures might still be found living in the unexplored West, was the first American to study and write in depth about fossils in his *Notes on the State of Virginia* (1785). His interest in the subject continued until his death in 1826, and for this, Jefferson is regarded as the father of vertebrate paleontology in the United States.[2]

In the nineteenth century's early years, two prominent members of the Academy of Natural Sciences, William Maclure and Thomas Say, were pioneers in the sciences of geology and paleontology. Maclure introduced the study of geology in America with his travels from Maine to Georgia, and he recorded his discoveries in an article and map for *Transactions*, the journal of the American Philosophical Society, in 1809.[3]

According to historian George H. Daniels, Say was the first American to point out the chronogenetic value of fossils.[4] In a paper for Benjamin Silliman's *Journal of Science and the Arts* in 1818, he explained how the fossil record could be used as a guide for dating rock strata.[5] He conceded that little was known about North American fossils but insisted that comparisons must be made with those of Europe—so ably described by Lamarck and other naturalists—before progress in this field could be made. Say recognized that America was "rich in fossils" and that the task of studying them required knowledge of all their different states, from the "unchanged specimen" (i.e., fossil) to its living counterpart.

In the 1830s, Academy member Richard Harlan eagerly entered this field in an attempt to bring American fossils to the attention of scientists abroad.[6] Harlan sent casts of a *Tetracaulodon* (mastodon) jaw to the Jardin des Plantes in Paris and to the Geological Society of London in 1830 or 1831, and in 1833 and 1839, he journeyed to Europe, where he met some of the leading lights of European geology, principally Richard Owen. According to historian Patsy Gerstner, the spirit of cooperation between Harlan and Owen "set an entirely new tone for the relationship of American and European paleontology."[7] In Owen's contribution "Fossil Mammalia" in *The Zoology of the Voyage of H. M. S. Beagle* (1840), Owen quoted Harlan extensively on the *Megalonyx*, the extinct giant sloth first described by Jefferson.[8]

Harlan pioneered the gradual professionalism of American science through his publications, with his technical writing style the antithesis of the anecdotal style of earlier naturalists such as Alexander Wilson and John Godman. Perhaps efforts like

8.1 Fossil horse teeth (*Equus* sp.) collected by Edgar B. Howard in Texas, 1933. ANSP Paleontology Department, #13669. Purcell photograph.

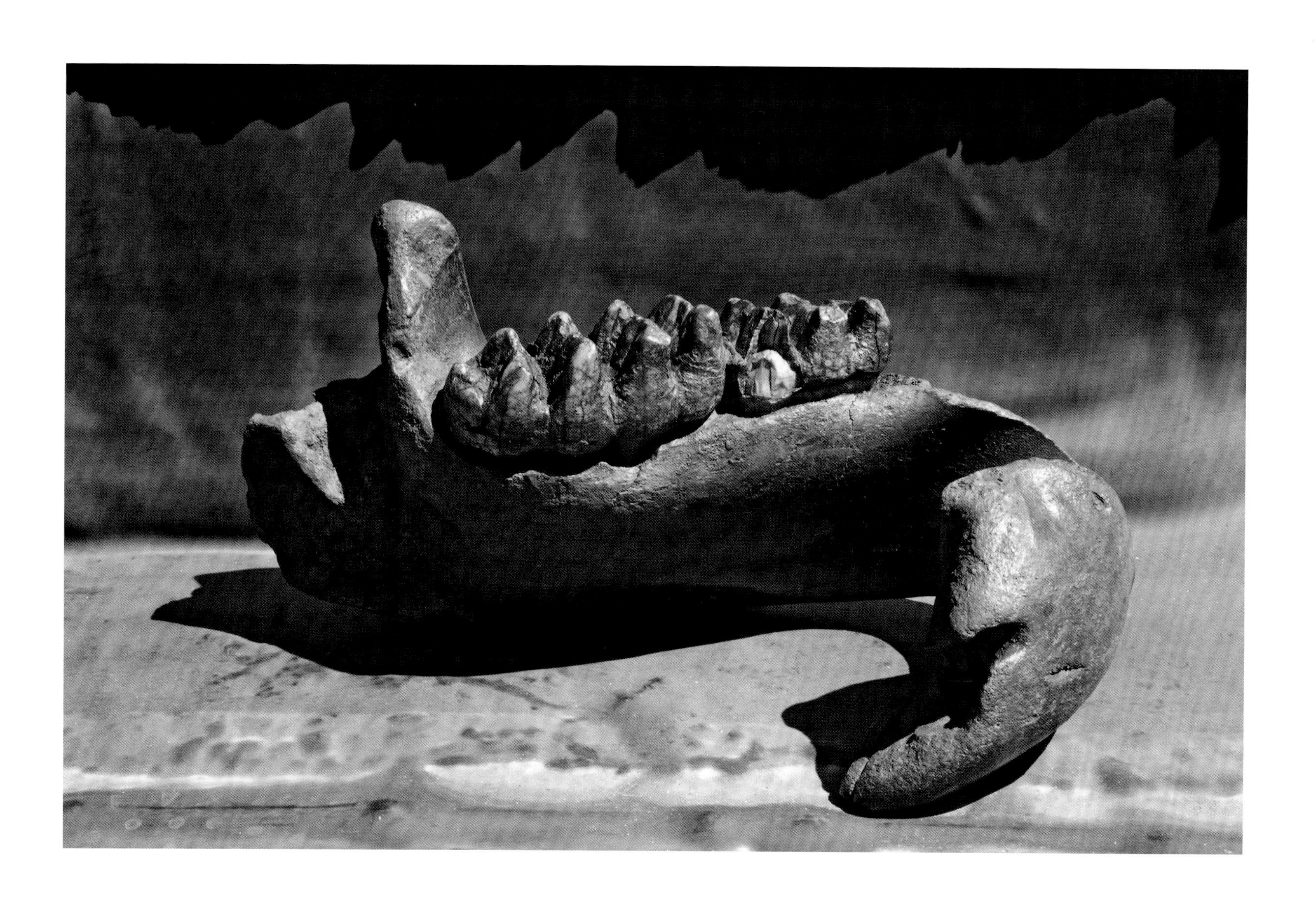

Harlan's can be characterized by saying that "the very first step through which science passes as it becomes professional is one of pre-emption, that is, it passes through a stage that makes it comprehensible to a few rather than to the many."[9]

In 1849, the American Philosophical Society recommended to its members transferring its small but historical fossil collection, including Thomas Jefferson's important specimens, to the care of the Academy, which had moved to new, larger quarters at Broad and Sansom Streets and had grown sufficiently to manage and provide for study of its own large collection of natural history specimens (figure 8.2).[10] The additional space would be increasingly necessary as many new fossils were added to the Academy's collections.

As the nineteenth century progressed, the federal government, following the lead of certain states, realized the necessity of accessing the geological resources of the country, to discover valuable minerals such as coal, iron ore, and building stone, as well as silver and gold. In 1839, Congress appointed David Dale Owen, son of the founder of New Harmony, where Say, Lesueur, and others from the Academy had gone to live in 1825, to survey some of the country's public lands. With this information in hand, the government intended to authorize the sale of mineral rights to private speculators. Owen, who may have been influenced to become a geologist by his father's onetime business partner, William Maclure, organized a team of surveyors who within a few years successfully surveyed some eleven thousand square miles in the upper Mississippi Valley.[11] Owen's explorations revealed, in addition to minerals, the great potential the West held for the discovery of fossil vertebrates.[12]

By mid-century, geological surveys mounted by the government to ascertain the most practical and economical route for a railroad from the Mississippi River to the Pacific Ocean, and the private enterprise for railroad building that ensued, disclosed a vast treasure house of amazing prehistoric creatures, buried in western canyons and basins and as yet unknown to the world. Thanks to these discoveries and the precedent set by earlier expeditions, the political and commercial expansion of the country continued to have a scientific component. Later, science would be pursued for its own sake.

The Academy of Natural Sciences, almost from its founding in 1812, was a key institution in advising and consulting on the government's involvement with western exploration. This participation began with requests for suggestions from the Academy concerning the scope of and expert personnel for the Long and Wilkes Expeditions. It gained even greater importance as the nineteenth century progressed, for the opening of the enormous western expanse of America brought the need for a great scientific inventory of the natural and human resources of the new territory.

Academy members Joseph Leidy (1823–1891) and Ferdinand Vandeveer Hayden, both physicians and intensely intellectual yet diametrically opposite in personality, would work together to understand the fossil record of the American West (figures 8.3 and 8.4). Leidy, a rising young star at the Academy, was reserved and methodical, while Hayden, a geologist and explorer who collected numerous western fossils and sent them to Leidy for description, was, though brilliant, also insecure, impatient, and aggressive.

Young Joseph Leidy, more interested in the study of natural history than in the practice of medicine, was strongly influenced by the Scottish geologist Charles Lyell, who visited Philadelphia in 1842 on a lecture tour. Lyell befriended the clever medical student, was impressed by his talent as an anatomist and illustrator, and encouraged him to leave medicine for paleontology.[13] In 1845, Leidy became a member of the Academy at the age of twenty-two. In a felicitous link to a founder of the institution, that same year, Leidy's first published paper appeared (in Boston), describing the anatomy of a mollusk first named by Thomas Say.[14]

Leidy was soon appointed Academy librarian. A year later, he was elected chairman of the Council of Curators, a position he held for the rest of his life.[15] A parasitologist as well as a mineralogist and paleontologist, in 1846 Leidy described an *Entozoon*, a small coiled worm in a cyst in a pig's thigh, which he identified as *Trichinella spiralis*, the worm that causes the dreaded, often fatal disease trichinosis in humans.[16]

A friend once noted with amusement, "never give Leidy anything for dinner that he can dissect." It seems that this friend had served Leidy nine turtles at a party, and instead of eating them, he had proceeded to dissect them, finding and naming three previously unknown intestinal parasites in the process.[17]

Leidy studied the morphology of creatures both living and extinct and ranging in size from microscopic organisms to mastodons and dinosaurs. Some fossil remains he gathered locally himself, but collectors in the field, like Hayden and, later, Edward Drinker Cope, sent him enormous quantities of fossil specimens from the west for description and classification. That he was a trained anatomist made possible his early entry into vertebrate paleontology. As Lyell had observed, Leidy was also a gifted artist. He drew exquisitely detailed renderings of the many subjects

8.2 Mastodon (*Mammut americanum*) jawbone collected by William Clark at Big Bone Lick, Kentucky, 1807. Jefferson Fossil Collection. ANSP Paleontology Department, #13315. Purcell photograph.

8.3 Joseph Leidy (1823–1891). ANSP Archives coll. 457.

he studied and wrote about (figure 8.4). Not willing to engage in theories, he carefully built up facts concerning the materials he worked with, preferring to describe, catalog, and arrange specimens in their proper orders. It was said that he could turn over any rock and name every creature under it, in addition to the rock itself and the mineral composition of the soil.[18]

Leidy began his work at a time when geology was no longer under the strong influence of the Bible—the dogma that God had created the earth in six days and every living thing whole and entire—because the meaning of the fossil record was becoming clearer with every new discovery. Lyell's theories on the link between the earth's history and life on earth, as expressed in his *Principles of Geology* (1830), affected Leidy's thinking, as it had Darwin's. Leidy accepted Lyell's advocacy of "gradualism" in biological and geological change, the idea that because the earth was formed gradually, layer after layer, in theory the various strata should contain transitional forms of life that arose and then became extinct.[19]

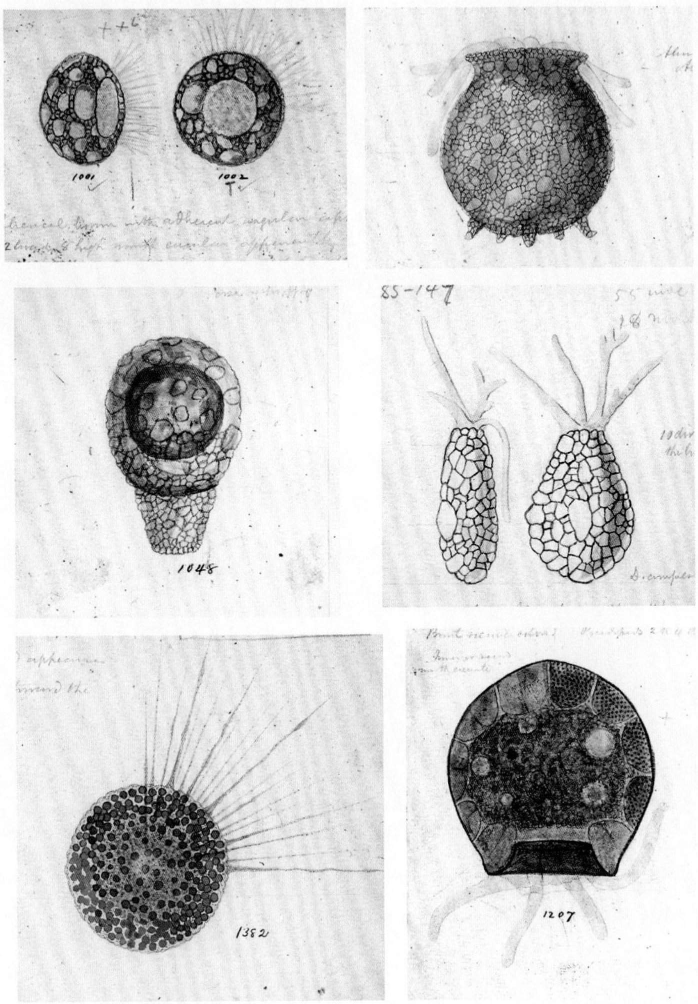

By the mid-1840s, Leidy had already received quantities of fossils to identify. The first one, sent to him from the Badlands of the Nebraska Territory, was the jawbone of a huge mammal related to the horse and the rhinoceros that he named *Titanotherium*.[20] This was the first fossil of any kind to reach a professional scientist in the East, and it marked the beginning of a new era in American paleontology as fossils continued to arrive for Leidy to examine.[21] He received fossils from numerous sources, including Spencer Fullerton Baird, assistant secretary of the new Smithsonian Institution. As the specimens accumulated, especially from Hayden and Fielding Bradford Meek (1817–1876), Leidy described many of them in the Academy's *Proceedings*. In 1853, he published *Ancient Fauna of Nebraska,* which included his descriptions of North American hoofed mammals, a saber-toothed cat, horses, turtles, and a prehistoric relative of the rhinoceros.[22]

Leidy was discovering an entirely new world in the virgin fields of the American West. It was not possible to base his studies on those of European paleontologists because, according to the eminent scientist Henry Fairfield Osborn, "every specimen represented a new species or a new genus of a new family, and in some cases a new order."[23] Excavations were easier in the West than in the East because nature had eroded the ground, exposing fossils in cliffs and canyons. Many fossils were visible on the surface, where only a knife was needed for excavation.

In an 1847 article, Leidy wrote about a miniature fossil horse, only three feet high, that had once existed in North America, although it had been extinct for millions of years. A cervical vertebra and the teeth of the horse had first been found in association with a deposit of mastodon bones in a tract extending from the Neversink Hills in New York to Bordentown, New Jersey, revealing to Leidy that the horse had existed at the same time as the mastodon.[24] On his voyage aboard the *Beagle* (1832–36), Darwin had found similar teeth concomitant with mastodon bones in South America.[25]

Leidy was the first to establish that horses had lived in North America prior to their reintroduction by the Spanish in the fifteenth century. After more horse teeth were discovered in the West, he named the species *Equus americanus,* though ten years later he had to change the name because it had already been assigned to a fossil in South America. Leidy changed the name of the fossil horse to *Equus complicatus,* demonstrating that, in spite of his serious side, in the midst of probable exasperation, he had a sense of humor. Eventually he identified five species of *Equus,*

8.4 Joseph Leidy's original drawings of *rhizopods* (protozoans). ANSP Archives coll. 3.

8.5 Prehistoric horse skull (*Anchitherium bairdii*) (now *Mesohippus bairdii*) described by Joseph Leidy and named for Spencer Fullerton Baird in 1850. ANSP Paleontology Department #10446. Purcell photograph.

8.6 Fossil horse teeth fragments (*Equus excelsus*) collected for F. V. Hayden on the Niobrara River, Nebraska. ANSP Paleontology Collection. Purcell photograph.

Fragments teeth of
Equus excelsus
Niobrara R.
For Dr. Hayden

which ultimately formed a significant part of the sequence in the evolution of the horse and, according to Leidy's biographer, Leonard Warren, became an important bit of evidence in support of Darwin's theory of evolution.[26]

In his 1847 article, Leidy questioned why the horse became extinct in the Americas when it had survived elsewhere in the world. He wrote in the Academy's *Proceedings*:

> At present [the fossil horse's] existence being fully confirmed, it is probably as much wonder to naturalists as was the first sight of the horses of the Spaniards to the aboriginal inhabitants of the country, for it is very remarkable that the genus *Equus* should have so entirely passed away from the vast pastures of the Western world, in after ages to be replaced by a foreign species to which the country has proved so well adapted.[27]

He speculated that the horse's disappearance must have been related to severe climate change. Darwin, in *On the Origin of Species* (1859), pondered the same mystery.[28]

Leidy wrote of another fascinating discovery in 1847, an ancestral camel that he named *Poebrotherium wilsonii*, in honor of Thomas B. Wilson, the Academy's benefactor. Jefferson Davis, secretary of war at the time (and future president of the Confederacy), suggested to him that it might be possible to introduce the modern camel to America as a beast of burden. Leidy readily agreed with the idea, since both the horse and the camel had lived and become extinct at the same time. The horse had thrived when reintroduced, so why not the camel?[29]

The experiment was indeed tried for a time after 1856, when army personnel stationed in the Southwest to protect settlers from Indian attacks had need of an animal able to withstand the harshness of the desert. Seventy-five dromedaries, with their Turkish, Greek, and Armenian drivers, were imported from North Africa to serve in a United States Camel Corps. But when additional camels were requested from the War Department, Congress was skeptical about the experiment and denied further funding. Nevertheless, for the next few years, camels continued to be used by survey teams, who preferred them to mules for long-distance travel, but in consort with other animals, the camels were a problem. One participant reported that the camels "frightened horses and mules and caused several accidents." Consequently, during the Civil War, the government's dromedaries were dispersed. Some were used for various purposes, such as road construction, while others escaped or were released from their stables to wander freely in the desert. But unlike the horse, the camel never established a wild population.[30]

Also in 1847, Leidy wrote about extinction and the great age of life: "The study of the earth's crust teaches us that very many species of plants and animals became extinct at successive periods, while other races originated to occupy their places . . . life may have been ushered upon earth, through oceans of the lowest types, long previously to the deposit of the oldest palaeozoic rocks as known to us."[31]

By the 1850s, when he had been selected to lecture at the University of Pennsylvania, there were powerful forces at the university who objected to his appointment because of his ideas on extinction. Leidy wrote to his friend Spencer Baird at the Smithsonian, "My being a comparative anatomist, naturalist etc. are all used as objections. I am shamefully abused as being an atheist and infidel . . . it has been positively asserted that I seek to make proselytes to infidelity, and that in my writings I have tried to prove that geology overthrows the Mosaic account of creation."[32] There are echoes in Leidy's words of the Academy's past difficulties in this realm. In the society's early years, its members had been castigated for banning the discussion of religion at meetings and were thus considered antireligious and labeled "atheists."

When, thirteen years later, Leidy read Darwin's earthshaking book, which laid out the workings of his theory of evolution, something that had puzzled Leidy for years, he wrote to Darwin, whom he had met once in London in 1848, "Night into day . . . I felt as though I had groped about in darkness, and that all of a sudden, a meteor flashed across the sky."[33] Darwin replied:

> I received a few days ago your note of Dec 10 & your most generous present of a whole bundle of your publications, which I value most highly & am extremely glad to possess. Your note has pleased me more than you could readily believe; for I have during a long time heard all good judges speak of your paleontological labours in terms of the highest respect. Most paleontologists (with some few exceptions) entirely despise my work; consequently approbation from you has gratified me much. All the older geologists (with the exception of Lyell whom I look at as a host in himself) are even more vehement against the modification of species than are even the paleontologists. . . . Your sentence that you have some interesting facts "in support of the doctrine of selection, which I shall report at a favourable opportunity," has

> delighted me even more than the rest of your note. I feel convinced that, though as long as I have strength I shall go on working on this subject, yet that the sole way of getting my views partially accepted will be by sound workers shewing that they partially accept them. I say partially, for I have never for a moment doubted, that though I cannot see my errors, that much in my Book will be found erroneous. Pray forgive this egotistical note & with cordial thanks for your letter & kind present, believe me Dear Sir, with sincere respect, your obliged Charles Darwin.[34]

Leidy's protégé in paleontology, Ferdinand Vandeveer Hayden, unintentionally provided confirmation of Darwin's theory of evolution, though he had not yet read *On the Origin of Species*, when he wrote his book, *Geology and History of the Upper Missouri* (1862).[35] Hayden, like Leidy, did not practice medicine but pursued in different forms of teaching and research his passionate interest in natural science, which he had discovered at Oberlin College in Ohio (1845–50). After further study at the Cleveland Medical School, he was introduced to James Hall, the state geologist of New York, who hired him to explore the Badlands in Sioux country (in what is today South Dakota). It was a dangerous place to be, but fortunately for Hayden, the Sioux regarded him as a harmless eccentric and left him alone, much as other Indians had been unconcerned with the botanist Thomas Nuttall years before. Impressed by Hayden's energy and enthusiasm, they called him "He who picks up stones running."[36]

In 1853, Hall sent Hayden, along with Fielding Bradford Meek (an Academy member based at the Smithsonian), to the White River Badlands in Dakota country (then the Nebraska Territory) to study the geology of the area and to collect fossil remains. Hayden sent many of his collections to Leidy in Philadelphia for identification. Together with Meek, Hayden worked out the cretaceous geological horizon of the Dakota country and a stratigraphic column for the upper Missouri region.[37] A few years later, collecting for Leidy along the Judith River on the upper Missouri River in Montana and in the Nebraska Territory, Hayden found teeth in Mesozoic beds that proved to be the first dinosaur material identified in North America.[38] Nominated by Leidy and Spencer Baird, Hayden was elected an Academy corresponding member in 1856.

8.7 Dinosaur fossil (*Palaeosyops paludosus*) collected for F. V. Hayden by Dr. Joseph Carson at Grizzly Buttes, Wyoming. ANSP Paleontology Department #10387. Purcell photograph.

Etheldred Benett (1776–1845)

ABOVE Etheldred Benett, artist unknown, private collection. Photo courtesy of Sir Henry Rumbold.

FACING PAGE Three wooden cards containing invertebrate fossil specimens assembled by Etheldred Benett, including echinoderms (top); gastropods, bivalves, and mollusks (center); and cnidaria (bottom), all from Bradford-on-Avon, Wiltshire, England. ANSP Invertebrate Paleontology Department. Purcell photograph.

Etheldred Benett, who should be regarded as the first female geoscientist, was born in 1776 into a wealthy family in Wiltshire, England. The countryside surrounding her home (the Vale of Wardour) included exposures of wonderful fossiliferous strata of Jurassic and Cretaceous age. Her brother-in-law, A. B. Lambert, was an enthusiastic fossil collector and, as a founding fellow of the Linnaean Society, well connected in the scientific community. These factors must have kindled Miss Benett's interest in geology and paleontology.

Never married and financially independent, Miss Benett enjoyed the freedom to pursue her interest in fossils. Most of her collecting (by her own hand or by quarrymen in her employ) took place within a ten-mile radius of her home in Norton Bavant. By 1810, she was corresponding with many of the leading paleontologists of the time about her activities and her collections. Specimens she collected were illustrated and published by James Sowerby in his seven-volume work *Mineral Conchology*. Some of her other fabulous specimens were illustrated by the notable paleontologist William Buckland in *Geology and Mineralogy* (1837).

More than just a collector, Miss Benett produced a remarkable bed-by-bed description of the quarry at Chicksgrove, which Sowerby published in 1816 (without her knowledge or consent). Her first version bore the note "There are Scales of fishes, ammonites and various other organic remains found in this quarry but I have not yet been able to ascertain to which of the beds they respectively belong and therefore could not insert them in their proper place." She continued to work on the quarry, and by the following year, she was able to revise her section to include all the fossils found in each bed.

Her greatest contribution to paleontology (after the remarkable specimens she amassed) was her monograph *A Catalogue of the Organic Remains of the County of Wiltshire*, published in 1831, first as part of *The History of Modern Wiltshire*, by Colt Hoare, and then by Miss Benett herself in a revised version that included eighteen lithographic plates of her own artwork.

After her death in 1845, her obituary was published in the *London Geological Journal*, written by no less a geologist than Gideon Mantell (1790–1852). Her marvelous collection was dispersed among many collectors and institutions. The most valuable specimens were purchased by Thomas B. Wilson on behalf of the Academy of Natural Sciences and presented to the museum in 1848.

—*Elana Benamy*
Curatorial Assistant
Department of Botany

Bradford,
Wiltshire.

The following year, heading back to the same fossil beds, Hayden wrote Leidy from Loup Fork, Nebraska Territory: "We have 11 wagons and 6 mules each and intend to give the Bad Lands a thorough Exploration. You may look for something nice on my return. Let me know if you hear of any one going up to the Judith region—I think no one will disturb it."[39] With other institutions entering the field of paleontology, competition for fossil discoveries was heating up.

Returning five months later aboard a Missouri River passenger packet, Hayden told Leidy jubilantly: "I have succeeded in getting all my collections of this year down safely after much trouble and anxiety—To you my collections from the Bad Lands will be the richest that you have ever had the privilege of seeing at any one time." He said that in an area of over 250 square miles he had found "a profusion of vertebrate remains most beautifully preserved." He wished Leidy would make the trip to South Dakota to join him.[40]

An earlier statement by Charles Lyell was proving to be truer than the great geologist could have known after his American visit in the early 1840s. Then he had been aware of only the fossils of New York State and New England and not of the vast treasure trove to be found in the West, yet he had presciently written: "We must turn to the New World if we wish to see in perfection the oldest monuments of earth's history, so far at least as it is related to its earliest inhabitants. Certainly in no other country are these ancient strata developed on a grander scale, or more plentifully charged with fossils."[41]

When, on the fateful day of 12 April 1861, Confederate guns fired on Fort Sumter, South Carolina, and the Civil War erupted, explorations of the American West were put on hold for some years, removing American paleontology from the world scene. By the time the war was official, Leidy had had ample warning of the dreadful conflict to come. His friend Francis Holmes, in Charleston, had written to him the previous January:

> Thanks for yr kind letters. I hope you did not think hard of me for this delay in answering. The truth is we are literally *in camp & armed to the teeth,* and everything has given way to the preparation for war. These are sad times for our once happy country, but *Black Republicanism* has driven us into this measure & now there must be a dissolution of the Union. *There must be a Southern Confederacy.* I have volunteered in a mounted troop. . . . Our young men are the best horsemen on this continent, familiar with our swamps, the use of the rifle & the gun, and with stout hearts & brave hands we will never flinch before a "*Lincoln force*" three times our number. . . . Remember your promise Leidy, "no matter what comes *we are friends forever.*"[42]

Long before, during the War of 1812, the international exchange between naturalists had taken precedence over the hostilities between nations, and, except when interrupted by captured ships on the Atlantic, American, English, and French practitioners had continued to cooperate in pursuit of science. But the Civil War was far more personal. Bitterness on both sides separated associates, friends, and families.

Like others at the Academy, Leidy was torn between his sympathies for the South and his loyalty to the Union. Many of his medical students at the University of Pennsylvania, where he had taught for eight years, were Southerners. He also had warm relations with some of his own medical school colleagues then in the South. Being a university professor, he was exempt from the draft, but his three younger brothers enlisted. Though not on the front lines, Leidy witnessed the horrors of war firsthand

8.8 Ferdinand Vandeveer Hayden (1829–1887) in Civil War Union uniform. ANSP Archives coll. 457.

when he was appointed an assistant army surgeon at the Satterlee General Hospital in West Philadelphia, where he served as a pathologist, performing autopsies. Later, as chief surgeon for the City of Philadelphia, he headed a group of physicians and surgeons who determined the fitness of draftees for war service. He also inspected and supervised hospitals around the city for the United States Sanitary Commission.[43] In the Union Army's report *The Medical and Surgical History of the War of the Rebellion*, Leidy told of treating soldiers wounded in the Battles of Gettysburg and Antietam.[44]

During the war, Ferdinand Hayden resumed his medical practice and served as an army surgeon and medical administrator (figure 8.8). Many other Academy members also became involved in war-related activities, and the society's regularly scheduled meetings were suspended until 1868.[45]

After the war ended, Hayden returned to his paleontological and geological survey of the West, sending most of the fossils he found to Leidy. By then it was an even more dangerous place. Encouraged by the Homestead Act and the recent discovery of silver and gold, prospectors flooded into many parts of the West, setting up tent cities, destroying the buffalo, and enraging the Indians, whose very existence was threatened by the loss of their land, food, and way of life. To protect the prospectors, the United States Army began to fight a guerrilla war against an enemy familiar with the landscape and able to endure great hardship.

In 1867, through his connections with Undersecretary Spencer Baird and Joseph Henry, secretary of the Smithsonian, Hayden obtained a job with the Department of the Interior directing an ambitious series of natural history surveys. He encouraged the federal government to substantially increase its funding for his Geological and Geographic Survey of the Territories, which attracted more attention and talent than any other scientific institution during the postwar years.[46] The principal objective was to search for coal and other commercially important minerals, but Hayden continued to collect fossils as well. In three seasons, from 1869 to 1871, he found one of the world's greatest treasure houses of fossils in the Eocene lake beds of the central Rocky Mountains, where the Green River flows south into Utah to join the Colorado River. Leidy later wrote a monograph (1873) on Hayden's work outlining a picture of the prehistoric landscape where his collections had been found: "The ancient lake-deposits now form the basis of the country and appear as extensive plains, which have been subjected to a great amount of erosion, resulting in the production of deep valleys and wide basins, traversed by the Green River and its tributaries."[47]

Hayden continued his surveys for the U.S. Department of the Interior for more than ten years. Recognizing the great scenic beauty of the West and wishing to share it with others, he invited the photographer William Henry Jackson (1843–1942) and the artist Thomas Moran (1837–1926) to accompany him on his journeys from 1870 to 1879. Jackson had previous experience in photographing the West, as he had been hired by the Union Pacific railroad to take photographs of thousands of views along the train's route (figures 8.9, 8.10). He exhibited his

8.9 "Horse drawn odometer." Photograph by William Henry Jackson (1843–1942) from the photograph album of the U.S. Government Surveys of Colorado, Wyoming, Utah, Idaho, and Montana, 1869–75. ANSP Archives coll. 34.

8.10 Ferdinand V. Hayden's surveying team. William Henry Jackson's photograph album of U.S. Government Surveys of Colorado, Wyoming, Utah, Idaho, and Montana, 1869–75. ANSP Archives coll. 34.

8.11 *Edward Drinker Cope* (1840–1897) by C. A. Worrall, 1897. Oil on canvas. ANSP Library coll. 286.

pictures of the wonders of the Yellowstone region in 1870 at the Capitol building in Washington, D.C., where his photographs so affected members of Congress that in 1872 they voted to make Yellowstone the first national park. Just three months later, Moran sold the government his large canvas *The Grand Cañon of the Yellowstone* (1872). His painting was the first landscape to hang in the Capitol.[48]

Inevitably, a host of other vertebrate paleontology enthusiasts soon invaded the field. None of them, however, was quite as ambitious, brilliant, and unscrupulous as Edward Drinker Cope at the Academy of Natural Sciences and Othniel Charles Marsh at Yale University (figure 8.16).

Cope was the son of a wealthy Philadelphia Quaker industrialist who expected his son to be a gentleman farmer, but Edward had other ideas after he graduated from Westtown, a Quaker school outside Philadelphia. As a youth, he spent many hours at the Academy, where, beginning in 1859, he recatalogued the entire reptile collection. Several years later, he worked at the Smithsonian Institution, but he had little formal university education other than a few courses in anatomy with Leidy at the University of Pennsylvania. Unfortunately, belying his Quaker ancestry, throughout Cope's life he was noted for a lack of decorum. A friend once observed that "no woman was safe within miles of him," and another recorded that "his mouth was the

8.12 *Dryptosaurus aquilunguis* confronts *Elasmosaurus,* in an illustration from Edward Drinker Cope's "The Fossil Reptiles of New Jersey," in *The American Naturalist,* 1870. ANSP Library.

most animal [vulgar talk] I've ever heard."[49] Apparently Cope's colleagues were able to overlook these faults in light of his scientific achievements, because he was elected a member of the Academy in 1861, secretary in 1865, and a curator in 1873.

Marsh, who did not grow up in a well-to-do family, nevertheless had a rich uncle, the financier George Peabody, who sent him to Philips Andover Academy and then to Yale. At both institutions he proved to be a brilliant student. After graduation (1860), he continued his studies of geology and mineralogy at Yale's Sheffield Scientific School under James Dwight Dana (who had been a member of the Wilkes Expedition) and Benjamin Silliman Jr. During the Civil War, both Marsh and Cope spent time in Europe studying geology, principally in Germany. Marsh, in particular, had a certain career in mind. His philanthropic uncle had proposed financing a science museum for Yale (this became the Peabody Museum of Natural History in 1866), and Marsh was being groomed to head the new institution. In the 1860s, like Cope, Marsh devoted his studies to vertebrate paleontology. A great impetus for this field of science was the publication of Darwin's *On the Origin of Species* in 1859, which gave respectability to theories of evolution. Fossils represented the science of the future for anyone interested in natural history.[50]

Soon after his return from abroad, Cope began to work with his usual whirlwind energy, always excessively proud of his accomplishments and quick to defend them. He had been fairly good friends with Marsh, whom he had invited to join him to see fossil sites in New Jersey and with whom he had extensive correspondence. But in 1868, a blunder with the reconstruction of a huge reptile fossil that had been sent to him from the West caused a serious rift with Marsh, and also with Leidy. When Cope assembled the skeleton of his sixteen-million-year-old sea creature and named it *Elasmosaurus platyurus,* he put the head on the wrong end. This strange animal was a kind of *Plesiosaurus,* a group well known to scientists by then (figure 8.12). Like all of its type it had an extremely long neck, but Cope thought his animal was different, and that its neck was its tail.

The following spring, when Marsh visited the Academy, Cope laid the fossil out for him. Marsh quickly pointed out Cope's mistake, much to the latter's annoyance. Even worse, Leidy, called in to settle the argument, agreed with Marsh. But most embarrassing of all, Cope had sent a paper on the subject to the American Philosophical Society to be published in its *Transactions,* expecting to wow the scientific world with his description of a creature that he believed was new to science. Advance copies had already been sent out to many of his friends and colleagues, including Marsh (figure 8.13).[51] The onetime friendship between Cope and Marsh quickly turned to bitter rivalry. Both men were willing to use ruthless, at times unscrupulous means to attain their goal of scientific immortality through describing, classifying, and naming vertebrate fossil remains.[52] They argued over the priority of their discoveries, used different names for the same species, and accused each other of stealing samples, collectors, even descriptions. One famous example of their intense competition occurred in 1877, when two Colorado

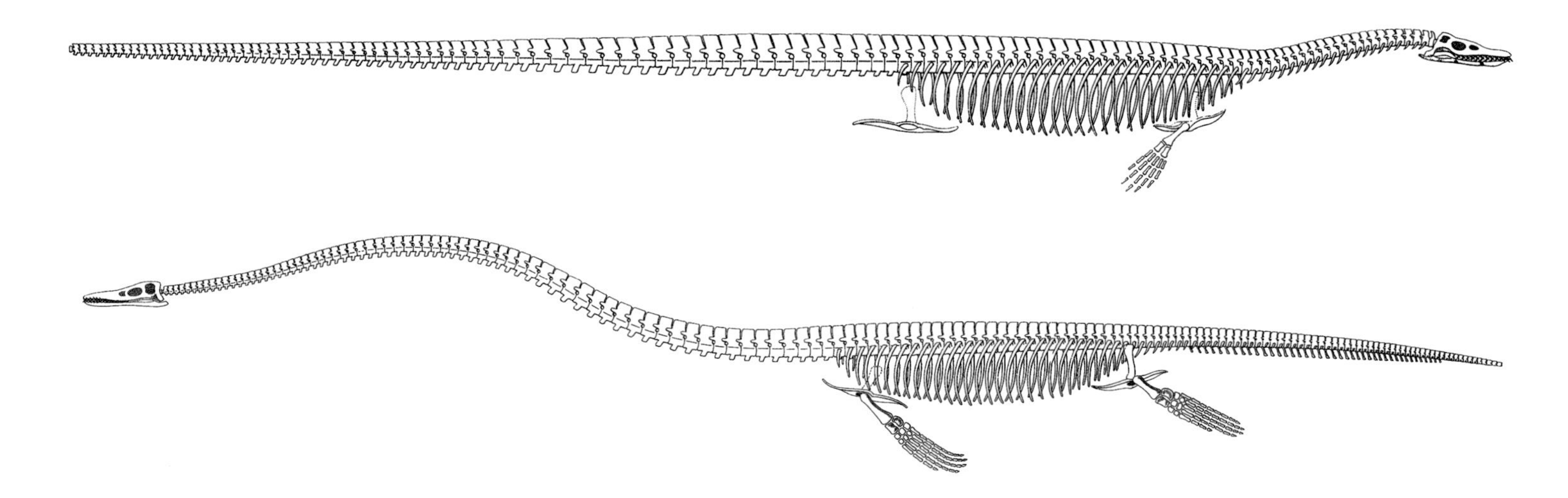

schoolteachers independently discovered exciting caches of dinosaur bones. One sent word to Marsh in New Haven, and the other alerted Cope in Philadelphia. Both men at once hired armies of diggers and raced to excavate and publish their finds. But even though the sites were hugely significant, the acrimony between Cope and Marsh robbed their work of some of its scientific importance.[53]

Their disagreements eventually had even larger ramifications for the Academy and its collections. According to Hayden's biographer Mike Foster, "By noisily impugning each other's integrity, they waged the most rancorous war in the history of American paleontology, which had repercussions on Hayden and his geological surveys."[54] Hayden was caught in the middle between Cope, Marsh, and Leidy. He had written to Leidy in 1870 promising all his fossil vertebrates to him, "his best and oldest friend."[55] But Cope managed to make Hayden break his promise of offering Leidy first choice of his discoveries. Cope and Marsh had inherited money that they used to buy fossils from Hayden and other collectors in the West, but Leidy could not afford to make such purchases with his modest teaching salary from the University of Pennsylvania and several other institutions.

Unable to compete with the wealth of Cope and Marsh, and uncomfortable with their increasingly hostile conflict, Leidy eventually abandoned western paleontology to pursue, and publish his great works on, protozoology and parasitology (figure 8.14). Before doing so, he published one last compilation of his studies of the West, *Extinct Mammalian Fauna of Nebraska and Dakota* (1869), in which he examined the relationship between Tertiary animals (tens of millions of years old) in Asia and North America and suggested faunal migrations between the continents.[56] Leidy resumed paleontological work in the late 1880s and published seventeen papers about Florida paleontology and major Pleistocene discoveries. His death prevented him from compiling a major opus.[57] In time, Hayden's sale of fossils to Marsh and Cope ended when Marsh began his own expeditions to the West and Cope lost his capital on bad investments.

Cope's first western trip occurred in 1871, when he went to excavate for fossils in the region around Fort Wallace, Kansas. His letters to his wife, Annie Pim, reveal an unsettling picture of animal destruction in the western lands that the newly built railroad was opening up for pioneer migration (figure 8.15). Many of the creatures that paleontologists were so eager to resurrect had lived for millions of years, but within a few decades some of the living animals of the West were being slaughtered at an alarming rate. Speaking of the bison, Cope wrote: "About the fort they appear in countless thousands; indeed there are millions they say between here and Denver. Carcasses and bones lie on each side of the R.R. track . . . remains of animals shot from the car windows."[58] (This statement makes one think of wolves being shot from helicopters in Alaska in our own day.)

Writing a week later from Fort Wallace, Cope alluded to his rivalry with Marsh. A year earlier, Marsh had been to the Bridger Formation in southern Wyoming with a group of twelve Yale students, an area that he had learned about from an 1869 paper of

8.13 Top: The suppressed illustration of Cope's *Elasmosaurus* with its head shown on the wrong end. Bottom: The published version of the plate with *Elasmosaurus*'s head shown on the correct end. Line drawing from Edward Drinker Cope, "Synopsis of the extinct Batrachia, Reptilia and Aves of North America." *Transactions of the American Philosophical Society*.

FACING PAGE 8.14 Tapeworms collected by Joseph Leidy, who pioneered the study of parasitology in the United States. ANSP General Invertebrates Collection. Purcell photograph.

Coll. Leidy
Taenia saginata Goeze
Homo.
Coll. Leidy
Taenia saginata Goeze
Homo.
Coll. Leidy
saginata Goeze, 1782

Leidy's about fossil mammals.[59] Cope wrote to Annie, "Indeed the stories I hear of what Marsh & others have found is something wonderful, and I can now tell my own stories, which for the time I have been here are not bad."[60]

After his return from this expedition, his colleague, a collector himself and Cope's eventual biographer, Henry Fairfield Osborn (1857–1935), asked him if the country around Fort Wallace was good collecting ground. "It was before I got there," Cope replied coolly.[61] The arrogance that made him so objectionable to many was evident even to his friends.

The following year, Cope was in that part of Wyoming where Marsh had already been and would return. In a letter to his brother, Cope gave an idea of his life in the wild: "We eat fishes from the streams and get occasionally a large grouse or sage hen which are very abundant. . . . We are still in the well-watered region of the heads of Ham's fork of Green River . . . then we strike Bitter Creek and follow it east into a howling wilderness, where water is scarce and bad, with grizzly bears plenty. I have 20 sp[ecies] of mammals (8 new)."[62]

Writing to his father in 1873, he attempted to assure him that he was not bothered by the ongoing feud with Marsh, who accused him of violating the canons of scientific practice. "As to the learned professor of Copeology in Yale, he does not disturb me, & as I promised in my last reply (last spring) I will not notice him again," he noted. "The longer he continues his course the more injury he inflicts on himself, as I find from various directions; the only harm he can do me is to frighten Hayden, and the risk of this is diminished or passed."[63] Presumably, Hayden was no longer collecting for Marsh.

In 1879, Congress established the United States Geological Survey, which marked the ascendancy of Marsh and the end of Hayden's collecting for the government. That year also marked the end of Cope's active fieldwork. Marsh had suggested Clarence King to serve as first director of the survey, and in 1882, the

8.15 *Buffalo Herd on the Upper Missouri* by Karl Bodmer, 1835. Watercolor on paper. Joslyn Art Museum, Omaha, Nebraska.

FACING PAGE 8.16 Othniel Charles Marsh (1831–1899) (back row, center) with Yale students on a western collecting expedition. ANSP Archives coll. 2011-003.

second director, John Wesley Powell, named Marsh the survey's paleontologist. This post gave Marsh access to fossil deposits, publication under government auspices, and a team of assistants (figure 8.16).

By this time, Cope had used up his inheritance on collecting and squandered the rest on bad mining investments. He was no longer able to afford costly travels out west or the purchase of fossils collected by others. He had hoped to secure a job teaching at Princeton when a chair in natural history was endowed in honor of Joseph Henry, but he was not appointed. In referring to the person to whom he had been recommended, Cope wrote his father that this professor "objected to my evolution sentiments, for those views are much condemned at Princeton."[64] Cope's theories of evolution, based on those of Lamarck, were another broken link with Leidy, who wholeheartedly supported Darwin.

Cope's quarreling over matters of priority and publications caused serious controversies at the Academy of Natural Sciences and at the American Philosophical Society. Constantly at odds with the Academy's president W. S. W. Ruschenberger over the way the institution should be run, he was called a "storm center" and a "law unto himself." Admittedly brilliant, he was still someone to keep at arm's length.[65] Cope was incensed that the Academy would not hire professional curators, including Cope himself, of course, and fought with other members who did not agree with him.

The scientific community at large, and especially the Academy, may have been most angered by Cope's feud with Marsh breaking out in the public domain. On 12 January 1890, the *New York Herald* published an article headlined "Scientists Wage Bitter Warfare" that laid out in detail the whole unpleasant story.

Thomas Jefferson Fossil Collection

Every specimen in the millions-strong biological research collections at the Academy has scientific value, but many are also tangible vestiges of the history of the study of natural science in North America. Among the most iconic specimens at the Academy are about one hundred fossils in the vertebrate paleontology collection that are the legacy of the dedicated effort and contemplative inquiry of Thomas Jefferson (1743–1826). The Thomas Jefferson Fossil Collection is composed mostly of the bones and teeth of Ice Age mammals derived from the Big Bone Lick site in Boone County, Kentucky, and includes the remains of mammoths, mastodons, giant ground sloths, bison, horses, and other animals.

Jefferson's fascination with such specimens was roused in 1796 when he was sent some large bones found in a cave in the western reaches of Virginia (now West Virginia). He had been keenly interested in cataloguing the animals of the region (especially the larger ones) for his *Notes on the State of Virginia* (1787). At the American Philosophical Society in 1797, Jefferson made a scientific presentation on the bones he had received in a paper titled "A Memoir of the Discovery of Certain Bones of a Quadruped of the Clawed Kind in the Western Parts of Virginia." He named the animal *Megalonyx*, "great claw," and suggested that it was a very large carnivore. Jefferson's large-carnivore hypothesis was soon supplanted as accounts of South American giant ground sloths reached North America, but such fossil remains continued to be one of his long-term intellectual pursuits.

This Big Bone Lick fossil site on the Ohio River in Kentucky was along a relatively busy trading corridor, and fossils had been collected there as early as 1739. Frustratingly, such objects were often treated as curiosities, and many were lost to scientific examination. Alternatively, fossils of mastodons and mammoths from Big Bone Lick and elsewhere were sent to Europe, where natural historians were fascinated by such North American megafauna. The science of paleontology was in its infancy, and the basic issues of diversity, extinction, and the meaning of fossils were being discussed. Jefferson wanted North Americans to be part of that conversation and consciously sought to build collections for the burgeoning American scientific community to consider.

Jefferson's enthusiasm for such a scheme was apparently widely known, as demonstrated by the two large fossil shark teeth (*Carcharocles megalodon*) William Reid of Charleston, South Carolina, sent to Jefferson in 1806. The fossils were accompanied by a message that reads in part, "Observing you attentive to Natural Philosophy as well as to other branches of science, I take occasion to present you with a fossil, which you may consider a curiosity, and not unworthy of your contemplation."

The lower molar of an American mastodon (*Mammut americanum*), collected by William Clark for Thomas Jefferson in 1807 at Big Bone Lick, Kentucky. ANSP Vertebrate Paleontology Department #13132. The tooth is of late Pleistocene age (approximately 18,000 years old).

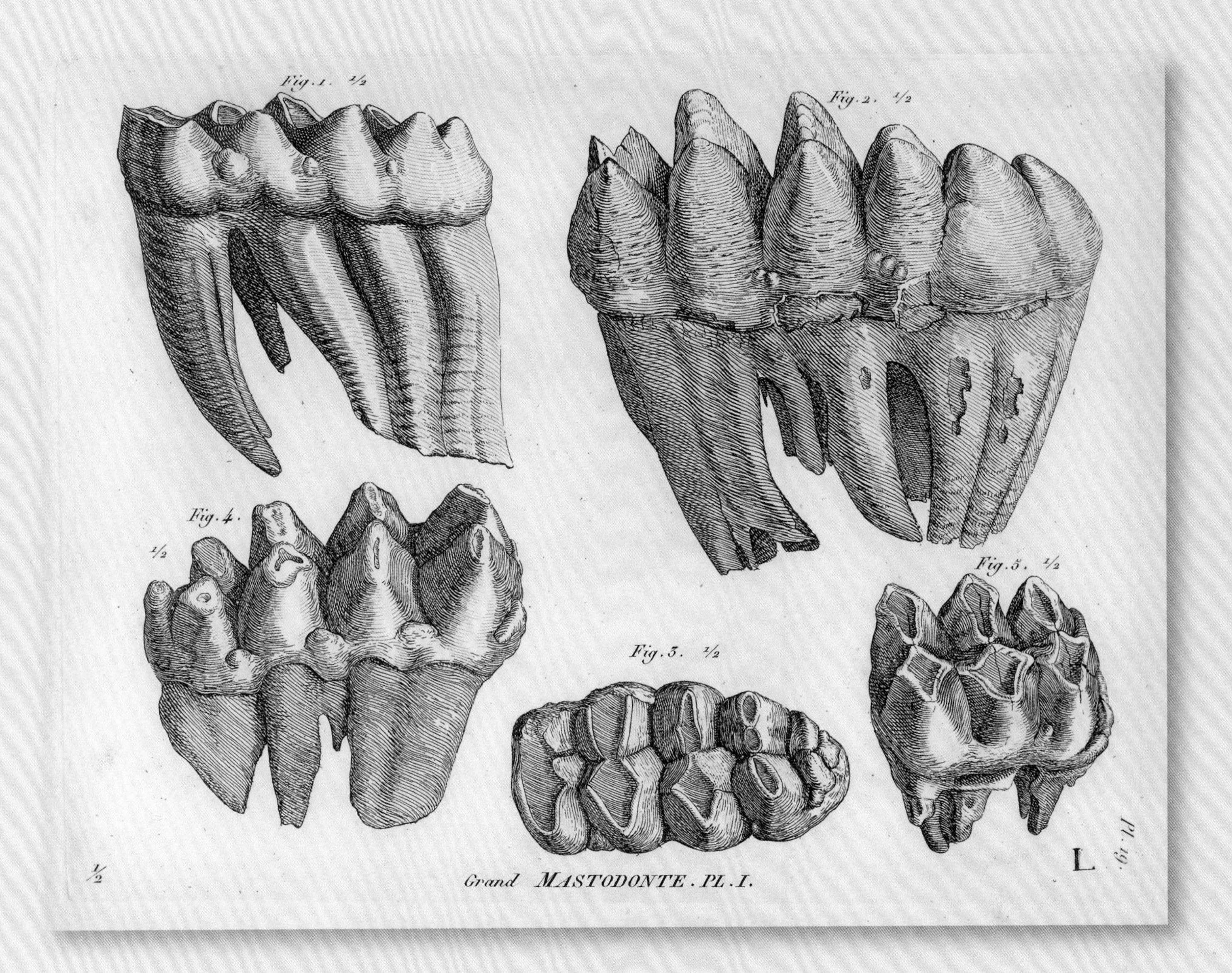

In 1807, soon after the return of the Corps of Discovery, Jefferson charged William Clark with traveling to Big Bone Lick to collect fossil remains from the site. This expedition was the source of the bulk of the material in the Thomas Jefferson Fossil Collection. Although Clark was disappointed on arrival by the scarcity of fossils remaining at the site, several months later, he shipped a collection of fossils to the White House, where Jefferson excitedly worked with Caspar Wistar in June 1808 to identify and organize the remains. Later that year, Jefferson sent part of the collection to the American Philosophical Society in Philadelphia. (This was transferred to the Academy of Natural Sciences about forty years later.) Jefferson gave the duplicates in his collection to the natural history museum in Paris.

—Ted Daeschler
Associate Curator and Chair
Vertebrate Biology

A plate of mastodon teeth from George Cuvier's *Recherches sur les Ossemens Fossiles*, published in Paris in 1825. ANSP Library.

ACADEMY OF NATURAL SCIENCES OF PHILADELPHIA, NO. 3626
Libinia canaliculata
Atlantic City.
Joseph Leidy.
Spider Crab
Libinia canaliculata
Atlantic City
N.J.

Cope pitted himself against the entire staff of the Geological Survey, officers of the Department of the Interior under which the survey operated, and Marsh's colleagues at Yale and numerous friends abroad. The final straw for Cope, which prompted him to cooperate with the journalist who wrote the article, had occurred the previous 16 December, when he received a letter from the Department of the Interior directing that he deposit his collections from the Hayden surveys into the United States National Museum. Enraged, he declared that he had collected his specimens at his own expense.[66]

William Pepper, provost of the University of Pennsylvania, was on Marsh's side and sought to remove Cope from his university post. Cope wrote to his friend Osborn a week before the article came out that "Marsh has made a dead set on Pepper the Pres[ident]t of the University here so as [to] secure my resignation or expulsion. Pepper is terribly frightened & yields everything to him."[67] A week later, he wrote again to Osborn, "Poor old Leidy has come out against me in the *Phila. Enquirer* [*sic*], just as he has always done."[68]

Leidy was sixty-eight at the time and would be dead within a year. He had encouraged and promoted Cope at the beginning of his career, but Cope's aggressive and pugnacious ways were offensive to him, and he gradually distanced himself from the younger man. Leidy was a giant of nineteenth-century natural science, every bit as large in his field as the dinosaurs he was the first to describe in America, but peaceful, retiring, and principled. His biographer, Leonard Warren, has written that "he is considered the founder of American vertebrate paleontology, parasitology, and protozoology and America's foremost anatomist, the person who revealed the power and versatility of the microscope (though few are aware of it)."[69]

Hayden predeceased Leidy by four years. An influential naturalist and innovative geologist, he was, perhaps most importantly, a masterful entrepreneur for science. He had supplied Leidy, Cope, and Marsh with numerous fossils that he uncovered and shipped to them, enabling and advancing their studies in paleontology. From his explorations of the Yellowstone region and other areas, he promoted and popularized the gorgeous scenery and wonders of the West.

Cope, after inheriting from his father, in 1876 bought two houses on Pine Street in Philadelphia, where he spent the rest of his life—one house for himself and his wife and daughter, and the other for his collections.[70] He never gave any of his huge accumulation of specimens to the Academy, though in the same year (1876), the institution had moved into a much larger building that could have accommodated them. He did propose selling his important finds to the Academy in 1884, but the price of $100,000 was more than the society could afford. In the late 1890s, Cope sold his entire collection to the American Museum of Natural History in New York, which helped boost that museum to the forefront of vertebrate paleontology, eclipsing the Academy of Natural Sciences in that field. For fifty years, the Academy had been the leader. It was there that the subject first developed as a specialized field of scientific inquiry in the United States.[71] Economics had changed the entire thrust of vertebrate paleontology at the Academy, forcing Leidy from the field and restricting opportunities for expansion.

There are vertebrate paleontology collections in other institutions that contain far more specimens and are far more systematically diverse than those of the Academy, but the Academy houses an abundance of historically significant specimens, especially types, in its relatively small collections.[72]

Cope died in 1897. He left the Academy his collection of recent fishes and reptiles and money to hire a vertebrate paleontologist, but no position was created.[73] What a curious man he was, for after all the trouble he caused Leidy, in his will Cope directed that his ashes be preserved in the same place as the ashes of "my esteemed friends, Dr. Joseph Leidy and Doctor Jno. A. Ryder."[74] Today, the urns containing the ashes of Cope and Leidy are together on the same shelf in the Wistar Institute in Philadelphia.[75] It is the large bronze statue of Leidy, however—that gifted and versatile man who was first in America to recognize the genius of Darwin—that stands in front of the Academy of Natural Sciences on Logan Circle.[76]

PTS

8.17 Spider crab (*Libina canaliculata*) collected by Joseph Leidy in Atlantic City, New Jersey. ANSP General Invertebrates Collection. Purcell photograph.

Notes

1 Silvio A. Bedini, *Thomas Jefferson: Statesman of Science* (New York: Macmillan, 1990), 97, 95.

2 Ibid. This is the title that would also be attributed to Joseph Leidy in recognition of his even more substantial contribution to this field.

3 William Maclure, "Observations on the Geology of the United States, Explanatory of a Geological Map," *Transactions of the American Philosophical Society*, 1809.

4 George H. Daniels, *American Science in the Age of Jackson* (New York: Columbia University Press, 1968), 221.

5 Thomas Say, "Observations on some species of zoophytes, shells, etc, principally fossil," *American Journal of Science and the Arts, more especially of Mineralogy, Geology, and the other Branches of Natural History including also Agriculture and the Ornamental as well as Useful Arts* 1, no. 4 (1818): 381–87.

6 Patsy Ann Gerstner, "The 'Philadelphia School' of Paleontology: 1820–1845," Case Western Reserve University, Ph.D. diss., 1967, 164.

7 Gerstner, "The 'Philadelphia School' of Paleontology," 186.

8 Charles Darwin, *The Zoology of the Voyage of H.M.S. Beagle, under the command of Captain Fitzroy, R.N., during the years 1832 to 1836* (London: Smith Elder & Company, 1840; Royal Geographical Society Reprint, 1994), 65–68.

9 Gerstner thesis, "The 'Philadelphia School' of Paleontology," 262.

10 Earle E. Spamer, Edward Daeschler, and L. Gay Vostreys-Shapiro, "A Study of Fossil Vertebrate Types in The Academy of Natural Sciences of Philadelphia: Taxonomic, Systematic, and Historical Perspectives" (Philadelphia: ANSP Special Publication #16, 1995), 19. The Academy's new building on the northwest corner of Broad and Sansom Streets opened in 1840 and was expanded by thirty feet in 1846–47. Jefferson's fossil collection was deposited at the Academy in 1849 and officially given to the institution in 1987.

11 Mary C. Rabbit, *The United States Geological Survey: 1879–1989*, U.S. Geological Survey Circular 1050.

12 Keith Thomson, *The Legacy of the Mastodon: The Golden Age of Fossils in America* (New Haven, CT: Yale University Press, 2008), 100.

13 Leonard Warren, *Joseph Leidy: The Last Man Who Knew Everything* (New Haven, CT: Yale University Press, 1998), 76.

14 Joseph Leidy, "On the Anatomy of the Animal of *Helix albolabris*; Say, *Proceedings of the Boston Society of Natural History* 2 (1845): 59; and Warren, *Joseph Leidy*, 53.

15 Warren, *Joseph Leidy*, 54.

16 Joseph Leidy, "Notice," *Proceedings of the ANSP* 3 (October 1846): 107–8.

17 Brian Burell, *Postcards from the Brain Museum: The Improbable Search for Meaning in the Matter of Famous Minds* (New York: Broadway Books, 2004), 179.

18 Burrell, *Postcards from the Brain Museum*, 183.

19 Ibid., 76.

20 Thomson, *Legacy of the Mastodon*, 109.

21 Warren, *Joseph Leidy*, 78.

22 Ibid., 80.

23 Henry Fairfield Osborn, *Cope: Master Naturalist* (Princeton, NJ: Princeton University Press, 1931), 23.

24 Joseph Leidy, "On the Fossil Horses of America," *Proceedings of the ANSP* 3 (September 1847): 262–66.

25 Darwin, *The Zoology of the Beagle*, part 1, 109.

26 Warren, *Joseph Leidy*, 77.

27 Leidy, "On the Fossil Horses," 262–63.

28 Ibid., 78.

29 Leidy, *Proceedings of the ANSP* 9 (1857): 210; and Warren, *Joseph Leidy*, 78–79.

30 Robert McCracken Peck, *Land of the Eagle* (New York: Summit Books, 1990), 205.

31 Osborn, *Cope: Master Naturalist*, 21.

32 Leidy to Spencer Fullerton Baird, 30 March 1853, ANSP; quoted in Warren, *Joseph Leidy*, 111.

33 Leidy's meeting with Darwin is in Warren, *Joseph Leidy*, 95. Leidy's letter to Darwin may have been destroyed in a fire at Darwin's home, Down House, in 1860. Excerpts are found in the mss of Joseph Leidy II, ANSP.

34 Charles Darwin to Leidy, 4 March 1860, Down, Bromley, Kent, ANSP archives, Leidy papers, coll. 1. Leidy's letter was probably destroyed in a fire at Darwin's home in 1860. Excerpts of the letter are found in the mss of Joseph Leidy II, ANSP, and David H. Wenrich, "Biography of Joseph Leidy," ms, ANSP.

35 Mike F. Foster, Ferdinand Vandeveer Hayden entry in *American National Biography* (New York: Oxford University Press, 1999), vol. 10, 374–76.

36 Thomson, *Legacy of the Mastodon*, 133.

37 William H. Goetzmann, *New Lands, New Men: America and the Great Age of Discovery* (New York: Viking, 1986), 177.

38 Warren, *Joseph Leidy*, 81; *Proceedings of the ANSP* 8 (1856): 72–73, 88–90. The Mesozoic era dates to some seventy-one million to thirty-three million years ago.

39 Hayden to Leidy, Loup Fork, 19 July 1857, Leidy Papers, ANSP, coll. 1.

40 Hayden to Leidy, aboard the *Florilda*, Missouri River Passenger packet, 13 December 1857, Leidy Papers, ANSP, coll. 1.

41 Thomson, *Legacy of the Mastodon*, 104; *Proceedings of the ANSP* 3 (1847): 315; and ibid. 4 (1848): 52.

42 Francis Holmes to Joseph Leidy, Charleston, 7 January 1861, Leidy Papers, ANSP coll. 1.

43 Warren, *Joseph Leidy*, 136.

44 Jane Pierce Davidson, *The Bone Sharp: The Life of Edward Drinker Cope* (Philadelphia: Academy of Natural Sciences, 1997), 27.

45 Edward J. Nolan, *A Short History of the Academy of Natural Sciences of Philadelphia* (Philadelphia: Academy of Natural Sciences, 1909), 15.

46 Foster, *ANB*, 375.

47 Joseph Leidy, *Contributions to the Vertebrate Fauna of the Western Territories; Report of the United States Geological Survey of the Territories*, 1873, pt. 1 (Washington, DC, 1873), 15–16, quoted in Thomson, *Legacy of the Mastodon*, 138–39.

48 The U.S. Congress purchased Moran's canvas in the spring of 1872 for $10,000. It hung in various public areas in the U.S. Capitol until 1950, when its ownership passed to the Department of the Interior. It is now on view at the National Museum of American Art in Washington. For more on Moran's involvement with the Hayden Survey, see Joni Louise Kinsey, *Thomas Moran and the Surveying of the American West* (Washington, DC: Smithsonian Institution Press, 1992), 42–67.

49 Burrell, *Postcards from the Brain Museum*, 183.

50 Thomson, *Legacy of the Mastodon*, 161.

51 Ibid., 162–63.

52 James G. Cassidy, *Ferdinand V. Hayden: Entrepreneur of Science* (Lincoln: University of Nebraska Press, 2000), 183.

53 See Mark Jaffe, *The Gilded Dinosaur: The Fossil War between E. D. Cope and O. C. Marsh and the Rise of American Science* (New York: Crown Publishers, 2000).

54 Mike Foster, *Strange Genius: The Life of Ferdinand Vandeveer Hayden* (Ireland: Roberts Rinehart Publishers, 1994), 278.

55 Cassidy, *Ferdinand V. Hayden*, 184.

56 Ronald Rainger, "The Rise and Decline of a Science: Vertebrate Paleontology at Philadelphia's Academy of Natural Sciences, 1820–1900," *Proceedings of the American Philosophical Society* 136, no. 1 (1992): 9.

57 Information kindly given the author by Dr. Thomas Peter Bennett, former president of the ANSP, microbiologist, and historian.

58 Edward Drinker Cope to his wife, Fort Wallace, Kansas, 21 September 1871, quoted in Osborn, *Cope: Master Naturalist*, 161.

59 Thomson, *Legacy of the Mastodon*, 180.

60 Cope to this wife, Fort Wallace, 21 September 1871, quoted in Osborn, *Cope: Master Naturalist*, 162.

61 Ibid., 27.

62 Cope to his brother, camp near Church Buttes, Wyoming, 28 July 1872, quoted in ibid., 186.

63 Cope to Albert Cope, 13 April 1873, quoted in ibid., 182.

64 Cope to Albert Cope, Haddonfield, 23 July 1873, quoted in ibid., 172.

65 Nolan, *A Short History*, 330, quoted in Warren, *Joseph Leidy*, 209.

66 Osborn, *Cope: Master Naturalist*, 402.

67 Cope to Osborn, Philadelphia, 5 January 1890, quoted in ibid., 409. Pepper was actually the provost of the university.

68 Cope to Osborn, Philadelphia, 14 January 1890, quoted in ibid., 411.

69 Warren, *Joseph Leidy*, 252.

70 Cope's houses were at 2100 and 2102 Pine Street in Philadelphia. They are now marked by a historic plaque.

71 Rainger, "The Rise and Decline," 1.

72 Spamer et al., "A Study of Fossil Vertebrate Types."

73 Ibid., 29.

74 Osborn, *Cope: Master Naturalist*, 32.

75 Burrell, *Postcards from the Brain Museum*, 183. John Ryder, like Leidy, Cope, and Marsh at the same time, was active in the field of comparative anatomy. The Wistar Institute is a venerable organization today principally devoted to the study of cancer. It was established in 1892 by Isaac Jones Wistar, a Union general in the Civil War and president of the Academy from 1881 to 1895.

76 The sculptor of the Leidy statue (1907) was Samuel Murray, a close friend of Thomas Eakins's. The sculpture originally stood at City Hall but was placed in front of the Academy in 1929.

Land Crab (Birgus latro)
From Flint Island,
South Pacific.
(C.D Voy!)

Coconut crabs (*Birgus latro*) collected on Flint Island (an uninhabited coral atoll four hundred nautical miles northwest of Tahiti in the Central Pacific) by C. D. Voy in 1875. ANSP General Invertebrates Collection. The

Chapter 9

"I Must Have Fame": Robert Peary Explores the Arctic

I cannot reconcile myself to years of common place drudgery and a name late in life when I see an opportunity to gain it now [by becoming an explorer] and sip the delicious draughts while yet I have youth and strength and capacity to enjoy it to the utmost.
—Robert E. Peary, February 1887

At the end of August 1890, as Philadelphia simmered in a heavy blanket of heat and humidity, at least one person in the city sought relief by thinking of a cooler place. Without bothering to cross town to look for himself, the man who later was hailed as one of the world's greatest explorers wrote to the Academy of Natural Sciences to ask "what the museum of the Academy contains in the way of Arctic specimens in any branch of Natural History and whether any of the Members are especially interested in Arctic studies."[1]

For Robert E. Peary (1856–1920), then a young engineering officer stationed at Philadelphia's Navy Yard, it must have been a rhetorical question. He surely knew of the Academy's past involvement with Arctic exploration, for he had read just about everything that had been written on the subject.[2] More likely, he was testing the waters to see if he could gain institutional interest and support for a plan he was concocting to break away from the a workaday life of a civil engineer. "I must have fame," he had written his mother in 1887, revealing the raw ambition that lay beneath his exploratory yearnings. "I cannot reconcile myself to years of common place drudgery and a name late in life when I see an opportunity to gain it now [by becoming an explorer] and sip the delicious draughts while yet I have youth and strength and capacity to enjoy it to the utmost."[3]

At his mother's expense, he had made a preliminary excursion into the unexplored interior of Greenland in 1886 to see just how far he could get across its frozen ice cap. Now he wished to launch a more extensive—and thus more expensive—expedition for which he would need help. Receiving an encouraging response to his initial inquiry from the Academy's corresponding secretary, Benjamin Sharp (1858–1915), Peary visited the Academy in January 1891 and then submitted a formal proposal outlining the goals of his expedition and seeking $6,000 in support.[4] He bolstered his case by quoting Dr. Hinrich Rink (1819–1893), the Danish geographer and geologist who was then arguably the world's greatest authority on Greenland. "No country appears to be better qualified to throw light on the problem of polar geography in general than Greenland," he quoted Rink as saying,

> Its northern extremity has not as yet been explored; here it disappears in regions which hitherto have braved the efforts of the boldest discoverers. . . . Here human inhabitants in their struggle for existence have advanced furthest toward the pole, the utmost limits of their abodes not being as yet pointed out with certainty. . . . No wonder that modern polar expeditions have considered the exploration of the northernmost part of Greenland one of their objects.[5]

9.1 Skull and bones of an Eskimo dog brought back from Greenland by the Peary Relief Expedition, 1892. ANSP Mammalogy Department #1870. Purcell photograph.

The centerpiece of his proposal was a desire to determine Greenland's northernmost point, but Peary wisely appealed to the Academy's many other areas of interest by promising to make "important additions to our knowledge of the geography, geology, ethnography, glaciation, meteorology and natural history of the Arctic regions" during a yearlong expedition. He also suggested that with the Academy's backing, he might be able to discover "the most practicable route to the [North] pole" during his trip.[6]

While this last goal had never been on the Academy's agenda, the gathering of new information and specimens from uncommon parts of the world was central to the institution's mission. In response to Peary's proposal, the Academy appointed a committee to work with Peary to refine the details of the expedition and to raise the necessary funds to carry it out. The two most interested and active members of the committee were the Academy's senior curator, Angelo Heilprin (1853–1907), a geologist who had just led a major Academy expedition to Yucatan and Mexico the previous spring (February–June 1890), and Benjamin Sharp, a physician and zoologist who had collected in the Caribbean and Hawaiian Islands and had been Peary's initial contact at the Academy.[7] These men did not want to accompany Peary on his grueling march across Greenland or endure the long, lonely wintering-over that it would require, but both were keen to accompany him as far as Greenland's northwest coast and to make as many natural history collections there as time and opportunity would permit. The two conflicting agendas were resolved by creating a two-tiered expedition, with two leaders and two sets of personnel. A small group of Academy members led by Heilprin would convey Peary, his wife, Josephine, and a team of five volunteers to their proposed winter quarters, and return south before the onset of the Arctic winter. From there, the following spring, Peary would take a sled-dog team across "the serene heights of the interior ice" to the northernmost tip of Greenland (if such a place existed). After solving what Peary considered "a geographical problem . . . second in importance only to the discovery of the North Pole" (whether or not Greenland was an island), he and his team planned to make their way south and secure passage home on an American whaling ship in the spring of 1892.[8]

In remarkably short order, the Academy succeeded in raising most of the funding Peary required. Three hundred dollars came from the institution itself, but, as was the Academy's usual practice, most of the money came from interested Academy members who contributed privately to support the trip.[9] Just before the expedition's departure, Lieutenant and Mrs. Peary were wined and dined at a "brilliant reception" in the Academy's library, which was "lavishly decorated with all the Arctic material the museum afforded" for the occasion.[10] At that time, Peary was presented with a forty-three-star American flag to carry on the Academy's behalf (figure 9.7).[11]

With great fanfare and pageantry, the combined West Greenland Expedition (under Heilprin) and North Greenland Expedition (under Peary) left the docks of Brooklyn on the steam-whaler *Kite* on 6 June 1891. Peary recalled the departure with understandable pride. "We were fairly off for North Greenland, and every ferryboat and steamer in the crowded East River knew it," he wrote.

> Scores of whistles bade us good-bye and *bon voyage*. All the way up the East River dipping flags gave us hail and farewell. The fleet of big Sound steamers passed us one by one, whistles saluting and the decks crowded with passengers waving handkerchiefs. At Flushing, and other points, many yachts saluted with their guns; and it was not until night hid us that the inspiring God-speeds of our friends and well-wishers were heard no more.[12]

The press made much of the fact that the expedition was to include its leader's wife, Josephine, for the presence of a woman was rare in any expedition and unheard of in previous Arctic exploration. Almost unnoticed at the time of departure was another member of the Peary team, who later became as well known as Peary when he challenged Peary's 1909 claim of precedence at the North Pole. Frederick Cook (1865–1940), making his first foray into the Arctic on the Academy trip, was an affable doctor from New York with degrees from Columbia and New York Universities (figure 9.3). Like most of the other participants, he was young, just twenty-six, and like them he had become a part of the Greenland venture by volunteering his services in response to news reports about the expedition. After a face-to-face interview with Peary and his wife in Philadelphia, Cook was signed on as the expedition's physician.[13]

The doctor's medical expertise was called upon far sooner than he or anyone else could have expected, for while still shipbound on the trip north, Peary suffered a serious injury that almost ended the expedition before it began. On 11 July, while crossing Melville Bay, the *Kite* encountered a heavy blockage of pack ice. As the ship's captain maneuvered the vessel backward and forward through the pack, the rudder jammed on a block of

9.2 Robert E. Peary (1856–1920), whose first scientific expedition to the Arctic was sponsored by the Academy of Natural Sciences and its members in 1891–92. ANSP Archives coll. 2011-003.

ice, wrenching control of the ship's wheel from the helmsman and lurching its heavy iron tiller into Peary's legs. Both of the bones of his right leg were broken between his ankle, and his knee. Mrs. Peary rushed to her husband's side, then watched as he was carried below deck by the anxious crew. Benjamin Sharp and Frederick Cook set the broken bones and administered whiskey and morphine to relieve the pain. The next day, Cook created a sturdy splint for the incapacitated explorer.

Despite an injury that would have caused almost anyone else to give up, Peary insisted on continuing with the expedition as planned. With several months before his scheduled ice travel, he figured he had time to heal. He was where he wanted to be and nothing was going to dissuade him from his goal.

When the *Kite*'s captain declared that his ship could travel no farther north without the risk of becoming icebound through the coming winter, Peary declared that their present position would be the site of his winter quarters (figure 9.4). Once a suitable place for the expedition's cabin was selected, Peary had himself carried ashore strapped to a wooden plank.[14] There, from the modest protection of a canvas tent, he oversaw the assembly of the partially prefabricated Red Cliff House, in which he and his team would live through the coming Arctic winter.[15]

A few days later, when Peary wrote his preliminary report to the Academy—to be carried back to Philadelphia by Angelo Heilprin on the departing *Kite*—he apologized for having to write in pencil instead of the customary ink. He did so, he explained, because he was "lying on my back" as the result of an "annoying accident to myself by which I am temporarily confined with a broken leg." The principal objectives of the expedition, he assured the Academy, would not be affected by the mishap.[16]

Peary's initial foray into Greenland in 1886 had given him an introduction to Arctic travel, but it was during the winter of

9.3 Dr. Frederick Cook (1865–1940), who served as the physician and anthropologist on Peary's 1891–92 North Greenland Expedition, would later claim to have preceded Peary to the North Pole. He is shown in Arctic expedition gear at left (in a studio photo) and in more formal attire at right. ANSP Archives coll. 457.

1891–92 that he really immersed himself in the methods of sled travel and terrestrial navigation that would become the hallmarks of his later career. Unlike many of his expeditionary predecessors, who looked down on the Inuit as anthropological curiosities who had not advanced into the modern age, Peary recognized that they were a resourceful and highly intelligent people who had learned to survive in the most challenging environment on earth.[17] With help from Frederick Cook, whom he had designated as the expedition's ethnographer as well as its physician, Peary tried to learn everything he could about Arctic survival from the group of Inuit he invited to settle near Red Cliff House.[18] By the time the last rays of sunlight left them and the long Arctic winter set in, no fewer than seventeen Arctic Highlanders and twenty-one Eskimo dogs had joined the Pearys and their five-person North Greenland Expedition team on McCormick Bay.[19] Despite the stark and inhospitable conditions outside, life inside Red Cliff House was surprisingly civilized. Peary noted in his memoir, *Northward over the Great Ice,* that on 26 November 1891, with the Academy's silk flag spread above them and dressed in "civilized attire," members of the party celebrated their good fortune with a special meal.

> With the temperature outside at −16½ F., we sat down in our comfortable little cabin to a tempting Thanksgiving dinner of broiled guillemot dressed with green peas, a venison pie, hot bisquit, plum pudding with brandy sauce, apricot pan-dowdy, apple pie, pineapple, candy, coffee, whiskey cocktail, and Rhine wine. . . . Later our Eskimo friends shared in our good cheer and the boys and natives amused themselves with games of strength until far into the evening.[20]

Because of Peary's isolation at McCormick Bay, he had no means of communicating with the outside world once the *Kite*

9.4 The steam-whaler *Kite,* which transported Peary's North Greenland Expedition and the Academy's West Greenland and Peary Relief Expeditions to and from the Arctic. Lantern slide. ANSP Archives coll. 145.

gave its farewell salute on 30 July 1891.[21] All that anyone knew was that five young men and one woman had been left in an extremely hostile place with an incapacitated leader and no hope of rescue for at least ten months. Previous exploratory disasters in the Far North, especially the highly publicized deaths of two-thirds of Adolphus W. Greely's Lady Franklin Bay Expedition to Ellesmere Island in 1884, further raised public anxiety about the fate of the explorers. During the winter of 1891–92, over the objections of some of its members, the Academy responded to the growing clamor for intervention by organizing a second Arctic expedition in as many years. The Peary Relief Expedition's name defined its primary purpose, but as with the West Greenland Expedition, the party hoped to continue the scientific research and collecting that had taken place in Greenland the previous summer. Once again, Angelo Heilprin would be in charge.

Unaware of the world's concern for their well-being, Peary and his associates had passed the winter exploring the area around McCormick Bay and preparing for their sled trip to the north. When the time came, Peary decided to simplify the journey by making it with only one companion. For this, he chose the best skier in his party, a twenty-year-old Norwegian named Eivind Astrup.

By the time the Peary Relief Expedition departed from New York at the end of June, a fully recovered Peary and his young assistant had been traversing the Greenland ice cap for more than a month and were nearing the northernmost point of their journey.[22] On 4 July they reached a towering 3,500-foot cliff overlooking a vast expanse of open water beyond which they could not pass. "Silently Astrup and myself took off our packs and seated ourselves upon them to fix in memory every detail of the never-to-be-forgotten scene before us," Peary later recalled. "All of our fatigues of six weeks struggle over the ice-cap were forgotten in the grandeur of that view. . . . I could now understand the feelings of Balboa as he climbed the last jealous summit which hid from his eager eyes the blue waves of the mighty Pacific."[23]

Embracing the symbolism of the date, and toasting his own success with a "thimbleful" of brandy, he "christened the great bay spreading its white expanse before us Independence Bay."[24] He named the cliff on which he and Astrup sat Navy Cliff and a nearby glacier Academy Glacier, in honor of his expedition's primary sponsor.[25] Peary put a message describing the expedition and its purpose into a corked bottle and placed it in a stone cairn he had built for the purpose. He then unfurled the American flag the Academy had given him and another from the National

FACING PAGE 9.5 Inuit artifact with walrus ivory carvings collected by Harry Whitney (1871–1936) in Greenland in 1909. ANSP Exhibition collection. Purcell photograph.

9.6 The flags presented to Peary by the Academy of Natural Sciences (top) and the National Geographic Society were flown over Academy Glacier on 4 July 1892. Photograph from *Northward over the Great Ice* (1898). ANSP Library.

> At twenty minutes past nine o'clock, when a short distance below Lincoln Park, the *Kite* was seen coming up [the Delaware] river in full holiday attire, her rigging gaily trimmed with flags, and the American ensign floating from the foremast, above the crow's nest. Her steam whistle was keeping up an endless series of shrieks in reply to the salutes from the numerous craft passing down the stream to the sea. . . . Every vessel that passed saluted. Boat-whistles shrieked and tooted, fog horns were blown and colors dipped. On the shores locomotives and factory whistles joined the din, cannon[s] boomed, sky-rockets hissed and exploded, and the thousands of people who blackened the wharves filled the air with their hurrah.[29]

Geographic Society and photographed the scene for posterity (figure 9.6). "How gloriously the brilliant colours sparkled," he later wrote, "as the wind from the mighty ice-cap spread them to the vivid sunlight and filled the air about the summit of the great bronze cliff with their laughing rustle!"[26]

Four weeks later, when he returned to his base on McCormick Bay after his grueling 1,200-mile round-trip journey, he was surprised to find Angelo Heilprin and the rest of the Academy's relief expedition there to meet him. Published accounts of the meeting make the reunion seem perfectly amicable, even joyous, but a reading between the lines of the archival record suggests that Peary may have felt insulted that anyone thought that he and his party needed rescuing.[27] No matter how generous and altruistic the motives of the relief expedition may have been, so ambitious an explorer as Peary was not eager to have his success diluted or to have his competence as a leader undercut by the implied need for help from a gaggle of tenderfeet. Josephine, by contrast, was delighted by the arrival of the *Kite* and, with it, news from her family and loved ones at home. The original plan had been for the North Greenland Expedition to make its way in open boats to southern Greenland and from there find a ride on a whaling ship to the United States. Now they would have the pleasure of traveling home directly from Red Cliff House in comfort.

Despite the loss of one member of the expedition, probably the result of a fall into one of the interior's many ice crevasses, Peary's party and the Heilprin relief team were greeted as heroes on their return to Philadelphia in the autumn of 1892.[28] An eyewitness described the excitement of the *Kite*'s arrival in Philadelphia on 24 September:

The Academy staged an even more elaborate welcome a few nights later with a reception at which "nearly 1200 people, representing every branch of professional and business life crowded the [museum's] library hall, and grasped in congratulations the hands of, Lieutenant Peary and his intrepid wife."[30] Contemporary newspaper accounts described the event as "the most successful and brilliant reception ever given by the Academy," with an attendance that "exceeded that of any previous entertainment of the kind."[31]

The extensive press coverage Peary received for his leadership of the North Greenland Expedition, enhanced by a number of reporters who accompanied both voyages of the *Kite,* helped to establish Peary as a certified Arctic explorer and enabled him to raise the funds necessary for future expeditions to the Arctic. Soon after his return, Peary called upon the influence of the Academy and its president, Isaac J. Wistar, a retired Civil War general, to help him arrange an additional leave of absence from the Navy so that he could return to Greenland for another expedition in 1893. Wistar agreed to help "on the understanding that the Academy will not be called upon for any money, its endowment not being lawfully available for this purpose, and will not be responsible for the risks to yourself and your companions."[32]

Peary remained forever grateful to the Academy for its support of his first major expedition and to Wistar for his effective intercession with the Navy on his behalf, but as Wistar had made clear, after 1892 Peary would have to turn elsewhere for his financial and institutional backing. While some Academy members continued to support Peary's efforts privately, the Academy as an institution had only indirect involvement with his continuing quest for the North Pole. Nevertheless, he went on to mount five more expeditions to the Arctic (1893–95, 1896, 1897, 1898–1902,

9.7 Peary's silk expedition flag, bearing forty-three stars, was the national flag for just a few days between the time Idaho and Wyoming became states (4 and 11 July 1890). Carried by the explorer to the north of Greenland, it was returned at an elaborate banquet in the Academy's library in September 1892. It was conserved in 2007. ANSP Archives coll. 145.

1905–1906) before claiming to have successfully reached the North Pole on 7 April 1909.[33]

In 2009, to commemorate the centennial of Robert Peary's North Pole quest, the American explorer Lonnie Dupre retraced Peary's route. He carried with him the green and yellow expedition flag of the Academy of Natural Sciences (figure 9.8). Like Peary's North Greenland Expedition flag of 1891–92, it represents the global reach of an institution dedicated to understanding the complexities of the natural world.

RMP

9.8 Lonnie Dupre, Maxime Chaya, and Stewart Smith display the Academy's expedition flag at the North Pole in 2009 after retracing Peary's travel route of a century before. Dupre photograph.

Notes

1 Letter to the ANSP from R.E. Peary, 22 August 1890. Coll. 145.

2 Peary would reference the Academy's collections in future correspondence. See letter to ANSP from R.E. Peary, 12 January 1891. Coll. 145.

3 Letter from R.E. Peary to his mother, 27 February 1887 quoted in Wally Herbert, *The Noose of Laurels: Robert E. Peary and the Race to the North Pole* (New York: Atheneum, 1989), 65.

4 The budget included $1,600 for provisions, $1,500 for equipment, $1,400 for instruments, and $1,500 for transportation. See Peary's "Outline of a Project for Reaching the Northern Terminus of Greenland via the Inland Ice," 1891 (hereafter referred to as Peary's Proposal), coll. 145, 8.

5 H. J. Rink, quoted by Peary in Peary's Proposal, coll. 145, 7, 8.

6 Ibid., 8.

7 In Nolan's unpublished history of the Academy (coll. 463, 416), he claims that the Academy's support of Peary grew from Peary's serendipitous meeting with Sharp on Peary's first visit to the museum in January 1891. Peary's earlier surviving letter to the Academy (coll. 145), cited in n. 1, indicates otherwise.

8 Peary Proposal, coll. 145, 7–8.

9 Although Peary later claimed that no money had come directly from the Academy, the records indicate otherwise. Other financial support came from the American Geographical Society, the National Geographic Society, the Brooklyn Institute, the *New York Sun*, and the World's Columbian Exposition (by way of Frederick Ward Putnam, curator of the Peabody Museum at Harvard). This last grant was for the purchase of ethnographic materials from the natives of Greenland for display at the Columbian Exposition. See James W. VanStone, "The First Peary Collection of Polar Eskimo Material Culture," *Fieldiana* 63, no. 2 (27 December 1972): 31–80.

10 Edward James Nolan, unpublished Academy history, ca. 1912. Coll. 463, 418.

11 The Academy minutes for 26 May 1891, report that the flag was donated to the Academy for the purpose by Hearstmann Brothers & Co. See coll. 502, 552. Returned by Peary in 1892, the flag was exhibited at various times over the ensuing years and became badly degraded. In 2007 and 2008, with support from the Geographical Society of Philadelphia and the Color Guard of the Pennsylvania Society of Sons of the Revolution, it was conserved and put back on exhibition in the museum.

12 Robert E. Peary, *Northward over the Great Ice* (New York: Frederick A. Stokes, 1898), vol. 1 , 43.

13 Cooke was paid $50 for his year of service. Bruce Henderson, *True North: Peary, Cook, and the Race to the Pole* (New York: W. W. Norton, 2005), 41.

14 Josephine Peary, *My Arctic Journal: A Year Among Ice-Fields and Eskimos* (New York: Cooper Square Press, 2002), 33; and Robert Peary, *Northward over the Great Ice*, 81.

15 The winter quarters was so named for "the striking red color of the mountainside" a mile to the east of the building. Robert N. Keely Jr. and G. C. Davis, *In Arctic Seas or The Voyage of the Kite* (Philadelphia: Rufus C. Hartranft, 1892), 133.

16 R.E. Peary report, July 29, 1891. Coll. 145, 8–10.

17 In his account of the second Grinnell Expedition of 1853–55, the American explorer Elisha Kent Kane, while generally admiring his Inuit hosts, nevertheless referred to them as "Arctic primates" and "indomitable savages." See Kane, *Arctic Explorations* (Philadelphia: Childs & Peterson, 1856), vol. 2, 24.

18 Peary's interest in the indigenous people of Greenland was reinforced by a $2,000 grant from F. W. Putnam at Harvard for the purpose of collecting information and artifacts to represent these people at the World's Columbian Exposition in Chicago.

19 Robert Peary, *Northward over the Great Ice*, 155.

20 Ibid., 162.

21 Peary did send a letter to the Academy (dated 15 April 1892) by way of a traveling Inuit from Cape York, but the letter did not arrive until long after Peary had returned to Philadelphia. For a transcript of the letter and Peary's comments about its long trip south, see Robert Peary, *Northward over the Great Ice*, 241–42.

22 The Peary Relief Expedition traveled aboard the steamship *Miranda* from Brooklyn, New York, to St. John's, Newfoundland, then transferred to the *Kite* for the rest of their journey.

23 Robert Peary, *Northward over the Great Ice*, 344–47.

24 Ibid., 349.

25 Unlike Peary Channel and a number of other features named by Peary during his trip, Independence Bay (or Fjord), Navy Cliff, and Academy Glacier are still considered valid geographical names.

26 Robert Peary, *Northward over the Great Ice*, 350.

27 Nolan's reference to this in his unpublished history of the Academy (coll. 463) and a formal letter from Peary denying press reports that he had ill feelings toward Heilprin both suggest that Peary must have shown at least some animosity toward the relief expedition.

28 For more on the disappearance of John M. Verhoeff, see Keely and Davis, *In Arctic Seas*, appendix 1, 461–81; Robert Peary, *Northward over the Great Ice*, 411–18, and Henderson, *True North*, 68–77. For a summary of the achievements of the expedition, see Clive Holland, *Arctic Exploration and Development c. 500 B.C. to 1915: An Encyclopedia* (New York: Garland Publishing, 1994), 372–73.

29 Keely and Davis, *In Arctic Seas*, 454–55.

30 Unidentified newspaper clipping, 29 September 1892, in ANSP mss. coll. 417, vol. 2, 77.

31 Ibid.

32 Isaac Wistar, quoted in introduction to Robert Peary, *Northward over the Great Ice*, 43.

33 There are many historians who dispute the veracity of Peary's claim.

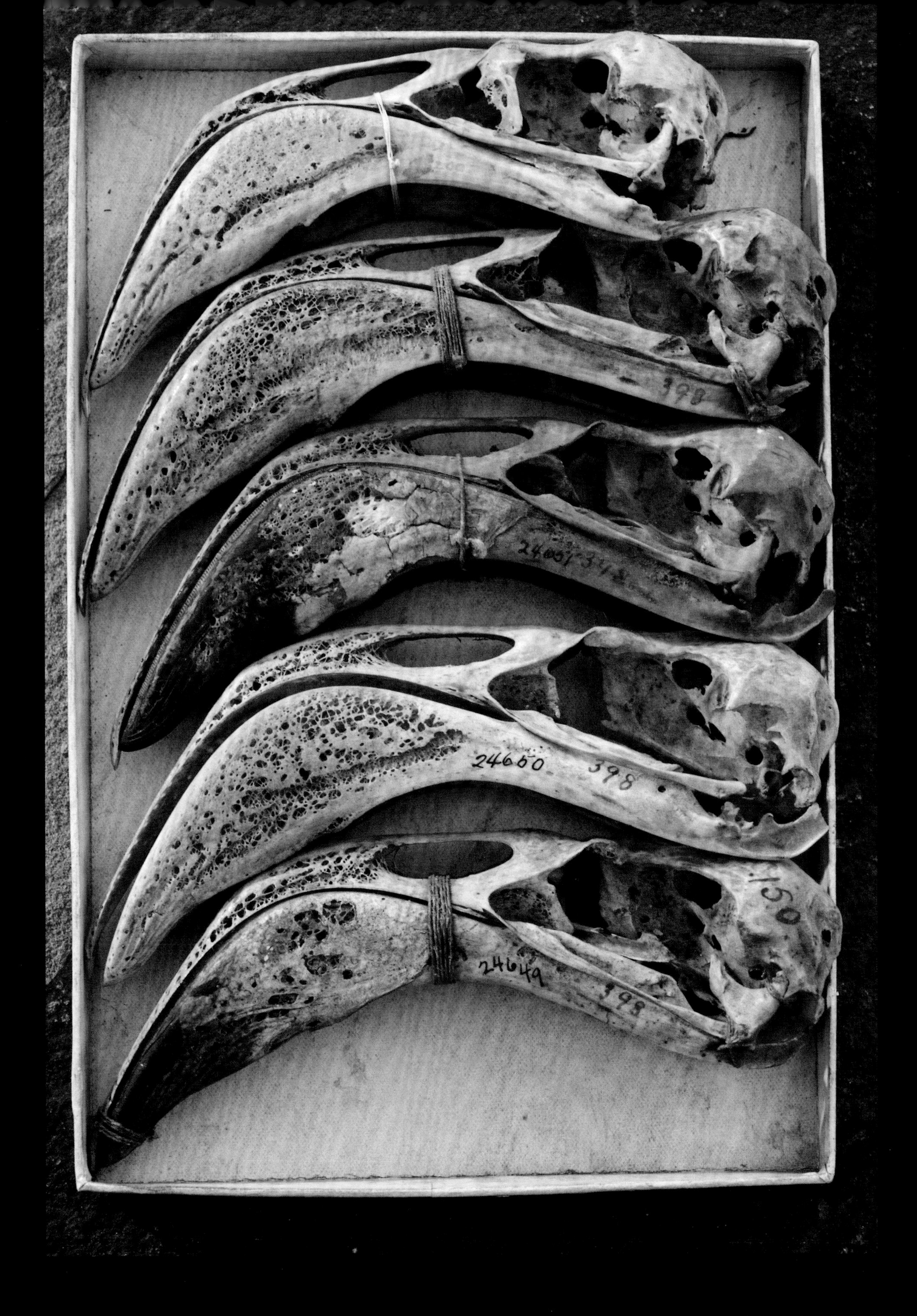

Skulls of the American flamingo (*Phoenicopterus ruber*) presented to the Academy by Thomas B. Wilson in 1846. ANSP Ornithology Department #24647–24652. Purcell photograph.

FOLLOWING PAGES Shed snake skins collected by George M. Feirer in 1942. ANSP Herpetology Department. Purcell photograph.

Thamnophis
S. sirtalis
9/1/42
Crotalus
viridis
oreganus
12/1/42
Crotalus
ruber
10/6/42
Ackistrodon
piscivorus
8/30/42
54"

CONSTRICTOR
C. CONSTRICTOR
10/10/42
PITUOPHIS
SAYI
10/7/42
Elaphe
obsoleta
confinis
8/29/42

Chapter 10

Early Man at the Academy

As a scientist, I must admit the evidence that man was born from the animal kingdom.
—Father Pierre Teilhard de Chardin, Academy of Natural Sciences Symposium, 1937

In 1937, as the Great Depression began to ease but ominous rumblings of war again threatened Europe, the Academy marked the 125th year of its founding in a most unusual way. Instead of celebrating its achievements in understanding flora and fauna, on which its members had focused their attention for a century and a quarter, it chose to explore the relatively unknown world of early human beings. At a weeklong symposium attended by most of the world's leading anthropologists, the Academy drew international attention by laying out the newest thinking on the beginning and evolution of humankind over time. The person who originated the idea and was behind every aspect of this remarkable gathering was Edgar Billings Howard (1887–1943).

After graduating from St. Paul's School and Yale's Sheffield Scientific School (1909) and spending a few years in business, Howard served as an officer during the First World War. Following the war and another stint in business, he decided to devote himself entirely to scientific research in the fields of archaeology, geology, and paleontology. Elected a member of the Academy of Natural Sciences in 1924, he was appointed research associate in 1931 and joined the board of trustees in 1935, the same year in which he received his doctorate in geology from the University of Pennsylvania. Howard's thesis was on early man in America.[1]

During the latter half of the nineteenth century, events related to the Civil War and the immense difficulties of Reconstruction so preoccupied the United States that archaeological activities in the country were overshadowed. By contrast, the situation in Western Europe was very different. Between 1850 and 1875, discoveries of large fossil bones in undisturbed association with stone artifacts led to the acceptance of Old World "Paleolithic Man." At cave sites in England and near France's Somme River valley, earth scientists established a basis on which to acknowledge that humans were contemporary with beasts of the last Ice Age.[2]

By 1880 in North America, hope and interest gave way to unbridled enthusiasm as enterprising men laid claim to early sites on the continent, which resulted in numerous "discoveries" of prehistoric humans.[3] Many of these finds were dubious and prompted skepticism and caution, especially the Smithsonian's eminent Czech anthropologist Ales Hrdlička (1869–1943), "whose reputation and his ability to intimidate grew with each confrontation, frequently stifling early man enthusiasts, preventing them from bringing forth new claims," according to archaeologists Anthony T. Boldurian of the University of Pittsburgh at Greensburg and John L. Cotter at the University of Pennsylvania Museum of Archaeology and Anthropology.[4]

When America entered the Great Depression in 1930, only one site of indisputable antiquity existed. This was near Folsom, New Mexico, where, in 1926–27, some stone artifacts were found

10.1 Discarded labels from the Academy's ethnographic collections. University of Pennsylvania Museum Archives. ANSP. Purcell photograph.

Dance Wands, with Horns of Mountain-Sheep and of the Cow. Cree (Algonkin) near Red River, Manitoba. A.H. Gottschall!
Knife-blade Arizona. Mrs. M.A. Haldeman.
ACADEMY OF NATURAL SCIENCES OF PHILADELPHIA, No.
FLUTES & WHISTLES. Used chiefly in the worship of the God of the Wind–Quetzalcoatl. Lamborn Collection. Ancient Mexico.
Plaited Sandals.
Dance Rattle of Pig's hoofs. (Tewa) San Juan Pueblo New Mexico. A.H. Gottschall!
Woman's Basket Cap. prob. Hupa. California. Mrs. John Reilly.
Umbilicus Pendent Assiniboin (Siouan) Milk R., Montana. A.H. Gottschall Coll.
Spoons made by steaming & pressing the horns of the Rocky-mountain goat; set with abalone shell & carved with heraldic crests. Haida, Queen Charlotte Islands A.H. Gottschall! British Columbia.
Hair-braid Wraps, Otter fur, beaded tassels. Comanche (Shoshonean) Cache Cr., Oklahoma. A.H. Gottschall Collection.
Ceremonial, of coyote skins, with feathers & cloth attached. Arikara (Caddoan) Upper Missouri R. N. Dakota. A.H. Gottschall.
ACADEMY OF NATURAL SCIENCES OF PHILADELPHIA, No. 4407. Dance Rattle made from carriboo hoofs. Yukon Valley. Alaska
Ornament; shell disc, dentalium & abalone shells. Frank Thomson Coll.
Necklace of trader's wampum, strung on raw hide. Ute (Shoshonean) Ogden R., Utah. A.H. Gottschall Coll.
ACADEMY OF NATURAL SCIENCES OF PHILADELPHIA, No. 15700
Frank Thomson Coll.
Xipe. Lamborn Collection
"Dance Skin" (wood-rat). Wichita (Caddoan) near Sugar Cr., Oklahoma.
Pipes, stone bowls and skin-covered stems. Cree (Algonkin) nr. Red River, Manitoba A.H. Gottschall!
Maskette for finger, and carving of a seal. Alaska. W.H. Jones.
Womans Basket Hat, Hupa (Athapaskan) N.W. California.
Xochi-quetzal Mexico. Lamborn Collection
Painted Dance-Armlets. San Juan Pueblo New Mexico. A.H. Gottschall Coll.
Beaded Bags for Trinkets. Crow, (Siouan) Little Big Horn River Montana. A.H. Gottschall.
ACADEMY OF NATURAL SCIENCES OF PHILADELPHIA, No. 24274. Ball-headed War-club, made of maple or "Sugar-tree" wood. Collected prior to 1858. Ojibwa (Algonquian). Michilimackinack, Mich. Wm. S. Vaux (James McBride Coll.)
Dance Pendents, bear-skin and claws. Arikara (Caddoan). Upper Missouri R. N. Dakota. A.H. Gottschall!
Xipe Lamborn Collection
Basket-Hat, woven double in two-ply twining. Such hats are usually of Haida make. Kelsimaht (Nootka), W. Coast
Dance-Skin (coyote)
Stone-headed War-Club Siouan tribe
Spoon-Mats, spruce-root; weave, two-ply twining; grass embroidery,

associated with extinct bison bones. Years earlier, in the summer of 1908, a flash flood had created a deep gully on a ranch outside this tiny town. The ranch foreman, George McJunkin, an African American born a slave in pre–Civil War Texas, found bison bones sticking out of eroded gully walls while checking fence lines after the flood. He showed his finds to a local blacksmith, who much later prevailed upon Jesse D. Figgins, director of the Colorado (now Denver) Museum of Natural History, to visit the site at Folsom. In August 1927, Figgins found a grooved spear point in direct association with a bison skeleton. The fieldwork Figgins supervised that fall confirmed the presence of humans in the New World at the end of the Ice Age. McJunkin, who had died five years earlier, never knew of the enormous importance of his discovery.

At that point in time, it was necessary to proceed judiciously in this field of archaeological inquiry. Edgar B. Howard, modest, genteel, diplomatic, yet intently energetic, was the perfect person to undertake the work. From an old New Orleans family that had relocated to Philadelphia, Howard had the considerable resources to pursue a scientific career. In the summer of 1930, nearly finished with his M.A. in anthropology at the University of Pennsylvania, he began his research with exhaustive surveys in caves in the foothills of the Guadalupe Mountains in New Mexico (figure 10.2).

Ostensibly he was studying the pre-Pueblo Basket Maker culture (ca. 1000 B.C.), but in fact he was eager to find evidence of much earlier human activity. Loren Eiseley, who would become a renowned anthropologist, was then a graduate student at the University of Pennsylvania. He worked for Howard as an assistant in the summer of 1934 and thought his mentor was unduly obsessed with finding the skeleton of Folsom Man.[5]

Accustomed to relaxing summers in Bar Harbor, Maine, Howard found the rigors of a field archaeologist daunting. He confessed to Charles Cadwalader, managing director of the Academy,[6] that the work was so very different from what he had naively anticipated. "The picture I had in my mind of this cave work was of a nice clean cave near a bubbling spring, on a good road, & with baskets & sandals hanging up on pegs or neatly piled in the corners, & well preserved mummies sitting huddled up around the walls! But it does not seem to work that way." He ended his letter, "Will let you know if I get anything worth while (besides experience)."[7]

In 1931, in another part of New Mexico, inside Burnet Cave—named for its discoverer, Bill Burnet, a "profane, blasphemous" local metalworker and relic collector whose shop motto was "We work iron and steal for a living"[8]—Howard found a grooved spear point of genuine antiquity similar to the one from Folsom. The stone tip was located beneath the floor of a sealed grotto near the charred bones of Ice Age musk oxen, caribou, camel, horse, and bison at the edge of a hearth.[9] He wrote to Horace Jayne, director of the University of Pennsylvania Museum, which was sponsoring his endeavor along with the Academy, from Raton, New Mexico, that he had "a large box of bones from the cave where the spear point came from, & I hope Barnum Brown will be able to identify them."[10]

Later that year, Howard invited Brown (1863–1963), an eminent paleontologist from the American Museum of Natural History, famous for discovering the first *Tyrannosaurus rex* skeleton, to visit the site. Howard should have been warned that Brown was unscrupulous in taking credit for himself, which he did in an article for the magazine *Popular Science*.[11] Howard wrote to Brown the same day with a diplomatic reprimand: "For myself I am not interested in the publicity of the work; but I *am* interested in seeing that the [University of Pennsylvania] Museum and the Academy be allowed to receive all the credit possible, in order to help to *build up* the public interest here in the institutions of Philadelphia."[12]

In the fall of 1932, Howard outlined to Cadwalader his ideas for reestablishing a new department at the Academy. "Our talks have covered a rather wide range," he said, "but mostly they have centred around the work I am so interested in, namely vertebrate paleontology and the probable existence of early man in North Amer[ica]. At first glance they do not appear to be very closely related; but, when, as has been happening in the last few years, extinct animal bones are found in association with human artifacts, a study of both subjects is essential."[13]

Howard said that the previous winter he had made a card file of all the paleontological material in the Academy's cases so that visiting scientists from other institutions could easily find the specimens they sought and could "use the Academy as it was once used for this purpose."[14] Many had expressed regret at not finding Joseph Leidy's type specimens available for study. Howard suggested reestablishing a Department of Vertebrate Paleontology, one that would work closely with the University Museum. He thought that if an archaeologist, a physical anthropologist, a geologist, and a paleontologist could visit all the important sites, as well as study material in museums and universities, they would be well positioned to report on all the evidence secured.

In his message to Cadwalader, Howard listed the various skeletal remains attributed to early man that had been uncovered in the Americas until then, none of which had been accepted

10.2 Edgar B. Howard in an unidentified cave, Guadalupe Mountains, New Mexico, 1937. Private collection.

because a tradition had grown up that "it just couldn't be." He said that this evidence, accumulating for several years, tended to strengthen the ideas of earlier investigators who believed that humans had lived in North America for a much longer period than at first supposed. "It may be too early to establish this department," he concluded, "but it is an opportunity to focus attention on the Academy and the University, and would result in securing the proper backing to undertake more extensive work and better exhibition of the material recovered."[15]

Accompanying this long memo to Cadwalader is a letter from Howard saying that he had hesitated to send it because he had just had "some very bad news." In the *American Journal of Physical Anthropology* for October–December 1931 was a notice of the formation of an International Commission on Fossil Man with an American Subcommittee. He said that this committee was composed of Barnum Brown, Chester Stock (from the California Institute of Technology), and others whom he had in mind for his Academy plan. "It looks as though it were all over but the shouting. I feel like a balloon that has been stuck with a pin. I can't understand why Barnum Brown never mentioned this to us. The only consolation I have is that it was an original idea with me, even though I was [a] year late getting it!"[16] Since ideas were being contemplated for celebrating the Academy's 125th anniversary, this disappointment may have planted the seed in Howard's mind for something more spectacular: an international symposium on early man.

When the excavations in Burnet Cave ended the previous summer (1932), Howard received word from a local artifact collector of another site containing fluted spear points and fossil bones near Clovis, New Mexico. In that part of the southwest, sand dunes mottled with yucca, sand sage, and tumbleweed extend one hundred miles across the flat, treeless Llano Estacado of eastern New Mexico and western Texas. Through the dune field runs a shallow valley in which the scrub blends with a carpet of buffalo grass and a profusion of wildflowers. This floral change marks the sand-choked headwaters of the Ice Age Brazos River, called Blackwater Draw, the setting of a former pond in this extinct river drainage. Dust Bowl winds of the late 1920s, which scoured millennia of water-deposited silt and clay, created a series of deflation hollows or "blowouts." Exposed in these depressions were bones of mammoth, bison, horse, and camel, with stone artifacts scattered among them.[17]

Three years earlier, a nineteen-year-old named James R. Whiteman had written to Alexander Wetmore, assistant secretary of the Smithsonian Institution, to report on finding this cache of "extinct elephant bones" and "warheads." Arrangements were made for Charles Gilmore, the museum's vertebrate paleontologist, to visit the site. Yet, after staying only an hour at Blackwater Draw, Gilmore declared that the spot was not interesting enough to warrant an archaeological dig.[18]

This was all very well for Howard, who was therefore free to excavate this most interesting site. He returned to Philadelphia brimming with plans for the next year's dig. But only three weeks after his departure, fate appeared in the form of the New Mexico Highway Department prospecting for road gravel near the very spot. Heavy machinery started pulling up mammoth bones and other evidence of extinct creatures. Some bones were put on display in Portales, New Mexico, while others were carted away by workers and onlookers, only to show up later on porches, in cupboards, and in garages. One local farmer made off with a hefty chunk of mammoth bone to use as a doorstop.[19]

Notified of this distressing event shortly after writing his long memo to Cadwalader and his letter of disappointment about the announcement of an international commission on fossil man, Howard rushed back to the Clovis site in mid-November. He telegraphed Horace Jayne at the University Museum: "Extensive bone deposit at new site mostly bison also horse & mammoth some evidence of hearths along edges will tie up permissions for future work & spend few more days investigating."[20]

In the summer of 1933, Howard and Whiteman canvassed the Blackwater Draw for extinct animal fossil bones in association with stone artifacts. Howard described one of his more curious finds to Cadwalader: "In the jaw of a mammoth, we removed the other day, I found his tail-bone! You might tell Dawson there were trained circus elephants down here 10,000 to 15,000 years ago!"[21]

Later that summer, Howard invited a group of eminent scientists who were attending an International Geological Congress in Washington, D.C., to visit the Clovis site and confirm his findings. Included in the group were John C. Merriam, director of the Carnegie Institution in Washington; Chester Stock, professor of vertebrate paleontology at the California Institute of Technology; Victor Van Straelen, director of the Royal Museum of Natural History in Brussels; and Sir Arthur Smith Woodward of the British Museum and Lady Woodward (figure 10.3).[22] Amid mammoth bones, the group saw spear points, fluted but larger and less refined than those that had been found earlier at the Folsom site.[23]

Coincidentally, that same summer, an article in a volume of essays on the origin and antiquity of American Indians suggested for the first time that a land bridge had probably existed between Siberia and Alaska and that a migration route south from Alaska

through the Mackenzie River Valley had opened in the wake of ice retreat. Within a few months, Howard and geologist Ernst Antevs put all the pieces together: twenty thousand to fifteen thousand years ago, the Bering Strait was dry land, and a corridor was open along the eastern side of the Rockies, removing all obstacles to migration from Central Asia to the Great Plains.[24]

In 1935, Howard traveled to the Soviet Union and visited museums in Leningrad (present-day St. Petersburg), where he examined archaeological collections from across the country amassed since the time of Peter the Great. None showed fluting like the Clovis points. From this, Howard deduced that fluting might have been an American invention—the *first* American invention.[25]

The Clovis venture of 1933 established the presence of mammoth hunters in the New World at the end of the Ice Age. Subsequent fieldwork by Howard's team (in 1936) revealed an even clearer image of these early big-game hunters, even though no human remains were found in association with the stone spear points. According to the renowned Paleolithic archaeologists Boldurian and Cotter, "Howard's work in the Blackwater Draw was among the first in American archaeology to employ an interdisciplinary approach, whereby the collaborative efforts of scientists from various disciplines resulted in an accurate reconstruction of the ancient lacustrine [lake] settings of these nomadic groups."[26] This interdisciplinary approach had been Howard's idea for the Department of Vertebrate Paleontology that he had proposed to Charles Cadwalader in 1932.

The Academy has had a long history of collecting archaeological, anthropological, paleontological, and ethnological artifacts. The *Minutes* records in 1815 that Gerard Troost presented "an artificial mummy found in the catacombs of Egypt and brought thence to Paris."[27] The first volumes of the *Journal* list many

10.3 Scientists and friends visiting Howard's excavation site, Clovis, New Mexico, 1933. ANSP Archives coll. 331.

ANTIQUITIES OF THE OUACHITA VALLEY.
GLENDORA, VESSEL NO. 132. (FULL SIZE.)

donated archaeological items: "Fossil remains from New Jersey, presented by S. Wetherill"; "fossil shells from near the Chesapeak from R. Randolph"; and "Fossils, etc. from Ohio, given by Correa de Serra."[28] Titian Peale added a large ethnological collection from the Wilkes Expedition in 1842. An article published in the Academy's *Proceedings* states that as late as 1915, Peale's collection was still regarded as "one of the best collections of Polynesian ethnica anywhere in the world. Others are larger, but none is so seriously representative of the period before foreign contamination had been introduced." Many of the items were brought back from Botany Bay in Australia, others from Samoa, Fiji, and Tonga.[29]

According to Academy president W. S. W. Ruschenberger in 1871, a Department of Ethnology had been established at the Academy in 1830. When Ruschenberger wrote his report, some collections of ancient Indian art had been deposited at various other institutions for lack of room.[30] Perhaps this transfer of artifacts set a precedent for what was to follow fifty-eight years later, in 1929.

The story begins with Clarence Bloomfield Moore (1852–1936),[31] a wealthy bachelor who amassed an important collection of Indian artifacts in the later part of the nineteenth century and the early years of the twentieth century. Born in Philadelphia, Moore graduated from Harvard in 1873 and then traveled throughout Europe and Central and South America, crossing the Andes on horseback and on foot. On a subsequent round-the-world trip, he visited India, Thailand, and Java. For a time, he was president of the Jessup and Moore Paper Company (his mother was the writer Clara S. Jessup Moore). A socialite in his younger days, he enjoyed big-game hunting, but an accident on an African safari that resulted in a serious eye injury turned him to less dangerous fieldwork: archaeological exploration in the southeastern United States. Moore wrote to Academy librarian Edward Nolan in 1897 that he would be unable to read one of his papers to the Academy membership because he was about to travel, and also, "since the severe accident to my eyes, I have been unable to sit in gas light which, I may say incidentally, is the reason I do not attend the regular meetings of the Academy."[32]

At the age of forty, Moore purchased a flat-bottomed steamboat, the *Gopher*, to explore the waterways of the American southeast, looking for potential archaeological sites. For the next twenty-five years, he explored every accessible site where Indians of the southern states had lived and buried their dead. Well organized, exacting, and accurate, he published his work in twenty-one large volumes, beautifully illustrated by the artist Mary Louise Baker (figures 10.4 and 10.5). Moore donated some of his material to the Wagner Free Institute in Philadelphia, but he deposited the major portion of his collections at the Academy of Natural Sciences. A very private person, Moore requested that the Academy staff not mention him or his work in any way.

Moore gave Indian artifacts to the Academy for many years. In 1893, the Academy *Proceedings* stated that "the collection of Indian remains, pottery and implements from the mounds of Volusia County, Florida, generously presented by Mr. Clarence B. Moore, is of exceptional interest, being largely unique. This collection has been labeled and arranged by Mr. Moore and his assistants in cases in the new room over the lecture hall, which has been open to the public since September 1st."[33] Several years later, the *Proceedings* reported: "Through the liberality of Mr. Clarence B. Moore, our museum has, for the first time in its history, been thrown open to the public on Sundays between

FACING PAGE 10.4 Glendora vessel with animal head (two views), painted by Mary Louise Baker for Clarence B. Moore, *Antiquities of the Ouachita Valley*. ANSP *Journal*, 1910. ANSP Library.

10.5 Glendora vessel, painted by Mary Louise Baker for Clarence B. Moore, *Antiquities of the Ouachita Valley*. ANSP *Journal*, 1910. ANSP Library.

the hours of one and five o'clock. This innovation has met with popular approval, the attendance of visitors being very large, and composed mainly of persons who would otherwise have been unable to view the collections."[34]

Many years later, George Gustav Heye (1874–1957) officially opened his Museum of the American Indian in New York City (1922). Heye (pronounced *High*), as passionate about Indian artifacts as Moore, offered to buy Moore's collection from the Academy for $10,000. A large man who stood six feet three inches and weighed three hundred pounds, Heye smoked huge cigars, and it was said that while traveling the United States in search of treasures, he drove his car at ninety miles an hour with his chauffeur as passenger.[35]

Moore, believing that his collections were not being properly appreciated at the Academy, was in favor of Heye's offer.[36] He had heard that the administration wanted to move his artifacts "upstairs" to make way for "stuffed animals," as he expressed it to Frank Keeley, Academy member and curator of mineralogy.[37] He not only feared that many of his items would be broken or stolen during the move to the upper level of the museum but knew they could not be easily seen in such an out-of-the-way place. He preferred that the collection be moved to the Heye Foundation in New York, a museum wholly devoted to the American Indian, where there was "no danger of its being crowded out by other fields of investigation." Moore confided to Keeley, "You are interested in archaeology and, therefore, set greater store by my collection than many others do. A relative of a member of the Council [of Curators] once said to me: 'Who the hell cares for Indian pots?'"[38]

Alarmed, the Academy's president, Richard Penrose, called a special meeting of the council to consider Heye's offer.[39] After due consideration, resolutions were adopted that "the acceptance of a unique collection is in the nature of a trust and that the deliberate disposal of any such collection would establish a precedent detrimental to the Academy." And further, "the Council regards the Clarence B. Moore Collection as one of the Academy's most valuable possessions, brought together at great expense and labor by a Philadelphian under the auspices of the Academy and described in the Academy's publications, and that it is the duty of the Academy to preserve it in Philadelphia, with which city it is so thoroughly identified."[40] And there the matter stood for six years.

But when Academy trustee Charles M. B. Cadwalader was named managing director in early 1929, events took a different turn. Cadwalader wanted to modernize the Academy's zoological exhibits and to provide space for the science of early man. Heye had renewed his offer to buy Moore's collections, and Cadwalader was strongly in favor of such a sale. From St. Petersburg, Florida, where he was spending the winter, Moore wrote to Henry Pilsbry at the Academy that he had long felt that many at the institution did not value his collection and should he die (he was seventy-seven at the time), it would be stored away "in undesirable quarters."[41] Consequently, on 1 April, he wired Cadwalader: "Under the circumstances I strongly advise sale of Moore collection to Heye Museum which is devoted entirely to American archaeology."[42] While still in Florida, in a letter to Heye at the time of Heye's meeting with Effingham Morris, then president of the Academy, and Cadwalader to seal the transfer, Moore revealed his opinion of the latter: "By this time your interview with Mr. Morris and Mussolini is probably underway. As the collection never cost the Academy one cent . . . and as the powers that be there are wild to get possession of the space . . .

10.6 Charles Cadwalader. ANSP Archives, coll. 457.

and know that I wish it placed where archaeology is the prime object of study I should think persons in control would be glad to get the sum named to purchase stuffed animals for the hall."[43]

Heye replied that he had gone to Philadelphia and met Cadwalader, whom he described as "a rather energetic young man but does not know, or does not claim to know anything about science. He wishes to run the Academy from a business point of view which makes it easier in a way for us to deal with them" (figure 10.6). He could see that Cadwalader wanted to get rid of archaeology at the Academy, but he had been frank in saying that he was worried about what people would say about allowing the collection to go to another city. "I told him that no one need know about it until long after the deal had been consummated," Heye said. But, he continued, Cadwalader had observed that Philadelphia was in some ways a small town and that the papers would seize upon anything that would seem detrimental to the city, so how could the collection be gotten out?[44]

Heye then suggested to Cadwalader that the collection could be packed and taken to a storage warehouse, where it would stay for a couple of months and afterward taken from Philadelphia without anyone knowing about it. Meanwhile, when inquiries were made as to what had become of the collection, the Academy could say that it was in storage.[45]

Three days later, Moore, back in Philadelphia, met with Cadwalader and reported to Heye, "Mr. C. . . . quite agrees with me that Philadelphia is not crazy over my collection and that its absence will scarcely be noticed." He said that storage was therefore not needed.[46]

Despite Cadwalader's sanguine predictions, the transfer of a valuable collection of Indian artifacts from the Academy to a museum in New York created a furor with several trustees and staff members, especially with H. Newell Wardle, assistant curator of the Archaeology Department, who had held her position for thirty years. A short time later, when other collections sent with Moore's artifacts to be studied at the Heye Foundation were returned, Wardle was horrified by their condition. In a rage, she wrote to the board of trustees: "These specimens went to New York carefully packed. They reappear in confusion. The Eskimo sled and Algonkian birch-bark canoe are badly crushed. The mass of objects, piled on the floor of the old museum, shows upon the surface, Pueblo pottery, California baskets, Northwest Coast carved wooden food-dishes, Eskimo sealing line, Plains Indian beaded cradle, quilled and feathered pouches, and Ancient Mexican stone statues. The latter are from the Lamborn Collection, which was expressly excluded from the transfer, but carried off notwithstanding."[47]

Several weeks after this letter, Wardle resigned. Citing a "breach of faith with the past," she registered her protest "against this blanket sale, secretly negotiated, and the proposed clandestine removal, without adequate time given for the security of deposited collections not the property of this Academy." A footnote to her letter stated that a copy had been sent to the press.[48]

Heye wrote to Moore on 9 May after the story appeared in the papers:

> It certainly was a hectic day, yesterday, at the Academy and I do not think I would have survived, or if I had I would not have been in as good shape, unless I had had that drink with you before I went there. I know it was a life saver. We had a succession of reporters there from eleven until three o'clock, with numerous telephone calls and also visits from old women that had never known the Academy was on the map, and it looked as if they had not been in town for several generations. Unfortunately, the transfer of the collection did not seem to stir up the feelings of the attractive young women, or flappers enough to have them visit the Academy. So, we missed what might have been a spot of brightness in a gloomy day.[49]

Sometime later, Wardle wrote a searing indictment of Cadwalader for *Science* magazine. In the article, "Wreck of the Archeological Department of the Academy of Natural Sciences of Philadelphia," she recounted the events that had taken place.

> In March [1929], the entire east end of the archaeological hall was ordered cleared. Archaeological and ethnographic material from thirty cases was sent to storage in the rather leaky old museum. This was to make space for proposed groups of sheep and goats. At this time, the managing director [Cadwalader] stated that the entire archaeological hall would be used for mammal groups and that no other exhibition space would be provided for the archaeological department, unless money were procured for it. The assistant curator [Wardle] called his attention to the half-million dollar bequest that had come with the R. H. Lamborn Mexican archaeological collection.[50]

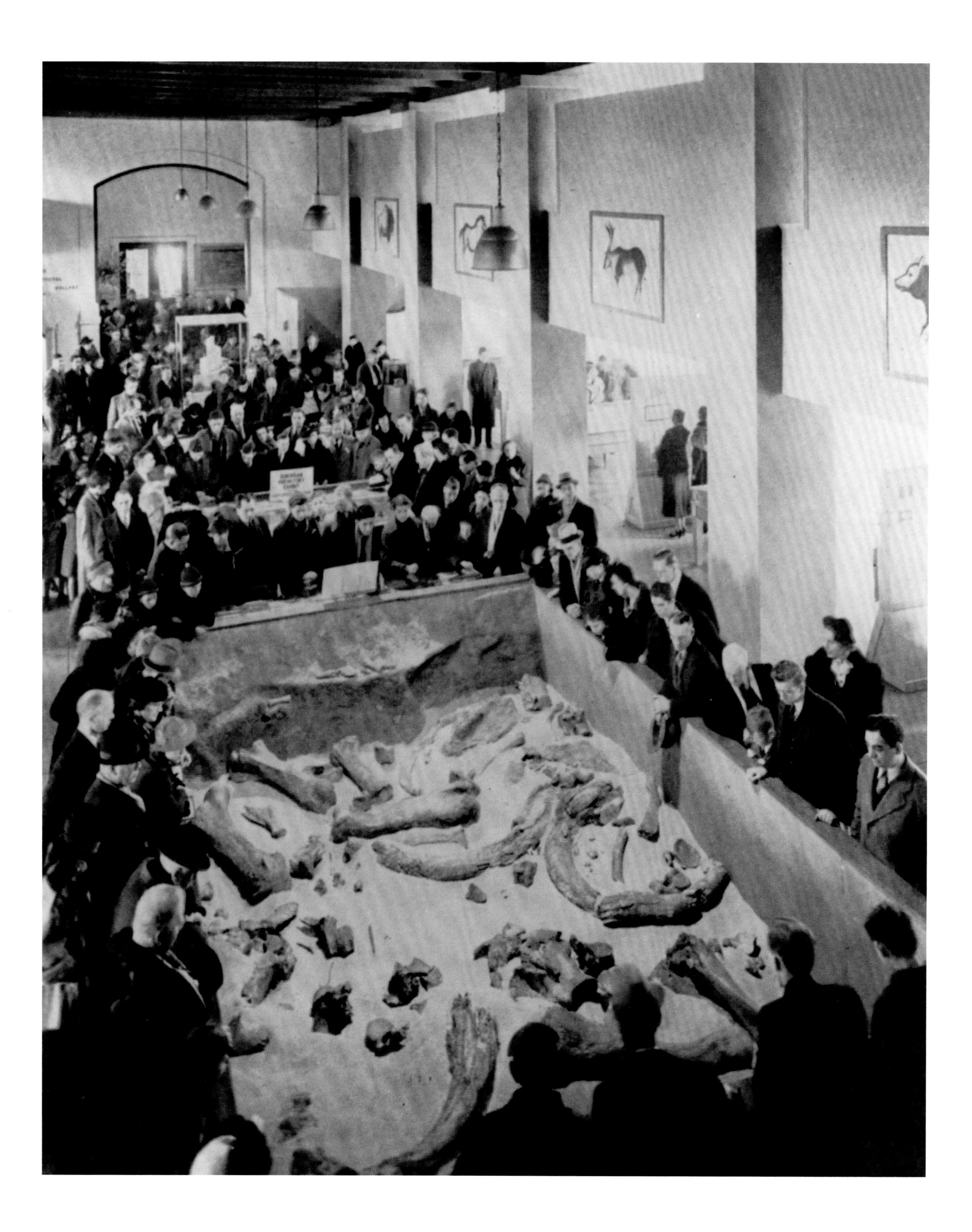

Wardle continued that when Cadwalader demanded to know who had told her that the Moore collection was to be sold, and she answered that Mr. Heye had told her, he reprimanded her for speaking to Heye and said, "Yes, the Moore collection has been sold, and with it goes all the American Indian stuff, North American, South American and Central American: that means all the Haldeman collection, all the Gottschall collection, all the Peary collection—everything that is not bound by the terms of gift. . . . Mr. Heye's trucks will be here next Monday morning."[51]

Summing up her sense of betrayal, Wardle concluded her article for *Science*: "If misguided trustees, chosen for their business ability to manage the financial affairs of an institution, have the power, without asking expert advice, so to wreck a scientific department, it shakes the foundations of confidence in every institution in America."[52]

The study of archaeology and anthropology at the Academy was not yet over, however, for as Clarence Moore's collections of American Indian artifacts left Philadelphia for New York, Edgar Howard's exciting finds of early man in New Mexico began to arrive. The decade of the 1930s was a difficult time for the country, with dust storms blanketing the Midwest and Southwest and the Great Depression taking its economic toll. Howard's letter to Cadwalader from Clovis in 1933 included both situations as he described the visit of Dr. Chester Stock, the archaeologist from California: "Unfortunately it was a terrible day, but, in spite of the dust storm, we showed him enough to help him form an opinion. . . . [Stock] said it was undoubtedly a new Pleistocine field & that Institutions would flock in here when & if they get any more money." It had been intense and laborious work with the temperature often 102 degrees in the shade.[53]

For the Academy, Howard's excavations were assuming increasing importance, and he was determined to set up an interdisciplinary department to study early man. He sent Cadwalader a memo (undated, probably 1934) with the idea of focusing attention on Philadelphia as a center for facts dealing with prehistoric humans in America and having the University Museum and the Academy act as a sort of clearinghouse for other institutions. He also proposed holding a symposium.[54] Cadwalader was agreeable to both ideas, and in a short time, the thought of an international symposium coalesced with the celebration of the Academy's 125th anniversary in 1937 (figure 10.7).

The Academy had great ambitions leading up to the symposium, which involved establishing the institution as a world-class center of research on the origins of the human race. To this end, several well-known scientists were hired and the Department of Geology and Paleontology reestablished. Benjamin Franklin Howell (1890–1976), associate professor of geology and paleontology at Princeton, employed on a part-time basis at the Academy, was to supervise the organization of some two hundred thousand fossil specimens collected the century before by Isaac Lea, Edward Drinker Cope, and Joseph Leidy. Hellmut de Terra (1900–1986), discoverer of prehistoric "shadow man" in the Indian Himalayas, was assigned to work at the Academy for a year through a grant from the Carnegie Institution in Washington, D.C., where he was a research associate. Edwin H. Colbert (1905–2001), a world-renowned paleontologist and dinosaur expert, was hired away from the American Museum of Natural History to work part-time at the Academy. In time, he would become a mentor to Stephen J. Gould, John Ostrom, Peter Dodson, and many others.[55]

The purpose of the symposium was to bring together eminent world authorities on prehistory and, as a memo to the Academy board stated, "to make possible the correlation of important new discoveries; to set up a special exhibit of recently found material relating to early man, including Siwalik primates, Indian Paleoliths, Pleistocene fossils from Java, Peking Man from China, and North American artifacts of Folsom and Clovis Man; to stimulate cooperation among anthropologists, paleontologists, and stratigraphers; and to focus the attention of scientists from other countries on the importance of the work being done in North America."[56] The committee in charge included the chairman, John C. Merriam, president of the Carnegie Institution of Washington; George Grant MacCurdy from Yale University; Edwin G. Conklin of the American Philosophical Society; Hellmut de Terra; and Edgar B. Howard, a trustee of the Academy and secretary of the conference.[57] Both Howard and de Terra believed that if the symposium was based on a wide variety of subjects, all of which dealt with early man, it would be the only time in the history of science that such a symposium had taken place.[58]

Preparations for the symposium, to be held 17–20 March 1937, began in earnest in the spring of 1936. A reconstruction of Clovis artifacts, such as fluted spear points and skin scrapers in association with mammoth bones that had been found in situ in 1936 by John L. Cotter, Howard's principal assistant at Clovis, was to be the foremost part of an exhibition in conjunction with the conference (figure 10.8). Cotter recorded that he "set to work improvising a portion of the Mammoth Pit [from Clovis] on the floor of the Hall of Man in the re-vamped main exhibit area of the Academy, transformed from Victorian décor of wood and

10.7 Attendees at the Academy's International Symposium on Early Man view a reconstruction of the Clovis excavation site in the Academy's first-floor exhibition hall, 1937. ANSP Archives coll. 2011-003.

glass cabinets around a central well (above which hung a whale replica) into an art deco designed décor. [The whale replica was removed.] My work was to position the major mammoth bones exactly as they had been found with relation to each other." Delaware River sand was the closest match he could find to the Clovis speckled sand. "Miraculously," he observed, "in those days, none [of the artifacts] was missing at the end of the Symposium."[59]

Howard wrote to Cadwalader that fall: "It is encouraging to know that my efforts, such as they are, have been appreciated, and I feel individually like I suppose one of the parties of the British Government feels when it has been given a vote of confidence by Parliament."[60]

Thirty-six world-famous scientists from Europe, China, Java, India, and South Africa were invited to give papers at this first international congress ever held on early man. Invitations were sent to geologists, archaeologists, physical anthropologists, professors, graduate students, and others. It was a fascinating story that was to be told of man's forebears: from Peking Man, who lived approximately a million years ago and was the oldest human of whom any substantial record had been found; the Java Ape Man; the Piltdown Man of England, who lived about five hundred thousand years ago (much later found to have been a hoax); the Neanderthal Man of Europe; the Folsom Man, who inhabited the southwestern part of North America probably ten thousand to fifteen thousand years ago; down to the Neolithic Man and the Swiss Lake Dwellers, who lived a mere eight thousand years ago.[61]

Telegrams from scientific luminaries began to arrive in October. On the 9th, Western Union brought word from Father Pierre Teilhard de Chardin in China that he was "probably coming."[62]

10.8 Edgar B. Howard and John L. Cotter preparing specimens for the Academy's early man exhibition, 1937. ANSP Archives coll. 2011-003.

10.9 (From left) Oswald Menghin, V. Gordon Childe, Gustav Heinrich Ralph von Koenigswald, and Pierre Teilhard de Chardin at the Academy's International Symposium on Early Man, 1937. ANSP Archives coll. 422.

Gustav Heinrich Ralph von Koenigswald wired from Bandoeng, Java, that he had booked his passage to Marseille and asked the Academy to arrange for the rest of his journey.[63] By mid-March, the arrivals began. Dorothy A. E. Garrod, from Newnham College, Cambridge, known for her important work on prehistoric man in the Near East and on Mount Carmel in the Holy Land, disembarked from the *Berengaria*. On the same ship was Kaj Birket-Smith of the Nationalmuseet, Copenhagen, a world authority on Eskimo culture. Oswald Menghin, professor of prehistoric archaeology at the University of Vienna, who had led expeditions in the Nile Valley and northern Africa, arrived on the *Europa*, accompanied by V. Gordon Childe, professor of archaeology at the University of Edinburgh (figure 10.9).[64]

The scientists were all requested to bring genuine artifacts or casts to be displayed in the Academy's Hall of Early Man. In a sign of the times that today would be considered taboo for its inference, the Academy stated in a news release: "Pre-historic Solo Woman, whose skulls Dr. von Koenigswald brought with him [from Java], can be recognized like living women by the lighter skull and smaller brain than that which her contemporaneous mate possessed." Without apparent irony, the next sentence read, "Well known women scientists and field workers who are taking an active part in studying the ancestry of the human race will be well represented among the delegates to the Academy's international symposium." The release continued: "Dr. Ruth Patrick [who would achieve world renown for her scientific accomplishments] will present in the 'Hall of Early Man,' an exhibit of Diatomes, minute unicellular algae that visitors to the Symposium will be able to study under microscope."[65]

Probably the best-known scientist to attend the symposium was Teilhard de Chardin (1881–1955), a French philosopher and Jesuit priest who trained as a paleontologist and geologist and took part in the discovery of Peking Man.[66] Consulting paleontologist to the National Geological Survey of China, Teilhard de Chardin was the first person with firsthand knowledge of Peking and Java Man to view the South African man-apes and the first to hypothesize that Africa rather than the Far East might be the birthplace of man.[67]

The symposium generated much excitement among the press, as 435 delegates from twenty-seven of the United States, Canada, England, Denmark, Norway, France, Austria, South Africa, China, Java, Hungary, Jamaica, and Australia attended meetings and roundtable discussions. A special newsroom was set up at the Academy with major newspapers and magazines represented: the *Philadelphia Inquirer*, *Evening Bulletin*, *New York Times*, *New York Herald-Tribune*, *Christian Science Monitor*, Associated Press, *Time*, and *Life*. A number of conference participants were questioned as to their views on primitive man (figure 10.10).

No doubt the views of the universally known Teilhard de Chardin were of special interest. It had been only twelve years since the famous Scopes Trial in Tennessee (1925), where John Scopes, a high school biology teacher, was tried for teaching evolution. According to a reporter from the *Philadelphia Bulletin*, after discussing Peking Man, or *Sinanthropus pekinensis*, whose beetle-browed, low-vaulted skulls had been found in a cave near Peking (present-day Beijing), and who Teilhard de Chardin judged to be three hundred thousand years old, the eminent priest was asked how he reconciled this scientific picture of man's early appearance with the religious one held by his church.

Teilhard de Chardin said that he had no difficulty at all with this apparent conflict. "The two most important events in the whole world were the very beginning of life and later the appearance of thought," he said. He suggested a sort of boiling point in evolution. "Just as water slowly warms up and suddenly changes into steam when it reaches the boiling point, so did life reach a critical point at which instinct changed into thought."[68] The *New York American* quoted him as saying: "Man must be considered as descended from the ape. In these discoveries there need be no difficulty for Christianity. As a scientist, I must admit the evidence that man was born from the animal kingdom."[69]

When reporting on the women scientists at the symposium, the tone was considerably lighter. "Dr. Ruth Patrick likes her tennis," stated the *Evening Ledger*, "but it is difficult to distract her attention from diatoms. She is now working on diatoms for the Brazilian Government and recently finished research on Siamese diatoms and their geographical distribution." The article continued that "two western girls, glowing with health and enthusiasm, 'bussed' it part of the way from Denver and found it great fun."[70] One of these "girls" was the curator and the other her assistant at the Colorado Museum of Natural History.

As president of the Academy,[71] Cadwalader invited several scientists to stay with him during the conference. Since they did not know him, he described himself as a fifty-two-year-old bachelor who lived in the country and traveled to work each day by automobile, received no salary from the Academy, got up when he was ready, went to bed when sleepy, and liked good food. "I am not partial to large and complicated dinners or other forms of entertainment, as I prefer to use up my energies in the Academy," he said.[72] Robert Broom, from Scotland, replied in kind. He said

10.10 These busts of prehistoric man, created by J. H. McGregor between 1914 and 1919, were exhibited at the Academy's International Symposium on Early Man, 1937. ANSP Exhibits Collection. K. Clark photograph.

he was seventy and married with a wife in South Africa. And in a sentence relevant to the late 1930s, in which a political storm of vast proportions was building abroad, he added: "Am neither a Nazi nor a Fascist nor a Communist—play a fair game of chess and checkers . . . but do not dance. My other failings you will find out in due course."[73]

During the symposium, the University of Pennsylvania invited delegates and visitors to a reception at which the keynote lecture, "The Pioneers: Leidy, Cope and Osborn," centered on the Academy's past in archaeology. Before the lecture, there was an academic procession and the awarding of honorary degrees to several of the symposium speakers, including Dorothy Garrod from Cambridge University.

The symposium was an enormous success. George Harrison, an Academy trustee and benefactor, wrote Cadwalader that he had "heard many say that it was the greatest thing of its kind that had ever been done in this country." Theodore McCown wrote from England: "Of the various scientific meetings which I have attended, this one was by far the most successful and the best organized. It has made a real contribution to science and to increasing the interest generally in what is a fascinating subject." And Frank B. Foster, another trustee and benefactor, observed: "I think the Symposium did more for the Academy than anything that has ever been done for it."[74]

Aside from all the success and the praise, in the following years there was no continuation of the study of early man at the Academy. And there was no follow-up to Howard's idea of establishing a Department of Archaeology and Anthropology at the Academy. It was nearly two years after the symposium, at the beginning of 1939, when Howard wrote to Cadwalader:

> I have been at a loss recently to discover just what the objectives of the Academy were supposed to be, and as a result it has become increasingly clear to me that no strong position could be maintained in formulating plans for the future, either of the department, or of any individual research, such as that in which I am interested.
>
> Constructive criticism offered by me towards the attainment of a definite policy is construed by you as complaint. This is somewhat discouraging in view of all the hard work that I have put in for the past several years in the interests of the Academy. Domination of all the activities of the Academy by one man, moreover, completely stifles all initiative of an individual nature, and blocks the formation of a comprehensive plan which I believe to be so badly needed by the institution.
>
> Therefore, after mature consideration, it seems best for me to seek some more congenial connection to carry on my work, but it will always remain one of the regrets of my life that we could not build together the greater Academy that was looming up in front of us after the Symposium, the opportunity to do which appears to have slipped past us.[75]

Howard's disappointment with Cadwalader for not moving the Academy ahead in a more scientific direction was reminiscent of Edward Cope's feelings toward W. S. W. Ruschenberger, when the latter was Academy president many years earlier.

John L. Cotter, who was for many years Howard's friend and colleague at the University of Pennsylvania Museum of Archaeology and Anthropology, said in his remembrance of Howard that Howard and Cadwalader had "a mutual antipathy." He recounted that "Howard's final break with the Academy came after a violent argument with Cadwalader, after which he stormed back to his office, cleaned out his desk, and left word with [his secretary] that he would never set foot in the Academy again."[76] And he never did. Sadly, four years later, at the early age of fifty-six, he died of a heart attack in California, where he had gone to live.

Since Howard's time, Clovis projectile points have been discovered all over North America, from Nova Scotia to Mexico, and similar stone points have been found in the Andes and Tierra del Fuego in South America. As Howard discovered, Clovis campsites typically contain the remains of prey that became extinct at the end of the Pleistocene, such as mammoths, mastodons, long-horned bison, and horses.[77] Because of his dedicated research into the origins of early man in America and his idea for, and involvement in, an international conference of like-minded scientists, Howard left an impressive legacy at the Academy. The symposium gave the Academy its day in the sun in 1937, with a historic meeting that brought recognition to the institution from around the world.

PTS

Notes

1 J. Alden Mason, "Edgar Billings Howard, 1887–1943," Society for American Archaeology: *American Antiquity* 9, no. 2 (October 1943): 230–34.

2 Anthony T. Boldurian and John L. Cotter, *Clovis Revisited: New Perspectives on Paleonindian Adaptations from Blackwater Draw, New Mexico* (Philadelphia: University Museum of the University of Pennsylvania, 1999), 1.

3 Ibid., 1.

4 Ibid., 2. Hrdlička was the first curator of physical anthropology at the U.S. National Museum, now the Smithsonian Institution Museum of Natural History.

5 Ibid., 5.

6 Effingham B. Morris was Academy president (1928–37) at the time, but he wanted someone else to handle the day-to-day operation of the society; thus the position of managing director was instituted.

7 Edgar B. Howard to Charles Cadwalader, Carlsbad, New Mexico, 18 August 1830, ANSP archives, Cadwalader papers, coll. 331.

8 John L. Cotter, "Recollections of Dr. Edgar B. Howard," University Museum Archives, June 1992.

9 Boldurian and Cotter, *Clovis Revisited*, 8.

10 Howard to Jayne, Raton, New Mexico, 4 September 1931, University Museum Archives.

11 Boldurian and Cotter, *Clovis Revisited*, 8.

12 Howard to Barnum Brown, Philadelphia, 15 March 1932, ANSP archives, Edgar B. Howard papers, coll. 331.

13 Howard memo to C. M. B. Cadwalader, October 1932, ANSP archives, Edgar B. Howard papers, coll. 331.

14 Ibid.

15 Ibid.

16 Howard to Cadwalader, Philadelphia, [Oct. 1932], ANSP archives, Cadwalader papers, coll. 331.

17 Anthony T. Boldurian, "Clovis Type-Site, Blackwater Draw, New Mexico: A History, 1929–2009," *North American Archaeologist* 29, no. 1 (2008): 67.

18 Boldurian and Cotter, *Clovis Revisited*, 11.

19 Ibid.

20 Howard to Horace H. F. Joyce [*sic*], Clovis, New Mexico, 16 November 1932, University Museum Archives. Also in Boldurian and Cotter, *Clovis Revisited*, 11.

21 Howard to Cadwalader, Clovis, New Mexico, 4 June 1933, ANSP archives, Edgar B. Howard papers, coll. 331.

22 Sir Arthur's keynote address to the congress triumphantly reviewed the proof that Piltdown Man was an early Pleistocene human. Piltdown Man was later found to have been a hoax. For more on the hoax, see Stephen Jay Gould, *The Panda's Thumb* (New York: W. W. Norton, 1980).

23 David J. Meltzer, *First Peoples in a New World: Colonizing Ice Age America* (Berkeley: University of California Press, 2009), 240.

24 Ibid., 241.

25 Ibid., 242.

26 Boldurian and Cotter, *Clovis Revisited*, 12–13. A touchstone in American archaeology, the Clovis site was designated a National Historic Landmark in 1981. In 2007, Blackwater Draw was nominated for the UNESCO World Heritage List. It could join Mesa Verde in Colorado and the Cahokia Mounds in Illinois as outstanding examples of cultural and natural heritage.

27 ANSP minutes, 31 October 1815, coll. 502. The *Journal of the ANSP* describes this donation as a "Glass model of a Mummy, found in the Pyramids of Egypt; brought thence by Denon [Napoleon's appointed director of the Louvre]," vol. 1, pt. 1, no. 6 (1817): 214.

28 *ANSP Journal*, vol. 1, pt. 1, no. 6 (1817): 214 and pt. 2 (1818): 501.

29 *Proceedings of the ANSP* 67, no. 2 (1915): 199–202.

30 W. S. W. Ruschenberger, "An Address: The Claims of the Academy of Natural Sciences of Philadelphia to Public Favor" (Philadelphia: Academy of Natural Sciences, 1871).

31 Biographical material for Moore comes from H. Newell Wardle, "Clarence Bloomfield Moore," *Bulletin of the Philadelphia Anthropological Society* 9, no. 2 (1956): 9–11; David Hurst Thomas, "C. B. Moore," *Anthropology, Science, and Archaeology* (1979): 30–33.

32 Clarence Moore to Edward Nolan, Philadelphia, 10 October 1897, ANSP, Clarence B. Moore papers, coll. 567.

33 *Proceedings of the ANSP* 45 (1893): 561.

34 *Proceedings of the ANSP* (1899): 539.

35 Lawrence M. Small, "A Passionate Collector," *Smithsonian Magazine*, November 2000.

36 Moore to Witmer Stone, Philadelphia, 22 June 1923, ANSP archives, Clarence B. Moore papers, coll. 102, file 4.

37 Frank James Keeley (1868–1949), mineralogist and microscopist, was an employee and a member of the Academy, elected to membership in 1894.

38 Clarence Moore to Frank James Keeley, 28 June 1923, ANSP archives, coll. 102, folder 4.

39 In 1925, the council's managerial responsibilities were replaced by a newly created board of trustees.

40 Resolutions of a special meeting of the ANSP Council, 25 June 1923, ANSP archives, coll. 6, folder 2.

41 Moore to Henry Pilsbry, St. Petersburg, Florida, 22 March 1929, ANSP coll. 102, file 3.

42 Western Union telegram, Moore to Cadwalader, St. Petersburg, Florida, 1 April 1929, ANSP coll. 102, file 3.

43 Moore to Heye, St. Petersburg, Florida, 11 April 1929, ANSP coll. 102, file 3.

44 Heye to Moore, 15 April 1929, New York, 15 April 1929, ANSP coll. 102, file 3.

45 Ibid.

46 Moore to Heye, 18 April 1929, ANSP coll. 102, file 3.

47 H. Newell Wardle, letter to the board of trustees, 6 May 1929, ANSP coll. 102, file 3.

48 Wardle to the Academy trustees, 25 May 1929, ANSP coll. 102, file 3.

49 Heye to Moore, New York, 9 May 1929, ANSP coll. 102, file 3.

50 H. Newell Wardle, "Wreck of the Archaeological Department of the Academy of Natural Sciences of Philadelphia," *Science* (2 August 1929): 119–20. Robert Henry Lamborn (1835–1895) was a wealthy industrialist and friend of Andrew Carnegie's, who collected ancient Indian pottery and other ethnological artifacts. He left his entire estate to the Academy of Natural Sciences for the advancement of biology and anthropology. ANSP biography file.

51 Wardle, "Wreck," 120.

52 Ibid. Wardle went on to work for the Museum of Archaeology and Anthropology of the University of Pennsylvania. For a complimentary obituary of Wardle, see J. Alden Mason, *American Anthropologist*, n.s., 67, no. 6, pt. 1 (December 1965): 1512–15.

53 Edgar Howard to Charles Cadwalader, Clovis, New Mexico, 4 June 1933, ANSP coll. 331.

54 Howard to Cadwalader, n.d. (1934?), ANSP coll. 331.

55 ANSP biography file.

56 ANSP memo to the board of trustees, 10 December 1936, Cadwalader files.

57 ANSP unprocessed collection, Cadwalader files, 1936. Folder labeled Symposium on Early Man.

58 Cadwalader to Kenneth Rose, 2 October 1936, ANSP folder on Symposium on Early Man.

59 Cotter, "Recollections."

60 Howard to Cadwalader, 9 November 1936, acc. 02:22, file 36, box 12.

61 *International Symposium: The Academy of Natural Sciences of Philadelphia*, ed. George Grant MacCurdy (Philadelphia: J. B. Lippincott Company, 1937), jacket text.

62 ANSP archives symposium file.

63 G.H.R. von Koenigswald, Western Union telegram, 24 October 1936, ANSP symposium file.

64 ANSP news release, 15 March 1937, Symposium file. Oswald Menghin (1888–1973) was a professor at the University of Vienna before World War II. His work on race and culture was used by the German Nationalist Movement of the 1930s. After the war, he was listed on the primary list of war criminals, but charges against him were dropped in 1956. He died in Buenos Aires, Argentina.

65 ANSP news release, 11 March 1937, ANSP archives, box 14, file 72.

66 Teilhard de Chardin's primary book, *The Phenomenon of Man,* set forth a sweeping account of the unfolding of the cosmos, which displeased certain officials in the Roman Curia and in his own order, who thought that his ideas undermined the doctrine of original sin. His writings were denied publication during his lifetime by the Roman Holy Office, but in 2009, the pope praised Teilhard de Chardin and his work.

67 Russell L. Ciochon, "The Search for Fossil Man in Asia," *The Explorer: Bulletin of the Cleveland Museum of Natural History* 18, no. 4 (1976): 7–17.

68 Reprint from the *Philadelphia Bulletin*, "Early Man News," Academy of Natural Sciences of Philadelphia [1937]. ANSP Cadwalader files, Symposium on Early Man.

69 At the time of the early man symposium, it was thought that man had evolved through a direct lineage from the apes. It is now believed that the linkage is much more complex, with certain lineages dropping off altogether (extinct), and new ones arising, much like an ever-branching tree.

70 Reprint from the *Evening Ledger* of Philadelphia, "Early Man News," Academy of Natural Sciences of Philadelphia.

71 President Effingham B. Morris retired in January 1937, and Charles Cadwalader was named president of the Academy.

72 Cadwalader letters to Dr. Robert Broom and Theodore McCown, 8 February 1937. Cadwalader unprocessed file, ANSP coll. 2009-034.

73 Dr. Robert Broom to Cadwalader, 3 March 1937, Cadwalader unprocessed file, ANSP coll. 2009-034.

74 Cadwalader unprocessed file, ANSP coll. 2009-034. Frank Foster gave the magnificent Lion diorama to the Academy.

75 Howard to Cadwalader, Philadelphia, 3 January 1939, ANSP, Cadwalader file.

76 Cotter, "Recollections."

77 Philip Kopper, *The Smithsonian Book of North American Indians* (Washington, DC: Smithsonian Institution Press, 1986), 37–38.

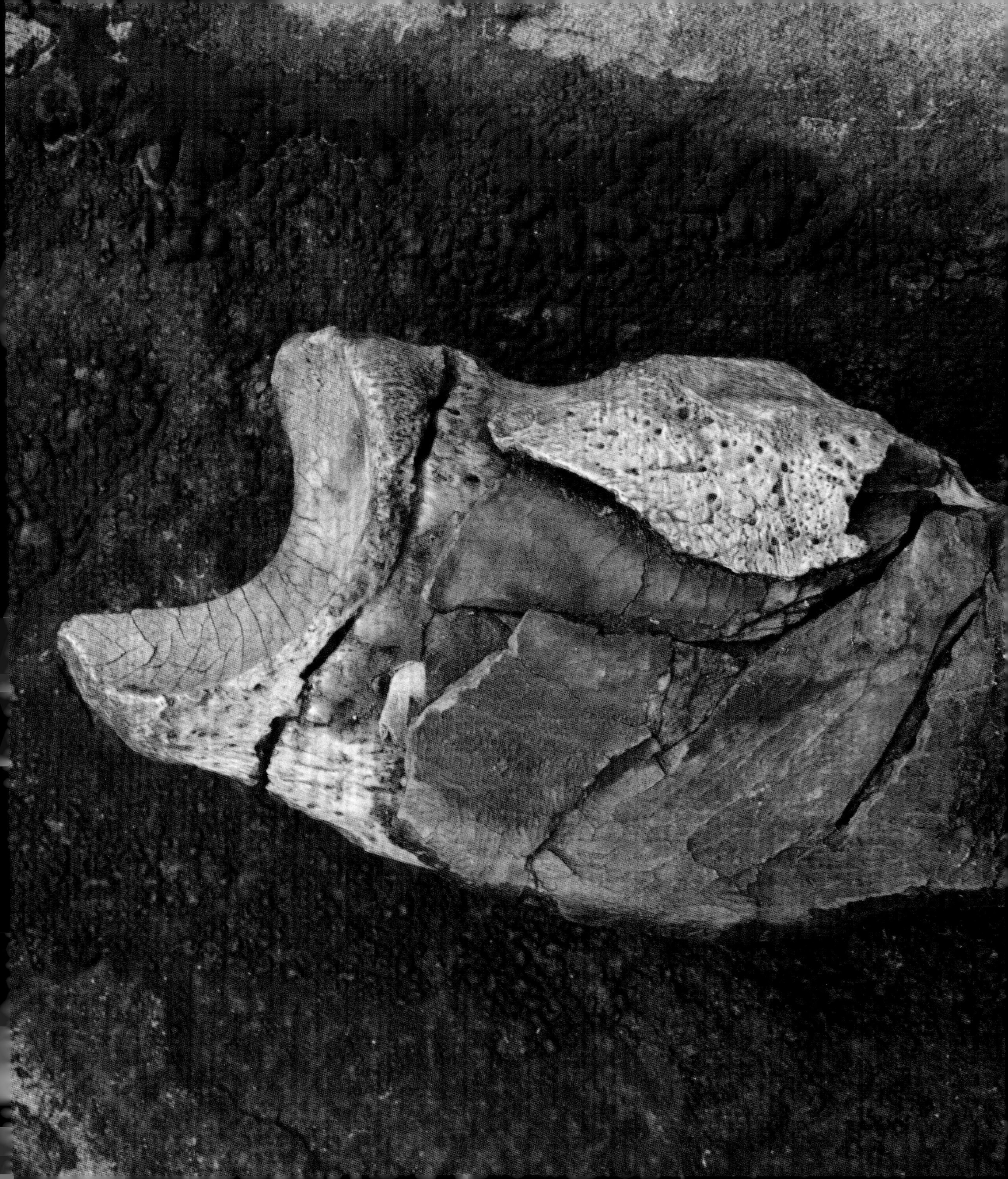

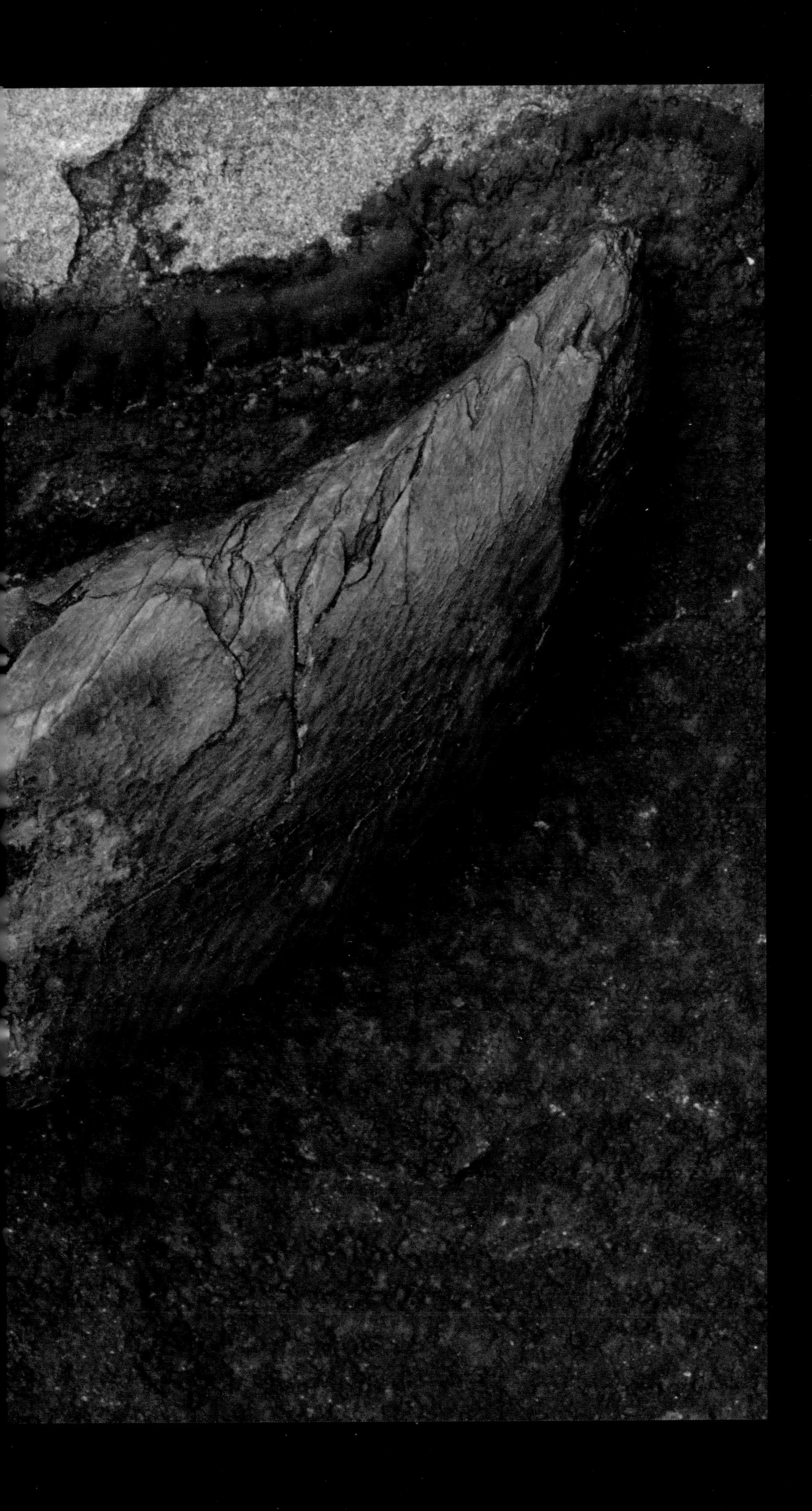

Fossilized claw of a giant ground sloth (*Megalonyx jeffersoni*) from Big Bone Cave, Tennessee. J. P. Wetherill Collection. ANSP Vertebrate Paleontology Department #12487. Purcell photograph.

Chapter 11

Volcanoes to Caverns: Exploring for Minerals

The heavens were aglow with fire, electric flashes of blinding intensity traversing the recesses of black and purple clouds, and casting a lurid pallor over the darkness that shrouded the world.
—Angelo Heilprin, Academy mineralogist, *Mont Pelée and the Tragedy of Martinique* (1903)

In May 1902, Angelo Heilprin (1853–1907),[1] the brilliant and courageous curator of mineralogy at the Academy of Natural Sciences, traveled to the Caribbean island of Martinique to investigate a devastating volcanic eruption that had occurred nearly a month earlier. In the book that grew from his research, *Mont Pelée and the Tragedy of Martinique* (1903), Heilprin tells of the fateful morning in 1902 when Mont Pelée exploded, swallowing the town of Saint Pierre in clouds of superheated poisonous gas and killing more than thirty thousand inhabitants in a matter of seconds. Accompanied by American journalist George Kennan, four local scientific investigators, and three boys to attend to their donkeys and equipment, Heilprin ascended the 4,583-foot mountain, later describing the experience as his party neared the crater:

> Across the steaming lake-bed, little mindful of its puffs of vapor and sulphur, we dashed to the line above which welled out the steam-cloud of the volcano, and almost in an instant stood upon the rim of the giant rift in whose interior the world was being made in miniature. We had reached our point. We were four feet, perhaps less, from a point whence a plummet could be dropped into the seething furnace, witnessing a scene of terrorizing grandeur which can be conceived only by the very few who have observed similar scenes elsewhere. Momentary flashes of light permitted us to see far into the tempest-tossed caldron, but at no time was the floor visible, for over it rolled the vapors that rose out to mountain heights. With almost lightning speed they were shot out into space, to be lost almost as soon as they had appeared.[2]

On his return, Kennan noted his admiration for Heilprin's utter disregard of imminent danger as they watched the raging, exploding Mont Pelée just a few feet away: "I must pay the highest possible tribute to Heilprin," he wrote. "He is modest and brave, a superb mountaineer and the nerviest and pluckiest man I ever knew. The ascent was the most terrifying experience of my life" (figure 11.2).[3]

Heilprin recalled that many years earlier, he had climbed Vesuvius, "gazed into the crater funnel, and watched the molten magma of the earth rise and fall," but "grand and inspiring though [Vesuvius] was," the scene could not compare in fury with Mont Pelée.[4]

He returned to Martinique for a second time in August of the same year to continue his studies and found that living conditions on the island had greatly improved since his first visit. A large number of refugees were back in their homes in Morne Rouge (Red Bluff), Grande Rivière, and other surrounding

11.1 Marble specimen resembling the outline of a cityscape, from Florence, Italy. Seybert collection, ANSP Mineralogy Department. Purcell photograph.

Keystone View Company,
Manufacturers and Publishers.
Meadville, Pa. St. Louis, Mo.
Copyright 1903, by B. L. Singley.
14344—Southeast from the Barren Slopes of Pelée above St. Pierre, Martinique, F. W. I., Professor Heilprin in Foreground.

320
Keystone View Company
Copyright 1903, by B. L. Singley. Made in U. S. A.
Manufacturers Publishers
Meadville, Pa., New York, N. Y., Portland Oregon, London, Eng., Sydney, Aus.
Natives Fascinated by the Fierce and Magnificent Sight of a Volcanic Eruption, Gros Morne, Martinique, F. W. I.

towns. The government, under the advice of a scientific commission, had ordered the repopulation of the deserted country, even though the volcano was far from dormant.[5]

Again Heilprin ascended Mont Pelée, which, he said, was raging with "a continuous roar that was simply appalling. I thought on my previous ascent to have heard something [very loud], but this time it was the old sound multiplied a hundred-fold. No words can describe it. Were it possible to unite all the furnaces of the globe into a single one, and to simultaneously let loose their blasts of steam, it does not seem to me that such a sound could be produced."[6]

While at dinner that night after the climb, Heilprin and others who were staying at the same inn saw a flash of lightning and heard a "dull thud." "We were out at once," he recorded:

> The heavens were aglow with fire, electric flashes of blinding intensity traversing the recesses of black and purple clouds, and casting a lurid pallor over the darkness that shrouded the world. Scintillating stars burst forth like crackling fireworks, and serpent lines wound themselves in and out like travelling wave-crests. The spectacle was an extraordinary and terrifying one, and I confess that it left an impression of uncomfortable doubt in our minds as to what would be the issue.[7]

In less than a minute, the "tornadic blast" from Mont Pelée had obliterated the town of Morne Rouge, its surrounding villages, and the rich vegetation Heilprin had so recently seen. An estimated two thousand people died in this second fatal explosion.[8]

On his visit to Martinique earlier in the year, Heilprin had traveled to the town of Morne Rouge and admired the lush countryside along the road, which "exhibited an enchanting display of tropical vegetation of palms, tree-ferns and bamboos, of heliconias, melastomes and rubber-trees—of giant foresters, cased in cables and creepers . . . of star-massed epiphytes and orchids, and great bursts of scarlet and blue blossoms." He had followed along "deep barrancas musical with their tumbling waters, and shrouded beneath an almost impenetrable maze of foliage."[9] Now all was devastation.

Heilprin's interest in the Caribbean continued a long tradition of West Indian investigations at the Academy, first recorded in a pioneering essay on the geology of the islands by William Maclure, published in the inaugural issue of the Academy's *Journal* in 1817. Maclure had explored most of these Caribbean islands with Charles-Alexandre Lesueur on his return journey from Europe in 1816. In his essay, Maclure detailed the rock formations of each one he had visited, including "Martinico" (Spanish for Martinique).[10]

From the Academy's beginning, minerals had been an important area of study at the institution. Gerard Troost (1776–1850), a Dutch scientist who had been trained in medicine, chemistry, and mineralogy, initiated the interest. Troost, who was born in s'Hertogenbosch (Bois-le-Duc), Netherlands, left his native town in 1795 when French revolutionary troops captured it. When an alliance was concluded that spring between the Netherlands and France, Troost served first as a foot soldier and then as a surgeon in the French-dominated Dutch army. By 1802, he was practicing pharmacy at The Hague and was well known as a mineral collector, which must have helped to secure him the job at the natural history museum in Paris where he became manager of the royal mineral collection of Napoleon's brother, Louis Bonaparte (king of the Netherlands from 1807 to 1810). There for three years, Troost was fortunate to be the protégé of Abbé René

FACING PAGE 11.2 Top: Angelo Heilprin (1853–1907), at left, pointing to Mont Pelée, Martinique. Bottom: Mont Pelée erupting with five people in foreground. Hand-colored stereographs. ANSP Archives coll. 147.

11.3 *William Maclure* (1763–1840), father of American geology, by Charles Willson Peale, 1818. Oil on canvas. ANSP Library coll. 286.

Just Haüy, the father of crystallography, who was established at the museum. A brilliant polymath with competence in Latin, Greek, Dutch, French, English, and German, Troost translated Alexander von Humboldt's famous 1807 treatise *Ansichten der Natur* (Views on Nature) into Dutch in 1808.

With the collapse of the Dutch monarchy imminent in 1810—Napoleon would annex the Kingdom of Holland to the French Empire, thus ending Louis's reign as king—Troost left France for the United States and settled in Philadelphia.[11] He brought with him a newly invented goniometer for measuring angles in crystals, a mineral collection, and a thorough knowledge of the latest methods in studying mineralogy.[12]

As the first president of the Academy of Natural Sciences (1812–17), Troost strongly influenced the society's study of minerals from the outset. Barely a month after the Academy's founding in March 1812, he proposed a mineralogical excursion to the mines of the Perkiomen Creek, some twenty-two miles outside Philadelphia.[13] In 1806, Adam Seybert (1773–1825), the first American trained in mineralogy, had announced the occurrence of "sulfhuret of zinc" (zinc blende) in this part of Montgomery County and had demonstrated that it would yield zinc metal, the discovery that led to the opening of the Perkiomen Mines.[14] In the years to come, Academy members would find many other fascinating specimens in these mines.

That summer of 1812, another founder, John Speakman, with an agreement from Academy members that he would be repaid, negotiated for the institution the purchase of Seybert's important collection of 1,825 minerals, including vials of ash and sulfur and samples of pumice and lava from Mt. Vesuvius (figure 11.4).[15] Seybert, a Philadelphia physician, had also studied at the École des Mines under Abbé Haüy.[16] Gerard Troost may have known Seybert through their mutual acquaintance with the brilliant cleric.

By the time of the transaction, Seybert had turned his attention to politics. Twice elected to the U.S. Congress, he nevertheless continued to work on his mineral collection and prepared a handwritten catalog of it shortly before his death in 1825. His son, Henry Seybert (1801–1883), gave his own collection of minerals and apparatus to the Academy about the time his father died.[17]

So important were Seybert and his collection to American mineralogy that Benjamin Silliman, newly appointed professor of chemistry and natural history (primarily meaning mineralogy) at Yale College, had journeyed south by stagecoach in 1802 on a five-month leave of absence to study mineralogy in Philadelphia, "then the scientific powerhouse of the nation."[18] Silliman, whose particular purpose was a visit to Seybert, carried with him a candle box containing Yale's entire mineral collection for Seybert to identify.[19]

The purchase of the Seybert collection enabled Troost to offer a series of lectures on minerals for Academy members and their friends in 1815, thus helping to raise the prestige of the fledgling institution. There may also have been an additional advantage. According to a later Academy librarian, Edward Nolan, "It is a tradition that the necessity of discharging the indebtedness incurred by the purchase of the Seybert minerals, was a bond of union during the first year or two of struggle and discouragement" (figure 11.5).[20]

The Seybert collection and its original, custom-built cabinet can still be seen at the Academy. It is the earliest American mineral collection to have survived intact.[21]

In another important transaction, in 1816, the newly elected Academy member Joseph Watson (d. 1841) purchased the mineral collection of Silvain Godon (ca. 1769–1840) for the society. Godon, born in Paris and also trained under Abbé Haüy, immigrated to the United States in 1807. François André Michaux (1770–1855), an eminent French botanist and one of the first corresponding members of the Academy, wrote to a friend about

11.5 *Dr. Edward J. Nolan,* Academy librarian, by Thomas Eakins, ca. 1900. Philadelphia Museum of Art. Gift of Mrs. Thomas Eakins and Miss Mary Adeline Williams, 1929.

FACING PAGE 11.4 Drawer of minerals from the Seybert Collection, purchased by Academy members in 1812. Cabinet dated 1825. ANSP Mineralogy Department #124.

Godon: "Mineralogy is for him what botany is for me. He is certainly one of the ablest mineralogists who have ever come to this country where he counts on locating, being very discontented with France. He is laden with introductions."[22] Godon arrived in Philadelphia bearing letters from, among others, William Maclure.[23]

Elected to the American Philosophical Society, Godon gave lectures on mineralogy around the Boston area[24] and attempted various scientific and business ventures that proved unsuccessful. By 1812, he was in debtors' prison and forced to auction off his minerals and personal effects in order to pay his creditors. Shattered by his reverses, Godon subsequently suffered a mental breakdown and spent the rest of his life in an insane asylum.

When Godon's collection was sold in 1812, Benjamin Smith Barton (1766–1815), the noted Philadelphia botanist who had been one of the scientists to train Meriwether Lewis before his western expedition with William Clark, outbid the newly established Academy. But three years later, when Barton died, Joseph Watson bought Barton's entire collection for the Academy. Included were Godon's minerals and those of Barton's father and older brother.[25]

In a sad twist of fate, in the same year that Watson bought the incarcerated Godon's collection for the Academy (1816), Governor Wilson C. Nicholas of Virginia asked Thomas Jefferson to name someone who could undertake a "mineralogical survey" of his state. Jefferson replied, "I have never known in the United States but one eminent mineralogist, who could have been engaged on hire. This was a Mr. Goudon [*sic*] from France, who came over to Philadelphia six or seven years ago. . . . It is long since I have heard his name mentioned, and therefore do not know whether he is still at Philadelphia, or even among the living."[26]

With the arrival of William Maclure, who had been elected to Academy membership in absentia in June 1812 but did not appear at a meeting until the spring of 1816, interest in mineralogy and geology at the Academy was given a further boost. Maclure has been called the father of American geology for his extraordinary feat, many years before his report on the Caribbean Islands, of completing a one-man geological survey of the eastern part of the United States, from Maine to Georgia and from the Atlantic Ocean to the Mississippi River.[27]

In 1808, Maclure produced the first geological map of America.[28] In his publication, he used terminology employed by many European geologists of the period for describing rocks: primary (rocks set in place by the Prime Mover), secondary (sediments much altered and containing fossils), diluvial (sediments put down at the time of Noah and the Great Flood), alluvial (later sediments created by erosion and deposited by rivers and seas). These terms, first employed by Abraham Gottlieb Werner (1750–1817) at the School of Mines in Freiburg, Germany, had gained wide acceptance in Europe. Like many of his counterparts, Werner believed that water had created planet Earth.[29]

A conflicting theory of the earth's formation was held by James Hutton (1726–1797), a Scottish genius who proposed that instead of progressing inexorably to an end, the earth might be in a rough sort of balance, the erosion of older rocks being matched by the uplift of new sediments and the products of forces such as volcanoes depositing ash and lava. Through this process, older rocks containing fossils, once in shallow seas, were raised up in mountains. Therefore, the earth was constantly renewing itself.[30]

One of the key questions of geology in the late eighteenth century was, how did rocks, which are made up of so many different minerals, form? Hutton believed it was the heat within the earth that was involved in mineralization, and most importantly, heat with extreme pressure. He was able to show that granite, an igneous rock, is formed from molten magma inside the planet and that it forms after, not before, the sediments surrounding it. Werner believed that all sediments had been formed by the universal ocean and laid down on primitive igneous rock.[31]

Werner's theory of water creating the earth, called Neptunism, and Hutton's theory, called Plutonism, produced much discussion among early geologists. According to science historian Keith Thomson, the only real answer to the question of the earth's formation is that both water and heat have been, and continue to be, involved in the formation and evolution of the earth's crust. It was not until the 1950s that scientists found evidence for plate tectonics, responsible for continental drift and the folding and faulting of massive portions of the earth's surface.[32]

At Maclure's first Academy meeting in 1816, he presented to the assembled members a huge (nine feet high by six feet across), expensive, and very important geological map of England, Wales, and part of Scotland, which had been published by William Smith the summer before. Using his own extensive exploration of the geology of European countries, Maclure also read at the meeting his paper on the "structure and original formation of Rocks," which was later published in the *Journal*.[33]

William Smith (1769–1839), an Englishman whose map earned him the popular title "Strata Smith," was a surveyor who drained marshes and built canals in England during the Industrial Revolution. His employment as a coal miner fueled his

fascination with geology and the minerals in rocks, and he studied the various sediments he saw in the mines. Smith realized that geology was a science requiring observations in three dimensions. He took copious notes on what he saw and came to the conclusion that all the rocks that had been laid down as sediments at a particular time had many of the same characteristics. They often contained the same fossils and always appeared in a vertical order, no matter where they were found. He believed that fossils would be the key to working out the order and to forecasting the succession of beds underground, and he further speculated that this predictability of strata could be a universal phenomenon.[34]

In the United States, because of Maclure's gift of Smith's map, Thomas Say no doubt knew of the Englishman's pioneering work. Presumably assisted by Smith's publication and from his own experience with geology, Say postulated that through knowledge of fossils in the sediments found in America, one could determine the ages of rock strata in the same way. He was the first in the New World to publish an article on this theory.[35]

At the time, for philosophical and religious reasons, there was immense interest in understanding the structure of the earth. The historian Simon Winchester has written that as the ideas of the eighteenth-century Enlightenment spread "in countless ways, both great and small, the faiths and certainties of centuries past were being edged aside, and the world was being prepared, if gently and unknowingly, to receive the shocking news of scientific revelation."[36] Thus William Smith's geological findings in England played a major part in preparing the way for Lyell and Darwin.

This awakening to the realities of the earth's structure was particularly concentrated in Philadelphia, the American center of science and culture in the late eighteenth and early nineteenth centuries. Fortuitous for the new science of mineralogy was Philadelphia's location along the eastern edge of the Piedmont Uplands. This complex metamorphic environment, the site of many igneous intrusions, was the storehouse for a wide variety of unique and beautiful mineral species, all within a hundred-mile radius of Philadelphia, two factors that created an ideal spawning ground for the development of mineralogy at the Academy.[37]

Aside from the philosophical side of geology, there was the competitive search for minerals for industry. Coal, iron, copper, zinc, gold, silver, and any number of other elements and

11.6 *Cave in the Rock [Ford's Ferry, Ohio]. Stalagmitic Chamber and Crystal Fountain—Magnificent Effects of Crystallization.* From a panorama of the Mississippi Valley by John J. Egan, c. 1850, tempera on lightweight fabric. St. Louis Art Museum. Eliza McMillan Trust.

compounds were essential to the advance of the American economy and society in the nineteenth century.

Enthusiasm for this burgeoning new science of mineralogy caused many Academy members and friends to donate minerals to the society's collections. The first issue of the Academy's *Journal* lists "specimens of native gold" from a Dr. Scott; "Nodular Iron Ore" from Miss Speakman; "Rock Salt from England" given by the architect William Strickland; minerals from Kentucky from J. R. Paxson; a tourmaline from Baltimore from James Griffith; bitumen from Havana donated by Dr. Mease; "minerals from the Missouri" collected by Academy founder John Shinn; and "a large mass of Rock containing Fossil Shells from near Bordeaux [France]" brought back by Captain Rush.[38] At the same time, essays on the subject of minerals were increasingly frequent. In the second volume of the *Journal*, delayed until 1821 because its editor, Thomas Say, was on the Long Expedition, seven out of eleven articles were devoted to mineralogy.

Since the area around Philadelphia was so rich in minerals, Academy members did not have to travel far to find them. Isaac Lea (1792–1886), one of the Academy's most active members, wrote one of the earliest papers on mineralogy published in America.[39] It focused on the minerals found in the vicinity of the city, including those from the mines of the Perkiomen Creek, where Troost had led the first Academy excursion. He wrote of blue carbonate of copper, "a beautiful dark blue colour . . . found in veins, with lead and zinc, in the old red sandstone formation [of the mine]"; of copper pyrites, "of a brass yellow colour and often externally iridescent, at Perkiomen, and on Chester creek, near a saw mill, three miles west of Chester, in Delaware county"; and of magnetic oxide of iron he had found "on the Schuylkill [River] in small quantities, of a dark iron black . . . strongly attracted by the magnet."[40]

Lea noted that he and Lardner Vanuxem (1792–1848), another Academy member and collector, had discovered silico calcareous oxide of titanium "imbedded in the hornblende rocks of the quarry at the end of the canal road." In speaking of "earthy minerals" in other localities, he wrote that zircon was "first discovered at the falls of the Delaware at Trenton" and "on the Brandywine [River] two mile[s] beyond Westchester; and beryl was found on Mr. C[harles Willson] Peale's farm, near Germantown."[41] "Near Westchester," he said, "Serpentine occurs very abundantly, and is used for common building stone. Colour, from light to dark green . . . fibrous asbestus is found in the serpentine rocks about one mile north of Westchester . . . it occurs also in very delicate fibres on quartz crystallized, in the hornblende quarry, end of canal road."[42] This same green serpentine would be used to face the Academy's building on Logan Square in the 1870s.

A century later, Robert Middleton, Academy curator of minerals, wrote of the importance of the museum's mineral collection, with its rich representation of specimens from eighteenth- and nineteenth-century sites around Philadelphia. He noted that the emphasis placed on Pennsylvania minerals, particularly those from the Philadelphia area, made the Academy's collection "unmatched in its coverage, containing superb specimens from numerous localities which are now inaccessible because of urbanization."[43]

In addition to Isaac Lea's paper in volume 1, many mineralogical firsts were published in the pages of the early Academy *Journal*. Troost's descriptions of new crystalline forms of minerals found in the United States contained the first crystallographic data, with crystal drawings, published in this country.[44] William Hypolitus Keating (1799–1840) and Lardner Vanuxem's definition of a mineral found at the then-defunct Franklin Iron Works near Sparta, New Jersey, was the first American description of a mineral as new. Keating wrote in the Academy *Journal*: "Mr. Vanuxeum has proposed to dedicate this mineral to Mr. Jefferson; I have readily assented to this proposal, and we now offer this mineral to the public under the name of *Jeffersonite*."[45]

Keating was the official geologist on the Long Expedition to the St. Peter's River in Minnesota in 1823 and a founding member of the Franklin Institute in 1824. In the late 1820s, he spent three years surveying mineral resources in Mexico.[46]

Troost and Maclure, through their professional competence, enthusiasm, and close contacts with French mineralogists, provided inspiration and support for the younger generation of Philadelphia scientists such as Keating and Vanuxem. The Maclure-sponsored 1825 excursion of Academy members to coal mines in Pennsylvania and New Jersey (discussed in chapter 2) to examine the anthracite being extracted for heating houses and for manufacture is a case in point. As the *Journal* evolved into an important publication on American mineralogy, the Academy became a solid base for the future growth of the science.[47]

In 1834, the Academy elected to membership a major figure behind the development of the Academy's mineral collection, the wealthy Philadelphian William Sansom Vaux (figure 11.7). Not only was Vaux influential in the erection and financing of the society's new building at the corner of Broad and Sansom Streets in 1839, but he also served as the unpaid curator of minerals for forty-four years, beginning in 1838, and as the Academy's

11.7 *William S. Vaux* (1811–1882) by Herman F. Deigendesch, [1873]. Oil on canvas. ANSP Library coll. 286.

11.8 Philadelphia Mineral Society excursion on a Toonerville trolley, Williams Corner, Pennsylvania, 30 September 1923. ANSP Archives coll. 2011-003.

vice president for more than twenty-one years, from 1860 until his death.[48]

An avid collector of minerals and archaeological specimens in the decades following the Civil War, Vaux traveled extensively in Europe and North America, amassing the important assemblage of nearly six thousand minerals he bequeathed to the Academy. This broad, general collection, with superb representation of both American and foreign occurrences, was then unmatched by any public or private collection in America.[49]

Until Vaux's time, mineral collectors had concentrated largely on specimens of a utilitarian nature, such as ores and soils that had potential commercial value. But as the century advanced, the search for minerals was more about knowledge for knowledge's sake and collecting for the sake of collecting (figure 11.8). The Vaux collection differed from previous additions to the Academy's Mineralogy Department because of its emphasis on the aesthetic appreciation of minerals, many selected for their great beauty rather than the products they could produce. It therefore held many educational possibilities and was ideal for attracting public interest.[50]

Vaux's nephew, George Vaux Jr. (1863–1927), inherited his uncle's enthusiasm for the Academy and for minerals. Elected an Academy member in 1892, he served at different times as the society's treasurer (1894–1929) and solicitor. In the 1920s, he cosponsored with the Academy four geological expeditions under the direction of the longtime associate curator of minerals Samuel George Gordon (1897–1952). These expeditions added many superb specimens to the institution's collections.

At a young age, Gordon attended classes at the Wagner Free Institute in North Philadelphia, where Edgar T. Wherry (1885–1982), professor of mineralogy and chemistry at the University of Pennsylvania, inspired him to learn about minerals.[51] Wherry recognized Gordon's intelligence and took a special interest in this bright Jewish youth from a poor section of the city. With Wherry's encouragement, Gordon applied for and won a Jessup Scholarship at the Academy.[52] Gordon, still a teenager, was assigned to the Mineralogy Department, working under the tutelage of curator Frank J. Keeley (1868–1949), an Academy life member since 1894.[53] At the age of nineteen, Gordon established the *American Mineralogist* (1916), which provided a needed outlet for the publication of specialized mineralogical research in the United States.[54]

In 1921, Gordon's first Academy expedition took him to Ecuador, Bolivia, Chile, and Peru, where he collected hundreds of superb specimens for the Academy. Two years later, "through the courtesy of the Danish Government and the Pennsylvania Salt Company," the Academy sent Gordon to the Julianehaab district of Greenland.[55]

After four weeks visiting mineralogical laboratories and examining collections in Paris, Brussels, London, and Oxford, Gordon, who had shipped his equipment earlier to England, joined a Norwegian steamer at Newcastle-on-Tyne. On his approach to Greenland, after a day of fog near Cape Farewell, he recorded: "The night was spent in dodging great icebergs, some of which we passed so closely as to be able to hear the waves beating against their sides." Soon after reaching Igaliko, where he saw many Norse ruins, Gordon engaged twenty-two men and women to carry his equipment and supplies to Narsaruk, near his selected excavation site. "While the coast of Greenland has a barren and forbidding aspect," he wrote, "the high rugged mountains of the interior, the wide and deep fiords, the glaciers at their heads, and the Inland Ice, present magnificent vistas that truly beggar description."[56]

When the expedition ended in October, a Norwegian ship, the *Skulda,* arrived to take Gordon's thirty cases of specimens aboard. "A short stop was made at Halifax," he noted, "and on November 11th we passed up the Delaware, marking the end of an exceptionally interesting and profitable trip."[57]

In 1925, Gordon journeyed to the Bolivian Andes (figure 11.9) and, with the intrepid vigor of a man in his twenties, descended into numerous mines. He shipped thirty-eight cases of minerals, ores, and rocks to the Academy, including the new minerals penroseite and trudellite, named for Richard A. F. Penrose (1922–26), then president of the Academy, and Harry W. Trudell, an Academy trustee, who, along with George Vaux Jr., defrayed the expenses associated with the analyses of the specimens.[58]

On his fourth Academy-Vaux expedition, Gordon collected specimens in the copper, gold, platinum, diamond, and chrome mines of Southwest Africa, the Transvaal, Northern and Southern Rhodesia (Zimbabwe), and the Belgian Congo (now the Democratic Republic of the Congo). He returned with 2,276 specimens of minerals, rocks, and ores and a section of the famous iron meteorite near Grootfontein. Among his rare finds were unusually large and perfect crystals of azurite, a rich blue mineral commonly regarded as one of nature's masterpieces. He had gone to a copper mine in Tsumeb, Southwest Africa (Namibia), in search of this treasure.

11.9 Samuel G. Gordon in South America, 1925. Hand-colored photograph. ANSP Archives coll. 2009-009.

As Gordon recorded in an article for the Academy's 1930 *Year Book*, "The copper mines of Tsumbeb are noted among mineralogists for extraordinarily perfect crystals of azurite. . . . Surface workings extend to a depth of 100 feet, while underground operations have reached a depth of 1,400 feet. I visited first the 8th level workings. The tunnels were narrow, and at each face were groups of naked Ovambos, working under the supervision of a mine boss."[59]

According to an unpublished manuscript in Charles Cadwalader's files, when Gordon descended eight hundred feet into the mine, he discovered "a small concavity in the ceiling of the stope faintly tinged with greenish blue, which looked promising. He began to work it, and uncovered a vug [a small unfilled cavity in a lode or rock] of azurite just as he realized that the available oxygen . . . was rapidly being exhausted. He worked desperately and finally got out the crystals and climbed down to the main galleries, where the air was better. He saw then that he had obtained the best [azurite] crystals yet found, immeasurably better than any previously discovered."[60]

Among nearly eighty mineralogical papers Gordon published are descriptions of his discoveries of eight new minerals, including the ones he named for his patron George Vaux Jr.: vauxite, paravauxite, and metavauxite.[61] Vauxite is a rare, sky-blue crystal occurring in small aggregates, and, in 1930, the Academy had "practically all of it." Paravauxite occurs in pearly, prismatic crystals considerably larger than aggregates of vauxite, while "metavauxite is the beauty of the three." It is found in vugs and "literally sprouts from the interior of the little caves in fountain-like sprays of dainty, resplendent needles, looking like a small, clipped aigrette. The spray sometimes terminates in delicate green."[62]

In addition to his field collecting activities, Gordon established a mineralogical research laboratory at the Academy, complete with chemical analysis apparatus, goniometers, and

FACING PAGE 11.10 Azurite specimen collected in Bisby, Arizona. Vaux Collection. ANSP Mineralogy Department, #17724. Purcell photograph.

11.11 Samuel G. Gordon with miners in an African mine. ANSP Archives coll. 2009-009.

microscopes. He also constructed a mineral exhibition hall, which in its day was the finest and most up-to-date in America.[63] In 1928, the Academy opened the first public exhibit of mineral fluorescence in the country.[64]

The Academy *Year Book* for 1931 stated that "No section of the Museum is more popular with visitors than Mineral Hall, and if sufficient funds were available for new cases, thousands of attractive specimens could be properly shown. . . . Plans were made for a series of educational exhibits to be installed in modern glass cases designed and built in the Academy."[65]

Four years later, Gordon wrote to the directors of the natural history museums in Cleveland, Chicago, Milwaukee, Buffalo, and Los Angeles and at the Century of Progress Exposition in Chicago offering to replicate the Academy's fluorescence exhibit, which he described as "famous" and "one of the most remarkable museum exhibits in the world." In his letter, Gordon explained that the Academy was organizing an expedition to acquire the fluorescent minerals needed to reproduce the exhibit and that the charge for a comparable display would be $491.75.[66] A receipt and some correspondence in the Academy's archives show that, for one, the Century of Progress Exposition did buy and install the fluorescence exhibit.[67]

In April 1938, the Academy opened a new, enhanced exhibit of fluorescent minerals, shown under both long and short wavelengths of ultraviolet radiation. The installation was what Gordon claimed to be the first completely "robot" exhibit in any museum. Activated by a visitor passing a pair of directional photoelectric cells, the mechanical "brain" caused the exhibit to go into a four-minute cycle: first, electric lights showed the mineral under ordinary illumination, and then the lights dimmed out and, at a cue, "the invisible became visible," and the minerals were shown fluorescing. (That is, energy in the form of short wavelengths of ultraviolet radiation vibrating rapidly are changed by certain minerals into rays of light of longer wavelength vibrating slowly enough to make the mineral visible.)

Of the forty thousand minerals examined by Gordon in the Academy's mineralogical laboratory in the 1930s, only a few showed this phenomenon conspicuously. According to Gordon, "Not all specimens of even the same mineral react [to light in the same way]. Those that fluoresce contain something, some impurity that activates them. *Willemite* and calcite from Franklin, New Jersey, contain small amounts of manganese as activators. Too much manganese inhibits them." After the ultraviolet lights in the exhibit were turned off, the minerals continued to glow, the afterglow known as *phosphorescence*.[68]

Franklin, New Jersey, the "fluorescent mineral capital of the world" and one of the sources for Gordon's display, is recognized today as the source of at least 250 minerals, of which at least 56 are fluorescent. Many of these minerals are found nowhere else in the world.[69]

In September 1939, the focus of the Academy's Mineralogy Department changed dramatically when the war that engulfed the world for most of the next decade began with Germany's invasion of Poland. Anticipating U.S. participation in the fast-developing global war, Gordon published an article in the October 1941 issue of the Academy members' magazine, *Frontiers,* applying his expertise as a scientist to issues of strategic concern. He warned that there were five minerals essential for defense that the United States did not produce: quartz (rock crystal), chromium, manganese, tungsten, and tin (which the United States acquired mainly from Southeast Asia). "Quartz, perhaps the most important of the strategic minerals, is the one which made the blitzkrieg possible," he wrote. "It is the essential unit of short-wave radio sets, permitting sending station and receiver to be accurately tuned with one another, whether the latter is a ship, airplane, submarine, or police car. The success of the blitzkrieg has been due to such remote control of rapidly-moving tank columns, dive-bombers, and parachute units, permitting closely coordinated effort at strategic points." "It is obvious," Gordon continued, "that our limited stockpiles of strategic minerals depend for replenishment upon imports from across the ocean. It is no wonder that some anxiety exists regarding future supply. At this writing teutonic hordes are moving toward the manganese of the Caucasus; while in Asia, the Japanese are threatening the tin of the Malays, operations which may mean a stranglehold upon much of the industrial world."[70] Gordon had every right to worry. Two months later, the Japanese bombed Pearl Harbor, forcing the United States into World War II and further restricting American access to needed mineralogical resources.

After the war, Gordon continued to encourage the professional and amateur study of minerals. But in 1948, he faced a shattering turn of events when he returned from a collecting trip to find, to his utter amazement, that his mineralogical laboratory had been eliminated in favor of a newly enlarged Department of Limnology. Shortly thereafter, the meager funding for the Mineralogy Department was phased out, and he was forced to resign. In his concluding remarks, a disheartened Gordon noted: "As far as the research program (field work and laboratory work is concerned), hardly a penny came from Academy sources. The Academy has merely furnished the space, light, heat, and the salary of

11.12 Hematite ("Bird wing") from Cumberland, England. Vaux Collection. ANSP Mineralogy Department #11409. Purcell photograph.

the associate-curator [Gordon], less than the lowest grade of a government worker. It is the unanimous opinion among mineralogists elsewhere that the associate curator has wasted his life in the Academy because of the lack of proper equipment."[71] Gordon's departure marked the end of the Academy's long-standing tradition of excellence in the field of mineralogy, and the department was without a curator for the first time in nearly 140 years.[72]

Since Charles Cadwalader was still Academy president at the time—his term would end in 1951—this episode has echoes of Cadwalader's treatment of Edgar Howard, the sale of Indian artifacts to the Heye Museum in New York, and the end of the Academy's Archaeology and Anthropology Department in 1939. Coincidentally, Howard and Gordon both died four years after leaving the Academy at almost the same early age: Howard was fifty-six, and Gordon was fifty-five. The fateful similarities of these two departments would be even more compelling in the following century.

After various heroic attempts to revive the Mineralogy Department over the ensuing years, in the early twenty-first century the Academy decided to sell most of the collections, despite vigorous protests from the natural history museum community. In the fall of 2004, an angry article appeared in *Mineral News* denouncing the deaccession. The writer stated that the Academy had built a substantial mineral collection since its founding in 1812 through the efforts of noted collectors and mineralogists. In addition, significant bequests were made by the Academy's many patrons, sometimes supported by large monetary endowments from their estates. "Famous collections and collectors, the likes of George Vaux (vauxite), Samuel Gordon (gordonite), and Edgar Wherry (wherryite), contributed greatly to mineralogy in general and to the Academy of Natural Sciences in particular," the article said. "Unfortunately, the current administration at the Academy has no appreciation for its mineral holdings, nor seemingly any respect for the wishes of its many donors, and it now seeks to sell off its holdings without full regard for the legality of its intended actions. Its judgment has apparently been clouded by the lure of money, and lots of it."[73]

One vocal critic wrote heatedly to the Academy's president, James Baker, in October 2004, that he did not think Baker's argument—that the Academy had not had a curator of mineralogy for twenty-five years—was a valid reason for the sale, nor was the caveat that most of the collection was not on view. "How much of the Smithsonian's collection do you suppose is 'on view'?" he asked. "Very little." He concluded that "early American mineralogy and the Academy are intimately linked" and that the Academy should celebrate the historical significance of the collection and its "links to so many important American scientists."[74]

Nevertheless, despite the writer's objections and those of others, two years later, on 20 October 2006, the Academy, under the jurisdiction of acting president Dr. Ian Davidson and the board of trustees, headed by Edward Montgomery, announced that it had sold "a portion of the mineral collection."[75] This portion was in fact the bulk of the collection, some nineteen thousand specimens sold to a consortium of three mineral dealers. Not included were the Seybert collection (1812), which contained a few minerals from the Lewis and Clark Expedition, a group of scientifically valuable type specimens to be transferred to other museums at no cost,[76] and the Vaux collection, which was under legal restriction by the terms of the Vaux bequest. The Academy expected to receive several million dollars from the sale of this historically important collection.

At a court hearing the following spring, Trina Vaux, the great-great-grandniece of William S. Vaux, upset that the Academy would betray the trust of her famous relative, submitted a plan to have Bryn Mawr College and the Wagner Free Institute of Science take the collection.[77] William Vaux had traveled the world to amass these thousands of minerals. Even some of the more common specimens had significant value because they were collected in mineralogical sites that no longer exist. And, according to Wendell E. Wilson, editor of the *Mineralogical Record*, because they had not been exhibited for more than half a century, although it was "a virtual time capsule from a gilded age long past, the collection might just as well have been reburied in the ground."[78]

Indications were that the court would rule against the sale of the Vaux minerals, and since no money would be made in the transfer that Trina Vaux requested, the Academy decided to keep the collection. The new Academy president, Dr. William Y. Brown, who felt strongly about deaccessioning any of the institution's collections, filed an affidavit in Orphans' Court to that effect. "The vast and seminal collections of the academy, including these, date from the dawn of science in this hemisphere. We have no higher priority than their stewardship."[79]

Even so, all the specimens given to the Academy by other donors and by William Vaux in his lifetime (exclusive of the collection he bequeathed the institution in 1882), those so laboriously—and often at great risk—excavated by Troost,

Maclure, Heilprin, Keating, Vanuxem, Gordon, and a host of others, and "rocks from Mts. Terror & Erebus collected on the Shackleton expeditions"[80] were sold.

Natural history museums must change at times, according to the dictates of the current administration, in order to accommodate popular taste, or the perception of it. Over the years, the Academy had shifted its focus from earth sciences (mineralogy) to the biological and ecological sciences. However, the institution had been brought up short by the fateful decision of the president and board of trustees in 2006 to sell off its historic mineral collection. With a new administration in place, it was important to change course.

A grant from the Pennsylvania Historical and Museum Commission in 2007 partially funded a project to reorganize, rehouse, and conserve the Seybert and Vaux collections, with an eye to a future display of some of the more extraordinary specimens. Purchase of new storage cabinets and supplies was the first step in the process of acquiring the twenty new cabinets needed. Ted Daeschler, associate curator and chair of vertebrate zoology, along with Ned Gilmore, collections manager for vertebrate zoology, and Fred Mullison, fossil preparator, took over the care of the historic collections. "You can't change what happened," Daeschler said, "you can only work to protect what you have left."[81]

PTS

11.13 Agate specimen from Brazil, William S. Vaux Collection. ANSP Mineralogy Department #11980. Purcell photograph.

Notes

1. Angelo Heilprin, the son of Michael Heilprin, a scholar of Polish origin who lived in Hungary, was an American naturalist, geologist, and traveler. He moved to the United States with his family at an early age but received his higher education in Europe at the Royal School of Mines in London, the Imperial Geological Institution of Vienna, and at Florence and Geneva. He was professor of invertebrate paleontology and geology at the Academy of Natural Sciences (1880–1900); curator of mineralogy at the Academy (1883–1892); professor of geology at the Wagner Free Institute of Science (1883–1890); first president of the Geographical Society of Philadelphia; and a fellow of the Royal Geographical Society of London. A talented painter, he illustrated some of his own works. Heilprin and his father are major figures in American Jewish history.
2. Angelo Heilprin, *Mont Pelée and the Tragedy of Martinique* (Philadelphia and London: J. B. Lippincott, 1903), 162–63.
3. Account of George Kennan, among the Heilprin letters purchased by the Geographical Society of Philadelphia and deposited in the Academy archives (2010), coll. 2010-287.
4. Heilprin, *Mont Pelée*, 164.
5. Ibid., 208.
6. Ibid., 220.
7. Ibid., 227–28.
8. Ibid.
9. Ibid., 194.
10. William Maclure, "Observations on the Geology of the West India Islands, from Barbadoes to Santa Cruz, Inclusive," *Journal of the ANSP* 1, pt. 1 (1817): 134–49.
11. James X. Corgan, "Gerard Troost, 1776–1850," in *The Tennessee Encyclopedia of History and Culture* (Nashville: Tennessee Historical Society, 1998; online ed., Knoxville: University of Tennessee Press, 2002), 2.
12. Robert G. Middleton, "Report on the Mineral Collection, Academy of Natural Sciences of Philadelphia," October 1988, 8, ANSP archives, coll. 2009-034. Middleton was the Academy's curator of minerals from 1981 to 1986.
13. ANSP original minutes, 18 April 1812, ANSP coll. 502. In the mid-nineteenth century, the Perkiomen mine produced the copper (that was alloyed with zinc from Franklin, N.J.) for the brass that went into the fabrication of the first weights for the just created National Bureau of Standards. The Perkiomen Creek is a tributary of the Schuylkill River.
14. Don Messenger, "A Museum Built on Rocks," unidentified printed article in Academy mineralogy file, coll. 2009-034.
15. Academy members subscribed $20 a share to repay Speakman for the $750 purchase of the Seybert minerals. Edward J. Nolan, *A Short History of the Academy of Natural Sciences of Philadelphia* (Philadelphia: Academy of Natural Sciences, 1909), 9.
16. William Maclure, *The European Journals of William Maclure*, ed. John S. Doskey (Philadelphia: American Philosophical Society, 1988), 247, n. 1.
17. Website of the *Mineralogical Record, Inc.*, biographical archive, 2010 (http://www.mineralogicalrecord.com). There is a note in the Academy's archives written by Samuel Gordon, longtime Academy associate curator of minerals, that "the chemical laboratory of Dr. Henry Seybert—1500 pieces of apparatus was deposited or given by Dr. [Samuel G.] Dixon [Academy president from 1895 to 1918] to the University [of Pennsylvania] and should be looked into." ANSP coll. 89, III, 28.
18. Geoffrey T. Hellman, "Onward and Upward with Science—Go On Investigators, Scrutinize!" *New Yorker* (3 November 1962): 142.
19. Messenger, "Museum Built on Rocks." For more on Silliman, see George Park Fisher, *Life of Benjamin Silliman: Late Professor of Chemistry, Mineralogy, and Geology in Yale College VI* (New Haven, CT: Yale University Press, 2007).
20. Nolan, *A Short History*, 9.
21. Included in this historic collection are some of the few surviving mineral specimens from the Lewis and Clark Expedition, which include bits of slag and a piece of pumice the explorers found floating in the Missouri River.
22. François André Michaux to Benjamin Vaughan, July 1807, quoted in J. C. Greene and J. G. Burke, "The Science of Minerals in the Age of Jefferson," *Transactions of the American Philosophical Society* 68, pt. 4 (1978): 32.
23. Ibid.
24. Silvain Godon, *Mineralogical Observations: Made in the Environs of Boston, in the Years 1807 and 1808* (1809).
25. Greene and Burke, "The Science of Minerals," 32–34.
26. Thomas Jefferson to Governor Wilson C. Nicholas, 19 April 1816, A. A. Libscomb and A. E. Bergh, eds., *The Writings of Thomas Jefferson* (20 v. Washington, DC (1907) 14: 486. Quoted in *Trans. of APS* (1978) 68, pt. 4, 33.
27. Maclure, *European Journals*, xv.
28. William Maclure, *Observations on the Geology of the United States of America; with some remarks on the effect produced on the nature and fertility of soils, by the decomposition of the different classes of rocks; and an application to the fertility of every state in the union, in reference to the accompanying geological map. Read as a Memoir before the American Philosophical Society, and inserted in the 1st vol. of their Transactions, New Series.* Philadelphia: printed for the author by Abraham Small, 1817.
29. Keith Thomson, *The Legacy of the Mastodon: The Golden Age of Fossils in America* (New Haven, CT: Yale University Press, 2008), Appendix A, 335–36.
30. Ibid., 63–64.
31. Online entry on James Hutton, *Chambers Biographical Dictionary*, ed. Una McGovern (Chambers Harrap, 2002).
32. Thomson, *Legacy of the Mastodon*, 64.
33. ANSP minutes, 4 June 1816, coll. 502; William Maclure, "Essay on the Formation of Rocks, or an inquiry into the probable origin of their present form and structure," *Journal of the ANSP* 1, pt. 2 (1818): 285–310. For more on Maclure's European explorations, see Doskey, *European Journals of William Maclure*.
34. Simon Winchester, *The Map That Changed the World* (New York: HarperCollins, 2001; Harper Perennial Edition, 2009), 74.
35. Thomas Say, "Observations on some species of zoophytes, shells, etc, principally fossil," *American Journal of Science and the Arts* 1, no. 4 (1818): 381–87.
36. Winchester, *The Map That Changed the World*, 11.
37. Jay Lininger, "Bob's Findings," October 2005, 6. http://www.meteorite-times.com.
38. *Journal of the ANSP* 1, pt. 1 (1817): 213–18.
39. Isaac Lea, "An Account of the Minerals at present known to exist in the vicinity of Philadelphia," *Journal of the ANSP* 1, pt. 2 (1818): 462–82.
40. Ibid.
41. Ibid.
42. Ibid., 476, 478. As one travels through this area of Chester County today, one sees many houses built with this lovely green stone.
43. Middleton, "Report on the Mineral Collection," 4.
44. Ibid., 10; Gerard Troost, "Descriptions of some new Crystalline forms of Phosphate of Lime and Zircon," *Journal of the ANSP* 2, pt. 1 (1821): 55–59.
45. Middleton, "Report on the Mineral Collection," 11. Lardner Vanuxem and William H. Keating, "Account of the Jeffersonite, a new mineral discovered at the Franklin Iron Works, near Sparta, in New-Jersey. Described and analysed by W. H. Keating," *Journal of the ANSP* 2, pt. 2 (1822): 257–77.
46. American Philosophical Society, online entry for William H. Keating.

47 Middleton, "Report on the Mineral Collection," 11.

48 George Vaux, "The Vaux Family's Scientific Pursuits," *Frontiers, Annual of the Academy of Natural Sciences of Philadelphia* 3 (1981–82): 57.

49 Middleton, "Report on the Mineral Collection," 13.

50 Ibid.

51 The Wagner Free Institute was established in 1885 by industrialist William Wagner to teach natural history to poor youths unable to pay for their education.

52 The Jessup Scholarship offered stipends to young men and women of unusual talent for work at the Academy.

53 Jay L. Lininger, "Mineralizing in Germantown: An Avid Pursuit in an Earlier Era," *Penn Minerals* (15 July 2010): 4–5.

54 Middleton, "Report on the Mineral Collection," 15.

55 *Year Book, The Academy of Natural Sciences of Philadelphia for the Year Ending December 31, 1923* (Philadelphia: Academy of Natural Sciences, 1923): 3.

56 Ibid., 3–4, 7.

57 Ibid., 11.

58 *Year Book* (1925): 62.

59 Samuel G. Gordon, "In Really Deepest Africa," *Year Book* (1930): 47.

60 Charles M. B. Cadwalader, as told to Paul Brown, "Field Collecting," unpublished mss. in Cadwalader files, submitted to the *Saturday Evening Post* in 1930 but rejected because considered "too scientific," ANSP coll. 2009-034, box 2, file 1, 19.

61 Vaux, "Scientific Pursuits," 57–58.

62 Cadwalader, as told to Brown, "Field Collecting," 22.

63 Middleton, "Report on the Mineral Collection," 15–16.

64 Ibid.

65 *Year Book* (1931): 16.

66 Samuel G. Gordon to various museum directors, January 1932, unprocessed Charles Cadwalader correspondence, ANSP coll. 2009-034, box 4, file 52.

67 Unprocessed Cadwalader correspondence, ANSP coll. 2009-034, box 4, file 52.

68 ANSP exhibits file, coll. 256. "The ultraviolet rays in the exhibit," according to Samuel Gordon, "are produced by two carbon arc health lamps, supplemented by quartz mercury-vapor lamps. Glass screens of remarkable composition shut out the visible rays, permitting only the invisible ultraviolet rays to shine through." Samuel G. Gordon, "Fluorescent Minerals: An Exhibit at the Academy of Natural Sciences of Philadelphia," ANSP coll. 256.

69 Online posting of *The Fluorescent Mineral Society, Inc.* (http://uvminerals.org).

70 Samuel G. Gordon, "Minerals for Defense," *Frontiers* 6, no. 1 (October 1941).

71 Samuel G. Gordon, "Conclusion," in "Operations of the Department," ANSP coll. 89.

72 Middleton, "Report on the Mineral Collection," 16. Also, Wendell E. Wilson, "Sam Gordon and the Tsumeb Azurite," *The Mineralogical Record* 38, no. 5 (1 September 2007). The department would be revived briefly in the 1970s and 1980s with the employment of Robert Middleton as a full-time and then a part-time curator.

73 Tony Nikischer, "Philadelphia Museum Attempts to Liquidate Historic Mineral Collection," *Mineral News* 20, no. 11 (November 2004); reprint by Robert Verish, "The Last in a Series of Articles: Museums Are Selling-off Their Collections—the Original Editorial," *Meteorite Times*, October 2005, 2.

74 John S. White, letter to Dr. James Baker, 6 October 2004, reprint in *Meteorite Times*, October 2005, 5.

75 Academy news release, 20 October 2006, ANSP website. The sale required permission from Philadelphia's Orphans' Court, which was granted.

76 Eight thousand specimens were relocated to other museums: the Pennsylvania specimens were acquired by the Carnegie Museum of Natural History; the Franklin Mineral Museum and the Sterling Hill Mining Museum took the New Jersey suite; the New York State Museum now has the New York suite; and the Museo di Storia Naturale in Milan took the Italian suite. "Notes from the Editors," *The Mineralogical Record* 38 (September–October 2007): 338.

77 "Minerals Make a Comeback at the Academy," *Natural History* (February 2009): 44–45.

78 Wendell E. Wilson, quoted in the *Philadelphia Inquirer*, 16 June 2007.

79 *Philadelphia Inquirer*, 16 June 2007.

80 Rocks presented by John H. McFadden in 1915. ANSP mineralogy file, Gordon's notes for 1933 annual report, coll. 89.

81 "Minerals Make a Comeback," 44–45.

Cambarus bartonii robust
Girard
328 [type]

Chapter 12

Academy Expeditions, 1928 to 1960

Governor Pinchot, who will try anything once, had iguana served for dinner; it is not at all bad.
—Henry Pilsbry, Academy malacologist, Galápagos Islands, 1929

Despite the world's economic crisis, the era of the late 1920s and 1930s was a time of unusual activity at the Academy. The museum's 1929 *Year Book*, without mentioning the catastrophic stock market crash in October, lists fifteen expeditions in the field that resulted in specimens collected in more than thirty countries and island possessions. Most of these journeys were privately financed.[1] New tax regulations gave further incentive for the sponsorship of these expeditions. The increase in participation and contributions to the Academy's research and public museum, encouraged by the Academy's president, Charles Cadwalader, helped to cushion the blow of the collapsing economy while simultaneously increasing the visibility of the museum.

Leading a period of rapid expansion of public exhibitions at the Academy were many impressive animal habitat displays, or dioramas, the first of which was completed in 1929. The idea for dioramas in the United States goes back to Charles Willson Peale, who pioneered the concept of creating painted backgrounds for the specimens in his museum. In the twentieth century, the concept was adopted and expanded by exhibit designers at the Field Museum in Chicago, the American Museum of Natural History in New York, and elsewhere.[2] The public's enthusiasm for habitat groups may have been stimulated by a growing awareness that wildlife and wilderness were finite and that fragile ecosystems had been targets for exploitation. The killing of great bison herds in the American West, the extinction of the Passenger Pigeon and Carolina Parakeet, and the needless slaughter of hundreds of thousands of egrets, terns, and herons for the millinery trade all contributed to a growing concern for wildlife and a wish to better understand it. In the Academy's habitats, the visiting public could see preserved and in realistic settings fascinating animals from parts of the world they might never have a chance to visit.

The first of the Academy's dioramas (no longer in existence) depicted the Rocky Mountain goat (*Oreamnos americanus*; not a true goat, but more accurately a goat-antelope), a species discovered on the Lewis and Clark Expedition of 1804–6 and described for science by the Academy's George Ord in 1815. Clement B. Newbold secured and presented the specimens from an expedition to British Columbia he had undertaken in 1927.[3] In that same year, Harry Whitney (1873–1936), an inveterate sportsman and explorer who contributed numerous specimens to the institution and would be elected to membership the following year, brought back three giant brown bears (*Ursus arctos*) from Kodiak, Alaska,[4] to serve as the focus of the Academy's second diorama. Plants, stones, and even soil were collected from the site to furnish an authentic context for the specimens. Paid for by subscription, the diorama was created in cooperation with the Bureau of Biological Survey in Washington, D.C.

12.1 The skull of a male musk ox (*Ovibos moschatus*) collected for the Academy by Harry Whitney on Clavering Island, Greenland, in July 1930. ANSP Mammalogy Department #13434. Purcell photograph.

Acad. Nat. Sci. No.
R.M. de Schauensee. Date
Sex
Philadelphia
Acad. Nat. Sci.

With support from Academy trustee Robert Ruliph Morgan "Ruli" Carpenter (1877–1949), whose many African expeditions contributed significantly to the Academy's collections, Whitney returned to Alaska in 1929 to collect Dall's mountain sheep and woodland caribou for the centerpieces of their own habitat groups. Cadwalader wrote: "The popularity of these realistic presentations with all visitors testifies to the wisdom of the plan to increase their number as rapidly as possible, especially in view of the fact that their artistic and educational value is immeasurable in advance of old methods of exhibit. The Academy should be as pre-eminent in this form of service as it is in the field of scientific research."[5]

In 1928–29, Rodolphe Meyer de Schauensee (1901–1984), the Academy's curator of ornithology, with his wife, Williamina, journeyed to Chieng Mai in the northern part of Siam (now Thailand), some 450 miles north of Bangkok, to collect little-known birds. Most of the collecting was done on the slopes of the 4,500-foot mountain Doi Souteb. "At 1,700 feet, the dry scrub was replaced by cool, green forest," Meyer de Schauensee recorded. "Close to the path a mountain stream dashed down over the rocks in a white torrent. Along its course we frequently saw the splendid Burmese Whistling Thrush (*Myiophoneus eugenei*) and occasionally the curious forktail (*Henicurus Schistaccus*) flitting over the rocks . . . gibbons sang frequently and once we saw a tiger. The forest seemed to vibrate with life." He added that "a great deal of credit is due to my colleague, Tyson Smith, who helped us in preparing skins."[6] In all, the expedition collected 376 bird skins. This small but important collection helped to establish the Academy as one of the world's leading repositories of Asian avifauna (figure 12.2).[7]

Collecting birds in South America was the major occupation of Melbourne "Meb" A. Carriker Jr. (1879–1965), "the most prolific and energetic collector of Andean and other Neotropical birds in the history of ornithology," according to an article in *The Auk*, the journal of the American Ornithologists Union. "His studies in the Neotropics began in Costa Rica and later extended to Venezuela, Colombia, Peru, Bolivia, and Mexico."[8]

In 1929, Carriker was hired as assistant curator of ornithology at the Academy, after he and his family moved to the United States from their coffee plantation in Colombia, so that their five bilingual children could attend American schools.[9] He then spent six months of every year collecting in South America.

His wife, Myrtle "Carme" de Flye Carriker (1893–1960), whom he married in 1912, traveled the entire length of western Colombia with him in the early years of their marriage and was his most accomplished bird skinner. On their first collecting

FACING PAGE 12.2 Black-backed Kingfishers (*Ceyx erithancus*) collected by Rodolphe Meyer de Schauensee in Siam (now Thailand), 1937–38. ANSP Ornithology Department. Purcell photograph.

12.3 Melbourne A. Carriker Jr. (1879–1965) beginning a collecting journey down the Rio Béni from Santa Ana, Bolivia. ANSP Archives coll. 900.

trip together, they climbed 17,500 feet into the Sierra Nevada of Colombia.

In 1916, the couple spent ten months in the Andes exploring between Norte de Santander and Bogotá and returned to their home in Santa Marta with 5,464 bird skins.[10] On that seven-month trip, the intrepid Carme Carriker added to all her other duties by taking along the couple's seventeen-month-old son, Melbourne. Carme was also pregnant with their next child. Her husband's friend and entomological colleague, K. C. Emerson, later wrote with admiration of her courage and stamina on this trip.[11]

Meb Carriker's employment by the Academy from 1929 to 1938 brought in his only regular salary.[12] His other income, aside from the years on his coffee plantation, came from the sale of bird skins to the Academy, as well as to other American museums, such as the Carnegie Museum in Pittsburgh, the Field Museum in Chicago, and the Smithsonian. According to his son, Meb Carriker is credited with collecting eighty thousand birds and mammals, but mostly birds, between 1902 and 1962. Few scientists have had the opportunity to collect and record information in the field over such a long period of time and during constantly changing political and economic situations. In 1982, a new bird species from Peru was named *Grallaria carrikeri* in his honor.[13] It is one of a dozen bird and animal species to bear his name.

In 1929, Gifford Pinchot (1865–1946), the first director of the U.S. Forest Service (under Theodore Roosevelt), twice governor of Pennsylvania, and an Academy trustee, took his wife, Cornelia, son Gifford Jr., an eleven-man crew, and five accompanying scientists on a cruise to the Panama Canal Zone and many tropical islands of the Pacific Ocean. They traveled aboard his 148-foot, three-masted schooner, *Mary Pinchot* (figure 12.4). "The Philadelphia Academy of Natural Sciences was with us," Pinchot wrote, "in the small, modest, and cheerful shape of Doctor Henry A. Pilsbry, one of the foremost living conchologists [and curator of malacology at the Academy]. None of us suspected, when we first saw him, what dynamite was concealed in that small package, or that we were in the presence of a great outdoor

12.4 The *Mary Pinchot*, Governor Gifford Pinchot's schooner, en route to the South Pacific Islands, 1929. ANSP Archives coll. 2010-004.

man." In a tribute to Pilsbry's accomplishments on the trip, Pinchot noted: "The largest and doubtless the most important series [of collections], with the most new species and even genera, and potentially the largest additions to our knowledge, were the land shells, work upon which has only just begun."[14]

In speaking of the Galápagos Islands, Pilsbry wrote: "In most of them the rocky shores swarm with ugly black iguanas from 18 inches to a yard long. They can be caught by hand without much trouble. Governor Pinchot, who will try anything once, had iguana served for dinner; it is not at all bad."[15]

Fascinated with the islands' tortoises, Pilsbry reported on capturing one: "I fell upon his shell and embraced him. I have wanted to see the famous Galapagos tortoise on his native islands ever since I read Darwin's 'Voyage of the Beagle' years ago. *Testudo ephippium*, for that, I believe, is his full name, replied to my caress with a hoarse hiss, pulled his head and legs in and settled down on the ground." In the same letter, Pilsbry expressed the affection he felt for his fellow Academy workers at home and the scientist's dedication to his chosen discipline: "I hope all are well at the Academy. Please give my regards to the staff. I am wild to see them all and to settle down to work on my Galapagos snails. I have a great collection of them. But I have to see the Marqueses!"[16]

On his return, Pilsbry named two newly discovered snails for his hosts—a singular honor in natural science. He gave the snails the genus name *Giffordius* and then designated the new species *pinchoti,* after the governor, and *corneliae,* after the governor's wife. Pilsbry had collected these snails on the mountainous Caribbean island of Old Providence, which lies between Costa Rica and Jamaica.

Africa, with its great variety of wild animals, was probably the most popular destination for the collecting trips of wealthy Americans during this period. If the safaris were taken under the auspices of a museum, avid sportsmen could justify their love of hunting with the knowledge that they were securing valuable specimens for education and research. In 1929, trustee Frank B. Foster commissioned the acclaimed American hunter Alfred Klein (1883–1944) to collect a group of five lions in Tanganyika (now Tanzania), in East Africa, for a habitat group for the Academy (figure 12.5).[17] Klein, a diminutive five feet four inches tall and weighing only 120 pounds, was considered the "mightiest lion-killer of modern times," according to his obituary in *Time* magazine. "There is something cruel about Africa," he once observed, "yet I have never known anyone who stayed there for any length of time who did not burn to go back."[18] After a year of preparation, the Academy's Lion diorama opened in 1931. In that same year, Foster would organize his own collecting trip to Indochina with his wife and his daughter, Elizabeth.

For the very rich, safaris of the 1930s were both comfortable and leisurely. Days in the bush routinely began with a wake-up call before dawn. Waiters in tribal costumes served morning tea in the client's sleeping tent, followed by a hearty breakfast laid out in a special mess tent nearby. Thus fortified, the safari party set out on foot, horse, camel, or motorized transport for the morning hunt. A picnic lunch, served on fine china, might consist of cold guinea fowl or sliced game meat accompanied by ham, paté, cheeses, and fresh-baked bread from special provisioners in Nairobi as well as from Fortnum and Mason's in London. A selection of fine wines, tea, or coffee accompanied the meal. After a siesta under a shade tree, the hunt would resume in the afternoon. At nightfall, when the hunters returned, hors d'oeuvres and drinks around the campfire would precede a hot bath or shower in a waiting canvas tub. The day was usually capped by a formal, multicourse dinner by candlelight under the stars.[19]

On a few safaris, there was a certain amount of hedonism and a heedless disregard for wildlife that seems shocking by today's standards. Elizabeth Foster's journal of hunting in India at the time states the extreme: "While I was so successfully wounding elephants and missing tigers, Daddy had been doing much better by the buffalo," she wrote (figure 12.6).[20] In an effort to explain such an attitude, Gertrude Sanford Legendre, who collected African animals for the Academy in 1927, recalled: "There was more game than anyone could possibly imagine . . . I was hunting for the sport of it, which, at the time, required no justification. Hunting to me was great sport and dangerous game was a special challenge. The plains game presented no particular risk, but lion, mgogo (buffalo), and elephant did. In those days, the African plains were full of game; no one thought there could be an end to it."[21] Theodore Roosevelt, on safari in Kenya, described the train trip from Mombasa to Nairobi as a journey through "the Pleistocene."[22]

A great majority of serious hunters, including Roosevelt, were deeply interested in wildlife and committed to collecting for educational purposes as well as for sport. Roosevelt's friend Prentiss N. Gray (1884–1935) shared these two interests. A successful industrialist and Academy trustee, he planned and led an expedition across Africa from Kenya to Angola in hope of

LONG SHOT AT ONE OF
4 LIONS ON DAY KILLED
3 THIS LION KILLED
WITH ONE SHOT COUGHED
AS THOUGH BLOOD IN
LUNGS
LION THAT WAS
WITH THE ONE THAT
CHARGED.
THIS BULLET BLEW
ITS HEART TO PIECES

securing a rare giant sable antelope (*Hippotragus niger variani*) for another Academy diorama. In March 1929, he and his party sailed from San Francisco to Mombasa, on the east coast of Africa. After traveling by train to Nairobi, the expedition set off with fifty-nine porters and two tons of supplies. Gray employed the renowned white hunter Philip Percival (1886–1966), who as a young man had been an assistant on President Theodore Roosevelt's grand safari of 1909, from which he brought back mammal collections for the American Museum of Natural History. He had subsequently led many other expeditions for the rich and famous, including George Eastman (1854–1932), founder of Eastman Kodak Company, and Gary Cooper (1901–1961), the leading man in such Hollywood epics as *Sergeant York* and *High Noon*. Five years after directing the Gray Expedition, Percival served as Ernest Hemingway's guide and would appear as "Pop" in Hemingway's book *The Green Hills of Africa* (1935).[23]

Witnessing thousands of wildebeests migrating northward across the Serengeti Plains in Tanganyika, Gray observed that "it recalled the stories we had heard our fathers tell of the bison on our own plains, before the westward movement of settlement practically wiped out this splendid American animal."[24] To twenty-first-century sensibilities, it seems ironic that after such a statement of nostalgia and acknowledgment of the consequences of overhunting, Gray went on to shoot elephant, rhino, buffalo, oryx, waterbuck, impala, zebra, and other game with enthusiasm. Even so, Gray and his friends, including Theodore Roosevelt, George Bird Grinnell, and Gifford Pinchot, were among the conservationist-sportsmen responsible for preserving vast wilderness areas in the United States and encouraging the establishment of the great game parks in Africa.

After collecting the desired number of museum specimens in East Africa, the Gray Expedition returned to the United States via a transcontinental journey to the west coast of Africa, traveling thousands of miles by automobile, train, and riverboat. On the way, expedition members followed much of David Livingstone's route, even passing the ancient mango tree where he and Henry Morton Stanley first met in 1871. As they traveled, they were uncomfortably aware that this same route had also been

FACING PAGE 12.5 Labels with bullets used by Alfred Klein on an African expedition sponsored by Frank B. Foster, 1929. ANSP coll. 2010-004. Purcell photograph.

12.6 Elizabeth Foster on safari in India, 1931. ANSP Archives coll. 2010-004.

the old Arab trade trail for ivory and slaves.[25] In Angola, they secured two giant sable antelopes for the Academy. This had been one of the principal objects of the expedition. Sadly, today, after decades of civil war and uncontrolled hunting, this species teeters on the verge of extinction.[26]

The giant sable antelope, revered as the national icon of Angola, was first reported in 1909 by a British engineer in charge of constructing a railway from Angola's Atlantic coast to the Belgian Congo. The animal was even then so rare that in 1922 Frank Varian persuaded the Portuguese colonial authorities to declare the giant sable "royal game," requiring special permission to shoot it. Quentin Keynes (1921–2003), a great-grandson of Charles Darwin's, took the first motion pictures of the giant sable in its natural habitat in 1954 and subsequently showed his historic sixteen-millimeter film of the animal at the Academy. From 1969 to 1970, Academy mammalogist Richard Estes and his wife, Runi, made a special study of this species of the antelope family in the Luando and Cangandala Parks of Angola before the country's independence from Portugal precipitated a devastating twenty-seven-year war. The Esteses found that the giant sable antelope was interbreeding with the roan antelope, further threatening the survival of the dwindling giant sable population. The last hope of bringing the giant sable back from extinction is now focused on captive breeding.[27]

Anticipating the modern era of photographic safaris, Gray took thousands of feet of motion picture film as well as still pictures of the animals the team encountered. "These photographs form one of the most important permanent records of the wild life of Africa ever made," wrote the editor of the Academy's 1929 *Year Book*. "With many species of these mammals becoming

12.7 An Ivory Gull flies past the *Effie Morrissey* in Arctic waters in a painting by Francis Lee Jaques (1887–1969). Oil on canvas, 1964. Collection of Thomas E. Lovejoy.

rarer yearly, the Academy is now the fortunate possessor of a record of what has well been called the closing scenes of the 'Age of Mammals'"[28] (figure 12.8).

The following year, 1930, under more modest circumstances, Academy curator H. Radclyffe Roberts collected waterbirds along the White Nile, and with the permission of the government scientist of the Sudan, the Academy secured a large number of specimens from the Nuba Mountains, the Blue Nile, and the Red Sea coast. The Rodolphe Meyer de Schauensee South African Expedition rounded out the Academy's efforts in Africa that year by collecting birds, mammals, reptiles, and fishes from the Kalahari Desert, while the Academy-Vaux Expedition explored the Belgian Congo's copper mines and the diamond deposits in the Transvaal, South Africa, bringing back hundreds of specimens of rare and beautiful minerals to be displayed in Mineral Hall.[29]

While his friends were enjoying the warmth and biological diversity of Africa, Harry Whitney journeyed to Greenland on the two-masted schooner *Effie Morrissey*, chartered by the Academy, to collect musk ox for another Academy diorama (figure 12.7). To his surprise and delight, he found the largest herd of these animals he had ever seen. "It was here that we captured 'Shannon'—the baby bull [musk ox] which is now on the farm of R. R. M. Carpenter, in Wilmington, Delaware," he recounted (figure 12.9). "We lassoed this little fellow without harming any of the other animals, and it was remarkable how quickly he became tame. In a few days after he was on board the *Morrissey*, he would follow one all over the vessel, like a dog."[30] Later, the party captured a young female musk ox they called Maureen. On an island a hundred miles to the south, Whitney and his team shot a group of five musk oxen—one bull, two cows, a yearling, and a calf—that served as the centerpiece for the museum's fifth North American diorama. The diversity may not have been as great as in Africa, but the wildlife was reassuringly abundant. "This was the first American expedition to that part of Greenland," wrote Whitney, "and I found it to be one of the greatest game countries I have ever seen. In the waters there were narwhal, walrus, seal and polar bear; in the rivers, plenty of salmon and trout; while on land, there were musk-ox, Arctic hares, ptarmigan, geese and ducks."[31]

By 1932, the effects of the Depression were beginning to take their toll on the Academy. A review of the year stated: "The 121st year of this oldest institution of its kind in America was marked by increased activity among officers, staff, and volunteer workers to offset, as far as possible, the handicap of hard times."[32] Nevertheless, there were many bright spots: a "magnificent collection of large butterflies from various parts of the world" that W. Judson Coxey gave to the museum; the large number of minerals, insects, and plants collected from the seldom-visited regions of British Columbia by Mrs. J. Norman Henry and her daughter Josephine; and the 975 birds, 200 mammals, and numerous mollusks brought back from China and Tibet by Brooke Dolan II (1908–1945) the year before, in 1931. Highlights of Dolan's mammal collection were the takin (*Budorcas taxicolor*), a huge, shaggy, cowlike creature, a relative of the musk ox that inhabits mountainous terrain up to an altitude of 4,500 meters, and the giant panda (*Ailuropoda melanoleuca*).[33] This animal, so rare and so beloved today, was described in a newspaper article at the time as "a grotesque bear-like creature."[34] Both were to become the focal points of dioramas in Asia Hall.

12.8 Prentiss Gray (1884–1935) and his wife, Laura Sherman Gray, filming in Africa, 1929. ANSP Archives coll. 457.

12.9 Musk ox calves Shannon and Maureen were brought back from Greenland by Harry Whitney in 1929. They were kept and bottle-fed at the Delaware farm of R. R. M. Carpenter. ANSP Archives coll. 2010-004.

None of the birds Dolan collected represented species unknown to science, but 155 species and subspecies and nine genera were previously unrepresented in the Academy's collections (figure 12.11). This was "one of the largest series of novelties added by a single expedition for many years," according to the Academy's curator of birds, Witmer Stone.[35]

On the 1931 trip, in addition to Gordon Bowles, who was doing ethnological work for the University of Pennsylvania's Museum of Anthropology and Archaeology, the Dolan Expedition included Hugo Weigold, a noted German ornithologist and director of the Provinzial Museum of Hanover, Germany; Ernst Schäfer, a brilliant young zoologist from Göttingen, Germany; and Otto Gnieser, a cameraman.[36]

In 1934, Dolan and Schäfer returned to western China and eastern Tibet. Dolan reported that the expedition, along with several tons of gear, began by steaming twelve days up the Yangtze River to Chungking (now Chongqing), China's innermost treaty post. For the first stage of the journey, he hired sixty yaks from a nomad prince at roughly twenty cents an animal per day. The only difficulty was that the yaks would not eat at night, so it was necessary to call a halt at noon each day and allow enough hours for grazing. Dolan recounted that during the trip he used "steamships, junks, rafts, coracles, suspension bridges, bamboo cables, airplanes, railroads, autocars, sedan chairs, rickshas, not to mention the usual horses, mules, and yaks" to transport his party and equipment.[37]

Aside from his numerous natural history collections, Dolan brought back thousands of feet of motion picture film recording his experiences on the high peaks of the northeastern Himalayas. Dolan's photography and his stated intention to conduct a year's

12.10 Harry Whitney (1871–1936), the scion of a wealthy New York family who settled near Philadelphia, made several trips to Greenland during the first three decades of the twentieth century. Many of the bird and mammal specimens he collected there were used for scientific research at the Academy, and some became centerpieces for the museum's North America Hall. ANSP Archives coll. 2010-004.

FACING PAGE 12.11 A series of Beautiful Rosefinch (*Carpodacus pulcherrimus*) specimens collected by Brooke Dolan in Tibet, 1934–35. ANSP Ornithology Department. Purcell photograph.

12.12 George Vanderbilt camping in Africa with Baron Bror von Blixen, James A. G. Rehn, and Harold T. Green, 1934. ANSP Archives coll. 457.

survey of the zoology of eastern Tibet prompted a London news service to comment: "It is gratifying to find that expeditions in out of the way parts of the world are concentrating more upon observation of animal distribution and habits than upon the unrestrained collecting which was often divorced from any sort of ecological observation."[38] Today, this focus is an important part of every serious collecting effort.

Although the country was still in the depths of the Great Depression, collecting for the Academy continued throughout the 1930s in many foreign countries and in various places in the United States. The institution was fortunate in having a number of patrons with the means to carry out its important fieldwork at a time when many other museums had been forced to suspend their overseas activities.

George Vanderbilt, the twenty-year-old son of Alfred Gwynne Vanderbilt, who had been lost in the sinking of the British ship *Lusitania* in 1915, financed and led a ten-month expedition to Kenya, Uganda, the Belgian Congo, the Cameroons, and the French Sudan beginning in 1934. He was accompanied by a friend, Vicomte Sosthenes de la Rochefoucauld; James A. G. Rehn (1881–1965), the Academy's curator of entomology; and Harold T. Green (1897–1967), director of exhibits. An unusual addition to the expedition's support staff was Vanderbilt's valet, who helped with the collections when not attending to his employer's personal needs. As their principal expedition guide, Vanderbilt hired the internationally famous white hunter[39] Baron Bror von Blixen, a Swedish resident of Kenya Colony and former husband of the famous writer Isak Dinesen, whose book *Out of Africa*, about the couple's struggles with a coffee plantation in Kenya, was made into a compelling movie starring Meryl Streep and Robert Redford in 1985. Donald Ker of Nairobi, Kenya, served as Blixen's assistant hunter (figure 12.12). Given his broad acquaintance with African big game and safari life, Baron von Blixen's task, in addition to guiding the hunters, was to handle "a four-truck safari and a score of native truck drivers, personal boys, trackers, gun-bearers, cooks, and skinners thousands of miles from their base. The trucks all were conspicuously painted on the roofs of their seat-housing for emergency airplane recognition."[40] Especially important among the five thousand birds, fifteen thousand insects, and numerous fishes and reptiles collected were the gorillas and okapis secured for new habitat groups.

The okapi was first discovered by Sir Harry Johnston, British governor of Uganda, toward the end of the nineteenth century. Formally named *Okapi johnstoni* in 1901, it is a long-necked, antelope-like animal the size of a horse, with striped

FACING PAGE 12.13 Harold Green on Vanderbilt African Expedition with a gorilla specimen collected for an Academy diorama, 1934. ANSP Archives coll. 2010-004.

legs and outsize ears, distantly related to the giraffe. According to the renowned explorer Roy Chapman Andrews, "The okapi is a veritable living fossil. For fifteen million years or more it has been quietly plodding up the hill of time changing with the years almost not at all. Today [1938] he is one of the rarest and shyest of all mammals, living deep in the gloomy forests of the Belgian Congo. Perhaps that is why he has held over from the Miocene Period."[41] Donald Ker wrote in his expedition notes that the okapi was "about as plentiful as the Loch Ness monster."[42]

The hunters were able to collect only two okapi specimens, a female and a young male three or four months old. In a report, Harold T. Green, the Academy's curator of exhibits, stated that it was not possible to obtain an adult male of the species because time was limited, but since the expedition had a permit from the Belgian government to collect the animal, arrangements were made with a Mr. Putnam to obtain and ship the skin and skeleton of a perfect male okapi. Unfortunately, however, Baron von Blixen held the permit and never allowed Putnam to have it or even to copy it.[43] Thus, to this day the Academy's diorama lacks a male okapi.

After Vanderbilt came down with fever and was transported to a hospital in Nairobi, Ker recorded, "Blixen, Rochefoucauld, Green, [Rehn], and I left on a foot safari across the Sanju River for gorilla . . . besides the five Europeans we had a flag bearer, a paramount chief [a Pygmy], two native chiefs, an armed guard of about seven or eight [men], four tipboys [?] with 32 native bearers, about 50 porters, our own servants, and the chief who owns 40 wives [and] brought along 10. We got 3 gorillas and the largest stood 5 feet nine inches high, a spread of 9 feet and was just too colossal being one mass of strength and muscle. They are extraordinarily human-like" (figure 12.13).[44]

While Ker makes no further mention of his feelings about killing gorillas, his friend Carl Akeley (1864–1926), an exhibit preparator at the American Museum of Natural History, who

12.14 Henry A. Pilsbry (right) oversees a worker on a trip with Francis W. Pennell collecting in Mexico, 1935. Glass lantern slide. ANSP Archives coll. 2009-013.

is considered the father of modern methods of taxidermy, expressed the pang of conscience he felt after shooting a gorilla for that museum in 1921: "As [the gorilla] lay at the base of the tree, it took all one's scientific ardor to keep from feeling like a murderer. He was a magnificent creature with the face of an amiable giant, who would do no harm except perhaps in self-defense or in defense of his friends. Of the two, I was the savage and the aggressor."[45]

In December 1934, after the Academy party had returned home, gorilla and okapi specimens in tow, Baron von Blixen wrote jovially to Rehn and Green wishing them a happy new year. "I feel that every good meal you have now," he said, "you will thank good [*sic*] you are not with the damn shot Blixen, who gave us nothing but duck and kudu!"[46] There is no mention of the okapi permit.

During this long and productive trip, Rehn amassed 14,000 insects, more than 1,300 birds and small mammals, and several thousand fishes, snakes, reptiles, and mollusks for the Academy's collections. Harold Green also considered the trip one of his most successful. Throughout his forty-year career at the Academy, Green traveled over two hundred thousand miles collecting animals, plants, and rock and soil samples, taking photographs and making sketches that he used in preparing the many dioramas he designed and built. The two that came from the 1934 Vanderbilt Expedition—the gorilla and the okapi—must have pleased him the most.

There were many expeditions in the Western Hemisphere in 1935. Cruising in West Indian waters among the Bahamas and the Virgin Islands on the diesel ketch *Antares*, Colonel Edwin M. Chance, accompanied by his wife and his son Britton,[47] led a party collecting fishes for the Academy and making a study of the habits of the swordfish and other large game fish. In the central highlands of Guatemala, Meyer de Schauensee spent two months collecting birds, fishes, and orchids and studying aquatic life in Lake Atitlán, five thousand feet above sea level and known to be deeper than one thousand feet. Academy trustees Brandon Barringer and Reginald Jacobs later flew from Philadelphia to join him.

Also in Central America were Henry A. Pilsbry (1862–1957), curator of malacology, and Francis W. Pennell (1886–1952), curator of botany at the Academy. These men spent time in the high plateau of central Mexico, Pilsbry searching for both fossil and present-day snails, and Pennell seeking certain species of plants (figure 12.14). The results of their research were used in completing a chart of the north-south migration of plant and animal life that followed the upheaval of the Isthmus of Panama some twenty million years ago.

S. Dillon Ripley (1913–2001), who was in charge of American Intelligence Services in Southeast Asia for the Office of Strategic Services (OSS) (the predecessor of the CIA) during World War II and later became secretary of the Smithsonian Institution (1964–84), joined the Denison-Crockett South Pacific Expedition bound for Dutch New Guinea in 1937, in charge of zoological collecting. The Academy sponsored the journey to learn more about the little-known birds of the region.

Ripley, a recent Yale graduate, had planned to study ornithology at Columbia when the invitation came from Charis and Frederick Crockett to join their proposed expedition. Such

12.15 Aboard the *Chiva*. Charis and Frederick Crockett, S. Dillon Ripley, and crew on the Denison-Crockett Expedition to Dutch New Guinea, 1937. ANSP Archives coll. 113 IV.

Acad. Nat. Sci. Philadelphia
PARADISEA MINOR
SALEM, VOGELKOP
Denison-Crockett So. Pac. Exped. 1936-37
D. RIPLEY coll.

an adventure was too good to resist, and Ripley postponed his graduate work, urged to go by his mentor at the American Museum of Natural History, the distinguished ornithologist Ernst Mayr. At the Academy, Meyer de Schauensee, curator of birds and a collector of Asian species from several previous expeditions, gave Ripley advice and encouragement and a copy of Alfred Russel Wallace's *The Malay Archipelago* (1869), described by Ripley as still the most up-to-date book on the East Indies.[48]

The party set sail from Philadelphia on 1 December 1936, aboard the sixty-foot auxiliary schooner *Chiva* (figure 12.15). Nine months and twelve thousand miles later, they reached Dutch New Guinea, having visited Tahiti, Samoa, and the Fiji and Solomon Islands.

In his book *Trail of the Money Bird* (1942), Ripley describes many species of exotic rare birds he collected, "but the most spectacular one of all," he wrote, "was a bird called the sickle-bill. They are big birds of paradise with tails nearly three feet long. They are soft velvety black all over, except for two curious butterfly-wing shields that rise out from each side of the breast. The bill is about three and a half inches long, thin and curved. I saw them only twice."[49]

After leaving his friends from the expedition, Ripley explored the island for several months on his own with two guides and various hired porters. He had heard tales that the Karoon people in the Tamrau Mountains were cannibals but dismissed this notion of their reputation as merely legendary, especially after a member of the tribe told him of a recent episode. It seems that a party of Chinese, trading for Bird of Paradise skins, happened upon the Karoon, who surrounded them and began plucking the arm of an especially fat man while drawing their fingers across their lips and making "smacking sounds." The Chinese fled in horror. Ripley reported that "the natives all thought this was a wonderful joke and have been laughing about it ever since."[50]

On the trip, Ripley collected more than three hundred rare bird specimens. Aside from the bird skins, he also collected and brought back to the United States eighty-seven live birds in forty-two specially made bamboo cages. On first surveying his accommodations aboard the small freighter, he wondered if the two-by-four storeroom allotted for provisions would be adequate for storage and preparation, as he would be "cutting up fish for the storks and herons, chopping bananas and melons for the paradise birds, parrots, and fruit-eating pigeons, pouring out grain for the ducks, pheasants, peacocks, and jungle fowl."[51]

Miraculously, after forty-nine days at sea, Ripley and his devoted mother, who had traveled to New Guinea to meet him and to accompany him on his return, as well as to help care for his birds, arrived safely in Boston harbor. "Dillon Ripley is much to be congratulated," wrote Ernst Mayr of his protégé's accomplishments, "for though hampered by fever, sometimes short of food, and having often to make very long marches, he managed to secure a really fine collection in a difficult country [Dutch New Guinea] which he was the first naturalist to explore."[52]

In September 1939, the war began that shattered the world well into the next decade. Although the United States was not involved for another two years, natural science at the Academy, by necessity, was confined to home base. On 7 December 1941, when the Japanese bombed Pearl Harbor, the United States entered World War II.

An editorial in *Frontiers* in the fall of 1942 stated: "Founded during the nation's first war after the Revolution, the Academy of Natural Sciences of Philadelphia has seen staff members go to the combat and supporting forces of every American national conflict. The service roster in the present war has been lengthening steadily since December 7, 1941."[53]

Brooke Dolan II, research associate in the Department of Mammals and an Academy trustee, joined the U.S. Army Air Forces and was assigned to the Office of Strategic Services, with the China-Burma-India theatre as his ultimate destination. Selected by fellow OSS officer Colonel Ilya Tolstoy, grandson of the great Russian writer, to accompany him on a special diplomatic mission to Tibet for President Franklin D. Roosevelt, Captain Dolan left for Lhasa, the Tibetan capital and seat of the Dalai Lama, in October 1942. The primary purpose of the mission was to find ways and land routes by which the United States could transport supplies to its Chinese allies after the Burma Road was closed and an American plane flying over the Himalayas crashed in Tibet, showing the difficulty of air transport.

Following a harrowing two-month trek over the mountains from Darjeeling, India, Dolan and Tolstoy arrived in Lhasa, where they were presented to the six-year-old Dalai Lama and his regent in the throne room of the Potala Palace (figure 12.17). They gave His Holiness a letter of greeting and a number of gifts from President Roosevelt. Dolan described the presentation of another gift that had special meaning because of its implied eternity:

> In addition to the President's gifts and our own, we are bestowing a certificate of life membership in the Academy of Natural Sciences on Gyalwa Rinpoche [the Dalai Lama]. The recipient so designated is a god constantly

12.16 Birds of Paradise, collected by S. Dillon Ripley in Dutch New Guinea on the Denison-Crockett Expedition, 1937. ANSP Ornithology Department. Purcell photograph.

reincarnated without death or cessation of existence, so that the membership is perpetual. The secretary, Surkhang, wrote reams of comment on this bestowal, saying that the priests would have to be consulted as to its acceptability since they might fear that the gift carried some obligation to the God-king. However they believed that it would prove acceptable.[54]

Permitted to journey on to western China, the mission ended after three months in Lanchow, north of Chongqing.[55] Nine months after his return to America, Dolan was back in China and off on a six-hundred-mile trek to the mountainous headquarters of Mao Zedong, Chou En-lai, and their small group of Communist followers. From there, he scouted with a small Chinese party, hiding in caves and tunnels to elude the Japanese who were searching for him. Subsequently, he helped rescue a B-29 crew and fighter pilot who had been forced down near Peking (now Beijing). Dolan did not return from his final OSS mission; he died prematurely in China in 1945.[56]

Although nothing came of the proposed land route for transporting supplies across Tibet to China, Dolan kept fascinating field journals of his encounters with the Tibetan people and their manners and customs. He also made detailed observations of the country's wildlife, with which he was already familiar from two previous scientific expeditions there on behalf of the Academy in 1931–32 and 1934–36. With Dolan's death the Academy lost one of its brightest stars.

In November 1945, the Academy's board of trustees stated in a resolution: "Due to [Dolan's] zeal, ability and generosity, the Free Museum [as the Academy was called at the time] acquired

FACING PAGE 12.17 The Dalai Lama (age six) on his throne in the Potala Palace, Lhasa, Tibet, 1943. ANSP Archives coll. 64. Dolan photograph.

12.18 Brooke Dolan and Ilya Tolstoy riding past the Potala Palace, Lhasa, Tibet, en route to China, 1943. ANSP Archives coll.64. Dolan photograph.

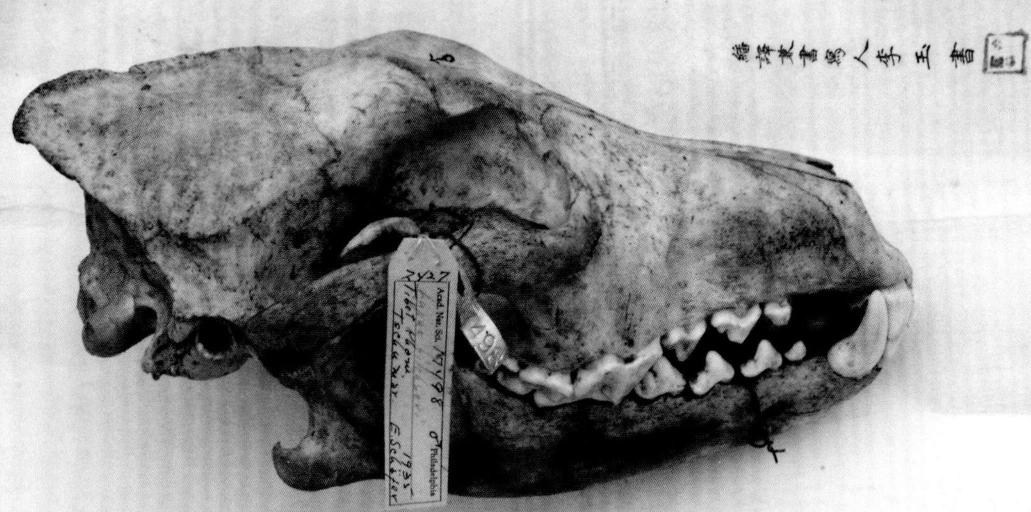

April 10, 1935 JYEKUNDO, Chinghai.

We four men of the party of the American expedition to China, Marion Duncan, Brooke Dolan, Americans, Ernst Schäfer German agree: — the magistrate of Jyekundo is not responsible for our safety or for the lives or property of ourselves or our men on the road across Chinghai to Hsining. Lee Yü Su, chinese (Chungking) agrees. We acknowledge he has requested us to return to Szechuan, he has advised we do not go thru Chinghai. Our Nanking passports do not give us specific right to travel in Chinghai, and the magistrate cannot help us in any way. If robbers, tibetans, or Ngoloks rob or kill us, that is not the affair of the Jyekundo magistrate and the government of Chinghai is not responsille.

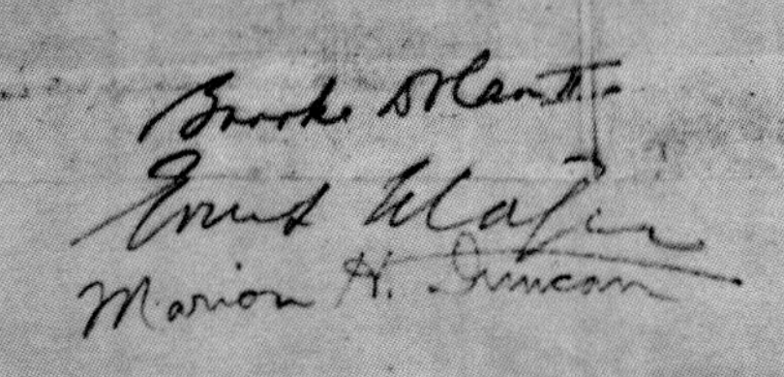

Brooke Dolan II
Ernst Schäfer
Marion H. Duncan

the Giant Panda and Takin habitat groups. His Asiatic expedition from 1934 to 1936 may be said to rank with the important field projects of modern times. More than 3,000 birds and 140 mammals were brought back. The Dolan birds made the Academy's Tibetan collections the best in the United States."[57]

For fifteen years after World War II, when the influence of European empires was waning, administrators of colonial island possessions were largely friendly toward American scientific expeditions. Up until 1960, there existed a window of relatively easy access to territories ruled by Great Britain, the Netherlands, Australia, and others.

Alfred James Ostheimer (1908–1983) and his wife, Ruth, with the cooperation of Henry Pilsbry, organized more than a dozen expeditions between 1949 and 1957, resulting in significant additions to the Academy's mollusk collections.[58]

An Ostheimer expedition in 1955 to the Pacific island nation of Palau (some five hundred miles east of the Philippines) produced twenty-six cases of specimens weighing 4,800 pounds for the Academy. Later, an Ostheimer trip to New Guinea, which included the Academy's malacologist Virginia Orr (1920–1986) and W. B. Dixon Stroud (1917–2005), an amateur who would become an important and innovative Academy trustee, succeeded in amassing possibly the best collections of any expedition of that era, with materials exceeding six hundred species and 6,500 lots.[59]

Because of changes in the ethics of collecting rare species of mammals, birds, amphibians, fish, and mollusks, as well as concerns for the environmental impact of their global loss, the practice of amassing huge collections of specimens of every kind has declined in recent years, but research using collections is still central to the Academy's mission. Of the eighteen million specimens cared for by the Academy of Natural Sciences, a great number are "types," meaning the individual from which the description was originally taken. These provide a historical record of biodiversity and are essential reference points for past and future taxonomic classification. The scientific data associated with these specimens provide an irreplaceable baseline that scientists use to better understand changing environments and the organisms that depend on them.

PTS

FACING PAGE 12.19 Wolf skull (*Canis lupis*, male) collected by Ernst Schäfer in Kham, Tschumar, Tibet, 1935. It is shown with the travel document carried by Brooke Dolan, Ernst Schäfer, and Marion Duncan absolving Chinese officials of any responsibility should the explorers be robbed, injured, or killed during their expedition. ANSP Mammalogy Department and ANSP Archives. Purcell photograph.

12.20 Brooke Dolan with "Miss Tick" (a Lhasa apso) in his saddlebag, Lhasa, Tibet, 1943. ANSP Archives coll. 64.

Notes

1 *Academy of Natural Sciences of Philadelphia ANSP Year Book for 1929*, 34.

2 Steve Quinn, *Windows on Nature* (New York: Abrams with the American Museum of Natural History, 2006); and Karen Wonders, *Habitat Dioramas: Illusions of Wilderness in Museums of Natural History* (Stockholm: Uppsala University, 1993).

3 Known as the Newbold British Columbian Expedition.

4 These bears live only on Kodiak Island in Alaska and are more closely related to Old World brown bears than to grizzlies.

5 Charles Cadwalader, *ANSP Year Book for 1929*, 3.

6 Rodolphe Meyer de Schauensee, "A Collection of Birds from Siam," *Proceedings of the Academy of Natural Sciences* 80 (1928): 554–55. Known as the Meyer de Schauensee–Smith Asiatic Expedition.

7 Rodolphe Meyer de Schauensee, *Proceedings*, 553–79.

8 Storrs L. Olson, "A Carriker Trilogy: Chapters in a Saga of Neotropical Ornithology," *The Auk* (January 2007): 357–61.

9 See Melbourne R. Carriker, *Vista Nieve: The Remarkable True Adventures of an Early Twentieth Century Naturalist and His Family in Colombia, South America* (Rio Hondo, TX: Blue Mantle Press, 2001), 257.

10 Barbara Mearns and Richard Mearns, *The Bird Collectors* (New York: Academic Press, 1998), 349.

11 Carriker, *Vista Nieve*, 137.

12 Although Carriker was not made an official Academy employee, Cadwalader told Stone he wanted Carriker "to consider himself a member of our staff so far as availing himself of the privileges, etc. of the Academy." He instructed that cards should be printed designating Carriker a "Field Representative, Department of Vertebrate Zoology, Academy of Natural Sciences of Philadelphia." See letter from Cadwalader to W. Stone, 9 August 1929, and Cadwalader to J. A. G. Rehn, 20 September 1929, Newbold file, Cadwalader papers, unprocessed collection, coll. 2009-034.

13 Carriker, *Vista Nieve*, 272.

14 Gifford Pinchot, *To the South Seas: The Cruise of the Schooner Mary Pinchot to the Galapagos, the Marquesas, and the Tuamotu Islands, and Tahiti* (Philadelphia: John C. Winston, 1930), 11, 492. Pinchot is regarded as the father of American conservation because of his great and unrelenting concern for the protection of the American forest (*United States Forest Society History*, Forest History Society, http://www.foresthistory.org). Conchology is the study of shells, while malacology is the study of the entire animal.

15 Pilsbry to Cadwalader, "On board 'Mary Pinchot,'" 17 July 1929, and in Pinchot, *To the South Seas*.

16 Ibid.

17 *ANSP Year Book for 1929*, 4–5.

18 Quoted from obituary of Alfred John Klein, *Time* (n.d.).

19 Brian Herne, *White Hunters: The Golden Age of African Safaris* (New York: Henry Holt, 1999), 71, 74. For more on the history of African safaris, see Bartle Bull, *Safari: A Chronicle of Adventure* (New York: Viking, 1988).

20 *ANSP Year Book for 1931*, 35.

21 Gertrude S. Legendre, *The Time of My Life* (Charleston, SC: Wyrick, 1987), 32.

22 Ibid., 31–32.

23 For more on Percival, see Emily Host, *Bwana Bunduki: A History of Early East African Professional Hunters* (Dannevirke, New Zealand: Quartz Publishers, 2007).

24 *ANSP Year Book for 1931*, 35.

25 Prentiss N. Gray, *The Livingstone Trail in Tanganyika, Belgian Congo, and Angola* (Missoula, MT: Boone and Crockett Club, 1929), 23. Reprint, Prentiss N. Gray, *African Game Lands: A Graphic Itinerary* (Boston: Office of the Sportsman, 1995), 131.

26 John Frederick Walker, *A Certain Curve of Horn: The Hundred-Year Quest for the Giant Sable Antelope of Angola* (New York: Grove Press, 2002).

27 John Frederick Walker, "Antelope from the Ashes," part 1, *Africa Geographic*, June 2010.

28 Special Reports on Expeditions of 1929: Prentiss N. Gray, "Across Africa from Kenya to Angola, The Gray African Expedition of the Academy of Natural Sciences of Philadelphia," *ANSP Year Book for 1929*, 34.

29 *ANSP Year Book for 1929*, 9.

30 Harry Whitney, "Where the Musk-Ox Ranges," *ANSP Year Book for 1930*, 35.

31 Ibid., 39.

32 *ANSP Year Book for 1932*, 7.

33 Ibid., 3, 5.

34 Theodore Roosevelt Jr. and his brother Kermit Roosevelt brought back the first giant panda specimen to the United States for the Field Museum of Natural History. It was put on display in 1930.

35 Witmer Stone, "Zoological Results of the Dolan West China Expedition of 1931, Part I—Birds," *Proceedings of the ANSP* 85 (1933): 165–222.

36 Ernst Schäfer was an eager and early recruit to the German SS. After the war, he insisted to the Americans that he joined only at the instigation of Himmler following his return from the United States in 1936, but his captured personal files confirm that he had applied for membership in 1933, immediately after Hitler came to power. Karl E. Meyer and Shareen Blair Brysac, *Tournament of Shadows: The Great Game and the Race for Empire in Central Asia* (Washington, DC: Counterpoint, 1999), 513.

37 Brooke Dolan II, "Road to the Edge of the World," *Frontiers* 1, no. 1 (October 1936): 5.

38 International Press Cutting Bureaus, 110 Fleet St., London (ANSP coll. 113, I-III, f. 6, news clippings).

39 The term "white hunter" was used at the time for nonnatives who came to Africa as hunters and guides for hire.

40 James A. G. Rehn, "Zoological results of the George Vanderbilt African Expedition of 1934—Part I, Introduction and Itinerary," *Proceedings of the ANSP* 88 (1936): 3.

41 Roy Chapman Andrews, "Giraffe Without a Neck: Meet the Okapi—a Living Fossil in This Amazing Planet," *This Week Magazine*, 17 April 1938.

42 Edie Ker, *The Anecdotes of Donald Ker—And a Bit More—As Remembered and Recorded by Edie Ker* (Alexandria, VA: 1988), 20. Today there are reportedly ten thousand to twenty thousand okapi in the wild.

43 Harold T. Green, "Report, " 4 January 1937, ANSP coll. 2010-004 (unprocessed), box 3, file 22a.

44 Edie Ker, *Anecdotes of Donald Ker*, 26–27.

45 Carl Akeley quoted in Quinn, *Windows on Nature*, 27.

46 Bror von Blixen to James A. G. Rehn and Harold T. Green, 13 December 1934, Batangafo, Africa (ANSP coll. 113, Expedition General).

47 Britton Chance would become an eminent biochemist at the University of Pennsylvania's Johnson Foundation.

48 Dillon Ripley, *Trail of the Money Bird: 30,000 Miles of Adventure with a Naturalist* (New York: Harper & Brothers, 1942), xii.

49 Ibid., 266.

50 Ripley, abstract, 2, coll. 113, folder 6.

51 Ripley, *Trail of the Money Bird*, 291.

52 Ernst Mayr (associate curator of the Whitney-Rothchild Bird Collection, American Museum of Natural History, New York) and Rodolphe Meyer de Schauensee, "Zoological Results of the Denison-Crockett South Pacific Expedition for the Academy of Natural Sciences of Philadelphia, 1937–1938: Part IV, Birds from Northwest New Guinea," *Proceedings of the ANSP* 91 (1939): 100.

53 Editorial from *Frontiers* (October 1942): 16.

54 "Across Tibet: Excerpts from the Journals of Captain Brooke Dolan, 1942–1943," *Frontiers* 2 (1980): 14.

55 Ibid., 1–45.

56 Meyer and Brysac, *Tournament of Shadows*, 548–49.

57 ANSP Resolution of the Board of Trustees, 8 November 1945, ANSP biography file.

58 John Ostheimer, "Alfred James Ostheimer, Shell Collector Extraordinaire: An Amateur Makes a Difference," draft article, 2010, 2.

59 Ibid., 10. The expedition to New Guinea took place in 1956. Virginia Orr married Robert Maes in 1963. See Robert Robertson, "Virginia O. Maes's obituary," *ANSP Proceedings* (1987): 527–32.

Cone of the sacred fir (*Abies religiosa*) collected in Ajusco, Mexico, November 1937, by J. H. Faull. ANSP Botany Department #13103 (gift of the Arnold Arboretum). The copper wire containment helps keep the seeds inside the cone. With a range that has been severely reduced by illegal timber harvesting in Mexico, the sacred fir provides critical wintering habitat for North America's migratory monarch butterflies (*Danaus plexippus*). Purcell

Chapter 13

Dioramas Defy the Great Depression

Don't you think that "biggest," "first," "only" etc. get a bit tiresome after a time? Like the MCZ [Museum of Comparative Zoology, Harvard], one of the charms of the Academy has always been that we haven't used too much bally hoo.
—Brooke Dolan to Charles M. B. Cadwalader, ca. 1932

At the end of North America Hall, on the first floor of what was once called the "Free Natural History Museum of the Academy of Natural Sciences," stands an enormous moose from Alaska's Kenai Peninsula.[1] To the millions of visitors who have viewed it through the years, it is the Academy's most iconic and beloved mammal specimen. Enlivened by its strikingly realistic setting, it not only represents the world's largest living deer but seems to embody the very essence of wilderness itself.

If encountered in life, the massive size, latent power, and unpredictable behavior of the moose would provoke understandable fear, but seen in the safe setting of a museum exhibition hall, for all but the very youngest, the intimidating giant stimulates curiosity and even affection. It inevitably draws visitors past the hall's other habitat groups for an irresistible closer look. In many natural history museums, mounted specimens do not benefit from close scrutiny, but the Academy's moose is a tour de force of modern taxidermy, which even after three-quarters of a century appears as lifelike as the day it was put on public view. Inconceivably large, and flawlessly presented in every detail, it reigns supreme over a reconstructed world of winter-worn spruce, stunted willow, and burned muskeg. The scene seems far larger than the seventeen-by-fourteen-by-eight-foot wood and Masonite shell in which it is contained. It is totally convincing, transporting, inspiring. And yet, for all its exacting accuracy, this reconstructed slice of nature says as much about human competition—and the ambitions of the Academy of Natural Sciences' managing director Charles Cadwalader—as it does about the species represented, Alaska, or the balance of nature.[2] If truth be known, the moose is a fraud.

When Cadwalader's cousin, Nicholas Biddle, returned from an Alaskan hunting trip in 1933 and presented the moose and other raw materials he had collected for a diorama, the Academy's managing director was delighted with everything but the size of the antlers. They were big, but not quite big enough for Cadwalader. Biddle may have shot the animal, but it was Cadwalader who was paying for its display, and he was determined that it would be the biggest and best in the world.[3] So, as the diorama artist Clarence Rosenkranz worked his magic on the background painting, and other members of the exhibits staff painstakingly re-created the rest of the scene, Harold Green, director of exhibits (who had been with Biddle when the moose was collected), and Louis Paul Jonas, the taxidermist commissioned to mount the animal, used their network of contacts in the museum and sporting worlds to secure the enormous substitute antlers that now grace the head of Biddle's moose (figure 13.2).[4] With the skillful use of his scalpel, and a few nuts and bolts to hold everything in place, Jonas replaced the impressive but uncompetitive sixty-nine-inch antlers that had come with Biddle's moose

13.1 Scale model of the Desert of Borkou diorama created by Jonas Brothers Studio and the Academy exhibition department. The habitat group was based on specimens collected on William K. Carpenter's expedition to French West Africa in 1955. ANSP Exhibits Collection. Purcell photograph.

with a superior seventy-plus-inch rack once borne by another animal.[5] Through the miracle of modern taxidermy, a hybrid supermoose was thus created to delight, instruct, and impress Academy visitors for generations to come. When the diorama opened to the public in the fall of 1935, no mention was made of the deception, but Charles Cadwalader could confidently state that the Academy was exhibiting the largest moose specimen in North America.[6]

Charles Meigs Biddle Cadwalader, whose full name reveals his filial connection to many of Philadelphia's oldest, wealthiest, and most aristocratic families, began his formal association with the Academy in 1925, when he was asked to chair a committee to improve the Academy's financial base by expanding its dues-paying membership. He was so successful in his recruitment efforts that, in 1927, he was elected to the Academy's board of trustees. A year later, he became the Academy's first managing director, serving under then-president Effingham B. Morris. Ten years later, he succeeded Morris as president and CEO of the Academy, securing almost total control of the institution for another fourteen years (figure 13.4). His tenure was transforming and left a legacy of public exhibition in the Academy's museum that extends to the present day.

At the time Cadwalader assumed the newly created position of managing director, the Academy, while still an institution with considerable scientific prestige, had fallen behind many of the country's other natural history museums in developing

13.2 The Moose diorama, created in 1935 with funds donated by Charles Cadwalader, has been a favorite Academy exhibit for generations. Sporting the largest antlers in any museum display, the specimen was originally less than it appears today. ANSP Archives.

13.3 Artist Clarence Rosenkranz (1871–1946) works his magic on the Moose diorama's background, while the star of the show waits to be installed. ANSP Archives coll. 2011-003.

13.4 Charles Meigs Biddle Cadwalader (1885–1959) served as managing director of the Academy from 1928 to 1947 and as president from 1937 to 1951. ANSP Archives coll. 457.

innovative exhibitions and programs for public education. Witmer Stone, whose resignation as the museum's director due to health considerations paved the way for Cadwalader's appointment, was first and foremost a scientist. During his four-year stint as director of the museum, he had also served as senior curator, which kept his focus on ornithology and other areas of zoological research, not on increasing membership or developing popular exhibitions. In this regard, Stone and Cadwalader could not have been more different. Although an enthusiastic sportsman who had contributed duck specimens to Stone's department, Cadwalader readily acknowledged his own lack of expertise in the field of science. His strength, the board believed, was in his experience as a businessman and savvy marketer who could cultivate his many social connections to enhance the Academy's position. Though bright, productive, and much beloved by his peers, Stone was quiet and socially retiring, preferring the isolation of his office or New Jersey's bird-filled marshes to the drawing rooms of Philadelphia's social and business elite. The two men maintained a respectful relationship, each recognizing the other's strengths and neither wishing to undercut the other's role in advancing the Academy's cause. Their different approaches to the institution's management did inevitably lead to conflict, however.

In August 1929, just nine months into his new administrative role at the Academy, Cadwalader sent a letter to Stone at his Cape May summer house to inform him of the launch of a major new ornithological initiative in Peru. For several months, Cadwalader had been trying to arrange an extended collecting trip for a wealthy young Princeton graduate named Clement Newbold, a family friend of Cadwalader's and a cousin of the Academy's revered treasurer, Arthur E. Newbold Jr. After unsuccessfully inviting several seasoned field collectors to accompany Newbold, Cadwalader secured the services of Melbourne A. Carriker Jr., arguably the greatest collector of neotropical birds of the twentieth century (figure 13.5). Carriker agreed to commit to a long-term project, to be financed through the Academy by private donations, principally from the Newbold family, up to six months in South America in the winter of 1929–30, teaching the aspiring young naturalist how to collect and prepare bird specimens. In his letter to Stone announcing his plans for the Newbold-Carriker expedition, Cadwalader enclosed a copy of a letter he had sent to the directors of the American Museum of Natural History, the Carnegie Museum, the Field Museum of Natural History, the Museum of Comparative Zoology (MCZ) at Harvard, and the U.S. National Museum of Natural History at the Smithsonian informing them of the upcoming expedition and requesting that they cede all future scientific research in Peru to the Academy.[7]

Although pleased by the news of Carriker's commitment and the ornithological specimens Stone knew that Carriker and Newbold were sure to secure, Stone was aghast at Cadwalader's audacious demand that other institutions keep their hands off

Peru. "I presume the letters have already gone," he wrote Cadwalader by return mail.

> If not, I should advise eliminating the last clause regarding leaving this field to the Academy. No such request has ever been made in such connection before so far as I know [and it] might arouse some resentment. It is not likely to do any good, should someone offer to make explorations in this field for one of the other institutions. We for instance did not hesitate to start West Indian work, even though [Thomas] Barbour [of the MCZ] was concentrating on it; nor to go into Siam when the Nat[ional] Museum had a man there—nor did they object![8]

As Stone had feared, Cadwalader's inexperience in scientific protocol did provoke a negative reaction from at least two of the institutions contacted.[9] The other recipients of Cadwalader's letter, though undoubtedly taken aback by his bold and unprecedented request, responded with polite reserve. It was a good learning experience for Cadwalader. From that moment onward, while still fiercely competitive on the Academy's behalf, the institution's champion advanced its preeminence by inviting others to cooperate instead of by trying to preempt their competing activities.

Briefly humbled by this example of his lack of experience, Cadwalader determined to learn everything he could about what made other natural history museums successful and to incorporate these lessons into new initiatives at the Academy. Devoting all of his time and personal resources to the Academy, he pursued an aggressive plan for improving the public museum while simultaneously stimulating a new wave of field research.

Though it began with an unfortunate misstep, the Newbold-Carriker effort in the high Andes of Peru proved something of a model for Cadwalader's future efforts. With the modest expenses of the enterprise covered by the Academy through private donations, and with young Clement Newbold in tow, the experienced Carriker was given free rein to organize an expedition that would secure thousands of rare birds for the Academy,

13.5 Melbourne A. "Meb" Carriker Jr. and his son, Mel, on a bird-collecting trip in Bolivia, 1934. ANSP Archives coll. 900.

Acad. Nat. Sci.
Philadelphia
Peru Exped. 1931
M. A. Carriker, Jr.
Length
Iris
Bill
Feet

making the museum's collection of high-elevation species one of the best in the world (figure 13.6).

Encouraged by this success, Cadwalader began to identify other wealthy amateurs with an appetite for adventure and a desire to help the Academy. He found that young, well-heeled sportsmen made especially good patron-participants for Academy expeditions, for they had the time, money, and energy to attempt trips that might be impossible for older, more established individuals. Thus, Clement Newbold, Brooke Dolan, and George Vanderbilt each led one or more overseas expeditions while still in their twenties.[10] William R. Carpenter made his first Academy-connected hunting trip—to Greenland—when he was just fifteen.[11] In all cases, these early associations with the Academy led to long-term relationships that were highly beneficial to the institution and rewarding for the individuals involved.

Against all odds, as the nation's economy spiraled into the Great Depression, Cadwalader found private sponsorships for a large number of collecting trips to many parts of the world, reinvigorating a research program that had grown moribund in some areas because of a lack of funds. Behind this renewed emphasis on collecting lay Cadwalader's determination not only to raise the Academy's profile and reestablish its preeminence in scientific research but to catch up to the country's other major natural history museums in creating dramatic displays to attract visitors. For several years, the National Museum of Natural History, the Field Museum of Natural History, the American Museum of Natural History, the Carnegie Museum, and others had been creating lifelike habitat groups to bring the natural world indoors.[12] Cadwalader's goal was to match them in presenting the charismatic wildlife of Africa, Asia, and North America to Philadelphians who might never see these species in their native haunts. Creating such displays, he believed, would help to reclaim the Academy's place among the leading science museums in the country.

The museum that Cadwalader inherited when he took up his office as managing director was little changed from the one Joseph Leidy—or even Thomas Say—had known. The Academy's

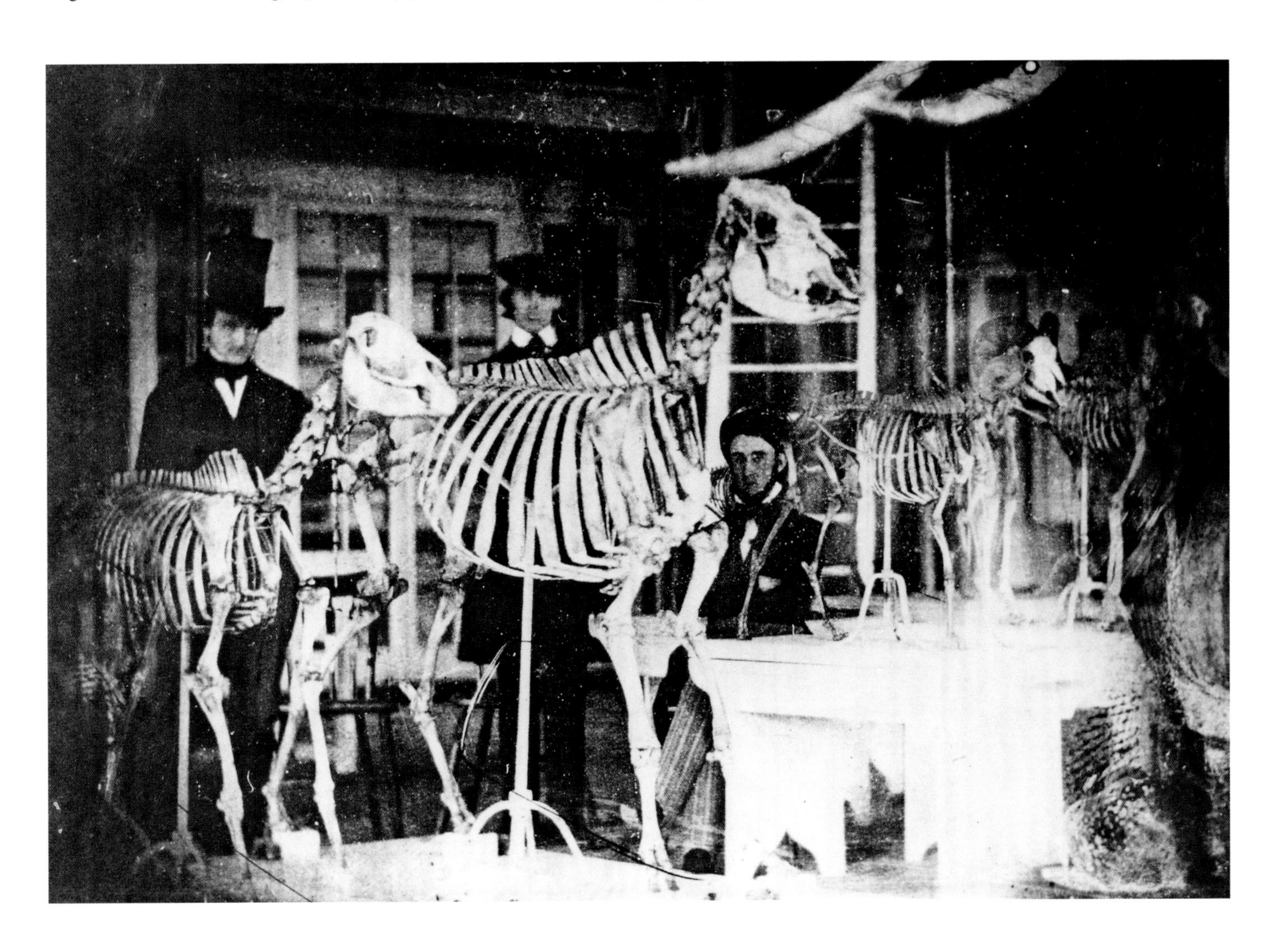

FACING PAGE 13.6 Hooded Mountain-Tanager (*Buthraupis montana*) specimens collected by Melbourne A. Carriker Jr. ANSP Ornithology Department. Purcell photograph.

13.7 Edgar Allan Poe (right), who spent time at the Academy doing research on mollusks; Joseph Leidy, a young medical student (center); and Samuel George Morton (left in top hat) were photographed together in the Academy's new building at Broad and Sansom Streets during the winter of 1842–43. This daguerreotype, possibly by Paul Beck Goddard, is the oldest-known photograph of an American museum interior. ANSP Archives coll. 49.

13.8 The Academy's Victorian exterior was changed to a more federal look in 1910 to conform with the City Beautiful movement and creation of the new Benjamin Franklin Parkway. This artist's rendering dates from the Cadwalader era. ANSP Archives coll. 49.

exhibition halls, filled with case after glass-fronted case of systematically arranged specimens, relieved by the occasional skeleton or stuffed skin mount, emulated the model of the best European natural history museums, which had served as unspoken models for the Academy in the nineteenth century. Whether in its Broad Street building (1840–76) or its "new" facility on Logan Square (after 1876), the Academy in the late nineteenth and early twentieth centuries confronted visitors with "hundreds of thousands of birds [uniformly mounted on little wooden stands] and ten thousand times ten thousand other beautiful, ugly, and interesting objects," all arranged in taxonomic order in endless rows of wood and glass cases (figure 13.7).[13] Except for the scope and volume of their contents, the presentation of such collections had changed little from the "book of nature" Charles Willson Peale had shown his museum guests in the 1790s, or what the Academy's members had proudly displayed in their Samson Street building in 1827. To the devoted naturalist, such encyclopedic displays were marvelous sources of information, but to the uninitiated, they could be bewildering, intimidating, or numbingly dull.

By the 1920s and early 1930s, such traditional methods of display seemed old-fashioned and irrelevant to a public less familiar with natural history and increasingly infatuated by speed and modernity. America of the 1920s saw passage of the Nineteenth Amendment, which granted women the right to vote, and the advent of Prohibition. It saw Charles Lindbergh cross the Atlantic and the first "talkies" make their way into movie houses. Georges Braque, Marc Chagall, Henry Moore, and Georgia O'Keeffe were changing the way people thought about art, while George Gershwin and Aaron Copland were reshaping contemporary taste in music. It was also a time of radical change in museum exhibitry and in the way information was conveyed. If the Academy didn't want to be left behind, it needed to change with the times. When the newly elected president, Effingham Morris, asked Cadwalader to serve as the Academy's managing director in 1928, part of Cadwalader's mandate was to move the Academy away from its nineteenth-century approach to exhibition and create a more vibrant and public-oriented museum. Through this, it was hoped, a broader base of financial support might be secured.[14]

The change Cadwalader was charged with making represented more than a new focus on public displays; it was part of a fundamental shift in the culture of the Academy. As a first step in this direction, in 1924, the Academy revised its charter and bylaws to move managerial responsibility for the Academy from its council of curators to its board of trustees.[15] Until that time, the curators had run the museum for the benefit of themselves, the Academy's members, and the scientific community at large.[16] Their intellectual control of the institution was nowhere more visible than in the public museum, where the Academy's research collections and public displays had been pretty much one and the same for more than a century. Arranged by the curators with the help of a professional taxidermist, these synoptic displays were not for the faint of heart.[17] Visitors were welcome to see them, of course, and without an admission fee, but to viewers who knew little about natural history, most of the museum would have been a closed book.

Thanks to an annual grant from Philadelphia's philanthropic Ludwick Institute, the Academy had been providing the public with a free lecture series on natural history since 1896, but that was the extent of the institution's educational outreach.[18] Lacking the time, inclination, or expertise to create interpretive exhibits in the museum, the curators preferred to let their participation in the Ludwick lecture series and their individual university teaching activities fulfill the Academy's obligations to public education. In the Academy's *Year Book of 1927*, Witmer Stone excused the institution's dearth of more engaging exhibits by citing the shortage of available resources for anything but science: "As heretofore," he wrote, "the funds available for the museum have been entirely absorbed in the maintenance of the building, the scientific staff, and the research being conducted by them."[19] Before the change in administrative structure and the board's specific focus on public outreach, it was not an excuse he would have thought necessary to offer.

The first recognition that the museum needed something to make it more accessible and appealing to the public had come in the spring of 1920, when the Academy's committee on policy recommended to the managing council (the curatorial body then still in control of the institution's policies) that "action be taken immediately to secure a first class museum man to take charge of the public exhibits and such other educational work as the Council may direct."[20] Unwilling to divert any of its research funding to public education, the council accepted the recommendation with the proviso that it would hire the needed museum professional "if and when the necessary funds shall be forthcoming."[21] Fortunately, the Ludwick Institute stepped forward to finance the effort, and by the fall of the following year, Stone had put out some inquiries to colleagues in other museums to see who might be available to rethink the Academy's exhibits and educational offerings. With the encouragement of the Cleveland Museum's director Paul Rea,[22] Stone hired Harold T. Green (1896–1967), an out-of-work salesman from Cleveland with no previous museum experience but with a passion for natural history and a determination "to leave no stone unturned in the search for a chance to get into natural history museum work."[23] It was an auspicious choice, for Green was to become a highly successful

13.9 Clarence Rosenkranz (with palette), Charles Cadwalader, and Harold Green review plans for a new diorama in Africa Hall. ANSP Archives coll. 2010-004.

exhibition designer and, after Cadwalader's arrival, the linchpin of the newly established Exhibits Department.

In his first few years at the Academy, Green devoted most of his time to coordinating the Ludwick Institute lecture series and garnering press coverage for the then quiescent institution. In addition to the Academy's senior science staff, who regularly taught the Ludwick courses, Green arranged for a number of out-of-town speakers. These included Barnum Brown, the first paleontologist to discover the fossil remains of *Tyrannosaurus rex*; George Cherrie, the explorer-naturalist who accompanied President Theodore Roosevelt on his famous "River of Doubt" expedition to Brazil in 1913; and the renowned ornithologists James Chapin and Robert Cushman Murphy, both of whom had made important discoveries of new bird species during overseas expeditions for the American Museum of Natural History.[24] Attendance at the lectures began to climb.

Green's early exhibit work included making casts of local reptiles and arranging seasonal displays of local birds. Though not sensational, and still in the spirit of the earlier displays, his small exhibits were demonstrably more interesting to visitors than the static exhibits that dominated the rest of the Logan Square building. After Cadwalader's arrival, Green's responsibilities were greatly expanded. Beginning in 1928, his primary exhibit focus shifted to coordinating the complex details of diorama design and installation (figure 13.9). He would ultimately oversee twenty-six such installations until his retirement in 1964. Between 1931 (when he went to Angola to collect the giant sable antelope) and 1952 (when he went to Oman and Muscat to create an Arabian tahr diorama, which no longer exists), Green participated in seventeen expeditions to gather materials for the Academy's habitat groups. His career was interrupted in 1942 by four years of military service in Australia, India, and Burma, where he worked for combat intelligence in the U.S. Army Air Forces.

Although Green possessed some painting and modeling skills when he arrived at the Academy and showed an interest in every aspect of museum exhibit making, Cadwalader knew that the exhibits director would need help transforming the Academy from what Cadwalader called "an old curio shop" to a modern public museum.[25] And so, as soon as he could raise the necessary funds, Cadwalader began to increase the size and capabilities of the exhibits staff. In 1929, he hired Joseph Santens, a man with "many years of experience in erecting habitat groups such as we have planned."[26] The following year, he added a recent college graduate, Harry Lance Jr., to the team. Like Green, Santens and Lance would get to participate in several of the museum's collecting expeditions, making trips to Greenland, South Africa, and Alaska and various parts of the United States to gather specimens for both research and exhibition. In the years to come, under Green's supervision, these men and others, joined by a number of talented female colleagues, devoted countless thousands of hours to meticulously preparing the habitat settings for dozens of dioramas on the first and second floors of the museum.

A critical member of Cadwalader's diorama team was Clarence C. Rosenkranz (1871–1946), a Minnesota-based muralist who had served as an expedition artist for the American Museum of Natural History in the 1920s. He was subsequently hired as a background painter for several of the dioramas in that museum's halls of Asian and African mammals.[27] Initially employed by the Academy as a freelancer in 1931 to paint the background for the Carpenter-Whitney caribou diorama for $750,[28] Rosenkranz was officially listed as the staff artist for the museum beginning in 1934.[29] He ultimately painted backgrounds for fifteen of the Academy's dioramas, sometimes working on as many as three a year (figures 13.3 and 13.10).[30] When Rosenkranz retired in the 1940s, he was replaced by Arthur August Jansson (1890–1960), an equally experienced and gifted artist who had painted the backgrounds for the Smithsonian's first two dioramas and for numerous habitat groups at the American Museum of Natural History. Jansson was first approached by the Academy to serve as an exhibition artist in the 1930s, but because of his many other commitments, he did not begin to work at the Academy for another decade. He ultimately painted the backgrounds for five of its most popular habitat groups.[31]

Rosenkranz and Jansson were among the very few artists who specialized in diorama painting. Like Charles Able Corwin (1857–1938), William R. Leigh (1866–1955), James Perry Wilson (1889–1976), Francis Lee Jaques (1887–1969), and a handful of other pioneering diorama artists of the period, Rosenkranz and Jansson had the skill to create a convincing illusion of depth and space by painting a two-dimensional scene on a three-dimensional (curved and concave) background.[32]

With the help of preliminary drawings and a three-dimensional scale model, Green would work with the background artist and the appropriate content experts to develop the plans for each diorama. These were submitted to Cadwalader, the board of trustees, and the donor for final approval. Once the money was in hand (or pledged), the work could begin. A space would be cleared and a shell built by the Academy's carpentry crew. With its inner surface coated with plaster, canvas, gesso,

13.10 Schematic designs for a proposed diorama, probably by Clarence Rosenkranz. ANSP Archives coll. 2010-004.

13.11 (ABOVE) AND FACING PAGE 13.12 Leaf molds. ANSP Exhibits Collection. Purcell photographs.

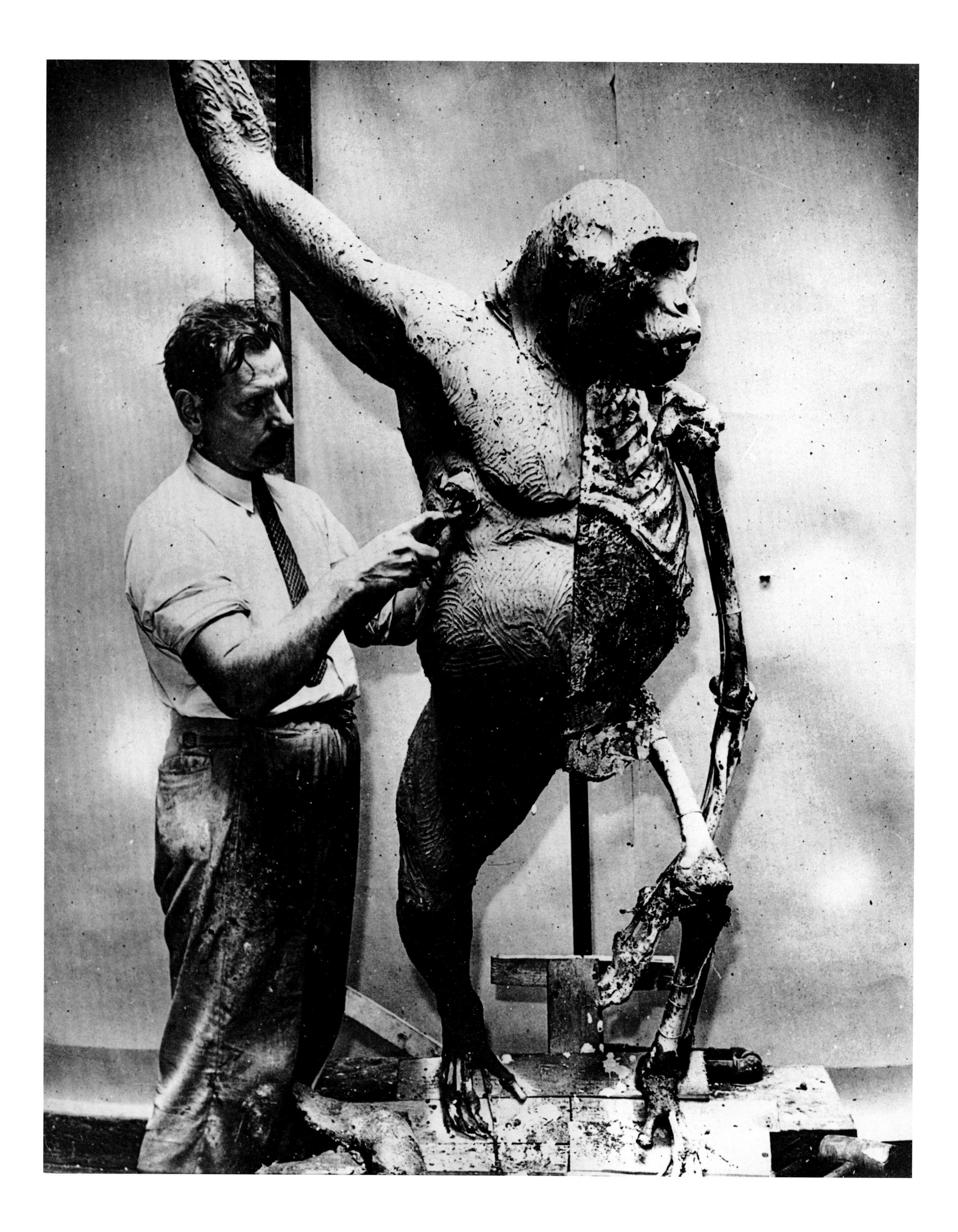

and a neutral priming paint, the empty diorama was ready to be transformed into another world.

The background artist would begin each diorama by working out the proper perspective for the eye of the museum visitor. Choosing an optimal viewing point from which to take in the scene (usually about three feet back from the center of the diorama's glass window and at a median eye height of about five feet two inches), he would sketch a horizon line in charcoal and then block out the rest of the composition.[33] With the basic outline established, he would turn to sketches, photographs, and samples of appropriate vegetation collected in the field to rough-in the rest of the scene, adding color and the specifics needed to underpin the final design. Over a period of many weeks, more details were added, and the landscape would take shape, starting with the sky and then moving with seamless skill to the horizon and below.

Where the background met the foreground, a second team of artists and preparators would take over, concealing the transitional junction and replicating the details of habitat with wax, clay, and facsimile soil.[34] This part of the installation required many months of work by the exhibition staff, as tens of thousands of leaves had to be made by hand from wax and paper, replicating the vegetation collected in the original habitat of the animals depicted. In some cases, actual leaves were cast in the field and then used to create metal molds (figures 13.11 and 13.12). In other cases, leaves were sculpted by hand, based on field sketches and herbarium specimens of the appropriate species, sometimes in different sizes. These were then used to create molds from which multiple copies could be made. Once the leaves were fabricated in either wax, paper, or plastic resin, the color was adjusted to match notes and color swatches made by Green or another member of the exhibits staff while in the field. When a sufficient number of leaves had been made (up to six thousand for the gorilla group), they were carefully wired onto actual twigs and branches to give the appearance of life.[35]

The ground was created from papier-mâché, gravel, sand, and leaf litter, all tinted to capture the soil types that had been collected or noted by the field team responsible for bringing back the specimens. Sometimes, shadows were painted or created by placing tinted sand beneath the animals to create the appearance of a natural source of light at a particular time of day. Such effects were employed to counterbalance the lights used to illuminate the diorama from above the proscenium soffit at the front.

After many months of preparation, the stage was set, and it was time to bring in the star attractions, with or without their original head gear. These were the animals that had been tracked and collected with such care from around the world by Academy members and staff. More often than not, the man who was central to preparing this critical part of the exhibition was Louis Paul Jonas (1894–1971).[36] While Jonas was never part of the Academy's full-time staff, his skill as a sculptor and exhibit designer perfectly complemented the in-house talent assembled by Green at the Academy. Born in Budapest, Hungary, Jonas moved to the United States at the age of fourteen to assist two of his brothers with a taxidermy business in Denver, Colorado. Three years later, on a trip to New York, he met Carl Akeley (1864–1926), the dean of American taxidermists, who was so favorably impressed by Jonas that he hired him to join the exhibits department of the American Museum of Natural History. While helping Akeley with the famous elephant group that still serves as the centerpiece for the Akeley Hall of African Mammals, Jonas extended his artistic training by studying sculpture with Herbert McNeil and George Bridgman at the National Academy of Design.

Beginning in 1930, a few years after establishing his own commercial taxidermy and exhibit design studio in Hudson, New York, Jonas was hired by the Academy to clean and tan the skins of large mammals collected on its far-flung expeditions and to mount specimens for display in its dioramas. Although other commercial taxidermists were sometimes employed by the Academy, the Jonas Studio mounted the specimens for all but ten of the thirty-one still-existing mammal dioramas produced at the Academy between 1930 and 1967.[37]

Like Akeley, Jonas was as much a sculptor as a taxidermist, determining a dramatic but realistic pose for each animal and then reconstructing its anatomical details in clay to serve as a base for the lifelike figure he was commissioned to create (figure 13.13).[38] He based his creations on photographs and personal observations of living zoo animals and on detailed measurements and plaster impressions taken from the animals as they were collected in the field and before they had been skinned. Once a full-size replica of the animal's body had been sculpted to his satisfaction, a plaster mold was made of the clay model, and from it a hollow papier-mâché form was made. The skin of the animal would be carefully fitted, stitched, and groomed over this form, making the hand of the taxidermist all but invisible. The resulting mounts, complete with accurately rendered muscles, bones, tendons, and veins, were a far cry from the taxidermied animals of an earlier era. These primitive precursors of modern mounts, several of which were still on view in the museum when Cadwalader began his renovations, were usually made of tanned

13.13 Louis Paul Jonas (1894–1971) sculpting one of several body forms for use in the Academy's gorilla diorama, ca. 1934. Once completed, the skeleton was extracted and returned to the mammal collection. The gorilla's skin was ultimately stretched over a papier-mâché cast of the body sculpture to create the final, lifelike display. ANSP Archives coll. 2010-004.

animal skins draped over a wood-and-wire frame, then literally stuffed like a pillow with sawdust, sisal, cotton, or excelsior. The formless blobs that resulted were neither convincing as living creatures nor useful as scientific teaching tools (figure 13.14). By contrast, the new taxidermy, pioneered by Akeley and further refined by Jonas and others, was as aesthetically pleasing as it was scientifically accurate. The combination was critical to the success of the diorama as a display technique.

Despite the noticeable improvement of the displays from the public's point of view after Cadwalader's arrival in 1928, not everyone was happy with the changes he was making at the Academy. In January 1930, Eleanor Carothers from the Zoology Department at the University of Pennsylvania, possibly speaking on behalf of her silent colleagues at the Academy, wrote President Effingham Morris to announce that she was maintaining her membership "under protest" because of what she believed was the undignified and unnecessary popularizing of the museum:

> Practically every Biologist in the country who is interested in the unique and honorable position which the Academy holds among our national scientific organizations must feel grave concern over the policies of those at present in control. I refer in particular to the attempt to imitate other museums in the matter of [habitat] group exhibits at the sacrifice of her scientific prestige.[39]

Cadwalader responded with a strong defense of his management and the new direction he had set for the Academy: "We are amazed that you should feel that the Academy is neglecting scientific work," he wrote. "You are apparently not aware that more scientific research is being done in the Academy today than in the past." He then cited an increased number of expeditions, publications, staff, and memberships within the last year, all of which he claimed were at an all-time high. "We would appreciate very much knowing just why an attractive exhibit illustrating

13.14 Bolivar, one of the largest Indian elephants ever taken into captivity, was a featured attraction in the Adam Forepaugh Circus and at the Philadelphia Zoo until its death in 1908. For the next ten years, its stuffed skin and skeleton stood side by side in the Academy's museum (above). Carl Akeley, Louis Paul Jonas, and others replaced this primitive style of taxidermy with more realistic and artistic mounts in the 1920s and 1930s. ANSP Archives coll. 49.

some particular phase of natural history should interfere in any way with the work of one of our scientists," he bristled.[40]

And so, to the displeasure of some in the scientific community, but with the full support of the board of trustees, Cadwalader continued with his changes. His plan was to offer not only the iconic mammals of each continent (elk, grizzly, bison, and so on for North America; lion, zebra, Cape buffalo, and others for Africa) but also some of the rarest (giant sable antelope, okapi, gorilla, panda, etc.) to give the Academy bragging rights over other museums. In this, he differed with some of his friends and patrons. "Don't you think that 'biggest,' 'first,' 'only' etc. get a bit tiresome after a time?' wrote Brooke Dolan in an undated letter to Cadwalader. "Like the MCZ one of the charms of the Academy has always been that we haven't used too much bally hoo."[41] Cadwalader was, of course, dependent on who would collect what, and while he was often successful at steering the expeditions toward species he wanted to represent, he was also realistic enough to accept what was offered. If it was the "biggest," "first," or "only" of its kind, so much the better. In the end, he could successfully justify exhibiting almost anything, as long as it was done on a scale and with the quality that would match or exceed the offerings of America's other great museums.[42]

As one dramatic habitat group after another replaced exhibit cases in the museum, the public responded just as Cadwalader had hoped. Between 1929 and 1935, attendance experienced a dramatic fourfold increase, from 40,000 to 181,000 visitors annually. In a 1936 promotional brochure, Cadwalader was able to boast of his success. "In 1929," he wrote, "not 70 people a day entered the Academy's doors. In 1935 people were coming at the rate of nearly 500 a day; while in that year some 30,000 school children went through the halls of the museum in classes."[43]

Adding to the public appeal of the dioramas was a series of temporary exhibits with which Cadwalader was able to garner good publicity, attract friends, and reinforce Philadelphia's historic role as the birthplace of American science. The two most ambitious of these were a 1931 national exhibit commemorating the two hundredth anniversary of the founding of the first botanical garden in the American colonies by John Bartram and the first national exhibition of the paintings (and related memorabilia) of John James Audubon, offered to commemorate the hundredth anniversary of the publication of *The Birds of America* (April–June 1938). For both exhibits, the Academy sought out for display manuscripts, paintings, and artifacts that were still in private hands (figure 13.15). The exhibits were accompanied by catalogs and gave institutional gravitas to the lives and accomplishments of the naturalists, who, though well known, were not yet the national icons they would become in subsequent years. Some of the items displayed, including Bartram's watch and fork and a copperplate used to create Audubon's "double elephant folio" of *The Birds of America*, were subsequently given to the Academy.

Smaller exhibits exploring the overlap of art, science, and the humanities included paintings of waterfowl by Richard E. Bishop (1935), paintings of the birds of Mount Desert Island by Carroll S. Tyson Jr. (1935), the art of early man (1937),[44] and African watercolors by Sanford Ross (1939). Each exhibit, while appealing to the general public and providing an opportunity to publicize the museum, was intended to help Cadwalader connect with another important constituency: the friends and acquaintances he was cultivating who he hoped would support expeditions, sponsor dioramas, and endow the operations of the museum. While the Academy's offerings had a populist emphasis (the museum was still free to the public), Cadwalader knew that the serious money he needed to support the institution was most likely to come from a relatively small group of individuals, and he worked tirelessly to capture and hold their interest in the Academy.

As the Great Depression took hold of the nation, there was considerable hardship at the museum despite Cadwalader's

13.15 This seventeenth-century pocket watch (by Hutchins, London) once belonged to the Philadelphia botanist and explorer John Bartram (1699–1777). It was included in the Academy's exhibition on Bartram and his garden in 1931. ANSP Archives coll. 15.

efforts. An across-the-board 10 percent salary cut was imposed on the Academy's staff in 1932, to help balance the budget. It was not restored for another four years, and then only thanks to a special fund-raising campaign.[45] At the time of the cut, Witmer Stone, the Academy's most senior curator, was receiving an annual salary of $4,000.[46] More junior scientists, such as James Bond and Rodolphe Meyer de Schauensee, were receiving $1,000 per year.[47] Many of the staff, including Cadwalader (by choice) and Ruth Patrick (beginning in 1934), worked as volunteers.[48]

Fortunately for the Academy, a number of government programs designed to relieve the nation's growing unemployment helped to bring as many as twenty-nine "artisans and laborers" to work at the museum in 1934.[49] Even though the Academy's conservative (and mostly Republican) board believed that "progress of scientific institutions comes best from private direction and support" and disapproved of "the strange philosophy 'let the Government do it,'"[50] they were not opposed to having artists assigned to work for free through the Public Works of Art Project, museum aides supplied by the Emergency Educational Council,[51] and dozens of other docents, laborers, and teaching assistants provided by the Works Progress Administration (WPA) throughout the Depression.[52]

Until the 1930s, with the exception of a few requests for building grants from the Commonwealth of Pennsylvania, the Academy had never gone beyond its own membership for financial support.[53] Public fund-raising was something many of the trustees found unseemly and distasteful. And yet Cadwalader saw it as an absolute necessity and made it his highest priority from the day he took office. Although he had done much to build the Academy's membership, he realized that dues would never be sufficient to keep the Academy running, let alone support the growth he knew was needed to achieve the greatness he envisioned. "The lack of funds is a very serious problem," he wrote a

13.16 In 1938 the Academy hosted a national exhibition on the life and ornithological contributions of John James Audubon (1785–1851). Included was the institution's subscription copy of *The Birds of America*. Seen here is plate 236, the Black-crowned Night Heron (*Nycticorax nycticorax*), engraved by Robert Havell Jr. in 1835. ANSP Library.

member who had been critical of his management style during his first six months as managing director:

> Scientists must receive proper wages just as other persons in order that they may live. You do not seem to realize the enormous costs of producing our publications and of housing the tremendous increase in scientific specimens now being received in our laboratories. The total number of scientists and helpers on our staff at this time is 47, which is the largest in the history of the Academy.[54]

With encouragement from the Academy's president, Cadwalader decided to use the Academy's 125th anniversary as an opportunity to mount a broad public appeal—the first in the institution's history. He did this by focusing on the importance of public education. In January 1936, Cadwalader began assembling a blue-ribbon advisory panel of community leaders, chaired by the Academy's treasurer Arthur E. Newbold Jr., to help determine "what the Academy should do and can do to serve the public and science in the future."[55] The group sponsored "an extensive survey of the Academy's responsibility and opportunity in connection with the teaching of the natural sciences in Philadelphia's schools and of ways and means to fulfill them."[56] Sixteen hundred replies came from teachers representing 205 public, private, and parochial schools and more than fifty-six thousand Philadelphia schoolchildren (figure 13.18). As Cadwalader had expected, they were overwhelmingly in favor of having the Academy offer more educational programs.

Not surprisingly, in response to the opinions expressed in the survey, public education was given top billing in the 125th anniversary campaign. However, to appeal to as many constituencies as possible, the fund-raising goals addressed some of the institution's other needs as well. The four objectives of the campaign reflected the Academy's continuing effort to balance its public and scientific priorities. The goals were to create a new Department of Education, to revive the Department of Geology and Paleontology, to create a hall devoted to "the Earth and its History," and to reclaim a "waste space above the present Mineral Hall" (present-day Dinosaur Hall) by adding two floors in what had been the dramatic, skylit free story of the museum's principal exhibition gallery.[57] The price tag for this ambitious plan had a fiscally responsible–sounding specificity: $3,548,475.[58]

Thanks to Cadwalader's many personal contacts and hard work, all four of these goals were eventually achieved. In 1937, W. Stephen Thomas, a dynamic young professional with a Harvard degree and previous experience at the Metropolitan

13.17 Harold Green (left) and his exhibits staff gather materials for the museum's bison diorama, Judith Basin, Montana, October 1933. ANSP Archives coll. 2010-004.

13.18 A school group from Girard College studies the Sonoran Desert diorama. ANSP Archives coll. 2011-003.

Museum of Art and the Newark Museum, was hired to head up the newly established Education Department.[59] At the same time, Edgar B. Howard, from the University of Pennsylvania, was appointed the unpaid acting curator of geology and paleontology,[60] while Helmut de Terra from the Carnegie Institution in Washington, D.C., Benjamin F. Howell of Princeton University, and Edward H. Colbert of the American Museum of Natural History in New York were all appointed associate curators in the reestablished Department of Geology and Paleontology. Under Harold Green's direction (with the young Arthur Newbold III serving as his assistant during a "gap" year between Harvard and the University of Pennsylvania Law School), work continued on a number of dioramas in the Africa and North America halls, and plans were developed for a comprehensive Hall of Earth History (opened in 1937) (figure 13.19). To add to the turmoil—and to the palpable feeling of activity and growth at the Academy—the promised floors were inserted into the previously unheated "no-man's land" above the Mineral Hall, permanently reducing the space available for exhibition but increasing the space available for specimen storage, offices, and laboratories.[61]

In the midst of this activity, Cadwalader launched *Frontiers*, a new members' magazine. With a title adopted from the Academy's successful 1936 fund-raising campaign, the publication was intended to help solidify the Academy's greatly expanded base of support and "serve as a means of communication between one of America's oldest and most active natural science institutions . . . and its members and friends."[62] The first issue of *Frontiers*, published in October 1936, contained articles by Brooke Dolan (on his recent expedition to Tibet), J. N. "Ding" Darling (on the conservation of migratory waterfowl), Samuel Gordon (on visiting the "ghost mines" of Philadelphia), and Witmer Stone (on his childhood in and around Philadelphia). It was an impressive start for a magazine that would grow in size and influence in the years to come. By 1939, it was being sent to 3,300 subscribers in forty-two states and had gained even broader attention through the republication of articles in a number of national magazines, including *Reader's Digest*.[63] Over the next four decades, it served as a publishing venue for many of America's leading natural history writers. Because of financial challenges faced by the Academy, the bimonthly magazine was reduced to an annual in 1979 and ceased publication in 1982.[64]

Through the 1930s and 1940s, there was much to report in *Frontiers*: new expeditions, new dioramas, new staff, and many other changes brought about by Cadwalader's irrepressible energy and vision. Described by *Time* magazine as the Academy's "spark-plug director,"[65] Cadwalader was relentless in pushing the Academy and its staff to new levels of achievement. Early in his tenure as managing director, he revealed his ambition for the institution when he wrote Witmer Stone that he wished the Academy's staff to "do three times as much work, only three

FACING PAGE 13.19 The Hall of Earth History, first envisioned in 1936, opened in November 1937. In order to provide local context, its entrance showed the geologic strata underlying Philadelphia. ANSP Archives coll. 49.

THE EARTH AND ITS HISTORY

RECENT
CAPE MAY
PENSAUKEN
PENSAUKEN
PENSAUKEN
QUATERNARY
PLEISTOCENE
BRYN MAWR
TERTIARY
CRETACEOUS BEDS
CRETACEOUS BEDS
CRETACEOUS
MISSING
JURASSIC MISSING
FT. WASHINGTON
FT. WASHINGTON
ARCOLA
YERKES
PERKIOMEN
TRIASSIC
MISSING
SILURIAN DEVONIAN CARBONIFEROUS MISSING
WISSAHICON
WISSAHICON
WISSAHICON
WISSAHICON
JACKSONBURG
BEEKMANSTOWN
ORDOVICIAN
ALLENTOWN
CHICKIES
SPRING MILL
CAMBRIAN
BALTIMORE
FRANKLIN
PRE-CAMBRIAN
BALTIMORE

STRUCTURE OF THE EARTH
ANIMALS OF THE PAST
GEMS OF THE WORLD
PRECIOUS METALS
WHAT IS A MINERAL
WHAT IS A FOSSIL
WHAT IS GEOLOGY
HISTORY OF THE HORSE
EARTH FORCES

CENOZOIC
MESOZOIC
PALEOZOIC
PICTURES NOT DRAWN TO SCALE

13.20 An Inuit figure created by Eda Kassel in 1937 as part of an ethnographic display in the Hall of Earth History. ANSP Exhibits Collection. Purcell photograph.

FACING PAGE 13.21 A Masai warrior by Eda Kassel, 1937. Painted plaster. ANSP Exhibits Collection. Purcell photograph.

13.22 Charles Cadwalader cuts a cake in the Library Reading Room as the Academy staff looks on. The May 1947 party honored Cadwalader's nineteen years of service as managing director. ANSP Archives coll. 457.

times as good, and five or six times more valuable."[66] While the comment was characteristically positive, it was uncharacteristically lacking in tact, for it implied that the long period of time during which Stone had served as the senior curator and director of the museum had been less than productive.

In November 1937, the Academy opened the new Hall of Earth History with a large anthropological section that helped to reinforce the leadership role the Academy had taken in this field with its symposium on early man the previous March. Before its new exhibits were installed, the room was emptied of its earlier displays and a false wall was installed to block all daylight from the Logan Square façade. Eight panels were erected to screen the mezzanine gallery to which the mineral collection was moved. According to a press release issued at the time of its opening, the new hall had a warm, terra-cotta color and internally lit exhibition cases that produced "a soft glow throughout the lower part of the hall."[67] As recalled by one of its designers, the room had been changed from a Victorian gallery to a contemporary hall of Art Deco design. Its contents traced evolution from the earliest emergence of life on earth to the present day. The earliest eras were illustrated with fossils drawn from the Academy's extensive collections, while the earth's human history was conveyed through specimens, casts, and models, some of which were borrowed from other museums.

To represent the diversity of living human populations, Eda Kassel, an Academy exhibits designer, created sixteen ten-inch models of human figures dressed in appropriate ethnic costumes (figures 13.20, 13.21).[68] These were "painted in natural colors and set off by a decorative map on which colors [were] used to designate the homes of the various races."[69] Kassel's anthropological sculptures paralleled similar work that had been done for the Field Museum of Natural History and the Century of Progress Exposition in Chicago in 1933 by Malvena Hoffman (1887–1966), though unlike Hoffman's life-size bronze sculptures, which were replicated for other museums, Kassel's were one-of-a-kind, hand-painted plaster models created specifically for the Academy's display.

Contained in a nearby case was a revolving cone with a spiral ramp on which much smaller sculptures of evolving humans trod ever upward, illustrating humankind's rise from its earliest-known form to the present time. The four-foot-high spiral, capped by evocative paintings of New York City's tallest buildings and a flying airplane, provided a dramatic bit of motion in a hall whose "modern" look transformed the Academy's public image.[70] Many elements of this exhibit, though altered slightly in 1957 with the insertion of several new dinosaur displays, remained in place until the hall was once again stripped bare for an "evolution and revolution" exhibition on natural history for the nation's bicentennial in 1976.

Although Cadwalader's presidency would extend for another four years and his association with the institution for another twelve, in June 1947, the Academy staff organized an elaborate surprise party to show their affection for "the Academy's most important and most prized type specimen, *Philadelphacuss Cadwaladerii Charlesiensis (Meigs-Biddle)*." Promising "refreshments, Academy style, fun, frolics, and maybe dancing," a "confidential" invitation described the event as

> A great gathering of curators, associate curators, assistant curators, associates of the associates and assistants to the assistants; botanical paster-uppers, exhibits tearer-downers, taxidermists, bursars, secretaries, assistant secretaries, just plain stenographers, heating engineers, photographers, maintenance personnel, jackgraham personnel, librarians, assistant librarians, cataloguers, associate cataloguers, scientific editors, unscientific editors, assistant treasurers, educators, docents, lecturers, monograph writers, microscope-peerers, microscope slide makers, and people who do the work around here.[71]

Pictures of the event show a somewhat somber-looking Cadwalader cutting an elaborately decorated cake on a white cloth–covered table with silver candelabra, vases of daisies and snapdragons, and the makings for tea and coffee near at hand. Given the formality of the setting (in the library reading room), the early hour (4:30 P.M.), and the nature of the refreshments (tea and coffee), it is unlikely than any dancing took place (figure 13.22). Frolics seem out of the question, but there was a big turnout (at least fifty people), and the event was undoubtedly a moving one for Cadwalader, who had fought through some strong opposition to create the Academy over which he then presided.[72] The occasion may have reminded Cadwalader of the Academy's far more jovial, pre-Prohibition centennial banquet, at which one speaker lauded Samuel Gibson Dixon as the man who "found the Academy serpentine and left it reinforced concrete," a reference to the massive physical changes that were made on the building during Dixon's presidency (1895–1918).[73] While Cadwalader did not change the physical footprint of the Academy during his twenty-two-year association with the institution, he can be fairly credited for bringing the Academy into the twentieth century.

RMP

Notes

1 Except for an admission fee charged from 1869 to the early twentieth century, the museum remained free to the public until 1953, when an admission charge of fifty cents for adults and twenty-five cents for children was imposed.

2 *Alces alces gigas.* The genus and species names, *Alces,* are from the Latin word for "elk," which is derived from the Greek word *alcae,* meaning "strong." The subspecies' name, *gigas,* means "giant."

3 Cadwalader committed the $10,000 honorarium he received from the prestigious Philadelphia Award in 1935 to pay for the moose diorama.

4 In a 1993 research report on the history of the dioramas, Keith Russell, collection manager of the Exhibits Department, wrote that yet another, even larger, set of antlers was given to the moose in April 1947; however, the author's research has not been able to substantiate this claim. See Keith Russell, *Dioramas of the Academy of Natural Sciences of Philadelphia*, ANSP Exhibits Department internal document, March 1993, 9.

5 According to Harold Green's careful records, the substitute antlers were "abandoned" or shed by their original owner, also from the Kenai Peninsula, thus a second animal was not killed for the display. See Harold Green papers in ANSP coll. 2010-004.

6 Harold Green's file on the Moose diorama gives detailed comparative measurements for the moose in dioramas at the American Museum of Natural History in New York and the Field Museum in Chicago. His files also reveal that the substitute antlers used in the Academy display were brought from Alaska in January 1935 by Fred W. Hollander. In his memoir, Biddle deftly avoids the issue, stating only that the goal of his trip had been to acquire a moose with a rack in excess of seventy inches for the Academy. See Nicholas Biddle, *Personal Memoirs* (Philadelphia: Sutter House, 1975), 235.

7 Charles Cadwalader to Witmer Stone, 9 August 1929, Newbold file, Cadwalader papers in ANSP coll. 2009-034.

8 Witmer Stone to Charles Cadwalader, n.d. (August 1929), Newbold file, Cadwalader papers in ANSP coll. 2009-034.

9 The two museums that objected were the American Museum of Natural History in New York and the Field Museum in Chicago; see letter from Frank M. Chapman to Charles Cadwalader, 10 August 1929, and W. H. Osgood to C. M. B. Cadwalader, 13 August 1929, both in Newbold file, Cadwalader papers in ANSP coll. 2009-034.

10 Newbold was twenty-one when he collected mountain sheep in Alaska for the Academy's first diorama in 1928. He was twenty-two when he went to Peru with Carriker in 1929. Dolan was twenty-three when he made his first expedition to China with the German naturalist Hugo Weigold, in 1931. George Vanderbilt led his first Academy expedition to Africa in 1943 when he was twenty; his second Academy trip, to Sumatra, was in 1947 when he was twenty-three.

11 The young "Billy" Carpenter was accompanied on his 1934 expedition to Greenland by Harry Lance Jr., from the Academy's Exhibits Department. The trip was led by the legendary Arctic captain Robert A. "Bob" Bartlett, who had commanded Robert Peary's ship the *Roosevelt* on his quest for the North Pole. Carpenter's father, R.R.M. Carpenter, paid for the expedition but did not participate. He said it was to provide his son with "a summer's outing and general 'conditioning.'" For records of this expedition, see ANSP coll. 316.

12 In 1889, Carl Akeley (1864–1926) created a diorama of a muskrat family at the Milwaukee Public Museum, which is often cited as the first modern museum diorama. In 1902, the American Museum of Natural History opened the Hall of North American Birds. This was the first museum hall entirely devoted to the habitat-group method of display. See Stephen Christopher Quinn, *Windows on Nature: The Great Habitat Dioramas of the American Museum of Natural History* (New York: Abrams in association with the American Museum of Natural History, 2006), 15–16. For Akeley's contributions to the development of dioramas, see Penelope Bodry-Sanders, *Carl Akeley: Africa's Collector; Africa's Savior* (New York: Paragon House, 1991).

13 Letter from Samuel Heaton to Mary [Heaton] Battey and Thomas Battey, 19 January 1869, following a visit to the Academy of Natural Sciences, private collection (courtesy of Daniel Snyder).

14 The ten-cent admission charge imposed in 1869 (prompted by the overwhelming response to the exhibition of *Hadrosaurus foulkii*) was removed by the first decade of the twentieth century. Since the museum was once again open to the public without charge during the Cadwalader era, the desire for improved exhibits was driven not by gate receipts but by a desire to raise the public profile of the museum and thus attract a broader-based membership. By the 1920s, membership, while still nominally by election, was more or less perfunctory and heavily dues oriented. Also, it was believed that more public use of the museum might stimulate more state and city support for the Academy.

15 Under the new charter and bylaws, the curators were still responsible for helping to direct the scientific work of the Academy, but the board of trustees was given responsibility for all other aspects of the Academy's operation. For a discussion of the changes, see R.A.F. Penrose, "Report of the President," in *Year Book . . . for the year ending Dec. 31, 1924* (Philadelphia: Academy of Natural Sciences, 1925), 6. See also broadside addressed "to the members of the Academy" signed by James A. G. Rehn, recording secretary, ANSP coll. 741.

16 Throughout most of the nineteenth century, the curators and the members were one and the same. As the institution and its collections grew in size, professional curators received modest salaries to help care for the collections.

17 David McCadden served as the Academy's principal in-house taxidermist from 1891 until sometime in the 1940s. Originally an employee of the Philadelphia Zoo who did taxidermy in the evenings and on weekends, he was invited by Angelo Heilprin to join the Peary Relief Expedition to Greenland in 1891, but he declined because of the loss of business a three-month absence would cause. McCadden was joined by Joseph A. Santens in 1929 and Harry Lance in 1930. McCadden, Lance, and Santens were excellent at handling scientific study skins and small specimens, but when the Academy began to need large mammals mounted for the museum's dioramas, these were usually sent out to the Louis Paul Jonas Studio in Hudson, New York.

18 The Ludwick Foundation was established by Christopher Ludwick (1720–1801), Baker General of the Army of the United States during the American Revolution, whose estate provided a trust "for the schooling and education gratis, of poor children of all denominations, in the city and liberties of Philadelphia, without exception to the country, extraction, or religious principles of their parents or friends" (Ludwick Foundation charter).

19 Witmer Stone, "Report of the Director of the Museum," *Year Book: The Academy of Natural Sciences of Philadelphia for the year ending December 31, 1927* (Philadelphia: Academy of Natural Sciences, 1928), 51. A similar excuse was given in Academy yearbooks throughout this period.

20 Report of the council meeting of 16 November 1920, ANSP minutes, vol. 16, coll. 502, 520.

21 Ibid.

22 See letter from Paul M. Rea to Witmer Stone, 5 November 1921, ANSP coll. 186.

23 Harold T. Green to Lawrence V. Coleman, 18 October 1921, ANSP coll. 186. At the time of his correspondence with Stone, Green was working for the YWCA and living in Woodstown, New Jersey.

24 The most frequent local speakers for the Ludwick lecture series during this period were J. Fletcher Street (from the Delaware Valley Ornithological Club), Spencer Trotter (from Swarthmore College), Philip Calvert (from the University of Pennsylvania), Alfred M. Collins (from the Geographical Society of Philadelphia), and Academy curators Henry Pilsbry, James A. G. Rehn, Wharton Huber, Witmer Stone, Samuel Gordon, Francis Pennell, Henry Fowler, and William Hughes. Other speakers from the American Museum of Natural History included G. Clyde Fisher, James L. Clark, W. D. Matthew, Frederick Morris, C. J. Albrecht, and S. Harmsted Chubb.

25 From "And the Work Will Be Completed," ANSP fund-raising brochure, 1936, 3.

26 Charles Cadwalader, report to the trustees, May 1929, Cadwalader papers in ANSP coll. 2009-034.

27 Quinn, *Windows on Nature*, 172.

28 For correspondence regarding the arrangements between Rosenkranz and the Academy, see Cadwalader papers, 1930, ANSP coll. 2009-034, box 8 (of 8), particularly letters dated 29 November and 2 December 1930.

29 Helen Winchester was listed as artist for Scientific Publications from 1923 to 1935. Their roles were quite different. Evidently Winchester was never involved with public exhibitions.

30 Rosenkranz painted the backgrounds for the following Academy dioramas: lion, musk oxen, caribou, tundra swan, sable antelope, giant panda, bison, takin, Kilamafeza waterhole group, pronghorn, okapi, cape buffalo, gorilla, moose, puma, and mule deer. It is possible that the background for the opossum diorama was taken from a larger, discarded Rosenkranz background painting.

31 Jansson was responsible for the background paintings for the Dall sheep, desert sheep and peccary, yak, kiang, and passenger pigeon dioramas.

32 Francis Lee Jaques, best known for his dioramas at the James Ford Bell Museum and the American Museum of Natural History, painted the Auodad or Barberry Sheep diorama for the Academy in 1950. Other artists who painted backgrounds for the Academy's dioramas include Charles A. Corwin (Rocky Mountain goat), Otto Bauer (Kodiak bear and bald eagle), Ruth Aspen (Arabian tahr), Jonathan Fairbanks, Jr. (Great Auk and Dinosaur Hall Mural), and Stephen T. Harty (polar bear, desert of Borkou, Colobus monkey, Pennsylvania black bear, and Pennsylvania bobcat).

33 The viewpoint calculation is given in Quinn, *Windows on Nature*, 150.

34 Some members of the Academy exhibits team who have worked on diorama preparation over the years include Frederick Stoll, Walter Kohler, Kelly Ash, Joan Anderson, Ruth Aspen, Pat Hagen, Eda Kassel (Newell), Charles Leibrandt, and Stephen Harty.

35 *Frontiers* 1, no. 5 (May 1937): 159.

36 The biographical information given here was obtained from an exhibition brochure "The Art of Louis Paul Jonas," published in conjunction with a traveling exhibition created by the Indiana State Museum, Indiana Department of Natural Resources, Indianapolis, 1981.

37 Other taxidermists responsible for the large diorama mounts at the Academy were C. J. Albrecht (Rocky Mountain goats and Stone's sheep), Joseph Santens (bald eagle and whistling swans), James L. Clark (musk oxen and giant sable antelope), Robert Rockwell (Barbary sheep and Arabian tahr), Fred Stoll (passenger pigeons), and Charles Liebrandt (Pennsylvania mammals). In an endorsement letter "to whom it may concern" dated 5 June 1956, Harold Green wrote that Jonas had already mounted some sixty-seven large mammals for the Academy and that he was then preparing an additional ten animals for the Academy's waterhole group. Harold Green papers, ANSP coll. 2010-004. Five dioramas created at the Academy between 1929 and 1967 no longer exist. These are Rocky Mountain goat (1929), Stone's sheep (1930), tundra swan (1932), snow leopard (1949) and Arabian tahr (1953).

38 Jonas usually started his work with the animal's actual skeleton to ensure accuracy. Once the taxidermy was completed, the skeleton could be deposited in the Mammal Department's research collection.

39 Letter from E. Eleanor Carothers to Effingham B. Morris, n.d. but January 1930, Cadwalader papers, ANSP coll. 2009-034, box 3, file 17.

40 Letter from Charles M. B. Cadwalader to Eleanor Carothers, 14 January 1930, ANSP coll. 2009-034, box 3, file 17.

41 Letter from Brooke Dolan to Charles Cadwalader, n.d., ANSP coll. 2009-034 If the letter was written after his 1931–32 expedition to Tibet, Dolan's remarks may have been intended to assuage Cadwalader's disappointment that the Academy's panda diorama (with specimens collected by Dolan and Ernst Schäfer) was only the second of its kind. It was preceded by three years by one at the Field Museum in Chicago, which used specimens collected by Theodore Roosevelt Jr. and Kermit Roosevelt in 1928 and was put on display in 1930.

42 A 1939 press release mentions the diverse habitats represented in the dioramas. It is not clear whether this had been Cadwalader's initial goal or was used after the fact to impose an educational logic on a disparate group of exhibits. See ANSP coll. 2010-004.

43 *Frontiers*, promotional brochure, ANSP (1936): 7.

44 While consisting mostly of high-quality reproductions, the *Art of Early Man* was an extremely popular exhibit nonetheless.

45 In a letter to staff dated 7 July 1936, Charles Cadwalader stated that the board of trustees had "secured sufficient funds to permit the restoration of the 10% salary cut [that was] established in June 1932 and made necessary in order that the Academy might balance its annual budget." Letter from Charles Cadwalader to Samuel Gordon, 7 July 1936, ANSP coll. 741.

46 Charles Cadwalader to Witmer Stone, 7 July 1936. ANSP coll. 186.

47 Charles Cadwalader to Witmer Stone, 16 August 1928. ANSP coll. 186.

48 As late as 1939, five years after Patrick arrived at the Academy, Cadwalader wrote the John Simon Guggenheim Memorial Foundation that "Dr. Patrick does not receive a salary from this Museum. We have never been able to include her on our salary list and thus all of her work is done on her own time." Letter from C. M. B. Cadwalader to Henry Allen Moe, Nov. 13, 1939, ANSP coll. 2010-051, 1939, box 1, Patrick file (38).

49 Charles M. B. Cadwalader, "Review of 1934," Cadwalader papers, ANSP coll. 2009-034.

50 From *Frontiers* (fund-raising brochure), ANSP (1936): 22.

51 C. M. B. Cadwalader, "Review of 1934," Cadwalader papers, coll. 2009-034.

52 See W. Stephen Thomas, Annual Report, ANSP Education Dept., 18 December 1936, ANSP coll. 2010-158, box 4 of 5.

53 The Academy had received the following funding from the state legislature: 1888–89, $50,000 for the construction of a new museum building; 1890–91, a second appropriation of $50,000 for the construction of a new museum building; 1907, $150,000 for the construction of a new fireproof library and auditorium; 1909, $60,000 for refacing the Academy's Race Street building and middle wing.

54 Letter from Charles M. B. Cadwalader to E. Eleanor Carothers, 14 January 1930, ANSP coll. 2009-034, box 2, file 17.

55 Letter from Effingham Morris to Horace H. F. Jayne, 3 January 1936, University of Pennsylvania Museum of Archaeology and Anthropology archives, Academy of Natural Sciences file.

56 Letter from Arthur E. Newbold Jr. to Effingham B. Morris, 4 May 1936, published in an Academy promotional flyer titled "An Experience Is Worth Ten Thousand Words," 1936.

57 A Hall of Microscopic Life was also proposed in the initial campaign launch, but this later went by the boards. See *Frontiers* (fund-raising brochure), ANSP (1936): 11.

58 From an Academy solicitation brochure titled "What Shall We Do With These 122 Years?" 1934.

59 The Education Department was officially established in 1936, but Thomas did not come on board until early 1937. He wrote about his role in "pioneering in a very new field" in the professional journal *Education*. See W. Stephen Thomas, "A Director of Science Education in a Large Museum," *Education*, March 1939.

60 Howard, a friend and St. Paul's schoolmate of Cadwalader's, had been "a volunteer member of the scientific staff of the Academy and also the University of Pennsylvania Museum." Letter from Charles M. B. Cadwalader to Childs Frick, 6 February 1935, ANSP coll. 331, folder 15.

61 The "no-man's land" reference is from *Frontiers* (fund-raising brochure), ANSP (1936): 20.

62 Effingham B. Morris, "A Note of Introduction," *Frontiers* 1, no. 1 (October 1936): 3.

63 C. M. B. Cadwalader, *A Museum at Work* (Philadelphia: Academy of Natural Sciences, 1940), 12.

64 *Academy Frontiers*, a new members' publication, was begun in 2009.

65 "Brutes and Scholars," in "Science," *Time*, 29 March 1937, 24.

66 Charles M. B. Cadwalader to Witmer Stone, 3 July 1931, coll. 186.

67 From a press release describing the new Hall of Earth History, ANSP coll. 2010-004.

68 Eda Kassel's married name was Newell, but at the time she created the sculpture, she was unmarried.

69 From a press release describing the new Hall of Earth History, ANSP coll. 2010-004.

70 The cultural/evolutionary spiral, painted by Clarence Rosenkranz, remained a popular exhibit well into the 1960s.

71 Invitation to the Cadwalader surprise party, 3 June (later rescheduled to 27 May), 1947, Cadwalader biography file. H. Radclyffe Roberts assumed the position of managing director of the Academy in January 1947. The Cadwalader party may have been organized as a formal acknowledgement of the transition and of Cadwalader's nineteen years of service as the first managing director (1928–47).

72 Cadwalader is reported "very much pleased with the desk set" that was given him at the party. Letter from John E. Bowers to A. Pomerantz & Co., 3 June 1947. Presidents records, ANSP coll. 2010-051, 1947, box 33, file 12.

73 Edwin G. Conklin, remarks made at the 21 March 1912, dinner commemorating the centennial of the founding of the Academy, *Proceedings of the Centenary Meeting* (Philadelphia: Academy of Natural Sciences, 1912), xxvii.

Skulls of the wandering albatross (*Diomedea exulans*), presented to the Academy by Thomas B. Wilson circa 1846. ANSP Ornithology Department #24531–24537. Purcell photograph.

FOLLOWING PAGES White-throated kingfishers (*Halcyon smyrnensis*) collected by Rodolphe Meyer de Schauensee in Siam (Thailand), 19 October 1937. ANSP Ornithology Department #130665, #14487, and #13388. Purcell photograph.

Acad. Nat. Sci. 130665
Philadelphia
R. M. deSchauensee
Acad. Nat. Sci. No 852 Sex
Date
Philadelphia

Chapter 14

Science and Celebrity: The Academy Goes Hollywood

The principal object, of course, of our visit there was to work with Mr. Ernest Hemingway, who provided the boat and all necessary equipment, in order that Fowler might make a study of the marlin swordfish. . . .
—Charles M. B. Cadwalader to Witmer Stone, about his trip to Cuba with Academy ichthyologist Henry Weed Fowler, August 1934

For an institution of such distinction and longevity, it is not surprising that a large number of persons of national and international acclaim have been associated with the Academy of Natural Sciences. From Henry David Thoreau, who attended a meeting in 1854 on the recommendation of Ralph Waldo Emerson, to pop star Michael Jackson, who made an off-hours visit in 1984 while in Philadelphia for a concert tour, the Academy has drawn celebrities from all walks of life.[1]

Two U.S. presidents (Thomas Jefferson and Ulysses S. Grant) and at least seven Nobel laureates (Wilhelm Röntgen, Marie Curie, Fridtjof Nansen, Thomas Hunt Morgan, Baruch Blumberg, Ernest Hemingway, and the Dalai Lama) are numbered among the Academy's corresponding members. John Frémont, also a corresponding member, former vice president Al Gore, Massachusetts senator John Kerry, former House Speaker Newt Gingrich, and General Colin Powell are among the presidential hopefuls and many politicians who have visited through the years, while several former heads of state—including Prime Ministers John Major of Great Britain and Brian Mulroney of Canada—have come to speak at the Academy, using the institution's stage as a place from which to offer their perspectives on international affairs.

Visitors from the world of literature have included Peter Matthiessen, John McPhee, and Stephen Ambrose (all Academy medalists), as well as George Plimpton, Patrick O'Brian, Hal Borland, Andrea Barrett, David McCulloch, Gerald Durrell, Bill McKibben, and Ernest Hemingway (figure 14.2 and discussed later in this chapter).

From the earliest years, important architects and fine artists have been associated with the Academy as well. William Strickland and John Notman designed buildings for the institution, and Charles Willson Peale, Rembrandt Peale, Thomas Sully, John Neagle, William Winner, James Lamdin, Samuel Murray, and Thomas Eakins made portraits of its early members. Sculptors Benjamin Waterhouse Hawkins, Charles R. Knight, Albert Laessle, and Jim Gary consulted with Academy curators and modeled their work from museum specimens. More recent artists with links to the Academy include Andy Warhol and Isamu Noguchi (both of whom were considered for sculptural commissions), Kent Ullberg (who created the *Deinonychus* dinosaur sculpture in front of the museum), and Signe Wilkinson (a Pulitzer Prize–winning cartoonist who worked for the Exhibits Department and *Frontiers* in the 1970s). Other artists who have spent time at the Academy are photographers Ansel Adams and Eliot Porter (both Academy medalists), Richard Barnes, Henry Horenstein, and Rosamond Purcell; painters Guy Tudor and Ray Troll (Academy medalists who based work on Academy specimens), Nelson Shanks (who was commissioned to paint a portrait

Participants in the Academy's far-flung expeditions often carried their equipment in specially adapted boxes, trunks, and shipping containers. What went out filled with supplies often returned filled with specimens. This trunk was one of twenty used by exhibits director Harold T. Green (1896–1967) on an expedition to Muscat, Oman, and Baasra in 1952. Purcell photograph.

H
TRAPS
CADEMY NAT.
SCIENCES
HILA. PA.
NOT WANTE

of board chairman and president Seymour Preston for the Academy in 2004), Albert Earl Gilbert, Richard Ellis (another onetime Academy employee); fish painters Flick Ford and James Prosek; and Walton Ford, a painter whose work has been inspired by the lives of a number of early Academy members.

Among the many notable scientists to have visited the Academy in recent decades are evolutionary biologists Ernst Mayr, Evelyn Hutchinson, Arthur Cain, Peter and Rosemary Grant, and Tim Flannery; anthropologists Margaret Mead, Don Johanson, and the Leakeys (Louis, Mary, and Richard); primatologists Jane Goodall and Dian Fossey; ornithologist Roger Tory Peterson; marine scientists Jacques Cousteau, Sylvia Earle, and Eugene Clark; astronauts Harrison Schmitt and Guion "Guy" Bluford; oceanographers Jacques Piccard and Bob Ballard; mammalogist George Schaller; malacologist Geerat Vermeij; entomologist E. O. Wilson; paleontologists Stephen Jay Gould and Jack Horner; botanist Peter Raven; and human genome pioneer Craig Venter.

The film and television industry, from which America often takes its cultural cues, has also found its way to the Academy, not just as a source of information and news, but as a destination of personal interest for its participants. The BBC's beloved natural history presenter Sir David Attenborough, an Academy medalist, not only spoke twice at the Academy but also filmed part of his 1985 documentary on John James Audubon, *The Million Pound Book*, in the Academy's Darwin Room, using Audubon specimens from the Ornithology Department and books and manuscripts from the library as props.[2] Newscasters Dan Rather, Jim Lehrer, and Chris Wallace have also visited the museum, while actors Lauren Hutton, Arnold Schwarzenegger, and David Morse have all made visits to the institution. Sylvester Stallone and his crew used the Academy as a staging area while filming parts of *Rocky* and *Rocky II* on the Benjamin Franklin Parkway. As a thank you, the actor donated the movie's turtle mascots Cuff and Link to Outside-In in 1977.

Since the 1930s, the Academy has been a source of information for the movie industry. It helped to pioneer the creation of documentary travel films, with Prentiss Gray and Brooke Dolan leading the effort in Africa and Asia, respectively. Gray made some of the first professional-quality films of African wildlife, and he argued persuasively that showing such films in natural history museums would add the important dimensions of sound and motion to the visitor's experience.[3]

Making and showing films about wildlife seemed a logical extension of the increasingly popular exhibition ideas the Academy was developing in the 1920s and 1930s. Making films about the people who studied natural history was a more novel concept. This was a genre successfully exploited by Martin and Osa Johnson, a Midwestern couple who made several appearances as guest speakers at the Academy in the 1930s (figure 14.3). The Johnsons managed to turn themselves into celebrities by making —and starring in—their own natural history travelogues. Following their lead, many others tried to do the same.

The Academy's direct involvement with the glamour of making motion pictures, and with the celebrity scene in general, gained new acceptance when managing director Charles Cadwalader was charged with raising the institution's profile. In two of the cases examined in this chapter, the opportunities came to the Academy. In the third, Cadwalader sought it out.

AMAZON ADVENTURE

The Academy's first starring role in the movie houses of the world came in 1931, when, along with Coca-Cola, popcorn, and Cracker Jack, moviegoers were able to enjoy a Pathé newsreel before each

14.2 Ernest Hemingway was one of many celebrities to help the Academy build its collections. Here he shows off some impressive specimens in Cuba, 1934. ANSP Archives coll. 707.

14.3 Martin and Osa Johnson made themselves international celebrities through their adventurous travel films. ANSP Archives coll. 775.

feature film.[4] According to the *New York Times*, the prefeature trailer for the first week of October gave moving-picture coverage of the first game of the World Series,[5] a blimp picking up mail from the tower of the newly built Empire State Building,[6] a visit to Japan by Charles and Anne Morrow Lindbergh, Maurice Chevalier's return to New York, and scenes from a scientific expedition to the Matto Grosso, Brazil.[7] This last segment proved so popular that "live" (silent) reports from the recently completed expedition became a regular offering in movie-house newsreels throughout the autumn. A clip in one of them showed two scientists capturing "a rare giant armadillo for the Academy of Natural Sciences of Philadelphia," and another showed the capture of a red wolf from the same expedition.[8]

The unlikely star of these zoological episodes was the Academy's normally retiring entomologist James A. G. Rehn (1881–1965), who had no idea he was going to spend five months of his life in the heart of the Amazon, let alone become a matinee showman in movie theatres across the country. His time of hardship, exhilaration, and fleeting fame grew from a cable sent to Charles Cadwalader from the Paraguay River in Brazil in early April 1931. Four months after the Matto Grosso Expedition had been launched, Academy trustee E. R. Fenimore Johnson, a principal sponsor of the enterprise, contacted the museum's managing director from the field with a belated request for an experienced naturalist. Through his invitation, Johnson hoped to help both the expedition and the institution on whose board he served. His message was at once generous and demanding:

> UNUSUAL ZOOLOGICAL OPPORTUNITIES HERE – STOP – OUR MAN NOT GOOD ENOUGH – STOP – IF YOU WILL SEND TRAINED COLLECTOR TO REPRESENT ACADEMY I PERSONALLY GUARANTEE DONATION COVERING ALL COST TO ACADEMY UP TO FIVE THOUSAND DOLLARS – STOP – ACADEMY TO CONTROL ZOOLOGICAL STATEMENTS [AND] OWN ZOOLOGICAL COLLECTIONS UNDER ARRANGEMENT SIMILAR TO OURS WITH UNIVERSITY [OF PENNA.] MUSEUM – STOP – DISPATCH SCIENTIST VIA SANTOS TO CORUMBA.[9]

Cadwalader responded immediately, accepting Johnson's offer. Four days later, he cabled Johnson again:

> SCIENTIST JAMES REHN SAILING NORTHERN PRINCE APRIL TWENTY-FOURTH ARRIVING SANTOS MAY EIGHTH WRITING AIR MAIL.[10]

No written records survive to describe the urgent discussions Cadwalader must have had with the Academy's senior staff to determine who among them would accept the enticing offer. His initial response to Johnson's cable suggests that the answer was not immediately apparent. As had been the case since Cadwalader assumed the role of managing director three years before, almost everyone at the Academy was already committed to summer fieldwork. Rehn, too, may have had other plans, but both he and Cadwalader knew that the Academy was unlikely to be offered such an opportunity again. The area to be explored was biologically rich and a blank spot on the map of scientific research. It was also a dangerous place, known to be populated by hostile Indian tribes (meeting with them was one of the expedition's objectives). This was the same area in which the famed English explorer Percy Fawcett (1867–1925) and his son had disappeared without a trace just six years before. But to make collections in one of the least studied parts of the planet was every naturalist's—and every institution's—dream. And so, with his wife's reluctant approval and the Academy's enthusiastic support, the scramble began to get Rehn and the necessary equipment to Brazil (figure 14.4).

In just three weeks, the fifty-year-old entomologist assembled more than two thousand pounds of supplies, meticulously documenting the contents of each of the twenty-five trunks he shipped to Matto Grosso. His nine-page inventory included guns, ammunition, navigational equipment, cameras, film, traps, cages, camping gear, clothes, and personal effects, including fourteen pairs of (used) socks, a Portuguese-English dictionary, a Colt army revolver, and a sombrero.[11] Never before or since has the Academy assembled such an exhaustive set of supplies on such short notice. Never before or since has a single collector required two thousand pounds of equipment for a single expedition!

As Rehn's eleventh-hour invitation to participate suggests, the Matto Grosso Expedition was neither conceived nor directed by the Academy.[12] It was, in fact, a partly commercial, partly philanthropic enterprise, intended "to create for posterity as complete a popular and scientific record of the human, animal and plant life and of the scenic and other features of the territory within a radius of 500 miles of Descalvados in Matto Grosso as the resources of the expedition permits."[13] The group planned to collect "zoological and museum specimens of the fauna and flora of the district" as well as "modern and archaeological native artifacts." The first would come to the Academy, the last to the University of Pennsylvania's Museum of Archaeology and

14.4 James A. G. Rehn (1881–1965), right, poses beside a dead jaguar with Mr. and Mrs. Floyd Crosby, Matto Grosso Expedition, 1931. ANSP Archives coll. 2010-129.

Anthropology, on whose board Johnson also served.[14] Above all, the organizers of the Matto Grosso Expedition wanted to make films so that their experiences and discoveries could be shared with the rest of the world. If these could help repay the considerable cost of the expedition, so much the better.

Conceived by Vladimir Perfilieff (1895–1943), a Russian-born artist and explorer, and Alexander "Sasha" Siemel (1890–1970), a Latvian "adventurer, linguist and professional hunter with twenty years' experience in the jungles of South America," the expedition was the first to use "sound picture" equipment to record the customs and languages of the indigenous peoples of South America.[15] An important silent partner in the enterprise was John S. Clarke Jr., the son of the founder of the Autocar Company and a University of Pennsylvania classmate of Johnson's, who was interested in exploring the educational use of film. Clarke helped pay for the enterprise and succeeded in lining up other backers among his wealthy friends in New York.[16] To ensure that the expedition's films would be of the highest quality, Clarke hired two exceptional photographers: Floyd D. Crosby and Arthur P. Rossi. Later that year, in vindication of Clarke's choice, the Philadelphia-born Crosby won an Academy Award in cinematography for the popular Hollywood film *Tabu: A Story*

14.5 E. R. Fenimore Johnson (1899–1986), with his fox terrier, Tupi, and the Sikorsky S-38 amphibious plane he provided for use during the Matto Grosso Expedition, talks with expedition ethnographer Vincenzo Petrullo of the University of Pennsylvania. ANSP Archives coll. 2010-049.

of the South Seas. He went on to be the principal cinematographer for dozens of other Hollywood films, including *High Noon* (1952).[17] Together, Crosby and Rossi provided the expedition with an unusually lively and extensive visual record of both still and motion pictures.[18]

The films Crosby and Rossi made during the expedition were intended to serve many purposes. Some were "purely scientific" and educational (i.e., documentaries), while others were aimed at a broader audience. Much effort was expended to create "a popular style talking moving picture featuring the experiences of one man [Sasha Siemel] hunting and traveling in Matto Grosso and living with the frontiersmen and aborigines."[19] This involved trying to capture on film Siemel's legendary prowess at killing jaguars by goading them into attacking him, which ended with them impaling themselves on a metal-pointed spear. It was hoped that this high-risk, low-tech method of aboriginal hunting would "provide the element of thrill which is so necessary if popular attention is to be successfully focused on scientific field work in these modern times."[20] Although Siemel was successful in carrying out many fatal cat impalements during the expedition, the action proved too quick and the conditions too difficult to capture on film.[21]

Rehn disapproved of such sport, noting that the speared animals were usually too badly mangled to serve as suitable museum specimens. He had not come to Brazil for theatrics. He had come for scientific specimens and whatever knowledge he could glean from his time in the field. After a long and difficult journey, made more complex by extensive flooding in the area, he established his headquarters in the expedition's recently built base camp. At the edge of its bustling compound was a menagerie of live animals, including two young adult jaguars, three jaguar cubs, two howler monkeys, two giant armadillos, two deer, a tapir, anacondas, toucans, guans, capuchin monkeys, and a giant river otter.[22]

Johnson had preceded Rehn in the field by several months and was there to welcome the naturalist when he arrived, along with a ten-seated Sikorsky S-38 amphibious plane that his father had chartered from Pan American Airlines for use by the expedition (figure 14.5). The handsome and impeccably dressed heir to the RCA Victor fortune also had a brown-and-white fox terrier named Tupi.[23]

There was plenty of precedent for including dogs on expeditions. On an early collecting trip to Florida, Academy president William Maclure and vice president George Ord took dogs to assist Thomas Say, Titian Peale, and themselves on their daily hunts. Over the years, other dogs played similar roles in catching game, pulling sleds, guarding camps, and even accompanying their masters on the lecture circuit after their return.[24] In some desperate cases, dogs served as emergency food for their owners.[25] Several ended up as specimens in the Academy's mammal collection.[26] But Tupi was not on the expedition to hunt. The high-strung (and by all accounts disagreeable) pet was an advertising prop and most likely the first dog brought on an expedition for purely commercial reasons. It was one of two fox terriers that were transported to Brazil from the United States by ship, train, and automobile, along with a well-trained pack of hunting dogs that helped the hunters track and tree jaguars, pumas, and other game. The photogenic hounds were featured in one of the Pathé newsreel stories.

Tupi was the visual successor to Nipper, the universally recognized mascot of the Victor Talking Machine Company that had built the Johnson family fortune. At the time, the iconic image of a quizzical dog listening to "His Master's Voice" through the funnel of a Victor gramophone was known the world around.[27] Even though his family had recently sold the company to RCA (for an estimated $40 million),[28] Johnson hoped to add to the logo's international cachet with some good publicity shots from the heart of South America. And so, somewhat incongruously, the little terrier appears in many of the photographs that document the Matto Grosso Expedition. Also captured in photographs by Crosby and Rossi are local Indians proudly displaying small metal castings of Nipper that Johnson presented to them, along with other trade items, in exchange for local food, specimens, and permission to film (figure 14.6).

More difficult for the photographers to record were the grueling conditions faced by Rehn, Johnson, and the others as they pursued the many noncinematographic goals of the expedition. After starting with weeks of cold, wet conditions in May, Rehn felt the thermometer climb to the uncomfortably higher tropical levels for which the region is famous. Heat and humidity thus provided a baseline for the many other hardships he endured, which included everything from hungry gnats and wounded jaguars to ravenous caimans and piranhas.[29]

Despite such challenging conditions, with the help of a native assistant named Pereira, Rehn was extremely successful in his collecting, ultimately bringing back to Philadelphia thousands of specimens in every discipline but his own.[30] "Insect work is a complete wash-out," he wrote Cadwalader in frustration. "It is mid-winter here now and the middle of the dry season, which means the resting season. There are no swarms of moths, almost

14.6 Near the Xingu River in the heart of the Amazon, two Tsuva men display the brass casting of Nipper, the RCA Victor mascot, presented to them by members of the Matto Grosso Expedition. University of Pennsylvania Museum Archives.

no butterflies, few beetles, hardly a grasshopper and little except countless gnats and mosquitoes out in the pantanal. I have a modest number of insects, but it is like trying to gather them at home at Christmas time."[31]

In addition to securing shells, fish, birds, mammals, and plants for the Academy's permanent collection, Rehn was asked to gather specimens and make photographs that would help Harold Green and the Exhibits Department create a jaguar diorama. In their confidential exchange of letters, Cadwalader and Rehn worried about just where such a diorama would go. At the time, there were only plans for African and North American diorama halls in the museum, but because creating a jaguar habitat group was a high priority for Johnson, and because Johnson was paying for the Academy's participation in the expedition, Rehn promised to get everything that would be needed for such a display. In this too he was successful.[32]

At summer's end, with supplies and finances exhausted and tensions running high after so many months of exhausting work, the expedition came to a suitably theatrical close when John Clarke was shot through the shoulder by an overly protective camera crew member while he was hunting a jaguar. Rehn reported that the accident occurred when "Crosby got rattled and had a nervous trigger finger *after* the jaguar was dead."[33] Had Johnson's amphibious Sikorsky not been on hand for a speedy evacuation, the loss of blood and near certainty of infection might have caused Clarke's death. Fortunately, a hasty exit was followed by lifesaving medical attention in Brazil and the United States. His shattered bones repaired, Clarke was able to continue his role as the financial manager of the Matto Grosso project.

The newsreels were released soon after the expedition's return, and some of the educational films made during the expedition were available for public showing in 1932. It was not until early the following year, however, that the full-length (six reels) feature film about the trip was ready for national distribution.[34] Distributed by Principal Adventure Pictures of New York, *Matto Grosso* was advertised as "the first sound and talking picture produced in the wilds of South America" and "one of the most unbelievable releases ever made."[35] Among the other travel documentaries offered by Principal at the same time were *Drifting Around with Lowell Thomas, Zane Gray's South Seas Adventure, Riding the Skies with Amelia Earhart*, and a round-the-world travelogue by Martin and Osa Johnson.[36] Interestingly, at least two other films appearing in movie theaters that year had strong Academy connections. They were *To the South Seas with Mr. and Mrs. Gifford Pinchot* (six reels), based on the 1929 expedition on which Henry Pilsbry had served as chief naturalist, and the somewhat more engaging *Devil's Playground*, a Bahamian fishing film starring Academy trustee George Vanderbilt, whom one smitten reviewer compared to Olympic swimmer and Tarzan actor Johnny Weissmuller.[37]

The film critic for the *New York Times* gave *Matto Grosso* a mixed review (figure 14.7). While he found the movie "distinguished by astonishingly fine photography," he considered the much-touted natural sound disappointing: "For their pains in lugging sound equipment into the up-country fastness, the cameramen are rewarded with a few animal noises and some native speeches," he observed. Its audio shortcomings notwithstanding, the *Times* ultimately proclaimed *Matto Grosso* "a definitely superior specimen of travel film."[38]

Unfortunately, *Matto Grosso* was up against considerable box office competition. In addition to the other nonfiction travel and adventure films distributed by Principal, Hollywood's commercial studios were creating fictional movies with similar adventure themes but on considerably larger budgets. The biggest and best-known of these was *King Kong*, which matched *Matto Grosso*'s exotic appeal but outdistanced it in drama. For every unusual animal offered in *Matto Grosso, King Kong* offered a special-effects-shop creation on a larger, more terrifying scale. Without dinosaurs, a giant gorilla, or a flirting (and hysterically screaming) Fay Wray, *Matto Grosso* simply could not match the box office draw of the Cooper and Schoedsack production.

The separate documentaries of Amer-Indian life filmed by Crosby and Rossi to fulfill the research and educational goals of the Matto Grosso Expedition proved more enduring. Produced by Ted Nemeth and narrated by Lowell Thomas, the films are still consulted today as the first visual and audio records of the Bororo, Tsuva, and Yawalapiti peoples.[39]

Intrigued by the filmmaking aspects of the expedition, Fenimore Johnson went on to pursue his own career in cinematography, specializing in the then-experimental field of underwater photography (figure 14.8). His pioneering documentary on the coral reefs of the Caribbean, *Undersea Garden* (1938), is acknowledged as the first of its kind ever made.[40] Johnson shared his knowledge of scuba diving and underwater photography with the U.S. Navy during World War II and, in the years to follow, helped to train the underwater teams that would evolve into the special forces group now known as the Navy SEALS. His company, Fenjohn, was the first in the United States to commercially manufacture camera housings for underwater use.

James Rehn was less enthralled with moviemaking, preferring to pursue the quiet, careful scientific research that had earned him a place on the Matto Grosso Expedition. Over the next several decades, he and his Academy colleague Morgan Hebbard made multiple trips to the American Southwest and elsewhere to study and collect insects. When Rehn died in 1965 at the age of eighty-three, he had spent sixty-four years working at the Academy. He left behind a distinguished record of more than 350 scientific papers and several definitive monographs.[41] The tens of thousands of insect specimens from Africa and Australia and North, South, and Central America that he and Hebbard contributed to the Academy's collections—including those from Matto Grosso—are still in frequent use.

14.7a AND b Filmmaking was the principal objective of the Matto Grosso Expedition. Here team members interact with their indigenous hosts. ANSP Archives coll. 2010-049.

During Charles Cadwalader's time as managing director, when opportunities for acquiring specimens came to the Academy, he always responded in the same positive but cautious manner, refusing to commit money up front but agreeing to have the museum's curators review the offerings on a specimen-by-specimen basis. If the Academy wanted the specimens and if funding could be found, he suggested, a purchase was likely to follow. To Cadwalader, expressing an interest in an expedition was a first step in a long, complicated dance that might or might not lead to an acquisition. There were many such dances. Many were remarkably successful.

Cadwalader was good at responding to opportunities coming from outside the Academy, but he was even better at initiating his own opportunities from within. Such efforts were behind many of the collecting trips dispatched to help create the dioramas with which he was reconfiguring the museum (see chapters 12 and 13). They were also responsible for securing assistance for the Academy's less visible research departments.

14.8 Academy trustee E. R. Fenimore Johnson (1899–1986) was a pioneer in the field of underwater photography. His documentary on the coral reefs of the Caribbean (1938) was the first of its kind ever made. ANSP Archives coll. 2010-049.

ERNEST HEMINGWAY

In 1934, using his network of well-connected sportsmen friends, Cadwalader sought out an unexpected ally to help with a scientific study of the fish of the North Atlantic then being conducted by the Academy's longtime ichthyologist Henry Weed Fowler (1878–1965).[42] Ernest Hemingway (1899–1961) and his wife Pauline were in Paris, on their way back from a ten-week safari in East Africa, when he received a letter from Cadwalader (forwarded from the office of Charles Scribner's Sons in New York). In it, Cadwalader asked if Hemingway would be interested in helping the Academy conduct some important research on saltwater fish near his Florida home. The goal, he explained, was to address "the lack of knowledge concerning the classification, life histories, food [and] migrations . . . of the sailfish, marlin, tuna and other large game fishes." He continued:

> We have here [at the Academy] a very large and comprehensive collection of fishes from all over the world, but specimens of the larger game fish are lacking. According to Mr. Fowler, no museum in the world has a comprehensive collection of these big game fish such as we are anxious to secure.[43]

For Hemingway, whose lifelong passion for fishing had taken a new focus on Atlantic game fish after his move to Key West in 1928, the Academy's request was intriguing. In a four-page, handwritten response, penned from the S.S. *Paris* en route to New York, Hemingway eagerly accepted Cadwalader's invitation, asking for more details about what fish specimens the museum already had and discussing at length the ways in which new fish would need to be collected and preserved.[44]

So enthusiastic about the project was Hemingway that he traveled to Philadelphia from New York to discuss it in person soon after his return to the United States.[45] During his 9 April meeting with Cadwalader and Fowler at the Academy, Hemingway extended an invitation for both men to join him for some fishing off the coast of Cuba that summer. Only by seeing live fish, fresh from the water, he claimed, would Fowler be able to appreciate the range of colors associated with the age, sex, and species of the fish he was studying.

Having heard nothing from the Academy for several months, Hemingway followed his Philadelphia meeting with a written invitation from Florida, suggesting a hotel in Havana where the two Philadelphians could arrange "a cool, pleasant, comfortable room with a bath . . . in a very pleasant situation, close to the old waterfront and overlooking the harbor and the Morro Castle for $2.00 a day. . . . It is much cooler in Havana now than in Philadelphia or N.Y," he assured his would-be fishing partners.[46] Hemingway closed his letter with an enthusiastic description of his brand-new fishing boat, the *Pilar* ("it is very comfortable, sea-worthy and has speed when it is needed"), and, even more enticingly, of a nine-foot, 119.5-pound sailfish he had just caught. "I fought him for 41 minutes and landed him," enthused the proud angler. "He was weighed on a tested scale . . . before eight witnesses and I believe he is the biggest ever caught in the Atlantic."[47] Convinced that Hemingway's offer of hospitality was genuine, the Academy representatives made their hotel reservations and were soon enjoying daily excursions into the Gulf Stream with their genial host.

Most of the information we have about the six-week sojourn comes from an article Hemingway wrote for *Esquire* magazine and from the dozens of enthusiastic letters exchanged between Hemingway, Fowler, and Cadwalader in the year following the trip.[48] Unfortunately, neither Fowler nor Cadwalader appears to have kept a diary of his time with Hemingway, but another member of the fishing party did. Arnold Samuelson (1912–1981) was an aspiring writer from the Midwest who, at the age of twenty-two, hitchhiked from Minnesota to Key West in the hope of meeting his literary hero. What Samuelson hoped might be a short interview with Hemingway turned into a yearlong period of employment, during which Samuelson helped the writer with various odd jobs in exchange for minimal pay and the experience of being near an admired mentor.[49]

Jealous of the amount of time and attention Hemingway was giving to Fowler and Cadwalader, and uncomfortable with the academic and social differences the visitors represented, Samuelson took an instant dislike to the two "scientificos" when they arrived, ridiculing the academicians for what he saw as stiff and condescending behavior (figure 14.9). In his account of their first meeting, Samuelson reveals his negative feelings about Hemingway's guests:

> E.H. met the scientificos at the hotel the night they came, and it was not until the next morning that we saw them, when they were walking down toward the dock, E.H. between them, so much bigger he looked like a father with two sons. One of them had white hair and the stiff-jointed walk of a man beginning to get old; the other, middle-aged and shorter, carried a big fishing rod and walked with his head back and his arms swinging stiff at the elbows. . . . The white-haired man said nothing. He was Henry Fowler, the ichthyologist; the short man was Charles Cadwalader, his boss, and out of politeness he [Fowler] let him talk to E.H. first. Cadwalader, short-legged, slightly pot-bellied, always wore the same club-room conversationalist expression on his freckled face, and when he talked to one person he spoke as if he were making a speech to a crowd or speaking for the benefit of those who might be trying to overhear, like a lecturer answering questions of people in his audience. . . . I had not yet been told that this bachelor philanthropist was the last of a distinguished line of money-making, money-hoarding Cadwaladers. It was not until later that I was told he kept twenty-seven servants in his house and was very much upset because an old woman intended to retire and it would be like losing one of the parts in

14.9 Charles M. B. Cadwalader (1885–1959), artist unknown. Pencil on paper. ANSP Archives coll. 457.

a smooth-running machine. This was the first man I had run into who had so many ancestors and so much money, and I had difficulty understanding him. He would not drink vermouth with us before dinner or wine with his meals or whiskey in the evenings, but would only drink bottled mineral water.[50]

Hemingway saw the situation quite differently. Clearly enjoying the company of his guests and the thrill of contributing new information to the scientific world they represented, he encouraged them to extend their visit from ten days to more than five weeks. Day after day, they spent time in the close quarters of the *Pilar* with only moderate success. Hemingway apologized profusely for his failure to produce more big fish during their time in Cuba. Less than twenty days after their return to Philadelphia, he wrote to his recent guests to invite them back for another visit. "I look forward to having you and Fowler with us again," he wrote Cadwalader in early September. "You know how much we thought of Fowler and, if you do not mind my saying it, I have never fished with a better sportsman or a more agreeable companion than you were. . . . Mrs. Hemingway and I both appreciate your invitation to stop with you some time in Philadelphia and look forward to it." In closing, he scrawled a further note of friendship and humorous quip for his scientific, non-drinking friend: "We'll split a bottle of scientific water!"[51]

Hemingway was less generous in his opinion of Samuelson:

He was an excellent night watchman and worked hard on the boat and at his writing but at sea he was a calamity; slow where he should have been agile, seeming sometimes to have four feet instead of two feet and two hands, nervous under excitement, and with an incurable tendency toward sea-sickness and a peasant reluctance to take orders. . . . If any more aspirant writers come on board the *Pilar* let them be females, let them be very beautiful, and let them bring champagne.[52]

14.10 Ernest Hemingway (1899–1961) admires a catch aboard the *Pilar*, 1934. His understanding of the game fishes of the Atlantic, communicated through the Academy, made significant contributions to the field of ichthyology. ANSP Archives coll. 707.

Despite losing an enormous striped marlin on his last day out, Cadwalader clearly enjoyed his experience. He wrote Hemingway to thank him for his "splendid hospitality and various kindnesses," describing the experience as "one of the finest holidays I have ever had."[53] More importantly, the trip had produced significant scientific findings for Henry Fowler. Five days after his return to Philadelphia, Cadwalader wrote Witmer Stone to report on his "successful stay in Cuba."

> The principal object, of course, of our visit there was to work with Mr. Ernest Hemingway, who provided the boat and all necessary equipment, in order that Fowler might make a study of the marlin swordfish. Although our expedition was only successful in landing one marlin, which weighed 420 pounds and was captured by our host, Ernest Hemingway, Fowler learned a great deal from him in regard to these large fishes. . . . [With this trip] he feels that he has accomplished something worth while.[54]

In the months that followed, Hemingway, Cadwalader, and Fowler exchanged many letters revolving around their shared interests in fish and fishing. Hemingway sent photographs of his angling triumphs and sometimes freshly caught fish. The latter, packed in ice, were accompanied by detailed information on their size and color and the conditions under which they were caught. One "small tuna shaped fish that looks, from length of fin to be an albacore" came with a suggestion of an alternate use if Fowler did not want to keep it for the Academy's collection: "If you don't want him as a specimen and he gets there in good shape, wash the salt out, cut the meat off both sides of the back bone and broil it, or cut the head off, make scores in the side of the fish and bake it."[55]

The dozens of post-trip letters that Hemingway sent to the Academy, some typed but many handwritten and some running up to six pages in length, are loaded with scientific information revealing the extent of his interest in natural history. They include descriptions of everything from whales ("I'd have given any damned thing if you and Fowler could have been here to see them") to migrating warblers.[56]

During the winter of 1934–35, Hemingway kept his Philadelphia friends informed of fishing conditions in Cuba, Florida, and Bimini (to which he invited Cadwalader and offered to pay to have Fowler join him for more fishing in 1935).[57] In return, Cadwalader wrote with pride about the Academy's activities during the same period. His reports on the Vanderbilt Africa Expedition—and Hemingway's responses to them—suggest that Hemingway was keen to participate in a future Academy expedition. Samuelson reported that the first discussion of this idea had occurred aboard the *Pilar*. His account helps to explain how a common love of field sports enabled the writer-sportsman and the Philadelphia socialite to get along so well:

> E.H. wanted Cadwalader to catch a marlin before he went back to the States and he extended his invitation from ten days to fifteen, then twenty and finally well into August [1934]. They had been on board a month now, and it was the last day of their stay. . . . [Hemingway and Cadwalader] talked about the deer and quail and prairie chickens up north and trap shooting and duck hunting in the old days. Cadwalader belonged to several gun clubs and had shot everything except big game and fished everything except big fish. He said he would like to have E.H. join his expedition to Africa and shoot a collection of gorillas and monkeys for the museum. Would E.H. consider it? Hell, yes, E.H. said, he'd shoot anything. Cadwalader wasn't positive about the expedition, though, because in the present economic crisis there was absolutely no assurance whatever [that he could find the necessary financial backers or that he himself could sponsor it].[58]

Hemingway later wrote Cadwalader that he intended to put aside some of the royalties from *The Green Hills of Africa* as an "expedition fund" to assist the effort.[59] Unfortunately, the African expedition the two men discussed never materialized, and Hemingway's life and travels took a different direction. One wonders just what might have emerged from an Academy expedition in which Ernest Hemingway had been a participant.

The relationship Hemingway developed with Fowler during their 1934 trip ran on a parallel but independent track. Hemingway continued to provide the ichthyologist with invaluable information and specimens long after their month of fishing. In return, Fowler helped Hemingway understand more about the sea life he was encountering in the waters off Cuba and Florida and during his forays farther afield. As a gesture of appreciation and respect, Fowler named a new species of fish *Neomerinthe hemingwayi* in honor of his fishing host and friend (figure 14.11).[60]

Hemingway appreciated the scientific immortality conveyed in Fowler's gesture, but he valued even more the stimulating exchange of ideas about fish and other forms of marine life that took place during Fowler's visit and subsequent exchange

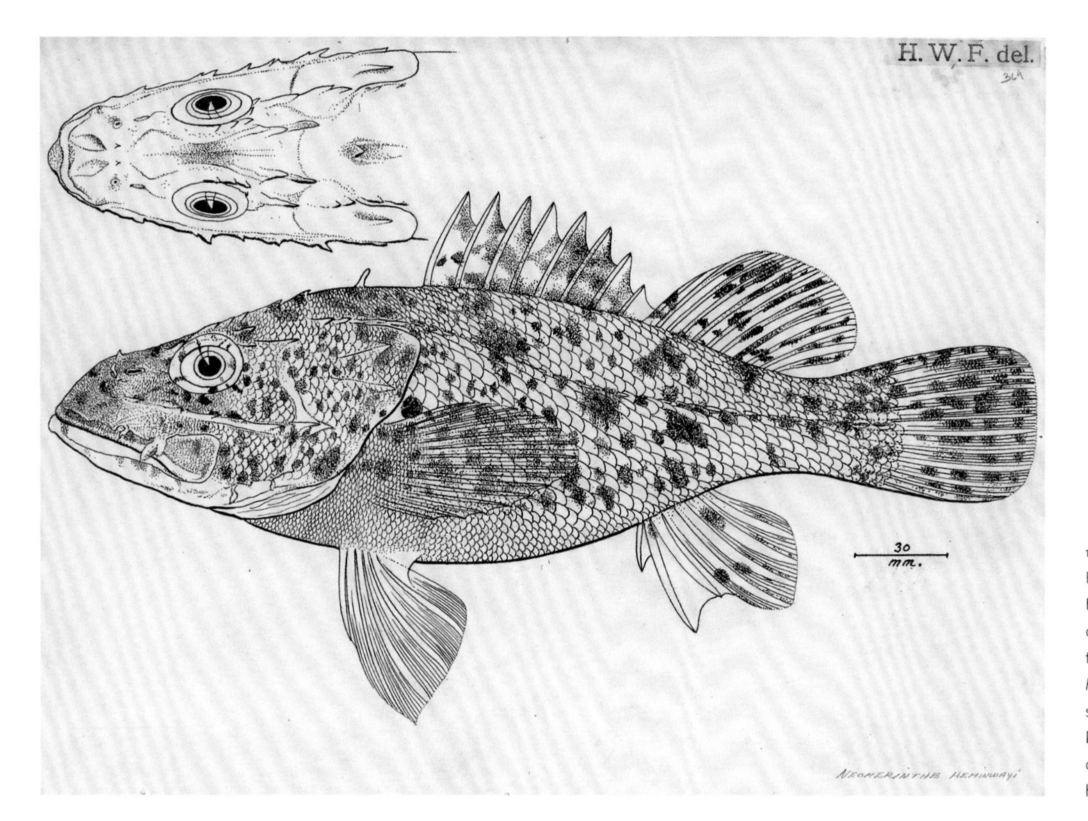

14.11 Academy ichthyologist Henry Weed Fowler (1878–1965) honored Ernest Hemingway by naming a new species of Atlantic sculpin after him. This is the drawing he made of *Neomerinthe hemingwayi*, the type specimen of which is still housed in the Academy's Ichthyology Department along with a number of fish collected for the Academy by Hemingway himself. ANSP Archives coll. 197.

of letters. In his *Esquire* article "Genio after Josie: A Havana Letter" (October 1934), he reported on Fowler's and Cadwalader's time aboard the *Pilar*. But it was in a later *Esquire* essay, "On the Blue Water: A Gulf Stream Letter" (April 1936), that he incorporated much of the ichthyological and marine knowledge he had gleaned from his discussions and correspondence with Fowler. This essay contained many of the elements that would resurface years later in *The Old Man and the Sea* (1952) in "the prose I have been working on all my life."[61] The scientific accuracy of Hemingway's writing in that classic novella and, to a lesser extent, in his posthumously published novel *Islands in the Stream* (1970) was undoubtedly shaped by the exchange he and Fowler enjoyed during and after their memorable weeks in Cuba.

JAMES BOND

While Ernest Hemingway and some of the figures associated with the Matto Grosso Expedition went to considerable effort to capture the public's attention through their writing and films, another Academy collector spent almost as much trying to escape it. James Bond was "happiest not only when inconspicuous, but if possible, invisible,"[62] yet he gained international fame when Ian Fleming stole his name for the fictional British secret agent 007. The real James Bond (1900–1989), who was affiliated with the Academy for more than fifty years, was a brilliant and sophisticated, but unassuming, gentleman who devoted his life to studying the birds of the West Indies. The last thing he expected—or wanted—was to become a universal symbol of sex, glamour, and intrigue.

Although Bond did not begin life thinking he would become an ornithologist, he always enjoyed outdoor adventures and field sports. His father may have inspired his interest in hunting and museum work with a 1912 collecting trip to Venezuela. When Bond was just twelve, Francis Bond (1867–1923), a wealthy Philadelphia banker and stockbroker, joined his childhood friend and schoolmate Stewardson Brown (1867–1921), curator of botany at the Academy, and Thomas Gillin (b. 1853), a surveyor and engineer, sometime farmer, and keen amateur ornithologist, on a three-and-a-half-month expedition to the delta of the Orinoco River.[63] From this trip, principally financed by the senior Bond, the three adventurers brought back for the Academy 850 sheets of vascular plants, 505 bird skins (representing 101 species), more than 1,000 insects, 49 trays of land and freshwater snails, and an undetermined number of fish, six of which were described as new species by Henry Fowler.

Although he was not close to his father, the young James Bond could not have helped but be impressed by the acclaim given his father's expedition (not to mention the live howler monkeys he brought back to their home in Spring House, Gwynedd Valley).[64]

Another strong influence in Bond's early life was his mother's brother, Carroll Sargent Tyson Jr. (1878–1956), a painter who had a great interest in birds. Together, Bond and Tyson later wrote a book on the birds of Acadia National Park on Mount Desert Island in Maine (1941), where the Tysons owned a summer house. Bond spent many happy holidays in Maine with his Tyson relatives, but he received most of his formal education in Great Britain, where his father moved when he remarried after the death of Bond's mother, Margaret Tyson Bond, in 1912.[65]

Educated at two of England's finest schools—first Harrow and then Trinity College, Cambridge, where he read economics—Bond returned to Philadelphia in 1922 unsure of his future.[66] Like his father, he began work as a banker, taking a job in the foreign exchange department at the Pennsylvania Company (later First Pennsylvania Bank). He soon grew tired of the drudgery associated with this position and dissatisfied with the long hours he was required to spend at his desk. He yearned for the outdoors and the chance to immerse himself in wild nature, which he had so enjoyed during his summers in Maine. So, in 1925, when his friend Rodolphe Meyer de Schauensee suggested that he join him on a trip to South America, Bond jumped at the chance (figure 14.12). Meyer de Schauensee, whom Bond had known before moving to England, had already made a successful trip to Brazil in 1924 to look for exotic animals.[67] The plan was for the two men to pay for the self-financed expedition by bringing back live animals for the Philadelphia Zoo and study skins for the Academy of Natural Sciences. They each borrowed $1,500 to make the trip and spent several months enjoying the magical diversity of the lower Amazon basin. Many of the animals they brought back were alive, including a black saki (monkey), a golden parakeet, and a twenty-three-foot anaconda.[68] "There were also two rare Hyacinthine Macaws, very much alive, who bit their way out of the cage on the ship on the way home," explained Bond's wife and biographer, Mary Wickham Bond. The macaws were valued at $300 apiece, and the two collectors "were horrified to see $600 worth of bird clutching the stern rail and going up and down with the swells. The cook rushed out with a fishnet and managed to recapture them."[69]

On their return to Philadelphia, Bond and Meyer de Schauensee sold their living specimens to the zoo and the skins to the

14.12 James Bond (1900–1989), right, and Rodolphe Meyer de Schauensee (1901–1984) solidified their lifelong friendship and love of birds with a collecting trip to Venezuela in 1925. ANSP Archives coll. 457.

Academy, eventually recouping their expenses. It was the beginning of a long though sometimes tenuous relationship between Bond and the Academy. In the years to come, he was given small amounts of money—sometimes up to $1,000—in exchange for his specimens, but he was never paid a salary. Even when he was named curator of the Ornithology Department in 1962, he was not put on the payroll.

From 1926 onward, Bond devoted himself entirely to the study of West Indian birds, personally collecting all but six of the three hundred native species of the region (figure 14.13). Writing in response to a biographical query in 1974, Bond's wife described some highlights of his career:

> [He] has climbed the 7000 foot La Selle Mountain in Haiti to pine forests, habitat of the White-winged Crossbill, and La Selle Thrush. [He has made] several trips to the Zapata Swamp, Cuba, collecting rail, wren, and finch peculiar to that region. [He] has explored remote islands, living on as little as twenty cents a day, slinging his hammock under wayside shelters, government rest houses, and in hospitable native huts.[70]

In the same document, Bond explained his general interest in science and his specific reason for studying the birds of the Caribbean:

> All my life I have been interested in natural history and as a young boy collected butterflies. . . . On my return [from the Amazon in 1926], I decided on my life work, a survey of the avifauna of the Antillean Subregion, and subsequently of extralimital islands of the Caribbean. My reason for undertaking this was not only the great number of endemic native species on these islands (birds not found elsewhere), but also the realization that probably more Antillean birds were in danger of extinction than in all the rest of the New World.[71]

14.13 James Bond was at his happiest when in the field. He is seen here on an early collecting trip in the Caribbean. ANSP Archives coll. 457.

14.14 The real James Bond (left) meets Ian Fleming, the author who stole his name, at Fleming's house in Jamaica, February 1964. ANSP Archives coll. 2010-045.

Bond would make his scientific reputation by theorizing —and then proving to a skeptical scientific community—that the birds of the Antilles, long considered neotropical (i.e., from South America), are in fact predominantly an old North American fauna.[72] In consequence, he was ultimately awarded almost every important scientific honor given in the field of ornithology.[73]

It was through his authorship of a field guide to the birds of the West Indies that he gained popular recognition. Illustrated by Don Eckelberry and Earl Poole, Bond's *Birds of the West Indies* was first published by the Academy of Natural Sciences in 1936. It was then republished in many subsequent editions by Collins (in the United Kingdom), and Macmillan and Houghton Mifflin (in the United States). For more than fifty years, it was the definitive reference guide for anyone interested in identifying the birds of the Caribbean.[74]

Among Bond's many devoted readers was Ian Fleming (1908–1964), an amateur bird-watcher with a house in Jamaica, who considered Bond's book "one of my bibles."[75] In a magazine interview in 1961, he explained, "There really is a James Bond, you know, but he's an American ornithologist not a secret agent. I'd read a book of his and when I was casting about for a natural sounding name for my hero, I recalled the book and lifted the author's name outright."[76] Confronted by Mary Bond through a tongue-in-cheek letter about his "brazen" theft of her husband's name, Fleming responded with a gracious apology and explanation. His letter, reprinted by Mrs. Bond in at least two books about her husband's adventures, reads, in part, as follows:

> Dear Mrs. James Bond,
> I will confess at once that your husband has every reason to sue me in every possible position and for practically every kind of libel in the book, for I will now confess the damnable truth.
>
> I have a small house which I built in Oracabessa in Jamaica just after the war, and some ten years ago, a confirmed bachelor on the eve of marriage, I decided to take my mind off the dreadful prospect by writing a thriller.
>
> I was determined that my secret agent should be as anonymous a personality as possible. Even his name should be the very reverse of the kind of "Peregrine Carruthers" whom one meets in this type of fiction.
>
> At that time one of my bibles was, and still is, *Birds of the West Indies* by James Bond, and it struck me that this name, brief, unromantic and yet very masculine, was just what I needed, and so James Bond II was born,

and started off on the career that I must confess has been meteoric, culminating with his choice by your President as his favorite thriller hero (see *Life* of March 17th.)

So there is my dreadful confession together with limitless apologies and thanks for the fun and fame I have had from the most extraordinary chance choice of so many years ago.

In return I can only offer your James Bond unlimited use of the name Ian Fleming for any purpose he may think fit. Perhaps one day he will discover some particularly horrible species of bird which he would like to christen in an insulting fashion. That might be a way of getting his own back?

Anyway, I send you both my most affectionate regards and good wishes, and should you ever return to Jamaica I would be very happy indeed to lend you my house for a week or so, so that you may inspect in comfort, the shrine where the second James Bond was born.

Yours sincerely,
Ian Fleming[77]

The Bonds did have a chance to visit Fleming at his Jamaican house, Golden Eye, in February 1964 (figure 14.14). As Fleming hosted them at lunch, Mary entertained him with tales of her husband's adventures and how his new double identity had affected their lives. As a parting gift, Fleming presented Bond with a prepublication copy of his most recent 007 thriller, *You Only Live Twice,* which he inscribed "To the real James Bond, from the thief of his identity, Ian Fleming, Feb. 5, 1964 (a great day)."[78]

While confusion with the fictional spy was often a source of embarrassment for Bond, his wife found the special service they received in hotels and restaurants quite appealing. In 1966, she summarized the story in a slender volume titled *How 007 Got His Name*.

In 1973 Bond's name was given a geographic immortality to match its literary one, when the British evolutionary biologist and Royal Society Darwin Medalist David Lack (1910–1973) identified the invisible boundary separating the avifauna of North and South America as Bond's Line.[79] Running between Grenada and Tobago, the theoretical line was like the better-known Wallace's Line in Indonesia, named in honor of Academy member Alfred Russel Wallace (1823–1913), who identified a similar separation between creatures of Asian and Australian origin running through the Malay Archipelago. Wallace, who with Charles Darwin is credited with developing the modern theory of evolution, was considered the nineteenth century's leading expert on the geographical distribution of animal species and is sometimes called the father of biogeography. "Wallace has a line and, after all your work in the West Indies, I don't see why you should not have one too!" Lack wrote Bond. "As you drew attention to this line, it would be most fitting [to have it named after you]."[80] While many more people have heard of James Bond, secret agent, the real James Bond much preferred this honor and the knowledge that his life's work has helped us better understand the birds of the tropical islands he loved.

RMP

Notes

1 Thoreau's Academy visit (recorded in some detail in his diary) came on 21 November 1854. He was hosted by Academy members William H. Furness and Elias Durand. The letter from Emerson introducing Thoreau to his "dear friend" Furness is dated 19 November 1854; see ANSP coll. 338. The reclusive Jackson, though staying at the Four Seasons Hotel next door, insisted on driving to the Academy in a windowless van. Surrounded by bodyguards, he entered the museum through the loading dock (service) entrance and toured its dark halls in his famous hat and sunglasses. Academy president T. Peter Bennett and VIREO director J. P. Myers were the only museum officials permitted to meet or even see the secretive singer during his hour-long visit. Personal correspondence with T. Peter Bennett, 27 September 2010, and J. P. Myers, 27 April 2011.

2 Academy fellow Robert Peck served as a content advisor to Attenborough and the BBC on this film.

3 Prentiss N. Gray to James Clark, 1 June 1927, ANSP coll. 186.

4 In 1931, Pathé had an exclusive arrangement with RKO and Trans-Lux cinemas.

5 In the first game, on 1 October, the Philadelphia Athletics beat the St. Louis Cardinals 6 to 2. Ultimately the Cardinals won the series in seven games, a rematch and reversal of fortunes of the 1930 World Series.

6 The Empire State Building was completed in May 1931.

7 "World Series in Movies," *New York Times*, 5 October 1931, 17.

8 Letter from David M. Newell (short reel director for the expedition) to Charles M. B. Cadwalader, 18 September 1931, coll. 2009-034, 1931, box 9 of 14, file 4.

9 Western Union telegram from E. R. Fenimore Johnson Jr. to Charles M. B. Cadwalader, 3 April 1931, coll. 2009-034, 1931, box 9 of 14, file 3; confirmed in letter from Cadwalader to Johnson, 14 April 1931, same location.

10 Western Union telegram from Cadwalader to E. R. F. Johnson, 8 April 1931, coll. 2009-034, 1931, box 9 of 14, file 3; confirmed in letter from Cadwalader to Johnson, 14 April 1931, same location.

11 Several versions of Rehn's equipment list exist in coll. 2009-034, 1931, box 9 of 14, files 3 and 4.

12 Although the current spelling for the region is Mato Grosso, we have used the spelling that was used in 1931.

13 From Fenimore Johnson, "A Description of the Matto Grosso Expedition," 8 January 1931, 1, coll. 2009-034, 1931, box 9 of 14, file 3. For more on the Matto Grosso Expedition, see Eleanor M. King, "Fieldwork in Brazil: Petrullo's Visit to the Yawalapiti," *Expedition* 35, no. 3 (1993): 34–43, and Alex Pezzati, "Where the Wild Things Are: The Mato Grosso Expedition, 1931," *The Codex* (newsletter of the Pre-Columbian Society) 10, no. 3 (June 2002): 11–16.

14 The designated collector for the University Museum was Vincenzo Petrullo, a graduate student in the University of Pennsylvania's Department of Anthropology.

15 Johnson, "A Description," 3. Already legendary in South America, Siemel gained greater international fame when, in 1931, Julian Duguid published his novel *Green Hell*, based on a 1929 trip across the Pantanal that Duguid and two other adventurers from England made with Siemel as their guide. In this book, Siemel was given the name "Tiger Man." This became the title of Duguid's 1932 biography of Siemel. In 1953, Siemel told his own story when he penned his autobiography, *Tigrero*. Siemel's second book, *Jungle Fury*, was published in 1954.

16 Among the financial backers of the expedition were Cryus McCormick, Francis L. Spalding, Sherman Pratt, E. R. Fenimore Johnson, and Mrs. William E. Green.

17 Crosby was the father of singer-songwriter David Crosby of the Byrds and Crosby, Stills, Nash, and Young.

18 King, "Fieldwork in Brazil," 37. See also coll. 2010-049.

19 Johnson, "A Description," 1.

20 Ibid., 3–4.

21 The film crew even built an enormous wooden corral with an elevated platform for the cameras in which to stage such a kill, but the terrified jaguar, which had been trapped and then released inside the corral, refused to attack.

22 Rehn describes the living specimens in two letters to Cadwalader, one dated 7 June 1931, and the other 20 July 1931. Both are in Cadwalader papers, coll. 2009-034, 1931, box 9 of 14, file 4.

23 For a description of the plane, see letter from James Rehn to Charles Cadwalader, 7 June 1931, Cadwalader papers,. coll. 2009-034, 1931, box 9 of 14, file 4.

24 Both Elisha Kent Kane and Academy member Isaac Israel Hays appeared on stage with their Inuit sled dogs to add color to their popular lectures on their travel experiences in the Far North.

25 Again, the reference here is to Elisha Kent Kane, though many other Arctic and Antarctic explorers are known to have eaten their dogs.

26 Curator Edward Drinker Cope built a large collection of dog skulls for comparative analysis. Elisha Kent Kane's dog "Tood la mink," from the second Grinnell Expedition (1853–55), was on display as a mounted specimen in the museum until 1940, when it was dismounted and placed in the mammal collection (specimen #618). Another mounted dog from Kane's expedition (specimen #661) is presently unaccounted for.

27 Painted in 1899 by Francis Barraud, a British landscape painter, and titled "His Master's Voice," the logo was adopted in 1901 as the trademark of the Victor Talking Machine Company, which was founded by Eldridge Reeves Johnson (1867–1945), the father of E. R. F. Johnson. Alex Pezzati, "Nipper in the Jungle," *Expedition* 47, no. 1 (Spring 2005): 6.

28 Eldridge R. F. Johnson obituary, *Philadelphia Inquirer*, 8 April 1986, 10-C.

29 James Rehn to Charles Cadwalader, 30 August 1931, 5–6, coll. 2009-034, 1931, box 9 of 14, file 4.

30 Rehn praised Pereira's ability to prepare specimens (he is "a wonder at skinning mammals," he wrote), but he admired his skills as a hunter even more. "He is a born hunter and the most wonderful trailer I have ever seen," wrote Rehn. Letter from James Rehn to Charles Cadwalader, 20 July 1931, coll. 2009-034, 1931, box 9 of 14, file 4.

31 Letter from James Rehn to Charles Cadwalader, 30 August 1931, 6, Cadwalader papers, coll. 2009-034, 1931, box 9 of 14, file 3.

32 The diorama was created and briefly installed near the entrance to North America Hall.

33 James Rehn to Charles Cadwalader, 30 August 1931, 5–6, Cadwalader papers, coll. 2009-034, 1931, box 9 of 14, file 4.

34 While copyrighted in 1932, the film appears not to have been given wide distribution until 1933.

35 Principal Adventure Pictures promotional brochure, coll. 2009-034, 1933, box 1 of 18, file 58.

36 Ibid.

37 Irene Thirer, "Vanderbilt Fish Film Has Thrills," *Daily News*, 26 December 1932, coll. 2009-034, 1933, box 1 of 18, file 58.

38 A.D.S., "In Brazilian Jungles," The Screen, *New York Times*, 14 January 1933.

39 The films, *Primitive Peoples of the Matto Grosso: The Bororo* and *Primitive Peoples of the Matto Grosso: Xingu*, were paid for by E. R. Fenimore Johnson. There are copies of each in the Academy's film archives.

40 In 2009, with a grant from the National Film Preservation Foundation, the Academy's nitrate print of this one-thousand-foot silent film was restored and transferred to 35mm polyester film and digitized to a Betacam DVD to make it more widely available.

41 This number includes 20 papers on mammals and 337 papers on orthoptera. See Maurice Phillips, "J.A.G. Rehn (1882–1965)," *Entomological News* 76, no. 3 (March 1965): 57–61. See also *Frontiers* (June 1950): 154–55.

42 For more on Fowler, see William F. Smith-Vaniz and Robert McCracken Peck, "Contributions of Henry Weed Fowler, with a brief early history of Ichthyology at the Academy of Natural Sciences of Philadelphia," in *Collection Building in Ichthyology and Herpetology*, ed. Theodore W. Pietsch and William D. Anderson, Jr., American Society of Ichthyologists and Herpetologists, Special Pub. no. 3 (Lawrence, KS: Allen Press, 1997), 377–89.

43 Charles M. B. Cadwalader to Ernest Hemingway, 6 March 1934, coll. 113, box 5, IV, #26.

44 Ernest Hemingway to Charles Cadwalader, 2 April 1934, coll. 113, box 5, IV, #26.

45 The meeting took place at the Academy on the afternoon of 9 April 1934. See related correspondence and telegram confirming his arrival in coll. 113, box 5, IV, #26.

46 Ernest Hemingway to Henry W. Fowler, 9 July 1934, coll. 220. The hotel he recommended was the Ambos Mundos Hotel.

47 Ernest Hemingway to Henry W. Fowler, 9 July 1934, coll. 220.

48 See Ernest Hemingway, "Genio after Josie: A Havana Letter," *Esquire*, October 1934, 21–22. The Hemingway-Academy correspondence can be found in collections 113, Box 5, IV, #26; 220; 707; and 2009-34, box 8, file 52. There is a discussion of some of these letters in Lawrence H. Martin, "Ernest Hemingway Gulf Stream Marine Scientist: The 1934–35 Academy of Natural Sciences Correspondence," *The Hemingway Review* 20, no. 2 (Spring 2001): 5–15.

49 Samuelson's account of his yearlong stay with Hemingway was published posthumously with a biographical foreword by his daughter, Diane Darby, in Arnold Samuelson, *With Hemingway: A Year in Key West and Cuba* (New York: Holt, Rinehart and Winston, 1984).

50 Samuelson, *With Hemingway*, 123–25.

51 Ernest Hemingway to Charles Cadwalader, 6 September 1934, coll. 113, box 5, IV, #26.

52 Ernest Hemingway, "Monologue to the Maestro," *Esquire*, October 1935, quoted in the foreword of Samuelson, *With Hemingway*, xi.

53 Charles Cadwalader to Ernest Hemingway, 31 August 1934, coll. 113, box 5, IV, #26.

54 Charles Cadwalader to Witmer Stone, 23 August 1934, coll. 186.

55 Ernest Hemingway to Henry Fowler, 9 October 1934, coll. 220.

56 Ernest Hemingway to Charles Cadwalader, 18 October 1934, coll. 113, box 5, IV, #26.

57 Ernest Hemingway to Charles Cadwalader, 29 March 1935, coll. 2009-034, box 8, file 52.

58 Samuelson, *With Hemingway*, 136–37.

59 "If I serialize it and make any money, will put a third of it aside as expedition fund," he wrote. Ernest Hemingway to Charles Cadwalader, 20 January 1934, coll. 2009-034, box 8, file 52.

60 The fish was a sculpin, collected seventy miles southeast of Cape May, New Jersey. *Neomerinthe hemingwayi*, Fowler [H. W.] 1935:42, *Proceedings of the Academy of Natural Sciences of Philadelphia*, vol. 87; ref. 1420.

61 Ernest Hemingway to Charles Scribner, October 1951, in *Ernest Hemingway: Selected Letters, 1917–1961*, ed. Carlos Baker (New York: Scribner's, 1981), 738. For a further discussion of Hemingway's interest in the sea and its influence on his writing, see Mark P. Ott, *A Sea of Change: Ernest Hemingway and the Gulf Stream, A Contextual Biography* (Kent, OH: Kent State University Press, 2008).

62 Mary Wickham Bond, *Far Afield in the Caribbean: Migratory Flights of a Naturalist's Wife* (Wynnewood, PA: Livingston Publishing, 1971), 59.

63 For a detailed account of this expedition see L. J. Dorr, "Muy poco se sabe de los resultados: Francis E. Bond's expedition to the Paria Peninsula and delta of the Orinoco, Venezuela (1911)," *Archives of Natural History* 37, no. 2 (2010): 292–308.

64 Willowbrook house (built 1906–7) was designed for the Bonds by the noted American architect Horace Trumbauer (1868–1938). It stood on a one-hundred-hectare estate with "a sweeping view of the countryside." It is today owned by Gwynedd Mercy College. See David R. Contosta, *The Private Life of James Bond* (Lititz, PA: Sutter House, 1933), 24–32.

65 James Bond had an older sister, Margaret (1897–1904), and brother, Francis (b. 1898). He was named for his great-grandfather, a physician and the American consul to Uruguay.

66 Bond spent a year as a student at St. Paul's School in Concord, New Hampshire, before moving to England.

67 Contosta, *Private Life*, 55.

68 James Bond to Robert Peck, 19 March 1985, author's collection.

69 Mary Bond, *Far Afield*, 21–22.

70 Biographical data entry for James Bond, 10 May 1974, from the personal files of Richard Estes.

71 Biographical data entry, 1974, 2.

72 Bond first presented his findings at the American Philosophical Society in December 1933; see Bond, "The Distribution and Origin of the West Indian Avifauna," *Proceedings of the APS* 73, no. 5 (May 1934): 342. See also Bond, "Origin of the Bird Fauna of the West Indies," *Wilson Bulletin* 60 (December 1948): 217.

73 Among Bond's awards were the Musgrave Medal of the Institute of Jamaica (1953), the William Brewster Medal of the American Ornithological Union (1954), the Joseph Leidy Award of the Academy of Natural Sciences (1975), and the silver medal from the Congresso Iberamericano de Ornithologico (1983). In 1987, he was voted an honorary member of the British Ornithological Union.

74 New editions of Bond's *Birds of the West Indies* were published in 1947, 1960, 1971, 1974, 1983, and 1985. See Contosta, *Private Life*, 99.

75 Ian Fleming to Mary W. Bond, 20 June 1961. This famous letter has been reprinted in a number of places. See Mary W. Bond, *How 007 Got His Name* (London: Collins, 1966), 21; Mary W. Bond, *To James Bond with Love* (Lititz, PA: Sutter House, 1980), 5; and Contosta, *Private Life*, 107.

76 Ian Fleming, from an interview in *Rogue*, February 1961, quoted in Mary Bond, *How 007 Got His Name*, 17.

77 Ian Fleming to Mary W. Bond, 20 June 1961, Bond collection, Free Library of Philadelphia, reprinted in Mary Bond, *How 007 Got His Name*, 21; Mary Bond, *To James Bond with Love*, 5; and Contosta, *Private Life*, 107.

78 Before her death Mary Bond gave this special copy of Fleming's book to the Free Library of Philadelphia. Sadly, in 2008, the library sold it at auction, where it realized $70,000. For a description of the book, see Profiles in History Auction Catalog #33, lot 103, Calabasas Hills, California, sale date: 11 December 2008.

79 Contosta, *Private Life*, 117–18.

80 David Lack to James Bond, unknown date, 1973, quoted in Mary Bond, *To James Bond with Love*, 24, and Contosta, *Private Life*, 117–18. The name was published (posthumously) in David Lack, *Island Biology Illustrated by the Land Birds of Jamaica* (Berkeley: University of California Press, 1976).

Scardinius erythophthalunus scardafa (cotype of *Leuciscus scardafa*) collected and described by Charles-Lucien Bonaparte (1803–1857) in 1837. Gift of Edward and Thomas B. Wilson 1850. ANSP Ichthyology Department #6212-68. Purcell photograph.

FOLLOWING PAGES Cleared and stained specimens of young horse-eye jacks (*Caranx latus*) from Dominican Republic, Santo Domingo. ANSP Ichthyology Department #81951. This is a species that was first described by Academy member Louis Agassiz (1807–1873) in 1831. Purcell photograph.

Chapter 15

Visions in Microscopes: Water Quality and the Environment

I was considered almost a woman of the streets for bringing corporate money to a place as hallowed as the Academy and for doing applied work instead of pure, basic research.
—Ruth Patrick

At the Academy of Natural Sciences' International Symposium on Early Man, held on the 125th anniversary of the Academy's founding in March 1937, a young woman Ph.D. from Kansas, only a volunteer amid a gathering of world-famous scientists, set up a microscope display of diatoms. Even if it was not clear how a look at these aquatic plants had much to do with the discussions of human evolution that were at the heart of the symposium, it was a toehold on the masculine world of science for Ruth Patrick (b. 1907). In time, her study of these microscopic organisms and their important place in the environment would make her known around the world.

From the earliest days of the Academy's existence, microscopy was an integral part of the institution's research, exhibits, and educational programs. One of the oldest books in the Academy's library is *Micrographia* (1665) by Robert Hooke, an inventor of the microscope (figure 15.2). The famous English diarist Samuel Pepys, one of the first to buy the work, recorded that on 20 January 1665, he went "to my booksellers, and there took home Hooke's book of Microscopy, a most excellent piece, and of which I am very proud." The next day, Pepys wrote: "Before I went to bed I sat up till two o'clock in my chamber reading Mr. Hooke's Microscopical Observations, the most ingenious book that I ever read in my life."[1]

There is ample evidence in the Academy's *Journal*, from 1817 onward, that the founding members regularly used pocket lenses and other optical apparatus in their studies. But until the mid-nineteenth century, microscopes were scarce in the United States. Paul Beck Goddard (1811–1866), Joseph Leidy's professor at the medical school of the University of Pennsylvania, had one of the very few in Philadelphia and introduced his clever student to the instrument.[2] Leidy, an American pioneer in the study of microorganisms, applied this marvelous piece of equipment to "a wide array of investigations in protozoology, microbiology, zoology, parasitology, paleontology, anatomy, pathology, histology, and entomology."[3] Leidy's 1853 publication *Flora and Fauna Within Living Animals* speaks to the detail of his microscopical observations. The book includes captivating botanical as well as zoological examples, illustrated with his beautiful and scientifically accurate figures.[4] Widely acclaimed as a landmark work, the book earned its author the title of father of American parasitology.

In 1856, a group of Academy members, including Leidy and Isaac Israel Hays (1832–1881), a future Academy president, formed the Biological Club for the sheer enjoyment of science. Their emphasis was microscopy. J. Cheston Morris, a member of the group, recalled many years later that they met every other

15.1 Freshwater mussel shells (*Elliptio complanata*) collected from the Choptank River, Talbot County, Maryland, by George M. Davis, September 1982. ANSP Malacology Department. Purcell photograph.

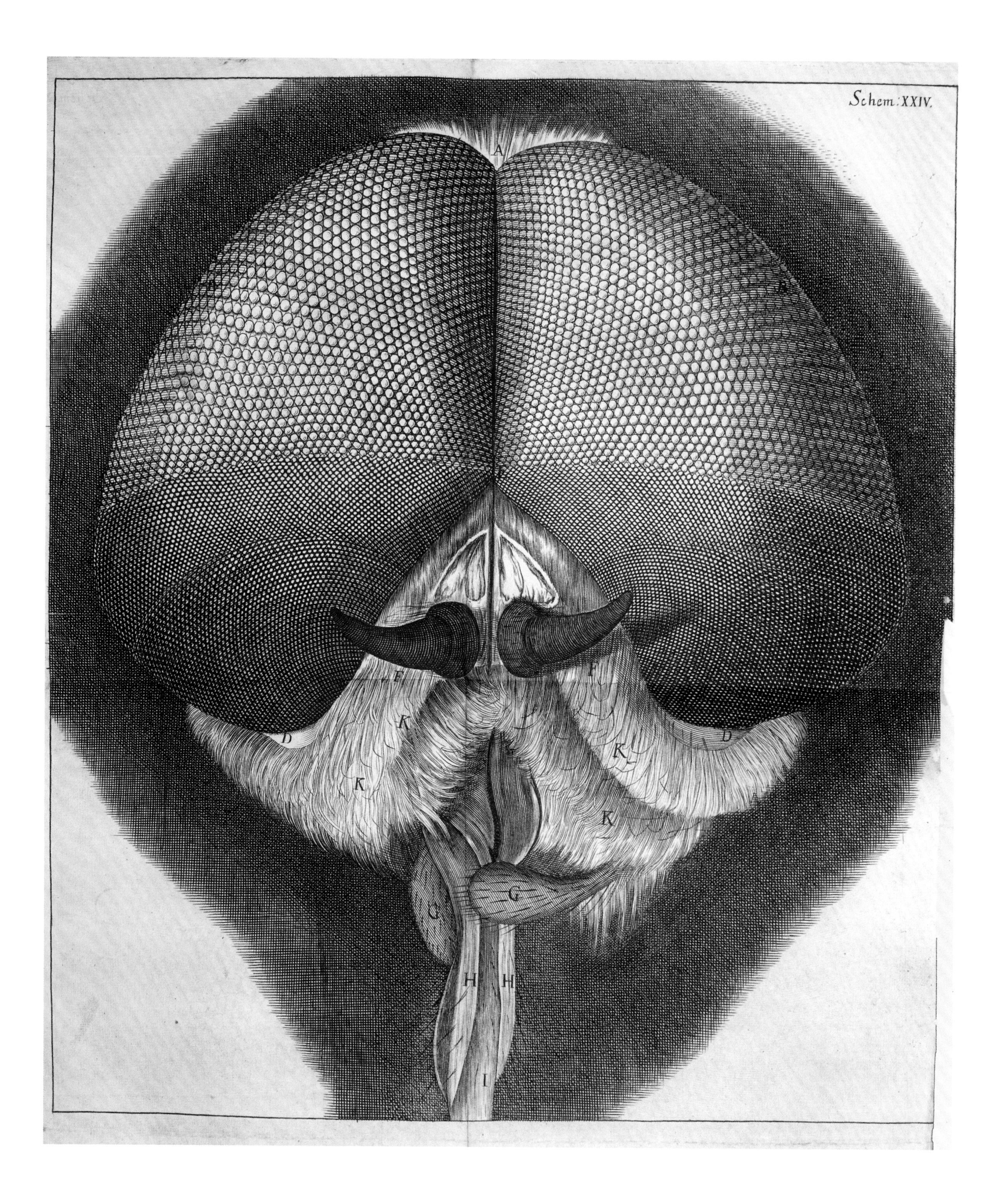

Schem: XXIV.
A
D
D
F
F
K
K
K
K
G
H
H
I

15.3 Three early books on microscopy, 1870, 1871, 1880. ANSP Library.

Friday evening at each other's houses. "The refreshments," he said, consisted of "crackers, and cheese, and ale, whiskey and tobacco." He added that one subject was taboo: business.[5] Morris's statement documents the changing times: at the first Academy meetings the founders had decided to ban any discussion of religion or politics. Now the forbidden topic was business. Perhaps politics were difficult to avoid, for through the club's formative and early years, the nation was gripped by internal strife and civil war. "We continued our bi-weekly sessions during all the 'troubles that tried men's souls' in '61 to '65," Morris noted.[6] For these biological enthusiasts, science forever rose above politics.

In 1858, Leidy and some twenty-five professionals, mostly medical doctors, founded the Microscopical Society of Philadelphia, which ten years later joined the Biological Club to form the Biological and Microscopical Section of the Academy.[7]

The increased use and availability of microscopes stimulated an enthusiasm for viewing all things through these instruments, from feathers to butterfly wings. Among the most popular items were diatoms, single-celled algae found in both freshwater and marine ecosystems, whose silica (a glasslike substance) cell walls make them durable and easy to observe by microscope (figure 15.4). Estimates are that 20 to 25 percent of all organic carbon fixation (the transformation of carbon dioxide and water into sugars by using light energy) on the planet is carried out by diatoms. This is possible because they contain chlorophyll. Diatoms are thus a major food resource for marine and freshwater microorganisms and animal larvae, and they are a major source

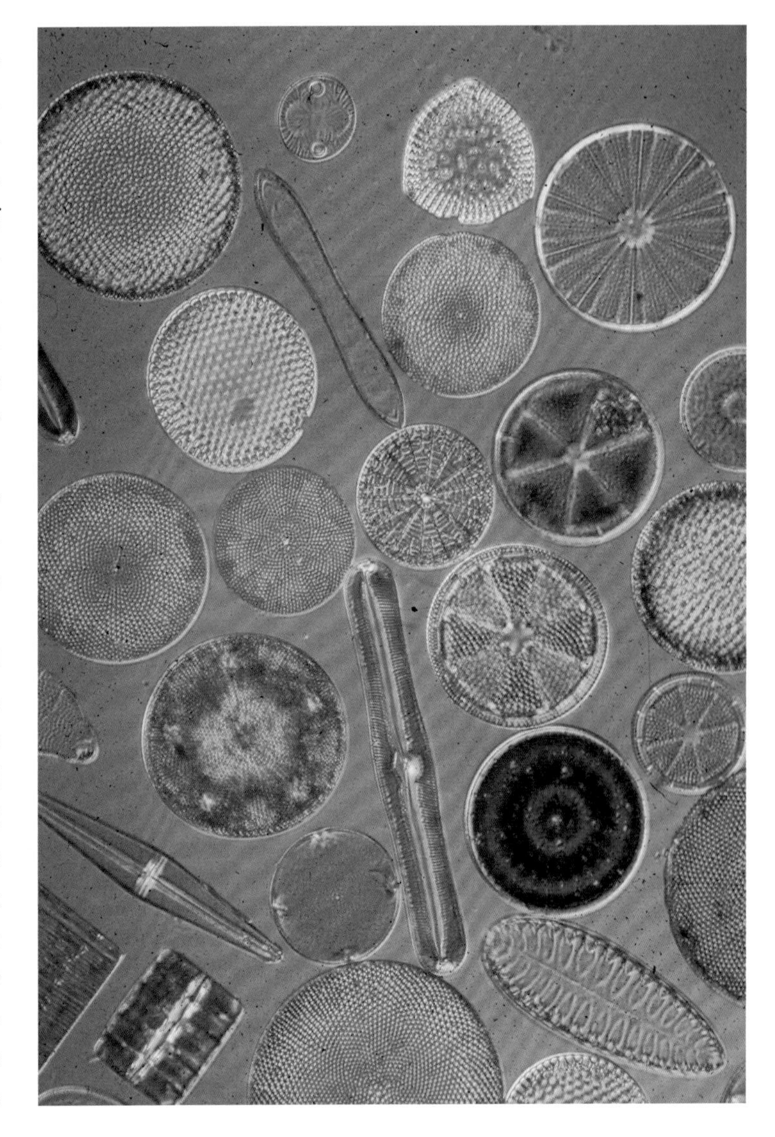

15.4 A collection of marine diatoms. ANSP photograph.

FACING PAGE 15.2 *Face of grey drone-fly* from Robert Hooke's *Micrographia: or Some Physiological Descriptions of Minute Bodies Made by Magnifying Glasses*. London, 1665. ANSP Library.

of atmospheric oxygen.[8] The largest of the many more than two thousand known species is one twenty-fifth of an inch long.

A 1942 article in *Frontiers* gives a more aesthetic description of this multiformed flora: "Although they belong to the lowest class of plants, the Algae, they are as wonderful to contemplate as are the vast distances of the universe. . . . Each plant consists of a single cell, encased in a shell of glass, architecturally more beautiful than a cathedral; structurally better than the finest works of man. This shell is indestructible."[9]

Christian Febiger (1817–1892), a wealthy businessman and director of railroads in Pennsylvania and Delaware, was the first Academy member to build a large diatom collection. In his researches, Febiger discovered that the diatoms illustrated in European journals were not like those he found in the Wilmington River near his home. He sent some of his slides to the diatomists of Europe, who were delighted with this new material. In time, the diatom slides Febiger sent became part of great European collections, and twenty species were named in his honor. After his death, his son Christian Carson Febiger gave most of his father's collection to the Academy, but he kept the "jewels" as *plantae exsiccatae* (herbarium specimens), dried plants that were bound in volumes. Fifty years later, Febiger's granddaughter, Mrs. Charles A. Fife, found these treasures amidst a collection of old books, recognized their importance, and gave them to the Academy.[10] A Philadelphia doctor and Academy member, Francis W. Lewis (d. 1902), was the first to write a paper for the Academy's *Proceedings* (1861) on the subject of living and recent fossil diatoms.[11]

So great was the popular interest in microscopic study by the 1870s that the Academy's Biological and Microscopical Section hosted a series of large exhibitions of the instruments at the museum. In 1879, the society's annual report stated: "The attendance at our meeting was probably larger than that of any similar meeting ever held here or in Europe. Nearly three thousand persons visited these halls during the evening. There were 137 microscopes lighted and set up in the library and adjoining rooms to show fascinating slides of various biological entities under magnification."[12]

By chance, the exhibit that year coincided with the publication of Joseph Leidy's last important monograph, *Fresh-Water Rhizopods of North America* (1879).[13] European scientists had studied rhizopods (single-celled organisms traditionally classified as protozoans, meaning "first animals") since the time of Hooke and von Leeuwenhoek, the seventeenth-century Dutchman who is also credited with the microscope's invention. Leidy's book marked the beginning of protozoan study in the United States. He wrote glowingly in his introduction: "The revelations of the microscope are perhaps not exceeded in importance by those of the telescope. While exciting our curiosity, our wonder and admiration, they have proved of infinite service in advancing our knowledge of things around us."[14]

In the late nineteenth and early twentieth centuries, many members of the Academy's Biological and Microscopical Section concentrated on the study of diatoms. One was Charles S. Boyer, a distinguished diatom scholar who amassed a valuable worldwide collection, published an important book,[15] wrote twenty-one papers on diatoms, and gave his collections to the Academy (figure 15.5).

Another, and a particularly important Academy member for many years, was Thomas A. Stewart (d. 1949). Elected to membership in 1900, he had a lifelong passion for microscopy, which led him to excel in microphotography. He was also intensely

15.5 Charles S. Boyer (1856–1928), the distinguished diatom scholar who gave his extensive collection to the Academy. ANSP Archives coll. 457.

interested in Wilhelm Conrad Röntgen's discovery of the X-ray (1895). The Prussian-born Röntgen (1845–1923), proposed for Academy membership in 1896 by Benjamin Sharp and Edward Nolan, won the Nobel Prize in Physics in 1901. According to an unsigned biographical manuscript in the Academy archives, "[Stewart] and Dr. Leonard were the first to bring x-ray to this country. In conjunction with Dr. Leonard and others, he was among the first to experiment as to its various uses."[16]

T. Chalkley Palmer (1860–1934), a chemist and Academy president (1926–28), and Frank J. Keeley (1868–1949), curator of mineralogy, were equally enthusiastic collectors of diatoms and actively exchanged materials with other collectors. At meetings, the members discussed new specimens, exhibited photographs of microscopical subjects, and donated materials to the institution. They also participated in field trips with the Academy's Mineralogical and Geological Section to southeastern Pennsylvania, New Jersey, and Delaware, where they collected recent and fossil diatoms along with minerals and other specimens.[17]

The camaraderie among the members of the Academy's Biological and Microscopical Section was celebrated at the group's fiftieth anniversary dinner in 1908. The lavish menu, typical of Philadelphia fare at the turn of the century, consisted of oysters, green turtle soup, Delaware shad, cucumber salad, quail, stewed snapper, sweetbreads, ice cream, "merangue," coffee, and bon bons.[18]

Years later, in the fall of 1924, the Academy abolished all scientific sections when the bylaws were changed and the board of trustees was established. The groups mentioned above were reorganized into an independent nonprofit entity called the Leidy Microscopical Club, which met regularly at the Academy for seventy years until it moved its meeting place to the suburbs in 1995.[19]

Diatom study at the Academy took a sudden leap forward in 1934 when Ruth Patrick moved to Philadelphia to be with her entomologist husband, Charles Hodge IV. After earning her Ph.D. from the University of Virginia, she volunteered to work

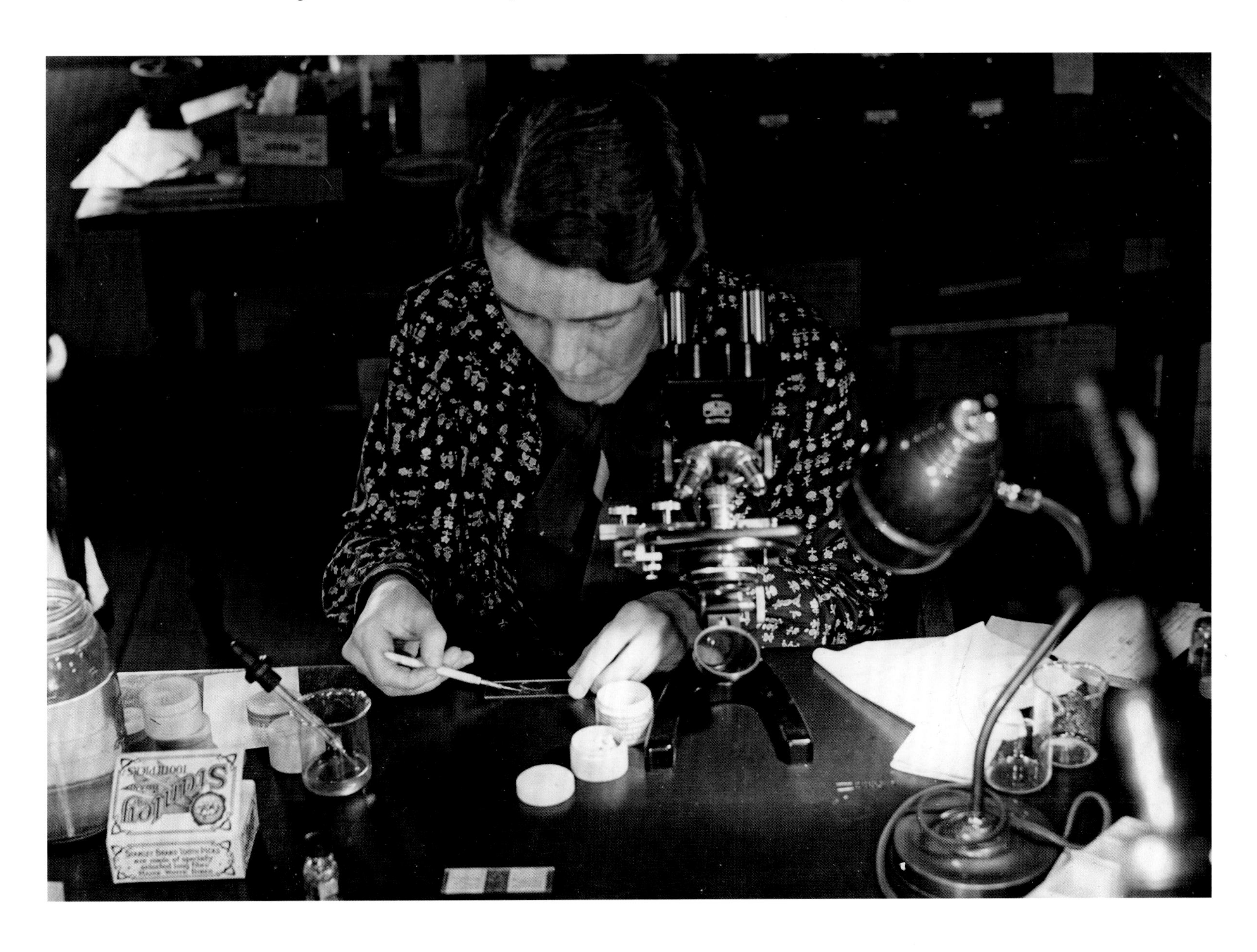

15.6 Ruth Patrick examining diatoms in the 1930s. ANSP Archives coll. 457.

with the famous diatom collections at the Academy. For an entire year, she wrote letters to Academy officials in an unsuccessful effort to gain access to the Academy's diatom collections, even though she had already coauthored and published two papers on the subject with her supervisor from the university, Ivey F. Lewis.[20] She wrote to the Academy's recording secretary, James A. G. Rehn: "You will remember I spoke to you some time ago about having access to the Boyer collection. As you know I am working on diatoms. I am particularly interested, at the present time, in distributional studies. I feel that being able to use the diatom collection at the Academy would be of great help to me. The National Museum, British Museum, Raffles Museum [in Singapore], and the Museum at Buitenzorg, Java, have cooperated with me in this work by sending me material from their collections."[21]

Still, there was no response to several more letters in the same vein until November, when Academy president Charles Cadwalader wrote to inform her that the Boyer collection belonged to the Leidy Microscopical Club and that he had taken up her "research problem" with Thomas Stewart, an active member of the club. Cadwalader suggested that she write to him "and see if that doesn't bring some result."[22] Apparently it did, for in spite of being a woman in the entirely male world of diatomists—and in natural science in general—she was allowed to use the collections. What may have convinced the members to cooperate was her doctoral dissertation, parts of which were published in the Academy's *Proceedings* in 1936.[23] In her paper, Patrick noted that others had studied the ecological conditions of various diatom habitats: "They have found that some forms are very specific for certain conditions, thus one might be able to foretell the conditions of the water by the occurrence of certain diatoms."[24] This concept would be the key to her future.

The following year, Patrick so impressed the members of the Leidy Microscopical Club with her display at the Symposium on Early Man that she was appointed curator of the club's collections (figure 15.6). Her duties, among other things, included preparing microscopes for the members' regular meetings. In 1938, she was named associate curator of the Academy's Department of Microscopy, under its curator, Thomas Stewart.[25] In later life, Patrick reportedly said: "Women weren't taken seriously. We thought we had to degrade our femininity by wearing knickers (knee-length pants), smoking small cigars and not curling our hair. Now it's much easier for women in the sciences."[26] Whether she ever submitted to these curious ideas for acceptance, her meteoric career had surely begun.

That same year (1938), Patrick published in the Academy's *Proceedings* a paper on the evidence of diatoms from the mammoth pit in Clovis, New Mexico, that Edgar Howard had discovered many years earlier (see chapter 10).[27] "Samples were collected from the various layers by Dr. Edgar B. Howard and submitted to the author for study of the diatoms present," she wrote. She thanked Howard for "his helpful advice and criticism during the progress of this paper."[28] Since Howard had spearheaded the Symposium on Early Man the year before, it was probably he who had encouraged Patrick to set up her microscope display at the event.

If Patrick felt discriminated against because of her sex, she would be reassured in late 1939 when she applied to the

15.7 Ruth Patrick collecting aquatic specimens circa 1970. ANSP Archives coll. 457.

Guggenheim Memorial Foundation for a grant and Cadwalader endorsed her with enthusiasm: "The personality and character of the applicant are excellent in every way," he wrote. "Her scholarship and capacity for independent research are also excellent. There is every reason to believe that Dr. Patrick will make a noteworthy contribution to the field of study proposed. Due to the fact that the Academy possesses a most important library of diatom literature, she has been able to read and refer to all of the original descriptions of the species in which she is interested."[29] In closing, he explained her status as a volunteer: "Dr. Patrick does not receive a salary from this Museum. We have never been able to include her on our salary list and thus all of her work is done on her own time."[30]

Born in Topeka, Kansas, in 1907, Ruth Patrick moved at a young age with her family to Kansas City, Missouri. Her father was a lawyer and banker by profession, but he was a scientist at heart, and his fascination with the natural world was an early and lifelong inspiration for his daughter. Walks in the woods to gather specimens of all kinds, especially water to be examined for diatoms, were followed by sessions with the microscope while she sat on her father's lap. By the age of seven, Patrick had her own microscope and looked at everything she could find, especially the endless shapes and forms of diatoms in various water samples. Her father taught her to prepare slides of these tiny plants and how to identify individual species. This was an unusual interest to inspire in a young girl at the time, but she credits her parents' steadfast faith as the key to her successful career.[31]

While pursuing her undergraduate degree at Coker College in South Carolina, Patrick did summer course work at such prestigious facilities as the Woods Hole Oceanographic Institution in Massachusetts and the Cold Spring Harbor Laboratory in New York, where she met her future husband. At Woods Hole she made the acquaintance of Ivey F. Lewis, the country's top algae expert and also the dean and chair of the Biology Department at the University of Virginia. Recognizing Patrick's intelligence and energy, Lewis suggested she become one of his graduate students. Patrick subsequently received both her master's and doctoral degrees from that university.[32]

Patrick's early research was as innovative and imaginative as her entire career would be. "I was working on my Ph.D. while at the Academy," she recalled. "My subject was diatoms of Siam [now called Thailand] and the Federated Malay States." She explained her method of obtaining her subjects:

> I got these diatoms by writing to museums around the world and asking for the tadpoles they had collected for research. I knew that tadpoles were vegetarians and that they ate a great many diatoms. The museums would send their tadpoles to me, and I would carefully draw the intestines out from each and put cotton back in where the intestines had been. This was of no consequence to the museums, because these tadpoles had been collected for taxonomic purposes and the organs were not of any use for this research. Then I returned the tadpoles.[33]

Soon after her arrival at the Academy, Patrick began reorganizing the diatom collections.[34] "My interest in diatoms wasn't so much in their taxonomy, but in what they indicated," she explained. "By studying the conditions in which diatoms lived, I discovered that the presence of different species often pointed to different types of water."[35] During her research, she found that some diatom species thrived in water that was heavily polluted with organic contaminants such as human sewage, animal manure, and crop fertilizers. Others flourished in pollutants of a different nature, such as mineral or chemical. By linking certain diatoms with specific water environments, it was possible to identify the sources of many different types of aquatic pollution (figure 15.7). For Patrick, diatoms provided a series of living litmus tests for environmental quality.[36]

In 1945, Patrick further solidified her reputation for original thinking and made a major breakthrough in her career when she delivered a paper at the annual meeting of the American Association for the Advancement of Science. In her talk, she explained her thesis that diatoms could tell a lot about the chemistry of water. In the audience was an executive from the Atlantic Refining Company, William B. Hart, who was responsible for waste disposal at the company's refineries. Hart recognized that Patrick's discovery could be essential in understanding and controlling water pollution. Though it would be another thirty years before the formation of the Environmental Protection Agency, Hart understood that water quality was rapidly becoming an important concern and that it was industry's responsibility to address its role in causing pollution.[37]

Some months later, Hart came to the Academy with an offer of funding to further Patrick's research. Some administration members were dubious about having a woman head the project, but Hart insisted that it be she or no one, so the study went ahead with Patrick in charge. Cadwalader had stated in his letter to the Guggenheim Foundation that "Dr. Patrick and I have held many

consultations and parlays on how best to advance her work. Through these consultations I have come to know her personally, and am certain in my own mind that she is a good scholar, [has] practical and common-sense, and [is] a most excellent diplomat."[38] Cadwalader was not only appreciative of Patrick's abilities but also prescient about her skills of communication and diplomacy, which had no doubt inspired Hart's enthusiasm. Henry A. Pilsbry, the Academy's principal malacologist, whose office was near hers and who became a good friend, encouraged her efforts. "Ruth," he once advised her, "always agree with the administration and then do exactly what you wish."[39] And so she did.

In 1947, with Hart's backing and Cadwalader's support, Patrick founded the Academy's Limnology Department and had the diatom collections placed under the care of her staff. The work of the department was focused on assessing the water quality in lakes and rivers by using a range of biological indicator organisms, including aquatic insects and fish, but especially diatoms.[40] One of the department's first projects, widely regarded as a milestone of environmental research, was a 1948 biological survey of Conestoga Creek near Lancaster, Pennsylvania (figure 15.8). In this study, Patrick broke new ground with her insistence that the diverse biological communities inhabiting a stretch of stream make up a single mosaic. This became known as the "Patrick Principle."[41]

An important addition to the Academy's Limnology Department was Charles W. Reimer (1923–2008), who was hired as an assistant curator in 1952. According to Marina Potapova, the current curator of the diatom herbarium, "Initially [Reimer] was responsible for translating diatom literature from German and other languages, for field work, and for the identification and enumeration of diatoms in samples collected in the course of numerous river and stream surveys across the country. He quickly became a renowned expert in diatom taxonomy and together with Patrick began working on a monumental flora of the United States."[42] This coauthored work, *Diatoms of the United*

15.8 Ruth Patrick and team assembled for the Conestoga Creek project, 1948. ANSP Archives coll. 2010-020.

FACING PAGE 15.9 Catherwood diatometers used for passively collecting aquatic specimens on glass slides. ANSP Archives. Purcell photograph.

CATHERWOOD DIATOMETER
Academy of Natural Sciences
of Philadelphia

15.10 Ruth Patrick and Robert Grant in the 1950s. ANSP coll. 2010-020.

States Exclusive of Alaska and Hawaii, was published by the Academy in two large volumes in 1966 and 1975. Reimer said of his choice to study diatoms as a career: "It's an unusual field. I was always interested in plants. I was raised on a truck farm. Ever since I was a kid I worked in the garden and orchard."[43]

Due to Reimer's presence, the Academy was one of the few places in the world where students could pursue diatom research at a professional level, and many did their dissertations under his tutelage. Jan R. Stevenson, codirector of the Center for Water Sciences at Michigan State University, recalled that Reimer was "a great teacher and mentor. He inspired passion in and camaraderie among us."[44] Reimer was also an adjunct professor at Drexel University in Philadelphia and a visiting professor in Chile and China. He officially retired as curator of the Diatom Herbarium in 1991 but continued his work at the Academy until his death.

Patrick's Limnology Department was a pioneer in multidisciplinary research in the 1950s and 1960s. Instead of specializing in one area, the department possessed expertise in all the major groups of aquatic organisms as well as the ability to analyze a stream's chemical and physical characteristics. The department was also noted for its technical innovations. The most useful of these was the Catherwood diatometer,[45] a device invented by Patrick in 1954, which passively collected diatoms, thus enabling scientists to accurately analyze the diversity and density of diatom communities and thereby assess the relative health of the bodies of water in which they lived (figure 15.9). A profile of Ruth Patrick in *Reader's Digest* (1962) observed that it was little understood in the 1960s that some forms of river life *use* industrial waste and sewage as food. The living organisms provide a river with a digestive system that keeps it clean. But if there is a waste overload, many healthy organisms are killed, undesirable ones take over, and the river can no longer clean itself.[46]

To assist with her river surveys, Patrick assembled a team of specialists, including a chemist, a bacteriologist, a zoologist, an ichthyologist, a specialist in the lower invertebrates like snails and clams, an expert in protozoa (the lowest form of animal life), an entomologist specializing in aquatic insects, and an algologist (one knowledgeable in aquatic algae).[47]

Many of the scientists Patrick assembled for the Academy's Limnology Department were eminent in their own right. John Cairns Jr., curator of the division for eighteen years (1952–65), was a member of the American Philosophical Society, a Fellow of the American Academy of Arts and Sciences and the American Association for the Advancement of Science, and a foreign member of the Linnean Society of London, among many others. He served on eighteen National Research Council Committees and authored 1,520 publications.[48]

Cairns joined the Academy's Conestoga River Survey team in June 1948 and experienced the long hours that leadership and research required, combined with "the joy and zest of being a member of a research team." "It was a major defining moment in my professional career," he recalled. "Under Dr. Ruth Patrick's mentorship, I discovered systems level research, team cooperation, grant and contract acquisition, and skills in administering a department while simultaneously carrying out research. No one in the department worked harder than Dr. Patrick. Our success came with a price and she paid it with grace and style. When I left the Academy in 1966 to begin research on rapid biological information systems, Dr. Patrick remained my mentor."[49]

To provide expertise in aquatic insects, Selwyn S. "Sam" Roback (1924–1988) came to the Academy in 1951. "He inhabits the world of tiny creatures and large words," noted a reporter for the *Philadelphia Evening Bulletin*. Roback said that sometimes fishermen seeing him in a stream would ask why he was scrounging for bugs. "Because I enjoy it," he said, "but that wouldn't satisfy them, so I just told them that someone was willing to pay me for it."[50] Roback was internationally known and brought further academic stature to the department, according to Clyde Goulden, who joined the department in 1966 to pursue testing for biological diversity in lakes.[51] From 1972 until 1978, Goulden headed the Limnology Department, after which he taught full-time until the late 1990s, when he embarked on a major new Academy project in Asia.

Robert Grant, who for many years collected diatoms for the Limnology Department and helped to coordinate river surveys for Patrick, recalled, as had Cairns, that she stressed a team effort (figure 15.10). "That is what set her apart," he said. Grant recalled that when the group was working in the field, she always reserved an extra room at the motel where they stayed so they could all meet to discuss the project. "It was a great opportunity to share our ideas and discoveries," he noted.[52]

As her reputation spread, companies such as DuPont and Sun Oil hired Patrick and her team of scientists to assess the water quality of streams where they operated plants. "This was heresy at the time," Patrick recalled. "I was considered almost a woman of the streets for bringing corporate money to a place as hallowed as the Academy and for doing applied work instead of pure, basic research." Patrick reasoned that she wanted to have an impact on the world and that there was no "better way to do that than by working with the people who could take the facts and put them to some positive use."[53] Her sphere of influence thus broke through traditional boundaries and reached well beyond the Academy's walls. She was instrumental in closing the gap between scientists and industry by being an advocate for the clear communication of science-based facts.[54] Patrick said the aim of the Limnology Department was "to produce good research and yet be completely self-supporting and if possible turn in an additional amount to the Academy's general income funds."[55] Her model proved remarkably successful, and the department grew dramatically during the 1950s and 1960s.

In the mid-1960s, Patrick made an innovative suggestion to William B. Dixon "Dick" Stroud (1917–2005) and his wife, Joan, who asked Patrick what they could do to help her work. Stroud was not a stranger to Patrick or the Academy. In 1956, he had accompanied an Academy snail-collecting expedition to New Guinea, where he spent two months diving for live shells.[56] Patrick's idea was to set up a research laboratory on the East Branch of the White Clay Creek, which ran through the couple's farm in Chester County, Pennsylvania. Thus, in the summer of 1966, the Stroud Water Research Center began its existence as an Academy field station in a hastily cleared space above the Strouds' garage.

Shortly thereafter, Patrick hired Robin Vannote, a young scientist working for the Tennessee Valley Authority, to head the lab. "Ruth and Robin set the tone," recalled Bernard Sweeney, the current director, who joined the Stroud Lab in 1972 while a first-year graduate student at the University of Pennsylvania. "They were constantly challenging, asking hard questions, never satisfied, always demanding another experiment. It's the same now. It's an intense and focused place."[57]

By the 1960s, it was evident that rivers across the country were in trouble. Massive weed beds clogged once-navigable waterways. The water had become too polluted to swim in, let alone to drink. A perfect example of the vital need for the work Patrick had set up in the Academy's Limnology Department occurred on 22 June 1969, when the Cuyahoga River, near its confluence with Lake Erie in Cleveland, caught fire, with flames leaping five stories high. For much of its length, the river was flanked by factories and refineries, especially between Akron and Cleveland, and by the time it flowed into Lake Erie, it had a thick coating of sludge and oil; it was so polluted that it was

completely devoid of life. The Cuyahoga had caught fire at least a dozen times before, with the worst outbreak in 1952. But the conflagration of 1969 gained worldwide attention when *Time* magazine wrote a story about it.[58]

The article also spoke of Lake Erie: "Some lake! Industrial wastes from Detroit's auto companies, Toledo's steel mills and the paper plants of Erie, Pa. have helped to turn Lake Erie into a gigantic cesspool."[59]

Identifying and ameliorating this kind of pollution was exactly what Ruth Patrick's Limnology Department was all about. The same year as the Cuyahoga River fire, the Stroud Center received a five-year grant from the Rockefeller Foundation to study the White Clay Creek. In this work, Stroud scientists established the importance of studying an entire watershed. Eventually a modern laboratory was built on property donated by the Strouds, with some of the White Clay Creek water diverted right through the building (figure 15.12).

In 1988, the Academy's Stroud Center extended its research internationally when its scientists helped establish the Maritza Biological Field Station in the Guanacaste National Park, Costa Rica, at the suggestion of the well-known tropical ecologist and University of Pennsylvania professor, Daniel Janzen. Surrounded by undisturbed tropical dry-forest, Maritza became a site where Costa Ricans and members of the international research community could study tropical stream, dry-forest, and cloud-forest ecosystems.[60]

Another important addition to the Stroud Center occurred in 1990 when Morris Wistar Stroud III (1913–1990), Academy trustee and brother of Dick Stroud, turned his 332-acre farm (formerly known as Georgia Farm) in West Chester into the Stroud Preserve. Three of Chester County's most important research and environmental organizations—the Natural Lands Trust, the Brandywine Conservancy, and the Stroud Center—formed a unique collaboration, and Morris Stroud bequeathed endowments to all three, with ownership of his farm going to the Natural Lands Trust and easements to the Conservancy. The land and water were to be managed for the benefit of science, education, and the environment. To the Stroud Center he granted perpetual use of the entire property for scientific research.[61]

In the early 1970s, the Limnology Department began to study the Delaware River, looking at the river in sections, especially the wetlands in the tidal portions, which extended from the river's mouth to the falls at Trenton, New Jersey. According to a 1976 article in the Academy's *Frontiers*, as far back as the early nineteenth century there had been a steady decline in the river's health. The Delaware's shad catch had passed its peak by 1820, and by 1870, the decline was so noticeable that attention focused on its causes: overfishing, dams, pollution, and the use of eel weirs and fish baskets, which blocked or killed large numbers of juvenile shad on their downstream migrations.[62] Thanks in part to millions of dollars spent by state and federal governments to develop and rejuvenate pollution laws, and with the guidance of the Academy, this situation was reversed by the 1980s, and once again shad returned in healthy numbers to the Delaware.[63]

In an effort to study the effects of electric power plants on the ecology of the Chesapeake Bay, in 1967 Patrick established the Benedict Estuarine Research Laboratory on the Patuxent River,[64] about thirty miles from its entrance to the bay. The largest estuary in the United States, the Chesapeake Bay is formed by the confluence of dozens of streams and rivers. These estuaries and their associated marshes are among the most productive natural areas on earth, acting as nurseries for thousands of maritime and freshwater mollusks, insects, fishes, amphibians, and birds. The wealth of life in an estuary depends on a rich biomass, from aquatic plants through algae and microscopic insect life, and along the food chain that eventually supports a human population.

15.11 Stroud Water Research Center team in the early 1980s. From left: Lou Kaplan, Peter Dodds, Kurt Dunn, Tom Bott, Robin Vannote, Cindy Dunn, and Bern Sweeney. ANSP Archives.

15.12 Part of White Clay Creek flowing through a Stroud Center laboratory. ANSP Archives.

15.13 Louis Sage collecting aquatic specimens on Chesapeake Bay from the Academy's research vessel *Joseph Leidy*. ANSP Archives.

The mission of the Benedict Lab, at first located in a converted motel, was to address the societal problems involved in balancing the ecology of the bay with the economy of the cities, towns, farms, and industries in its watershed and their supporting installations and waste products. The lab's scientists were supported by contracts from the state of Maryland, the U.S. Navy, the Baltimore Gas and Electric Company, and the Maryland Oceanic and Atmospheric Administration.[65] There was a "huge amount of work," according to Richard Horwitz, an ichthyologist who joined the Academy in 1976. "The impact of the increasing population surrounding the bay, new kinds of biological information coming in, and changing methodology in collecting, produced a lot of data."[66]

As the federal government became more concerned about water quality, it called on Ruth Patrick and the Academy to advise on appropriate legislation to address the problem. The Clean Water Act, enacted in 1972, established goals for eliminating the release of high amounts of toxic substances into water by 1985 and ensuring that surface waters would meet standards necessary for human sports and recreation by 1983.

In 1983, when the Academy's Limnology Department was renamed the Patrick Center for Environmental Research in honor of Ruth Patrick's fiftieth year with the institution,[67] philanthropist Mrs. Jefferson Patterson donated a part of her farm on the Patuxent River to the state of Maryland to be used as the site for an Academy laboratory. The project was to be the responsibility of the Academy, the Maryland Historical Trust, the University of Maryland's Chesapeake Biological Laboratory, and the Calvert Marine Museum.[68]

Louis E. Sage became director of the Benedict Laboratory in 1973 and built on a number of the power plant and river survey projects started by Patrick (figure 15.13). In 1982, he moved to Philadelphia to head the Limnology Department and was principally responsible for raising the substantial funds needed to build the new Benedict Laboratory (figure 15.14). Large projects on the Chesapeake Bay involved monitoring the environmental effects of the Potomac Electric and Power Company's electric power plant and, later, the Baltimore Gas and Electric Company's nuclear power plant at Calvert Cliffs, Maryland, where the scientific questions were different and the work was on a more diffuse scale.

The Academy's longest-running river survey project began in the early 1950s, when the Limnology Department was called upon to determine the effects of thermal discharge from DuPont's nuclear plant on the Savannah River ecosystem in South Carolina and Georgia. The Atomic Energy Commission oversaw the work because the plant was producing hydrogen bomb materials. The Academy continued to monitor the river for fifty years.

Patrick retired in the early 1970s, but she remained active in the Academy. Her primary focus was on finishing her set of books *Rivers of North America*, but she remained involved in

15.14 Benedict Estuarine Research Laboratory. Artist's rendering. ANSP Archives coll. 49.

numerous other projects, particularly the dredging and management of the Savannah River.

Because of the Academy's experience with nuclear issues, when the Three Mile Island nuclear meltdown occurred in March 1979, Pennsylvania governor Richard Thornburgh called Ruth Patrick within hours to seek her advice on how to monitor the extent of environmental damage. Seven years later, while under contract with the Pennsylvania Power and Light Company to study the Susquehanna River, Academy scientists detected fallout from the Chernobyl nuclear power plant disaster in Ukraine, even before the Soviet Union acknowledged the full extent of the accident. The scientists followed it up the food chain for ten years. Not only was radiation found in aquatic organisms, but it passed into higher animals when, for example, deer ate mosses on the riverbank.[69]

Aside from the power plant work, in the early 1980s, the Benedict Laboratory developed other research programs. The lab was a leader in environmental studies of arsenic (Jim Sanders), mercury (Cindy Gilmour), and oxygen depletion in the bay (Denise Breitberg). The facility had large outdoor tanks (mesocosms), which were used for a variety of studies.[70]

By 1999, the Stroud Center had evolved from a makeshift laboratory into a multibuilding research and education facility with twenty-three full-time employees and a multimillion-dollar annual budget. From a focus on one small tributary, White Clay Creek, the center's range had expanded to its own foreign research station in Costa Rica and other programs that spanned the globe. "None of this could have happened without the support of the Academy, which inculcated the Stroud Center and nurtured its growth," wrote James Blaine in the Stroud Center's history. "Above all, the affiliation with one of the oldest and most respected scientific institutions in the Western Hemisphere provided the Center with a mantle of credibility that proved invaluable to the fledgling organization."[71]

By the late 1990s, however, the Academy and the Stroud Center had reached a crossroads. The center's growth, financial independence, and physical distance from the Academy, as well as the increased funding through research grants and long-term contracts with public agencies and private corporations, pointed to the inevitable. In the spring of 1999, at the Stroud Center's request, the Academy's trustees authorized the center to become a separate nonprofit corporation, and that fall, the organization became fully independent.[72] According to Horwitz, the split allowed both organizations to pursue their own research without overlapping each other's work, which was a potential problem. But relations continued to be completely amicable between the Academy and the Stroud Center, with the scientists sharing information and data.[73]

At this time, river surveys began to decline because many companies established their own in-house capabilities for evaluating water quality. As regulating standards became routine, surveys became more formulaic, more cut-and-dried. Factories in the South, where labor had been cheap, were closed, and new ones were built overseas, where labor was even cheaper—and environmental protection less stringent.[74]

After a difficult period, the Benedict Laboratory ceased operation in 2003, when the program direction changed. Part of the problem involved commercial consulting firms undercutting its work, but the reason for the lab's demise was complicated by many factors that are still controversial. After a lengthy negotiation to ensure the continued operation of the facility, the building was given to Morgan State University in Baltimore, Maryland.

With David Hart as director, the Academy's Patrick Center continued to play a vital role in studying the human impact on aquatic systems in 1996. In addition to developing studies on black flies—the insects' distribution and drift—and maintaining river surveys, Hart built several group projects involving

15.15 Academy scientists Mike Kacher (right) and Barry Vance monitor the environmental impact of Pennsylvania Power and Light's nuclear plant in Berwick, Pennsylvania. ANSP Archives coll. 2010-301.

FACING PAGE 15.16 Ruth Patrick and team assembled for Guadalupe River project in Texas, 1947. ANSP Archives coll. 2010-020.

ACADEMY OF NATURAL SCIENCES OF PHILADELPHIA
LIMNOLOGY DEPT.

15.17 Roger Thomas and team examining mussels from the Delaware River. ANSP Archives coll. 2010-30. Partnership for the Delaware Estuary photograph.

river restoration and river ecology. These included studies on the effects of dams and dam removals on river systems. The work was relevant to Pennsylvania because of the many old milldams in the state and the push to remove them for ecological and safety reasons. Patrick Center operations also emphasized the importance of planting trees along streams and rivers, where they would take up nutrients such as fertilizers, a project initiated by Bern Sweeney with a major National Science Foundation study on riparian trees.

In 1989, the center received a grant from the William Penn Foundation to evaluate water quality, channel form, stream erosion, and the presence of insects, fish, reptiles, and amphibians in Philadelphia's Fairmount Park. There was also terrestrial work to be done in the park, which brought in Alfred "Ernie" Schuyler, longtime curator of the Academy's Botany Department. The Academy helped to create a master plan for the park's ecosystems between 1998 and 2001.[75]

The Patrick Center, among its many projects, has continued (since 1947) to evaluate the overall impact of plant operations on aquatic communities in a series of field studies on the Guadalupe River and the Victoria Barge Canal in Texas for DuPont, Union Carbide (Dow Chemical), and BP Chemicals (figure 15.16). Academy scientists have made assessments of total mercury concentrations in fishes from rivers, lakes, and reservoirs in New Jersey and surrounding estuarine waters. Current research focuses on nutrient and containment chemistry along with biological indicator development. Impacts from human development in watersheds from the east coast of the United States to the west coast of South America are areas of study, as are the threats of climate change in the wetlands surrounding the coastline of the region.[76]

David Velinsky became Academy vice president and director of the Patrick Center in 2005. Studying the ecological history of nutrients in the Delaware River basin and the effects of tidal marshes (freshwater and brackish) on nutrient chemistry has been a major focus for the department. Another important project has been to develop a research program for studying the ecological effect on Pennsylvania's streams and groundwater of drilling for natural gas in the Marcellus Shale formation.[77] A preliminary study found that populations of salamanders and aquatic insects, animals sensitive to pollution, were 25 percent lower in streams with the most drilling activity. In testimony before Philadelphia's City Council, Velinsky said of the early findings: "This suggests there is indeed a threshold at which drilling—regardless of how it is practiced—will have a significant impact on the ecosystem."[78]

In the summer of 2010, in a joint effort to study the Delaware River, Academy scientists and the Partnership for the Delaware

Estuary found seven species of mussels, including two native species long thought to have disappeared from the river, indicating that the Delaware may be cleaner than previously supposed. "It was very, very surprising to find mussels like this in an urban waterway," noted Roger Thomas, staff scientist at the Academy, where mussels and crayfish have a long history of study. "We think it's an exciting discovery" (figure 15.17).[79]

Diatoms are still very much a part of the Academy's investigations. Donald F. Charles, the Patrick Center's diatom ecologist, uses paleolimnological approaches for inferring change in the biology and chemistry of lakes and assesses the perturbations in aquatic ecosystems caused by municipal and industrial effluents, disturbances in land use, and climate change. "As a diatom ecologist," says Charles, "the Academy provides me with the opportunity to put my ideas into practice because I can tap into the Academy's historical expertise with diatoms, as well as the excellent collections and lab facilities."[80]

For her innovative work on water quality and the environment, Ruth Patrick has received numerous awards throughout her long career, most notably the John and Alice Tyler Prize for Environmental Achievement in 1975 (considered the equivalent of a Nobel Prize in ecology), the Benjamin Franklin Award for Outstanding Scientific Achievement from the American Philosophical Society in 1993, and the National Medal of Science presented by President Bill Clinton in 1996. In addition, Patrick has received lifetime achievement awards from the American Society of Limnology and Oceanography and the National Council for Science, the Gold Medal from the Royal Zoological Society of Belgium, and election to the National Academy of Sciences. In spite of all this recognition, her sense of humor was intact when once, during a short hospital stay, her colleague Bob Grant asked whether she wanted him to bring her mail. "Only the awards and honors," she replied.[81]

It must have pleased her when she was elected the first woman chair of the Academy's board of directors (1973) after her early struggles to be recognized in a man's world and, again, when she was elected the first woman on DuPont's board of directors. She was a director of the Pennsylvania Power and Light Company and an advisor to presidents Lyndon B. Johnson on water pollution and Ronald Reagan on acid rain. Appointed to the Academy's Francis Boyer Chair of Limnology in 1973, Patrick taught botany and limnology for thirty-five years at the University of Pennsylvania, and over time, she was awarded twenty-five honorary doctorates at various universities. During all this, she and her husband found time to raise a son, Charles Hodge V (a pediatrician), entertain friends, and faithfully attend the First Presbyterian Church in Philadelphia. She has also written more than two hundred scientific papers and a number of books on the environment.

Because of her long tenure at the Academy, some of the stories surrounding Patrick reach almost legendary proportions. One involves her service to military intelligence during World War II, when the U.S. Navy captured a German submarine far from its base and officials came to Patrick to identify a collection of diatoms scraped from its hull. The story goes that she recognized the species as indigenous to the Norwegian fjords, thus helping the Allies trace and eventually counter U-boat activity in the North Atlantic. In subsequent tellings, the submarine became Japanese and the diatoms were from the Bay of Japan;[82] later, as the Cold War replaced earlier conflicts, the submarine became Russian and the diatoms were from Cuba. Since the assessments involved national security, the true story will forever be cloaked in mystery. Perhaps all three versions are true.

"Ruth was a pioneer in so many ways," said David Hart, who succeeded Sandy Sage as director of the Patrick Center (1997–2005). "She was thinking about how her research could help society solve real-world environmental problems twenty-five years before we even had a Clean Water Act. She was building an unheard-of partnership between environmental groups, government agencies, and corporate heads. She was exploring how all of those partners could come up with better solutions to the problems they faced."[83]

Edward O. Wilson, from Harvard's Museum of Comparative Zoology, wrote in 1983 to congratulate Patrick on the celebration of her fifty years at the Academy: "On this occasion you should also be celebrated as one of the pioneers of modern ecology. . . . I am aware more than most how few first-class scientists preceding us were willing to commit themselves to ecology in the 1940s and 1950s. You led the way, set high standards, and made it exciting, and for that we owe you a great debt."[84]

In 1983, Thomas E. Lovejoy, then director of World Wildlife Fund–U.S., who had worked with Patrick at the Academy in the early years, summed up her extraordinary career with an apt and touching allusion to Alice in Wonderland: "The world is a better place for your having followed your curiosity down through your father's microscope."[85]

PTS

Notes

1 Samuel Pepys, quoted in unpublished remarks by Dr. Thomas Peter Bennett in a talk given at the Academy. The Academy's "Science Smorgasbord" lecture series, 20 April 1977. Presidential file under Thomas Peter Bennett lectures, coll. 2010-051, series 4, box 83.

2 Leonard Warren, *Joseph Leidy: The Last Man Who Knew Everything* (New Haven, CT: Yale University Press, 1998), 28.

3 Ibid., 69. The University of Pennsylvania has commemorated Leidy in the Leidy Laboratory of Biology housed in a building on the campus. There is a picture of the Leidy lab in George E. Thomas and David B. Brownlee, *Building America's First University: An Historical and Architectural Guide to the University of Pennsylvania* (Philadelphia: University of Pennsylvania Press, 2000), 239.

4 Joseph Leidy, "Flora and Fauna within Living Animals," in *Smithsonian Contributions to Knowledge,* vol. 5. (1853).

5 J. Cheston Morris to Joseph Leidy II (Leidy's nephew), 24 February 1909. ANSP coll. 553, Leidy Microscopical Club, papers and correspondence, 1909–1980.

6 Ibid.

7 Edward J. Nolan, *A Short History of the Academy of Natural Sciences of Philadelphia* (Philadelphia: Academy of Natural Sciences, 1909), 22.

8 F. E. Round, R. M Crawford, and D. G. Mann, *The Diatoms* (Cambridge: Cambridge University Press, 1990), http://www.tolweb.org/Diatoms/21810.

9 C. Bernard Peterson, "A Niche for Amateurs," *Frontiers* 6, no. 5 (1942): 133–35.

10 Ibid., 134.

11 F. W. Lewis, M.D., "Notes on New and Rarer Species of Diatomacea of the United States Sea Board," *Proceedings of the ANSP* 13 (1861): 61–73.

12 Biological and Microscopical Section, report of annual exhibition for 1879, ANSP coll. 295.

13 At his death, Leidy had two unpublished manuscripts nearly ready.

14 Joseph Leidy, *Freshwater Rhizopods of North America* (Philadelphia), and introduction to *Report of the United States Geological Survey of the Territories,* vol. 12 (Washington, DC: Government Printing Office, 1879).

15 Charles S. Boyer, *Diatomacae of Philadelphia and Vicinity* (Philadelphia: J. B. Lippincott, 1916); and Charles S. Boyer, *Synopsis of North American Diatomaceae* (Philadelphia: J. B. Lippincott, 1927).

16 "Dr. Thomas Summerville Stewart," undated, unsigned manuscript, possibly written by Ruth Patrick, ANSP coll. 262, file 6. I have been unable to find "Dr. Leonard" in order to substantiate this statement.

17 Marina Potapova, "The ANSP Diatom Herbarium: An Important Resource for Diatom Research," *Proceedings of the ANSP* 160 (November 2010): 3–12.

18 General papers of the Biological and Microscopical Section, correspondence 1868–1922, ANSP coll. 296, file 4.

19 The Biological and Microscopical Section and the Mineralogical and Geological Section were reconstituted as the Leidy Microscopical Club (later Society) in 1925. ANSP coll. 295, file 8. In 1995, the Leidy Microscopical Society moved its meeting place from the Academy to the Northminster Presbyterian Church in Fairless, Pennsylvania. The society removed its instrument collection from the Academy in 2005.

20 I. F. Lewis, C. Zirkle, and R. Patrick, "Algae of Charlottesville and Vicinity," *Journal of the Elisha Mitchell Scientific Society* 48 (1933): 207–23; and E. C. Cocke, I. F. Lewis, and R. Patrick, "A Further Study of Dismal Swamp Peat," *American Journal of Botany* 21 (1934): 374–95.

21 Ruth Patrick Hodge to Dr. James A. G. Rehn, 10 March 1934, ANSP coll. 262, file 6.

22 Charles M. B. Cadwalader to Dr. Ruth Patrick Hodge, 10 November 1934, ANSP coll. 262, file 6.

23 Ruth Patrick, "A Taxonomic and Distributional Study of Some Diatoms from Siam and the Federated Malay States," *Proceedings of the ANSP* 88 (1936): 367–470.

24 Ibid., 467.

25 Potapova, *Proceedings,* 4.

26 ANSP Ruth Patrick biography file.

27 Ruth Patrick, "The Occurrence of Flints and Extinct Animals in Pluvial Deposits Near Clovis, New Mexico: Part V—Diatom Evidence from the Mammoth Pit," *Proceedings of the ANSP* 90 (1938): 15–24.

28 Ibid.

29 Charles M. B. Cadwalader to Henry Allen Moe, Secretary, John Simon Guggenheim Memorial Foundation, 13 November 1939. ANSP archives 2010-051, Presidents: Cadwalader, box 1, Patrick file.

30 Ibid. After more than ten years as a volunteer, Patrick became a full-time employee in 1945 with a salary of $1,000 a year.

31 Barry Lewis, "The Patrick Principle," *University of Virginia Alumni News* (Fall 2003): 21.

32 Lewis, "The Patrick Principle," 22.

33 ANSP biography file. ANSP website, "Learn About the Academy: Who's Who: Dr. Ruth Patrick," 18 September 2006, 3 of 5 (http://www.ansp.org/research/pcer/np/biography.php).

34 Potapova, *Proceedings,* 5.

35 Lewis, "The Patrick Principle," 21.

36 Bayard Webster, "Wading in the Name of Science," *Scientists at Work: The Creative Process of Scientific Research,* ed. John Noble Wilford (New York: Dodd, Mead, 1975; reprint 1979), 183.

37 Marion Steinmann, "Rivers of America: The Source Is Ruth Patrick," Rockefeller Foundation, 1983. ANSP biography file.

38 Draft of a letter from Cadwalader to the Guggenheim Foundation, 9 November 1939. See note 29.

39 Ruth Patrick, "The Academy at the Forefront of Science," manuscript in ANSP coll. 262, file 7, p. 8.

40 Potapova, *Proceedings,* 6.

41 Lewis, "The Patrick Principle," 21.

42 Potapova, *Proceedings,* 6. Ruth Patrick and Charles W. Reimer, *Diatoms of the United States Exclusive of Alaska and Hawaii,* Monographs of the Academy of Natural Sciences of Philadelphia, no. 13, vol. 1, 1966; vol. 2, 1975.

43 *News of Delaware County,* 17 May 2006. ANSP biography file.

44 Charles W. Reimer (1923–2008), ANSP website.

45 The diatometer was named to honor the Catherwood family, whose foundation had provided financial support for some of Patrick's aquatic research.

46 James Poling, "She Takes the River's Pulse," *Reader's Digest* (August 1962), condensed from *National Civic Review* (July 1962): 2. ANSP archives, coll. 417c.

47 Poling, "She Takes the River's Pulse," 2.

48 John Cairns Jr. biographical sketch (last updated January 2010), http://www.johncairns.net/aboutme.htm.

49 Communication from Dr. John Cairns Jr., University Distinguished Professor of Environmental Biology Emeritus, Department of Biological Sciences, Virginia Polytechnic Institute and State University, Blacksburg, Virginia, November 2010.

50 "It was love at first bite for Philadelphia entomologist," *Philadelphia Evening Bulletin,* 31 August 1981. ANSP biography file.

51 Author interview with Clyde Goulden, November 2010.

52 Interview with Robert Grant, October 2010.

53 Patrick's remarks quoted in Lewis, "The Patrick Principle," 23.

54 Ibid., 23.

55 "The Limnology Department" by Ruth Patrick, Chairman, typed manuscript, ANSP coll. 262, file 2.

56 The expedition was funded by Alfred James Ostheimer, who had financed several previous Academy trips to the Pacific area.

57 Bernard Sweeney, quoted in James G. Blaine, *Stroud Water Research Center: A Portrait, 1967–2000* (Stroud Water Research Center, 2000).

58 "Environment: America's Sewage System and the Price of Optimism," *Time*, 1 August 1969, article at http://www.time.com/time/magazine/article/0,9171,901182,00.html.

59 Ibid.

60 Blaine, *Portrait*, 30–31.

61 Ibid., 26–27.

62 Neal Foster, "A Tale of Two Rivers: The Thames and the Delaware," *Frontiers* 40, no. 4 (1976): 36–41.

63 For more on the history of shad in the Delaware River, see John McPhee, *Founding Fish* (New York: Farrar, Straus and Giroux, 2002).

64 The Patuxent River is the largest river whose watershed lies completely in the state of Maryland. Part of the river is a fifty-two-mile-long tidal estuary.

65 "Checking on the Chesapeake," *Academy News* 5, no. 3 (Fall 1982).

66 Interview with Richard Horwitz, November 2010.

67 This name change also related to the establishment of the Ruth Patrick Fund for Research in the Environmental Sciences, which was the first endowment.

68 Chronology of Patterson Park Site Development, ANSP archives (1986).

69 Interview with Sandy Sage, November 2010.

70 Communication from Richard Horwitz, 17 January 2011.

71 Blaine, *Portrait*, 34.

72 Ibid.

73 Horwitz interview.

74 Horwitz interview.

75 R. J. Horwitz, B. W. Thompson, C. Cianfrani, N. C. Coulter, W.C. Hession, J. E. Pizzuto, A. Rhoads, and A. E. Schuyler, "Master Plan for Natural Lands Restoration in Fairmount Park, Philadelphia, 1999–2001," 3 vols. For more on Fairmount Park, see David Contosta and Carol Franklin, *Metropolitan Paradise: Wissahickon Valley, 1620–2010* (Philadelphia: St. Joseph's University Press, 2010).

76 Communication from David Velinsky, 20 January 2011.

77 Sandy Bauers, "Study Sees Threat in Shale Gas Drilling," *Philadelphia Inquirer*, 12 October 2010.

78 Ibid.

79 Kathy Matheson, "Mussel Find Suggests Delaware River Health," *Philadelphia Inquirer*, 26 November 2010.

80 ANSP biography file, unpublished manuscript by Don Charles.

81 Interview with Robert Grant, November 2010.

82 This part of the story is told by Thomas E. Lovejoy in "National Security, National Interest, and Sustainability," Environmentally and Socially Sustainable Development, Latin American and Caribbean Region (The World Bank, Washington, DC), 511. ANSP Ruth Patrick biography file.

83 Lewis, "The Patrick Principle," 24.

84 Dr. Edward O. Wilson to Dr. Ruth Patrick, Cambridge, MA, 4 November 1983, ANSP archives, coll. 262, "Ruth Patrick Jubilee," file 1.

85 Dr. Thomas E. Lovejoy to Ruth Patrick, Washington, DC, 31 October 1983, ANSP archives coll. 262, "Ruth Patrick Jubilee," file 1.

Skull of a quagga (*Equus quagga quagga*). ANSP Mammalogy Department #6317. This subspecies of the plains zebra became extinct in 1883. This specimen, from South Africa, was donated to the Academy by Edward Drinker Cope (1840–1897). Purcell photograph.

6317

Chapter 16

Regrouping and Looking Forward in the Postwar Years

It's difficult to be sure of anything, but that's the plan at the moment and we plan to go ahead with it.
—Ruth Patrick, speaking of the Academy's expansion plans, 1 June 1975

In the closing days of World War II, Ernst Schäfer, the German explorer, zoologist, and politically savvy associate of Heinrich Himmler, Hermann Göring, and the Nazi high command, left Schloss Mittersill, the fifteenth-century castle that had been his headquarters as director of the Sven Hedin Institute for Inner Asian Research since 1943.[1] He was dressed in civilian clothes, his SS uniform no longer providing the respectful deference he had enjoyed since joining the Third Reich's elite inner circle in 1933.[2] Making his way to the command center of U.S. forces in the area, he insisted on seeing the American general in charge of the occupying troops. "Who's asking?" barked one of the soldiers guarding the doors of the improvised headquarters. "It is I, Ernst Schäfer, the zoologist and explorer," Schäfer replied in the English he had learned during two Academy expeditions to Tibet. "I am a life member of the Academy of Natural Sciences of Philadelphia."[3] To Schäfer's amazement and disgust, the Academy's name and his offer to surrender Schloss Mittersill and all of its intellectual resources to the control of that Philadelphia institution had absolutely no effect on his subsequent treatment.[4] He was arrested, jailed, and eventually tried at Nuremberg for complicity with the Nazi regime. Thanks to the testimony of many whom he had helped during the war, Schäfer was ultimately exonerated by the tribunal.[5]

Halfway around the world, as hostilities ceased in the Pacific theatre, U.S. general Douglas MacArthur met with Japanese emperor Hirohito (Emperor Shōwa) (1901–1989). To break the tension of their first encounter, the emperor is alleged to have started his conversation with MacArthur with a friendly question: "How's my friend Pilsbry in Philadelphia?"[6] Hirohito, an enthusiastic collector of shells, was referring to the Academy's curator of malacology, Henry A. Pilsbry (1862–1957), with whom he had exchanged publications and correspondence before ascending the Chrysanthemum throne.[7] Pilsbry had earned Hirohito's respect not only by his enormous contributions to the study of Japanese shells but also by the central role the Academy had played in helping to develop the field of malacology in Japan during the late nineteenth and early twentieth centuries (figure 16.2). Hirohito undoubtedly intended his reference to an American colleague and a period of friendlier Japanese-American relations as a gesture of reconciliation.

Unfortunately, although MacArthur's grandfather had designed Philadelphia's city hall, the American general knew nothing about Pilsbry or the work of the Academy of Natural Sciences, so the reference went unappreciated. In researching the subject after the meeting, however, his aides may have come to understand how much the Academy and its associates had contributed to the war effort.

Henry Crampton (1875–1956), a friend and colleague of Pilsbry's, had provided useful information on the flora, fauna, and human populations of island communities in the South Pacific

16.1 A flying fox (*Pteropus sp.*) collected on Biak Island, Korrido, by S. Dillon Ripley during the Academy's Denison-Crockett Expedition to the South Pacific in 1937. ANSP Mammalogy Department. Purcell photograph.

based on his many expeditions to that part of the world in the decades leading up to the war.[8] Crampton had been focused on the evolution of land snails, but his "ground truthing" and photographs of specific islands proved invaluable to the OSS in planning Allied activities in the area long before topographical information became available to anyone online.

An even larger contribution had been made by another Pilsbry colleague, H. Radclyffe Roberts (1906–1982), who applied his twelve years of experience studying insects at the Academy to create an atlas of the mosquito species responsible for carrying malaria through many areas in which American troops were stationed. According to a contemporary press report, Roberts's guide enabled "our military experts to combat a wily foe which some have described as potentially more dangerous than all the regiments of the Reich and Nippon" by helping "those charged with protecting our soldiers, sailors and Marines from disease" to identify malaria's vectors.[9] Published by the American Entomological Society and distributed in loose-leaf form by the War Department, the atlas went through several editions during and after the war and helped to save countless thousands of lives.

Roberts and many other Academy staff members returned from military service with a changed perspective on the vulnerabilities of civilization. As the focus of the museum's public exhibits shifted from war-related topics (*A Victory Exhibition: Mineral and Vegetable Resources Essential to National Victory* had opened in January 1942, and another, *The Raw Materials of the Atomic Bomb,* ran for several months in 1945) to more peaceful subjects, Roberts asked his fellow curators to develop plans for safeguarding the research collections in the event of future conflict. In December 1947, a confidential document was prepared with instructions to move the type specimens and the most important historical collections to secure locations in the event of another war.[10] It was recommended that the collections should be dispersed and secretly transported by private automobiles and trucks to "bank vaults in small towns (Oxford, West Chester, Doylestown, Boyertown, Mount Holly) and buildings

16.2 Under the tutelage of the Academy's curator of malacology, Henry A. Pilsbry (1862–1957), the Japanese shell dealer and artist Yoichiro Hirase (1859–1925), seen here with some of his students and colleagues, encouraged Japanese scientists to focus on the study of mollusks. It was to this long and fruitful Academy influence that Emperor Hirohito referred in his first meeting with Gen. Douglas MacArthur. Nishinomiya Shell Museum, Hyogo, Japan.

16.3 The Academy's renovated 1876 building seen from Logan Circle. ANSP Archives coll. 49.

on larger private estates, say twenty miles from Philadelphia." The document further specified that "Any place selected for the purpose should be located not less than ten miles from any industrial plant and not less than five miles from any airfield."[11] Fortunately, although the plan was revisited during the Cuban missile crisis, it was never necessary to put it into effect.[12]

With provisions made for safeguarding the Academy's existing collections, the museum could continue its efforts to expand them. Overseas collecting expeditions, suspended during the war years, resumed quickly after the war, beginning with projects in areas considered safe and friendly to American interests (British East Africa, Canada, Mexico, and the Caribbean). Some were tied to diorama creation, while others were more purely scientific in nature.

In 1947, Roberts assumed the managing director's role from Charles Cadwalader, who remained the Academy's president through 1951. An accomplished scientist with degrees from Princeton and the University of Pennsylvania, Roberts kept a lower public profile than his predecessor, focusing his attention more on science than fund-raising or public events. During his tenure as managing director (1947–72), the Academy saw the continuing professionalization of the scientific staff. Volunteer curators who had been self-financed were slowly replaced by competitively paid professionals who were often recruited from outside Philadelphia. Over time, this shift created a subtle but increasingly significant distance between the Academy's trustees and professional staff. Some, like James Bond, Rodolphe Meyer de Schauensee, Morgan Hebbard, and Roberts, who were part of or had married into the upper tiers of Philadelphia society, were either independently wealthy or able to find funding from wealthy individuals who enjoyed their association with the Academy's scientific enterprise. The younger, non-Philadelphia-born scientists who joined the Academy through a more traditional academic track augmented their meager salaries with teaching appointments at the University of Pennsylvania, Princeton, Swarthmore, Haverford, or Bryn Mawr, while they sought further financial support from private foundations or the federal government.

The National Science Foundation (NSF), created by Congress in 1950 "to promote the progress of science; to advance the national health, prosperity, and welfare; to secure the national defense,"[13] became an important source of funding for the Academy's research staff. The highly competitive, peer-reviewed nature of NSF funding reinforced a narrowing of focus in scientific research, and so exacerbated the already growing distance between the research and the exhibition and educational activities of the institution.

When John W. Bodine (1912–1991) became the Academy's first paid president and CEO in 1963, the Academy was funding its programs and expensive fixed costs in a variety of ways. The Limnology Department, under the leadership of Ruth Patrick, was paying for itself and helping with the Academy's overhead

primarily through consulting contracts with industry and government. The Systematics and Biodiversity group (the section that includes most of the "ologies" and holds most of the collections) was increasingly dependent on NSF support and restricted endowment, while the public museum and education section was drawing its support primarily from corporate and individual philanthropy, earned revenues (including admission fees, beginning in 1953),[14] and modest, but ever-decreasing city and state appropriations.

Because of the increasing expense of salaried positions, the institution encouraged volunteers to assist with educational and exhibit activities. In the 1950s, a spirited group of women, dubbing themselves the "snake ladies," began to give live animal demonstrations inside the museum (figure 16.4), while another group, called the "roadrunners," took live animal programs to schools, hospitals, and nursing homes around the city and beyond.[15] Their efforts significantly augmented the educational offerings of the museum and helped to expand the Academy's reach into the community.[16]

In 1959, the Junior League of Philadelphia, a philanthropic women's organization, helped to create a new children's area within the museum. "They are not only planning it," explained the Academy's museum director Kenneth Prescott in an interview with the *Philadelphia Inquirer*, "they are carrying it out—carpentering, painting, making props. They are working like beavers."[17] The Academy's Junior Museum, which opened in 1960, proved a successful precedent for subsequent child-friendly spaces at the museum, including the Please Touch Museum (which was started at the Academy in 1976 and moved into its own facility in 1978) and, later, Outside-In (established at the Academy in 1979) (figure 16.5).[18] It was part of a larger, nationwide movement toward educational offerings for younger audiences, developed in response to the postwar baby boom and the perception that the United States lagged behind the Soviet Union in science education, brought on by the Soviets' launch of Sputnik in 1957.[19] Although prompted by financial need, the reliance on amateurs to teach and build exhibits was a return to the Academy's tradition of amateur involvement during the previous century.

This is not to say that the Academy's museum was without professional educators or that the public and scientific arms did not interact or support each other. Where there was intersection, the results could be inspiring. Jerry Freilich, director of the Academy's Live Animal Unit from 1971 to 1976 and now a research scientist with the National Park Service, attributes his lifelong career in natural history to the Academy's Junior

16.4 The Academy's energetic volunteer "snake ladies" helped to expand the museum's offerings during the 1950s and 1960s. ANSP Archives coll. 2010-301.

Curators program,[20] in which he first enrolled at the age of ten. "I was a student at the Academy in the summers of 1958, 1959, and 1960," Freilich recalls:

> The Director of Education in '58 and '59 was Bill Overlease who I regarded pretty much as God. He was enthusiastic, loved nature, and knew exactly how to talk to us kids. Overlease led the classes, introduced the guest speakers, and conducted various lessons in the museum halls. But the "Big Magic" was that each week we were introduced to one of the curators of the museum and given a behind-the-scenes visit to their department. I cannot express how exciting, how overwhelming, and how completely wonderful these visits were.
>
> The curators came to our classroom and gave a talk on what they did. But the "Real Magic" occurred when they escorted us in a line upstairs, past those doors marked "Staff Only" and into this "Other World." WOW!!!!!!! That was the Real Thing!!! (figure 16.6).
>
> Academy offices may look different today. But I'll bet that in essential qualities they are still the same. Strange smells. Dusty tarps covering big bones on high shelves. Large tables with trays of spectacular shells, bugs, or fossils. Piles of books with foreign titles and wall pictures of extraordinary creatures from far-away lands. What treasure! This was complete love at first sight for me.
>
> These meetings with the scientists were so important because they made us aware of the larger world of science behind those magical doors. They made us aware of the Science Gods and what they looked like. And it gave us a specific goal to aspire to. . . . I wanted to be one of them.[21]

For children not participating in the Junior Curators program, the principal means of interaction with the Academy was the free lecture series sponsored by the Ludwick Institute (offered continuously from 1896 until the 1970s) or the formal classes offered by the Academy's teacher-naturalists (both paid and volunteer).[22] From 1935 to the early 1990s, organized school groups coming to the museum from the School District of Philadelphia were instructed by one of two school district teachers who were on call at the Academy throughout the week.[23]

Those wanting to learn something about natural history by studying outdoors could join any number of local field trips. The Expeditions for Everyone program, the first of its kind in Philadelphia and one of the first in the country, was begun by Stephen Thomas in 1937.[24] It was suspended briefly during the Second World War but then resumed and ran successfully until the turn of the twenty-first century.

16.5 The popularity of the Junior Nature Museum led to the creation of Outside-In, the Academy's museum within a museum designed especially for young audiences, in 1979. ANSP Archives coll. 2010-003.

16.6 Ichthyology curator Henry Fowler (1878–1965) spent an astonishing and highly productive sixty-eight years at the Academy (1894–1962). Meeting such renowned scientists in their laboratories was one of the many privileges offered young students in the Junior Curators program. ANSP Archives coll. 2011-003.

Lorene McClellan (1905–1998), the first coordinator of this program, recalled the fun of the trips and the eclectic nature of its participants:

> We started modestly with perhaps one or two trips a month to local areas favored by PTC [Philadelphia Transportation Company] bus and trolley [such as] the Pennypack, Carpenter's Woods, Wissahickon, Tinicum Marsh (before it was a preserve and we had to be cautious of [gun] shots), Bartram's Gardens, Morris Arboretum, Horticultural Hall, etc. . . . We usually had two leaders on the early expeditions, mostly culled from ANS staff. . . . From the beginning with local half-day trips in the city, we extended, by chartered bus [to] full day and overnights, finally week-end trips.[25]

Interestingly, the field trip program was not initially intended for Academy members. Rather, it was part of an outreach to the community, attracting large numbers of people from professions ranging from business and health to "the technical trades . . . including a bookbinder, a cigar maker, a painter, an upholsterer, a toolmaker, and a waitress—in all, about one fifth of the whole."[26]

One of the most popular and enduring Expeditions for Everyone leaders was a member of the Exhibits Department who agreed to lead field trips on the side. Stephen Harty (d. 2003), who joined the exhibits staff in 1956 and remained at the Academy for the next twenty-five years, made the program national (to Florida and California) and international (to Costa Rica and Panama) while building a loyal support group he named the "Rare Bird Branch" (figure 16.7).[27]

When not leading field trips, Harty was involved with many other aspects of the Academy's museum. He left some permanent marks on the institution. As Harold Green's chief preparator from 1957 until Green retired in 1964, Harty was responsible for creating the black bear and bobcat dioramas in the Hall of Pennsylvania Mammals and painting the backgrounds for several of the dioramas in Africa, Asia, and North America Halls.[28] During a period of internal flux that followed Green's long (forty-four-year) tenure, during which the museum had six exhibit directors in twelve years, the affable Harty provided much-needed stability and continuity in the Exhibits Department. Like Green, he worked well with both the scientific and the educational staff.[29]

Harty was responsible for many museum exhibits but not all.[30] When the Hall of Earth History was updated to include dinosaurs in 1957, the demands of an enormous wall mural (twenty-two by twenty-four feet) required the services of a freelance artist.

For this project, the museum was fortunate in finding Jonathan Leo Fairbanks, a painter who combined the requisite artistic talent with a deep knowledge of natural history. Fairbanks, then a graduate student at the University of Pennsylvania and the Pennsylvania Academy of the Fine Arts, was asked to re-create a Cretaceous swamp with two large duck-billed dinosaurs happily munching on aquatic vegetation while being stalked by a distant but still menacing *Tyrannosaurus rex*.[31] The image was intended to commemorate the one hundredth anniversary of the discovery of the first (nearly) complete dinosaur skeleton in the world (*Hadrosaurus foulkii*, discovered in 1858) and to highlight and help to interpret a recently acquired skeletal mount of the closely related *Corythosaurus* (duck-billed dinosaur) that was to become the centerpiece of the hall (figure 16.8).

Although the bones of the *Hadrosaurus foulkii* and other dinosaurs were put on view nearby, the *Corythosaurus* remained the only mounted dinosaur on display at the museum for the next thirty years.[32] During that period, it assumed such an iconic place in the public's perception of the Academy that the skeleton was given a nickname (Dolores) and was used for a while as the museum's logo, appearing on Academy letterhead, museum guides, and promotional literature.[33] At the height of Cold War tensions in the 1960s, it was added to the list of important specimens considered for removal from the museum in case of nuclear attack.[34]

16.7 Exhibits preparator Stephen Harty debates a bird call with trip participant Evelyn Kramer during an Expedition for Everyone field trip, May 1974. J. Freilich photograph.

FACING PAGE 16.8 This mural by Jonathan Leo Fairbanks was on view in Dinosaur Hall from 1957 to 1998. ANSP Archives coll. 2011-003.

Fairbanks's mural, made under the direction of the Academy's paleontologist Horace G. Richards and botanist Edgar T. Wherry, was such a powerful image that it became as much a focal point for Dinosaur Hall as the skeleton of the creature it portrayed (figure 16.9). For three decades, until the hall was redesigned in the 1980s, Fairbanks's dinosaur depictions conveyed a living image of Cretaceous life to millions of museum visitors and helped to turn the Academy into Philadelphia's "Dinosaur Museum." Eventually, new thinking about dinosaur posture and locomotion made the Fairbanks interpretation obsolete. The long-familiar scene was given a new interpretation in 1986 as part of the museum's *Discovering Dinosaurs* installation. Using the outdated postures of the dinosaurs as an example of evolving thinking in paleontology, the mural became an instructional display used in juxtaposition to a remounted *Corythosaurus*.[35] In a later renovation of *Discovering Dinosaurs* (in 1998), the Fairbanks mural was finally removed, bringing to a close its forty-year presence in Dinosaur Hall.[36]

Another knowledgeable and talented artist associated with the Academy's Exhibits Department during this period was Richard Ellis, who offered his services to the financially strapped museum free of charge through much of 1961 and 1962.[37] Together, Harty and Ellis created a number of small, temporary exhibitions, including one for the Academy's sesquicentennial celebration (1962).[38] "We had a great story to tell, and some wonderful specimens to show," recalls Ellis, "but not much of a budget to work with. I hand-lettered two large signs in the front hall announcing the anniversary. The rest of the exhibit was displayed against pegboard using stick-on labels made with Prestype."[39]

The Academy was reluctant to spend money on its exhibits for, even as it looked back on its remarkable history, it was looking forward to its next 150 years, which many insiders believed would more likely than not take place in a new or completely reconfigured museum. The growing activities of the Limnology Department and an increased demand for exhibits and educational programs since World War II had prompted calls for expanded facilities as early as 1959. A concurrent initiative by the city to develop new cultural and commercial activities along the Delaware River attracted the attention of the Academy's board of trustees and prompted a study exploring the advantages and disadvantages of building a new museum facility in that location.

At a meeting of the Academy's administrative council on September 13, 1960, the Academy's director, Kenneth Prescott, reported "that the City proposes to redevelop the waterfront area to include a new marina, historic ships museum, science center, etc. The Academy is being considered as a focal point of the cultural-scientific section."[40]

Adding momentum to the prospect of moving to the waterfront was a proposal by Pennsylvania's Department of Forests and Waters to create a "conservation museum" in conjunction with the Academy at that location. In 1961 and 1963, the state legislature appropriated $190,600 to develop plans for this museum, which it was estimated might cost the commonwealth up to $6 million.[41] In the meantime, the Academy undertook a comprehensive study of its present and future space needs. The Turner Construction Company estimated that the cost of renovating the Logan Circle building would be almost as high as building an entirely new facility ($7,250,000 vs. $7,300,000), adding further support to the relocation plan.[42]

With all signs favoring an Academy move, at the annual meeting in 1964, in the presence of several hundred members, Mayor James Tate formally—if prematurely—welcomed the Academy to Penn's Landing.[43]

Because of the enormous expense associated with the proposed move, the idea languished. More studies were made and a tentative architectural plan was prepared, but discussions continued throughout the 1960s without resolution.[44] As committee after committee reviewed the options, some Academy staffers

FACING PAGE 16.9 Longtime Academy teacher Lorene McClellan (1905–1998) uses *Corythosaurus* to teach a class on prehistoric life. It was the erection of this iconic duck-bill dinosaur and the accompanying mural in 1957 that caused the Academy to become known popularly as the "Dinosaur Museum." Collection of Liz Drayton.

16.10 The Academy's museum has provided formal education programs for school groups since 1935. ANSP Archives coll. 2010-301.

offered alternatives to moving the entire institution. One proposal suggested that "the Academy recognize the dichotomy of function within its activities and that it consider physical separation of the research departments from those of teaching and education." This plan argued for moving the systematics collections, laboratories, and staff to the proposed Delaware River site, while leaving the Education, Exhibits, and Limnology departments at Logan Circle, where they would have plenty of room to expand.[45]

Whether it included part or all of the Academy, the institution's contemplated move to the waterfront came as unwelcome news to the University of Pennsylvania. At a meeting with the Academy's board chairman, Lane Taylor, and past president George Clark on 5 August 1970, the university's president, Gaylord Harnwell, and provost, David Goddard, tried to encourage the Academy to move closer to the university. Land "along Market Street somewhere near 37th Street," they assured Taylor and Clark, could be made available for "about $250,000 an acre."[46] It was the fourth time in a hundred years that the university had tried to effect a move to West Philadelphia by the Academy.[47] "I can well recognize that the cost of land may seem a barrier," Goddard wrote Taylor a few weeks after their meeting, "but I believe the Academy's relationship with the University would be such a uniquely important one that some of us may be willing to approach major foundations on behalf of the Academy with its goals for physical relocation, recognizing that the support by the foundations will be based more on the intellectual relocation than a purely physical one."[48] In a follow-up letter, Goddard conveyed the university's incoming president Martin Meyerson's "strong support" for an Academy move closer to Penn and expressed his own view that a move to Penn's Landing would "make more difficult the maintenance of the Academy as a first rate institution."[49]

To further confuse the situation, at about the time the Academy was weighing the relative benefits of a move to the Delaware waterfront or to West Philadelphia, Sam Smith, a Henry Houston heir with extensive land holdings in Roxborough, offered to give, rent, or sell the Academy land for a new building site near

16.11 The Fidelity Mutual Life Insurance Company building at 25th and Spring Garden Streets, designed by the architectural firm of Zantzinger, Borie, and Medary, was built in 1926. It was purchased by the Academy in 1972 and occupied by many of the scientific departments the following year. Philadelphia Museum of Art.

the recently established Schuylkill Valley Nature Center on the edge of Fairmount Park.[50]

In late 1970, a newly reorganized Penn's Landing Committee informed the Academy that the city's earlier offer of up to five acres of rent-free land would be significantly reduced, that the building it might occupy would be a high-rise, and that the space available for parking would be far less than originally discussed. The Academy, stunned by the change in conditions and wearied by years of fruitless negotiations with the city, retained the firm of Jackson Cross to provide a comprehensive review of new site possibilities and to revisit the projected costs associated with each.[51]

The Academy's board of trustees had high hopes for the institution's future when it hired William A. Marvel, a professional educator and foundation director, as president and CEO in 1971. At an elaborate inauguration ceremony with a formal dinner for 250 invited guests, Marvel articulated his ambitious vision for the Academy. This included a new building; additional and expanded outlying research stations; dramatically expanded educational programming, through which the museum would involve itself "simultaneously with the needs of people in the ghettos as well as the Main Line, in the revitalized inner-city of Society Hill, as well as in communities at the outskirts of the Greater Philadelphia area"; "the only scientifically based public aquarium between Coney Island and Miami, Florida";[52] strengthened systematics departments; and a Limnology Department "joined by new dimensions of work in terrestrial ecology."[53] All of this should, could, and would be funded, he assured his audience, by greatly expanded city, state, and federal support. A collection of congratulatory messages from Mayor Frank Rizzo, Governor Milton Shapp, senators, congressmen, EPA chief William Ruckelshaus, former president Lyndon Johnson, and incumbent president Richard Nixon seemed to support Marvel's optimistic view of government interest in what he called "the grand old gentleman of Logan Circle."[54]

In referencing the nineteenth-century chair in which he was ceremonially seated during his installation, Marvel said it made him think of "the throne of an English king or the stool of an African chief." He joked that he hoped it would not "really be the 'hot seat.'"[55]

With prospects of a move to the Delaware waterfront dimming, the newly installed president urged exploring other locations for the Academy. A board committee looked into creating a new building on the open ground across 19th Street from the Academy (then an open-air parking lot that was subsequently developed as the Four Seasons Hotel). They also explored an unspecified location in Franklintown, where sponsors were said to be "very anxious for the Academy to become a major factor in that development."[56] Still another idea was to build a new Academy headquarters on an open plot of city-owned land directly behind the Franklin Institute, retaining the original 19th Street building for exhibitions and educational activities.[57] Then, in the spring of 1972, talk turned to the Fidelity Mutual Life Insurance Company building at 25th and Spring Garden Streets, adjacent to the Philadelphia Museum of Art (now the Perelman Building of the Philadelphia Museum of Art) (figure 16.11).[58] The advantages seen in this building were its ample space, its suitably institutional location, the opportunities it afforded for further growth, its fund-raising potential, its immediate availability, and its relatively reasonable cost ($15 to $18 per square foot vs. $40 to $45 per square foot to build a new building).[59]

In October 1972, after more than a decade of agonizing deliberation over the best course of action, the Academy announced that it had "made a definite decision concerning a new home for its research, scientific collections, museum and education programs."[60] In explaining the plan to the public, Marvel prefaced his remarks by reiterating the inadequacy of the Academy's present building: "In giving the Academy the first accreditation awarded in Philadelphia," he said, "the American Association of Museums noted, in effect, that its inadequate physical plant is a major barrier separating the institution from the bright future it is destined to achieve."[61] He then laid out a three-phase plan for a projected $15 million expansion:

> Phase I: Purchase ($1.7 million) and remodeling ($1.7 million) of the Fidelity Mutual Life Insurance Company building . . . to house most of the scientific research and vast specimen collections (for a total of $3,400,000); Phase II: Construction of a new wing on the Fidelity building to provide the additional space needed for the library and Limnology Department ($2,500,000); and Phase III: Construction of a new museum either on present Logan Circle property or at a new Parkway site now being discussed with the Mayor and other city officials ($9,000,000).[62]

Within a year of the purchase of the building, the details of phase III of the plan were revealed with a second stunning announcement: The Academy wished to sell its Logan Circle building and acquire from the city five and a half acres of open

16.12 In 1973, the Academy announced its intention to build "the newest and most modern museum on natural history and the environment in the nation" on a plot of city-owned land near the Schuylkill River and the Philadelphia Museum of Art. Unable to raise the $7 million needed for the project, the board soon abandoned the idea and redirected its energies to renovating the Academy's 1876 building on Logan Circle. Artist's rendering. ANSP Archives coll. 2010-227.

land between the Benjamin Franklin Parkway and the Schuylkill River (just above the Park Towne Apartments), where it would build a new museum from scratch.[63] "On a location so dramatic and suitable, this museum and educational center would add further luster to the Parkway, Philadelphia's cultural spine," enthused Academy officials. "It will be the first new major cultural institution to be built here in 50 years, and also the only major natural history institution in the nation now creating a new environmental museum. This presents a rare opportunity for Philadelphia to reinforce its historic role as a leader in the arts and sciences."[64] Although this plan had been in the works since the summer of 1972, because of the "delicate diplomacy involved in determining whether the City-owned land parcel for the new museum [could] be obtained," it had been kept "highly confidential" by the president and board.[65] Nevertheless, a great deal of detailed planning had already gone into what the Academy's new museum would look like and how it would be run.[66]

Promising to give Philadelphia "the newest and most modern museum on natural history and environment in the nation," the Academy offered to create a building that would have been revolutionary for its day, incorporating wind, solar, and hydroelectric power, recycled water, and methane gas (produced in the anaerobic decomposition of sewage and organic wastes) to generate its own heat (figure 16.12). In an eight-page outline of the "ecological principles" the Academy wished to demonstrate in its new museum, engineer Philip Steadman stated that, among other things, its design would help "ease the problem of the rapid exhaustion of fossil fuel supplies and lessen the environmental damage resulting from their use."[67]

The proposed two-and-a-half floor, 130,000-square-foot building (occupying 53,000 square feet of land) promised to serve as a teaching model in itself. "There are ways in which the architecture and engineering of the new museum can teach lessons about the relation of man to his environment and to the natural world," explained Steadman. "In this perspective, the building can itself become an exhibit quite as much as the displays which it houses.[68]

Although the only such museum proposal in the country,[69] the Academy's cry for energy efficiency was perfectly in tune with the political and economic climate of the time. Skyrocketing fuel costs and new calls for energy independence brought on by a six-month OPEC oil embargo (October 1973 to March 1974) gave compelling relevance to the Academy's call for new ways of thinking about energy use. Unfortunately, the concurrent drop in the U.S. stock market made raising the necessary funds for the building more challenging than anyone had anticipated. To further frustrate the Academy's plans, the city of Philadelphia, which had already disappointed the Academy by its failure to deliver on its proffered Penn's Landing property, now refused to release the land needed for the proposed Schuylkill site without imposing significant rental fees and other layers of control. As negotiations with the city stalled, the Academy became increasingly stressed trying to operate its two downtown facilities (not to mention its satellite operations at the Stroud and Benedict laboratories).

With the financial situation growing worse, the president, who had once joked about his chair becoming a hot seat, discovered the truth in his humor. His ambitious plans for the Academy thwarted by obstacles he had failed to foresee and was unable to overcome, William Marvel resigned the presidency a little more than two years after he had arrived.[70] His departure signaled the end of the dream of a new museum.

On June 1, 1975, the *Philadelphia Inquirer* ran a story on its front page announcing what everyone at the Academy already knew: "The Whole Earth Museum is Dead":

> Less than three years after it bought the building, the Academy of Natural Sciences has decided to sell its research center on Fairmount Avenue at 25th Street. But there is more to the decision than just a sale. It also marks the abandonment of a grandiose set of plans that included the building of a new $7-million "whole earth museum" next to the Museum of Art and the development of an expanded research center in the building that is being sold.[71]

The paper went on to explain that the Academy's board had decided to "rebuild and expand its cramped, dingy, 99-year-old home on Logan Square," though the decision to do so was far from unanimous. Ruth Patrick, then chairing the Academy's board of trustees, reflected the tenuous nature of the decision in her response to a reporter's question about the institution's "consolidation" at 19th Street and the Parkway. "It's difficult to be sure of anything," she stated, "but that's the plan at the moment and we plan to go ahead with it."[72]

Having laboriously moved millions of plants, birds, mammals, and insects up the Parkway in 1973, the staff was reluctant to reverse the process and leave its spacious new quarters at 25th Street. A rift between the board, the administration, and the professional staff over the real estate debacle prompted the first-ever petition from the scientific staff. On January 30, 1976, seventeen of the Academy's leading scientists and other key personnel addressed the board of trustees and Milton Wahl, the managing director and interim CEO, in a formal document expressing their grave concern that "the wellbeing and future growth of our institution will be adversely affected if the decision to dispose of the Fairmount Building and return to the Nineteenth Street location is implemented."[73] The petition was an embarrassing vote of no confidence in the administration on the way in which it had reached its decision.

In the series of meetings that followed, the staff explained its conviction that there was not sufficient space to continue its work in the Logan Circle building. In response, the board agreed to raise the funds necessary to provide better facilities in an eight-story research wing next to the Academy's existing Logan Circle building.[74]

Although few staff members agreed that the sale of the Fairmount building and a return to 19th Street and the Parkway was desirable ("It seems to me that we might just be trading a white elephant for a white dinosaur," commented Patrick's colleague Charles Reimer),[75] financial realities left no choice. And so, still filled with many of the Academy's collections and administrative offices, the spacious but decaying Art Deco building at 25th and Fairmount was put on the market.

Into this cauldron of discontent stepped the Academy's twenty-third president, with a mandate to restore the institution's flagging finances and morale. Thomas Peter Bennett (b. 1937), a biochemist with a Ph.D. from Rockefeller University, had had extensive teaching and administrative experience at Harvard and Florida State Universities. His appreciation of the scientific research conducted by the Academy and his experience in dealing with restive faculty enabled him to calm the turbulent waters. He moved quickly to heal the rifts between departments

16.13 A publicity photo for the public launch of the Academy's new research building fund drive shows (left to right) George E. Bartoll III, chairman of the Fund for the Future campaign; Minturn T. Wright, chairman of the board of trustees; Ruth Patrick, honorary chairman of the board; and Thomas Peter Bennett, president of the Academy. ANSP Archives coll. 49.

and divisions by promoting what he called the "one Academy concept." This was given manifest form in a series of exhibits and educational programs titled "Festival of Fishes" (April–June 1977), which the *Philadelphia Inquirer* described as generating "an atmosphere of bustle and enthusiasm at the Academy . . . that is a marked contrast to the pessimism that . . . prevailed during a difficult period in the early 70s."[76] Bennett also took charge of the fund-raising for and construction of a new eight-story, 50,400-square-foot research building, as well as a massive renovation of the Academy's deteriorated 19th Street facility.[77]

What he was not as experienced at handling were the complicated political intrigues that would accompany the sale of the Fidelity Mutual building. While city officials had indicated that, if all else failed, Philadelphia itself might buy the building, these discussions appeared to be going nowhere. With the expense of running two buildings draining the Academy's coffers, and no other buyers in sight, in 1978 the board decided to speed the process of sale by offering the burdensome building at public auction. An interested party soon emerged: the Reverend Sun Myung Moon (b. 1920), the founder and leader of the worldwide Unification Church, considered the Academy's building ideally suited to be the headquarters for his organization.

The neighbors, alerted to Moon's interest by the news media, were horrified by the idea of living near an organization whose young converts were well known for their noisy bell-ringing and chanting and for their relentless proselytizing. When residents voiced their objections to the press, their elected officials, and the office of the mayor, a political scramble ensued.

Mayor Frank Rizzo and the city's managing director, Hillel Levinson, responded to the neighbors' pressure and ultimately obtained city council approval to preempt the church purchase by agreeing to buy the building from the Academy. "We had an emergency board meeting the morning the auction was scheduled to take place," recalled Bennett. "With documents in hand from City Hall confirming the city's desire to buy the building at a price we had discussed long before Reverend Moon arrived on the scene, we felt we had no choice but to sell to the city."[78] So, only hours before the auction, the building was withdrawn from sale.

Reverend Moon and members of the Unification Church were outraged, claiming the Academy had no right to make a "secret deal" with the city.[79] The Academy and the auctioneer, Louis Traiman Auction Company, with advice from their own and the city's legal counsel, felt otherwise. The Philadelphia press covered the controversy for several days.[80]

It took many months for the city to pay the Academy for the building, far longer than it took the Academy to move its millions of specimens back to new places in the old building in a logistical tour de force dubbed "Operation Shoehorn" by the staff and supervising architect.[81]

For the next several years, Peter Bennett oversaw the completion of the new research wing, the construction of a new auditorium, and the renovation of a badly deteriorated museum building whose cornerstone had been laid by W. S. W. Ruschenberger more than 106 years before.[82]

The $12 million project was made possible by money realized from the sale of the Fairmount building ($1 million, $700,000 less than the original purchase price), a federal Economic Development Administration grant obtained through the city of Philadelphia ($4.5 million), corporate gifts, foundation support (including a grant from the Pew Memorial Trust for a new auditorium), and contributions from the Academy's members.[83] The new research building was officially opened in April 1979 with a festive party at which Missouri Botanical Garden director Peter Raven spoke about the importance of the Academy's many contributions to systematics and environmental research.

Despite periodic calls to reduce the Academy's expenses by eliminating one or more of its functions, Bennett and board chairman Minturn Wright believed that all parts of the Academy were necessary and that striking a balance in the Academy's three-part mission—research, exhibition, and education—was critical in attracting the public support needed to maintain and

16.14 A variety of summer programs, internships, and other educational offerings for underserved students helped reach new audiences in the 1970s and 1980s. The most successful and long-running of these, Women in Natural Science (WINS), was—and continues to be—a free, after-school and summer science enrichment program for girls. Thanks to the mentoring they receive at the Academy, many WINS students have gone on to college and successful careers in science. ANSP Archives coll. 2010-222.

expand the institution. Under Bennett's direction, new financial management systems were put in place, the divisions of Systematics and Evolutionary Biology and Environmental Research engaged in new levels of activity, and the public side of the museum began to offer a robust new menu of classes and public lectures.

One of the successful activities that grew from this period of renewal and outreach was the George Washington Carver Science Fair, which attracted submissions from the best and brightest science students, grades four through twelve, from private, public, parochial, and charter schools in Philadelphia County. Jointly sponsored by the Academy, the School District of Philadelphia, the Archdiocese of Philadelphia, and Temple University, the fair has stimulated the participation of more than thirty-four thousand students since its founding in 1979. Many participants have gone on to regional and national competitions and to professional careers in science.

In 1982, the Academy reinforced its decades-long commitment to youth education with a new focus on underserved girls (figure 16.14). The program, Women in Natural Science (WINS), was established with the encouragement of Carole Chew Williams, an Academy trustee and science curriculum specialist with the School District of Philadelphia, and developed by Russell Daws, the Academy's coordinator of education. Its objective was to provide a free after-school and summer science enrichment program for girls from households with financial challenges. The still-flourishing program places special emphasis on families headed by single parents, grandparents, nonparent guardians, and foster parents. Each year, twenty-five to thirty students are chosen to participate.[84] As its founders had hoped, many WINS students have gone on to successful science-based careers.

The Explorers Lecture Series, which flourished in the late 1970s and throughout the 1980s, focused on a broader and mostly adult audience. It attracted a roster of distinguished speakers to the Academy, including, among others, author and naturalist Gerald Durrell, gorilla expert Dian Fossey, anthropologists Don Johanson and Mary Leakey, Smithsonian secretary (and onetime Academy collector) S. Dillon Ripley, explorer and Charles Darwin descendant Quentin Keynes, marine biologists Sylvia Earl and Eugene Clark, *Kon Tiki* navigator Thor Heyerdahl, writers George Plimpton and Peter Matthiessen, paleontologist Stephen Jay Gould, chimpanzee researcher Jane Goodall, whale communications pioneer Roger Payne, and bird-watching guru Roger Tory Peterson. Each talk offered an associated reception and dinner. These events, combined with the Women's Committee galas, annual Wildlife Art Exposition, and periodic medal presentations, helped to reestablish the Academy's profile on Philadelphia's social scene and added luster to membership on its Women's Committee and board. This, in turn, enabled the Academy to attract more contributions than it had at any time in its history.

With the creation of a new Hall of Changing Exhibits in 1981, the Academy for the first time was able to create and host large temporary exhibitions, adding a critical new dimension to the otherwise static museum.[85] Ironically, the inaugural exhibit harkened back to some of the displays created in the museum's first Hall of Earth History during the 1937 Symposium on Early Man. The exhibit *Ice Age Art* came from the American Museum of Natural History. Its opening was heralded by a preview reception and dinner at the Plaza Hotel sponsored by the Women's Committee (figure 16.15).[86]

The Academy's next big exhibit, *Treasures of the Academy,* highlighted specimens and stories from the Academy's extensive collections. Designed and beautifully installed by the museum's gifted exhibits director, Norman Ikeda, the exhibit demonstrated a new level of sophistication for its in-house creations and

16.15 Additions to the Academy's facilities created a new commons room and a changing exhibits gallery for the museum in 1981. Seen here at the opening of *Ice Age Art* are board chair Howard P. Brokaw, Women's Committee chair Patricia Tyson, and President T. Peter Bennett. ANSP Annual Report.

reinforced the public's understanding of the Academy as a scientific institution with a distinguished past and a promising future.

A second in-house exhibit opened in the fall of 1982 and solidified the museum's reputation for creating its own beautiful and content-rich exhibits. *A Celebration of Birds* focused on the life and extraordinary art of the American bird painter Louis Agassiz Fuertes (1874–1927). Drawing on an important collection of original paintings recently acquired from the U.S. Fish and Wildlife Service,[87] and funded in part by a grant from the National Endowment for the Arts (a rarity for a natural history museum), the exhibition was the first national traveling exhibition ever organized by the Academy. Its three-year tour eventually took it to the American Museum of Natural History, the Field Museum of Natural History, the Los Angeles Museum of Natural History, and three other venues in Texas, Georgia, and Connecticut.[88]

The final phase of the Academy's resurrection from the period when it was planning to abandon the Logan Circle building concerned the quasi-permanent displays in the museum. An informal public opinion survey in June 1977 identified the dioramas, the dinosaur, and the mummy (an incongruous hold-over from the abandoned Department of Anthropology) as the three most popular attractions in the museum. While some visitors found the dioramas "dusty," most found them "exciting," "fantastic," "realistic," "impressive," and so lifelike that "they looked like they would move the next second."[89] Comments about the museum's only dinosaur were unanimous in their praise and in their desire for more. "Please bring back—even expand the dinosaur exhibits," commented a teacher. "All kids love them and they'll get into other exhibits through this interest." "Our teachers were disappointed to find the first floor dinosaur and fossil exhibits [reduced]," commented another. "Your museum was [and is] the only place [in Philadelphia] to view these real specimens."[90]

The visitors were correct in noticing a reduction in dinosaur-related exhibits, for the old Hall of Earth History had been dramatically altered in 1975 in anticipation of a huge influx of out-of-town visitors with an interest in American history during the nation's bicentennial. The Academy had made a misguided decision to replace its popular paleontological displays with something timely for the bicentennial and had hired the New York

16.16 Five years in the making, *Discovering Dinosaurs* opened to great acclaim in 1986. New dinosaur displays, including this cast of *Tyrannosaurus rex*, helped attendance soar. ANSP Archives coll. 2010-301.

design firm of Ramirez and Woods to create a modular exhibit on the Academy's history in the hall. *Evolution and Revolution in the Natural Sciences* included a few scientific specimens and artifacts, but it was dominated by large backlit transparencies of natural history subjects. Designed and written for adults, for the modest cost of $100,000, the display proved disappointing for younger visitors and failed to attract the projected attendance.[91] Malacology curator George Davis declared the exhibit "a total disaster" and called for its removal "as soon as possible."[92] Other curators were less outspoken, but all agreed that the exhibit failed to capture the significance of the Academy and its work.[93] Some minor alterations were made to respond to the criticism, but the exhibit remained a disappointing and unwelcome addition to the museum's offerings for a number of years.

With the Academy's financial resources already overstressed and with higher priorities elsewhere, a quick replacement of *Evolution and Revolution* was not possible. But the need for a different kind of exhibit was clear, so even as a series of small traveling and temporary in-house exhibits appeared in other parts of the museum in the late 1970s and early 1980s, planning for a complete overhaul of Dinosaur Hall began.

With a $150,000 planning grant from the National Endowment for the Humanities, an ambitious exhibit titled *Discovering Dinosaurs* became the focus of intense activity at the Academy beginning in 1981.[94] Dennis Wint, director of the museum, and Hollister Knowlton headed up the planning team that included paleontologists, historians, educators, and display experts from across the country.[95] The intent of the exhibit was to show the excitement of scientific discovery by stressing recently developed and still-open questions about dinosaur anatomy, posture, and behavior, rather than to replicate the more traditional, didactic displays found in other museums (figure 16.16).

The humanities component of the exhibit focused on how dinosaurs have captured the human imagination ever since Benjamin Waterhouse Hawkins put the first articulated dinosaur skeleton of *Hadrosaurus foulkii* on display at the Academy in 1868. Dinosaur cartoons, children's toys, and scale models of the life-size dinosaurs displayed at the New York World's Fair were exhibited along with more purely scientific subjects. These included a life-size cast of the perpetually intriguing *Tyrannosaurus rex*, the late Cretaceous ceratopsian *Chasomosaurus belli*, and a new dinosaur species (*Avaceratops lammersi*) discovered by University of Pennsylvania paleontologist Peter Dodson on a joint Academy-Penn dig in Montana in 1981.[96] Also featured in the hall were two skeleton casts of *Deinonychus antirrhopus*, the first examples of the species ever shown in a museum. Depicted in a large commissioned sculpture that still stands in front of the museum,[97] these predatory dinosaurs became even more famous seven years after their first appearance at the Academy when they were enlarged and animated as Velociraptors in Steven Spielberg's 1993 film version of Michael Crichton's best-selling book *Jurassic Park* (figure 16.17). Interviewed about the movie and his interest in bringing dinosaurs back to life, Spielberg acknowledged developing his fondness for dinosaurs through visits to the Academy from his childhood home in nearby New Jersey.[98] It was hoped that the new $2.5 million dinosaur exhibition might have a similarly inspiring effect on future generations of museum visitors.[99]

When *Discovering Dinosaurs* opened to the public in January 1986, Harvard's paleontological guru Stephen Jay Gould declared it the best dinosaur exhibit in the country, and the public flocked to see it. "We tried to provide something for everyone," recalls Knowlton, "with interpretive copy that would appeal to people of all ages, and I think we succeeded."[100] Museum attendance

16.17 Artist Kent Ullberg (left) and paleontologist John Ostrom stand before Ullberg's bigger-than-life sculpture of *Deinonychus antirrhopus* (a dinosaur discovered by Ostrom), created to mark the Academy's 175th birthday in 1987. ANSP Archives coll. 2010-222.

climbed from the normal 130,000–150,000 visitors per year to a record-breaking 326,664 in 1986, and memberships soared.[101]

By contrast, a second permanent exhibit on earth history that opened five years later was far less successful. The $1.2 million installation *What on Earth* featured a large globe and three-dimensional road-cut, a working laboratory, and a small display of Academy minerals. Despite a simulated earthquake component and the appealing visual effect of the exhibit, it never captured the public's attention the way *Discovering Dinosaurs* had. It was removed without fanfare in 1994 to make way for an exhibit of living butterflies.[102]

Following the extended periods of uncertainty that plagued the Academy during the 1960s and early 1970s, the late 1970s and early 1980s were years of great optimism and positive activity. Renovation of the Logan Circle building, the construction of the new research wing and auditorium, the opening of Outside-In, the creation of the institution's first major traveling exhibit, and the installation of *Discovering Dinosaurs* combined to make the Peter Bennett era among the most dynamic periods of the Academy's growth. In 1986, when Bennett announced his intention to leave the Academy to head up the University of Florida's state museum, an editorial in the *Philadelphia Inquirer* noted that "When he moves on to his new post in Florida, [Bennett] will leave behind a healthy academy with [a] more-than-doubled endowment and a reinvigorated national reputation."[103]

This period of renewal and growth was followed by another period of expansion in the late 1980s and early 1990s, during which Keith S. Thomson, a former director of the Peabody Museum and dean of the Graduate School of Arts and Sciences at Yale, oversaw a successful $15 million capital campaign and two further additions to the Academy's infrastructure: a much needed $7.2 million library wing (which opened in 1992) and a $4.8 million aquatic research facility in Patuxent, Maryland (which opened in 1994) (figures 15.14 and 16.18).[104] Sadly, just as the Academy appeared to be moving toward financial security and increased scientific capability in its third century, a financial scandal rocked the institution.

In the early 1990s, board chairman Marvin Heaps introduced the Academy to John G. Bennett Jr., a secretive but seemingly generous philanthropist. Bennett (no relation to the former president) had established an organization called the Foundation for New Era Philanthropy in 1989. Wealthy donors were invited to support an array of Christian educational institutions by placing their money with the foundation for varying lengths of time, during which the funds would be matched by other anonymous donors. Once enough money had been collected, the foundation would direct double the amount of the original donation to the charities of the donors' choice. This mysterious and miraculous multiplying effect generally took from three to six months to achieve. No one had ever heard of anything quite like it.

Heaps, a personal friend of John Bennett's, with a shared interest in supporting fundamentalist Christian causes, saw the charity as a potential boon to Academy fund-raising. In 1993, he convinced John Bennett to allow the Academy to invest in New Era, in the hope that it too could double its money. Since the city had just cut its annual $250,000 appropriation to the Academy, Heaps reasoned, the institution needed to take advantage

16.18 The Academy's old library stack building was demolished to make way for a new $7.2 million library and archive wing. Books were stored off-site until the new facility opened in 1992. ANSP Archives coll. 2010-222.

of New Era's secret donor base and leveraging capabilities in order to meet its budget shortfall. The Academy's chief financial officer, unable to find appropriate financial records to support the foundation's claims to legitimacy, objected to putting Academy funds at risk in this way, but he was overruled.[105] In early 1993, $250,000 was turned over to John Bennett, and within six months, $500,000 was returned. It all seemed too good to be true. It was.

Over the next three and a half years, the Academy would give a total of $2.7 million to the Foundation for New Era Philanthropy.[106] Not all of it would come back. When, on a tip from someone investigating suspicious banking practices by New Era, Prudential Securities Corporation called in a $44 million loan it had made to the foundation, John Bennett was unable to pay. On 15 May 1995, in a front-page story, the *Wall Street Journal* exposed John Bennett's Ponzi scheme to the world. It was the largest financial scandal in the history of American charities.[107]

The Academy was not alone in its belief that the Foundation for New Era Philanthropy would help it build a bright future. With the notable exception of the Philadelphia Museum of Art, almost every other charitable institution in Philadelphia had invested in and lost at least some money to New Era, but the Academy was the first and, proportionally, the largest investor in the city, and, as such, it was the focus of a great deal of bad publicity.[108] Among the 1,100 individuals and organizations in Philadelphia and throughout the country to become ensnared in Bennett's nefarious scheme were the Boy Scouts of America (whose board Heaps also chaired), the Franklin Institute, the Philadelphia Orchestra, Haverford College, the University of Pennsylvania, Princeton University, Harvard University, the Nature Conservancy, the Environmental Defense Fund, One to One Partnership Inc., Planned Parenthood, the United Way, Stanford University Medical School, and Yale Law School.

From 1989 until its collapse in 1995, the Foundation for New Era Philanthropy collected between $350 and $500 million from well-meaning individuals and institutions seeking to increase their financial security. Even such business-savvy investors as Laurence Rockefeller, John C. Whitehead (a former cochairman of Goldman Sachs & Co.), Sir John Templeton, and former U.S. Treasury secretary William Simon were taken in by the scam.[109]

When the case was brought to court in the spring of 1997, John Bennett faced eighty-two federal counts of wire, mail, and bank fraud, filing false tax returns, and money laundering.[110] At first, he tried to defend his actions by claiming he had been possessed by "unchecked religious fervor," but the judge would not allow this defense. In the end, Bennett pleaded no contest to all the charges. Though federal sentencing guidelines suggested twenty-two to twenty-seven years of jail time for his crime, the sympathetic judge gave him only twelve.

Because the Academy had benefited from some of John Bennett's early money doubling, it was required by the court to turn back money it had received from New Era. Far more damaging than this financial loss, however, was the shaken confidence of the donor community, membership, staff, and board, from most of whom the New Era transactions had been intentionally concealed. In an effort to explain its involvement and control the damage, the Academy sent a letter to its members explaining that its actions had not been as irresponsible as they seemed, that "until May 1995, every gift [to New Era] had been doubled in six months," that "no Academy membership fees were turned over to the Fund," and that "no gifts were submitted without the donor's written permission."[111] In the end, an honest admission of failure seemed the best defense. "We took what we thought to be prudent measures and precautions," said the letter. "We looked—carefully—before we leapt. But we were conned just the same."[112]

In the decade following the New Era crisis, the Academy suffered through a turbulent period of unusually rapid change in administrative leadership. From 1995 to 2010, the institution saw eight different presidents and acting presidents (see the list at the back of the book), each of whom struggled to balance the budget while continuing to provide educational services to the community and remaining globally competitive in scientific research. As one strategic plan replaced another with each new administration, the staff, members, and even the board grew confused about what direction the institution should take.

In its 2007 annual report, titled "Celebrating Our Past; Embracing Our Future," the Academy announced that its board of trustees had unanimously readopted its original mission statement, which, "crafted by our founders in 1812 and set forth by the Pennsylvania Legislature in 1817, is 'the encouragement and cultivation of the sciences.' Short, simple, and sincere. By recommitting ourselves to it," concluded the report, "we acknowledge that much of our history is just as valid, relevant, and exciting as it ever was."[113]

The board's recommitment to the institution's core values was accompanied by the launch of a comprehensive master planning process in which, for the first time since the 1876 building

was designed, every aspect of its much-altered infrastructure was analyzed. President William Brown described the critical review undertaken by the architectural firm of SaylorGregg as a sort of archaeological dig through which the history of the institution itself could be read.[114]

In the spring of 2009, in a beautifully presented, multivolume report filled with useful statistics and graphics, the planning team set out an inspiring vision of what the Academy could become (figure 16.19). The projected price tag for this vision of the future was a sobering $200 million. Realizing such an ambitious architectural plan, acknowledged Brown, would require "a multi-phased approach, as well as a multi-faceted fund-raising strategy." Writing at the depth of the greatest economic recession since the Great Depression, Brown continued:

> Although the economic climate of the day seems less than favorable for such a momentous undertaking, the Academy has a strong tradition of being able to accomplish great things in uncertain times. We were founded in the war-torn year of 1812, constructed our first building in the wake of the Bank Panic of 1837, and moved to our current building as the country was still recovering from the Civil War.[115]

"As the oldest natural history museum in America, the Academy was once the finest natural history museum in America," concluded Brown. "The work outlined in this master plan will restore our ability to make that claim."[116] Three months later, the plan's spiritual architect and principal cheerleader announced that he would be taking another job in Massachusetts. "We are disappointed by Bill Brown's decision to leave the Academy," commented board chairman James Macaleer, "but we are determined to keep the institution moving forward. Just how much of what is in the master plan we will be able to realize, and when, remains to be seen."[117]

RMP

16.19 Creative ideas for a glass atrium and dramatic new exhibit space were offered in the Academy's 2009 master plan by SaylorGregg Architects. ANSP/SaylorGregg Architects.

Notes

1 Christopher Hale, *Himmler's Crusade: The Nazi Expedition to Find the Origins of the Aryan Race* (Hoboken, NJ: John Wiley & Sons, 2003), 324.

2 Ibid., 58.

3 From an interview with Ernst Schäfer by Robert Peck, December 1986.

4 Hale, *Himmler's Crusade*, 336, and ibid. Schäfer's surprise and hurt over this was still evident forty years after the event.

5 During his hearing, Schäfer's attorney presented affidavits from more than forty witnesses who stated that Schäfer had cooperated with resistance groups and assisted Jews and Polish scientists persecuted by the Nazis. On the strength of this evidence, the denazification tribunal cleared him of all charges in June 1949, awarding him an "exonerated" classification. Heather Pringle, *The Master Plan: Himmler's Scholars and the Holocaust* (New York: Hyperion, 2006), 307–8.

6 This story and quotation are third hand, recounted to Pilsbry and then to Ruth Patrick, who referenced it in a talk she gave on the occasion of the unveiling of her portrait at the Academy, 9 November 1978; see Ruth Patrick, "The Academy at the Forefront of Science," unpublished lecture notes, coll. 262, file 7, 7.

7 Letter to Pilsbry, 21 December 1925, coll. 459. Hirohito became emperor of Japan on 25 December 1926. In October 1949, a letter from the Imperial Household reached Pilsbry, presenting the American with a copy of the emperor's book *Opisthobranchia of Sagami Bay*. Letter to H. A. Pilsbry from Takanobu Mitani, Grand Chamberlain to H.M. The Emperor of Japan, 28 October 1949, coll. 459.

8 While Crampton was more closely associated with Columbia University and the American Museum of Natural History in New York, much of his personal collection of shells and his personal papers are now at the Academy. See ANSP coll. 755.

9 Don Fairbairn, "Philadelphia Scientist Aids Army War on Mosquitoes," newspaper clipping (n.d.), ANSP biography file.

10 H. R. Roberts, J. P. Moore, and P. P. Calvers, "Report to the Council of the Academy of Natural Sciences," 2 December 1947, Managing Directors Files, coll. 2009-034, box 4 of 4, file 19.

11 Ibid.

12 "No Shelter for Dinosaur," *Philadelphia Inquirer*, 26 October 1962, ANSP coll. 417.

13 National Science Foundation founding document quoted on the NSF website, http://www.nsf.gov.

14 Admission fees in 1953 were set at fifty cents for adults, twenty-six cents for children, twenty cents per child for non-Philadelphia school groups, Philadelphia school groups and members free. The Academy was the last of the city's major museums to charge admission fees.

15 Many, but not all, of the "snake ladies" and "roadrunners" were also members of the Academy's Women's Committee.

16 Beginning in 1956, the live animals used in these programs were kept in a cinder-block building with indoor and outdoor cages built in a courtyard behind the museum. Their variety grew from small, easily handled animals to large, showy ones, including a miniature donkey, a wallaby, a red wolf, a coyote, and even a puma (which lasted less than a year before becoming too difficult to handle and moving on to a zoo). Though operated with the help of many volunteers, the live animal unit was headed up by a series of paid professionals from 1956 onward. In the late 1960s, an indoor facility was opened in the Academy's basement near the fish department. Among the early directors were Forrest Strawbridge, Harold Torboss, Pat Brodey, Jerry Freilich, and John Robinson.

17 Kenneth W. Prescott, quoted in Katherine Dunlap, "Junior League Planning Natural Science Project," *Philadelphia Inquirer* (Society News and Fashions Section) 4 October 1959, ANSP coll. 417. According to the extensive press coverage of the time, more than 120 members of the Junior League were directly involved with creating the Junior Museum.

18 During the first several years of its existence, under the direction of Portia Sperr and an independent board of trustees, the Please Touch Museum occupied two rooms at the back of Dinosaur Hall. The museum was so successful that it soon outgrew the available space and moved into its own facilities in 1978. At that time, the Academy created Outside-In to serve the same young audience (one- to seven-year-olds). It started in the old Please Touch spaces on the first floor and then moved to the third floor of the museum, taking over much of what had been the Audubon Hall of Birds (created in 1942).

19 With a few exceptions—most notably the Junior Explorer program established in 1938—the Academy's educational offerings until the late 1950s were almost entirely aimed at adults.

20 The Junior Curators program had a precedent in the Junior Explorers program, established in 1938 by the first director of education, W. Stephen Thomas. See Anon. (but probably W. Stephen Thomas), *Learn to Be an Explorer*, ANSP, 1938.

21 Jerry Freilich to Robert Peck, September 2010, personal correspondence.

22 The Ludwick Institute continues to support the Academy's education programs but not the lecture series, which was sponsored by a number of different sources through the 1970s, 1980s, and 1990s.

23 The three longest-serving teacher naturalists in this city-funded program were Lorene McClellan, Jean Doris, and Frank Marinaro. For more on Doris, see *Academy News* 16, no. 2 (Fall 1993–Winter 1994). The program was cancelled by the School District of Philadelphia for financial reasons in the 1990s.

24 The program was originally titled "Expeditions in Philadelphia." See W. Stephen Thomas, "Expeditions in Philadelphia," *Journal of Adult Education*, June 1939.

25 Lorene McClellan McDonald to Stephen Harty, 15 April 1977, coll. 2010.160, folder 1. McClellan served as a teacher-naturalist at the Academy from 1935 to 1973.

26 This information resulted from a survey of participants, summarized in Thomas, "Expeditions in Philadelphia."

27 Established in 1975, the Rare Bird Branch ultimately consisted of several hundred loyal field trippers and quite a few honorary members invited by Harty to join.

28 According to his own résumé, Harty was responsible for the museum's Great Auk diorama (with background by Jonathan Fairbanks Jr.), desert of Borkou, black bear, colobus monkey, bobcat, tiger, polar bear, and tropical insect displays. Coll. 2010-129.

29 The Academy's exhibit directors have included Harold Green (1921–64), Gilbert E. Merrill (1964–65), Winslow M. Shaughnessy (1967–70), Beverly Mowbray (1970–75), Christopher Ray (1975–79), Norman Ikeda (1979–84), Lawrence P. Hutchinson (1985–87), Raylene Decatur (1988–90), Mark Driscoll (1990–96), J. Willard Whitson (1998–2006), and Barbara Ceiga (2006–2011).

30 For Harty's visual and written review of his own accomplishments at the Academy, see coll. 2010-129.

31 Fairbanks was paid $3,700 for the job based on the square foot, with a bonus for finishing on time. Personal correspondence between Fairbanks and Peck, 20 December 2010.

32 Excavated in 1927 by the famous fossil-collecting Sternberg family, the *Corythosaurus* was purchased by the Carnegie Museum in Pittsburgh, which in turn traded it to the Denver Museum of Natural History. The Academy purchased it from the Denver Museum (in unmounted form) for $2,000 in 1954. Because the Academy had no vertebrate paleontologist on staff at the time, the dinosaur was mounted by professional preparators from the American Museum of Natural History in New York (George Whitaker and Walter Sorenson). Also helping with the 1957 mount were Academy exhibit preparators Charles Leibrandt and Pat Cole.

33 The dinosaur's nickname was selected through a citywide naming contest. At various times the *Corythosaur* was also called "Cory." Two later Academy mascots (created with fabric costumes and worn by a series of long-suffering volunteers) were dubbed "Philasaurus" and "Logan C. Dinosaur," respectively.

34 "No shelter for Dinosaur," see note 12.

35 Fairbanks offered to repaint the mural, altering the dinosaurs' posture, but his offer was declined. Personal correspondence between Fairbanks and Peck, 20 December 2010.

36 It was cut into several panels, with each part representing a stand-alone scene. These were transported to the Bruce Museum of Arts and Science of Greenwich for inclusion in a traveling exhibit, *Designing Dinosaurs: Solving Prehistoric Puzzles* (September 1998–January 1999).

37 Personal communication from Richard Ellis, 16 July 2010. Following his time at the Academy, Ellis went on to a paid position in the Exhibits Department at the American Museum of Natural History and then a distinguished career as a natural history writer, lecturer, and wildlife artist, specializing in subjects relating to marine life.

38 For photographs of these displays, see coll. 2010-129, box 1.

39 Personal communication from Richard Ellis to Robert Peck, December 13, 2010.

40 Minutes of the Academy Council, 13 September 1960, coll. 502. The following spring, it was reported that the city might give the Academy a site "between Chestnut Street and Market Street, with ample room for growth and development of facilities," Minutes of the Academy Council, 4 April 1961, coll. 502.

41 Letter to Stephen S. Gardner from John W. Bodine, 24 June 1969, coll. 2010-051, box 122. The conservation museum idea was still being discussed at Academy meetings in the fall of 1977. See Minutes of the Administrative Council, 4 November 1977, 3, coll. 2010-213, box 1, file 5.

42 Minutes of the Women's Committee, 10 June 1965, coll. 786, file 12.

43 Letter to Stephen Gardner from John Bodine, 24 June 1969, coll. 2010-051, box 122. The Academy's president, Albert Linton, had declared publicly as early as 1962 that the Academy "envisions a new home on Philadelphia's Nautical Mile, the proposed Delaware River waterfront development, for the Academy has again outgrown its space limitations." M. Albert Linton, sesquicentennial address, reprinted in *The Academy of Natural Sciences of Philadelphia: 150 Years of Distinguished Service* (New York: Newcomen Society in North America, 1962), 25.

44 The designs were created by the Academy's architect, Robert Geddes. See coll. 2010-227.

45 "An Alternative Plan for Rehousing and Reorganizing Academy Departments," unsigned but possibly by Rad Roberts, 14 January 1969, coll. 2010-051, box 122.

46 Letter from David Goddard to Lane Taylor and George Clark, 24 August 1970, coll. 2010-051, box 122, folder 1.

47 Previous offers had been made in 1875, 1889, and 1890. See Chapter 7.

48 Letter from David Goddard to Lane Taylor and George Clark, 24 August 1970, coll. 2010-051, box 122, folder 1.

49 Letter from David R. Goddard to Lane Taylor, 16 September 1970, coll. 2010-051, folder 1.

50 Minutes of the Site Committee, 22 June 1971, coll. 2010-051, box 122. The minutes indicate that the offer had been made some time (possibly years) earlier but was once again on the table.

51 Minutes of the Site Committee, 21 May 1971, coll. 2010-051, box 122. In a reversal of earlier thinking, the firm also was asked to explore the expense of renovating the original 1876 building should the Academy decide to remain on Logan Circle.

52 Baltimore's National Aquarium was not established until 1981.

53 William W. Marvel, *An Institution for All Seasons (Inaugural Address)* (Philadelphia: Academy of Natural Sciences, 1972), 9–12, coll. 2010-218.

54 Ibid., 1. The reference here is to Philadelphia's Academy of Music, long known as "the grand old lady of Locust Street."

55 Ibid., 1.

56 Memorandum from William Marvel to Messrs [Henry] Mirick, [Lane] Taylor, and [Elias] Wolf, 20 January 1972, coll. 2010-051, box 122, folder 5.

57 See "Highly Confidential" memorandum from William W. Marvel to board of trustees, 24 August 1972, coll. 2010-051, box 48, folder 7.

58 The Fidelity Mutual Life Insurance Company building was designed by the architectural firm Zantzinger, Borie and Medary. It was built in 1926. For a review of all the other proposed sites, see "Highly Confidential" memorandum.

59 For a more detailed argument in favor of the purchase, see "Highly Confidential" memorandum.

60 Press release, "Academy of Natural Sciences Plans for Its New Home," 10 October 1972, coll. 2010-051, box 122, folder 1.

61 Ibid.

62 Ibid.

63 This plan was built on the assumption that the Logan Circle building could be sold for "$3,000,000 or more"; see "Highly Confidential" memorandum. Such a sale (and zoning issues relating to commercial development of the site) was not viewed favorably by the city administration.

64 "A New Museum for the Academy of Natural Sciences," 14 November 1973, coll. 2010-051, box 122.

65 See "Highly Confidential" memorandum.

66 For an artist's rendering of the proposed building, see coll. 2010-227.

67 Philip Steadman, "The New Academy Museum: An Exhibit in Energy Conservation," 10 September 1973, coll. 2010-051, Box 122.

68 Ibid.

69 Steadman's paper on the innovative aspects of the Academy's proposed museum cited several examples of domestic and institutional applications of the systems discussed, but it concluded that the Academy's proposed building "is the only structure of size, significance, and public prominence" that would make use of all these energy-conserving and environmentally conscious systems. Steadman, "The New Academy Museum," 8.

70 According to the *Philadelphia Inquirer*, "despite having no immediate job prospects, he [Marvel] felt compelled to resign" in March 1974. Larry Eichel, "The Whole Earth Museum Is Dead," *Philadelphia Inquirer*, 1 June 1975, 1. Coll. 2010-051, box 122.

71 Ibid., 1.

72 Ruth Patrick as quoted in the *Philadelphia Inquirer*, 1 June 1975. According to the *Inquirer*, the decision to sell the building was made at the Academy's board meeting on 29 April 1975. Patrick later gave the meeting date at 14 April. In any case, according to her memo to the board of 7 February 1976, there were two dissenting votes. Unhappiness with the state of affairs had prompted the resignation of Academy president William Marvel in March 1974 and the resignation of several Academy trustees. See coll. 2010-051, box 122.

73 Memorandum to Milton H. Wahl from Academy staff, 2 February 1976, coll. 2010-051, box 122.

74 Memorandum to Frank B. Gill and Clyde E. Goulden from Milton Wahl ("my last effort to explain the basis for the Trustees decision to sell the Research Building at Fairmount Street"), 3 March 1976, coll. 2010-051, box 122.

75 Charles W. Reimer, addendum to memorandum from Academy staff to Milton H. Wahl, 2 February 1976, coll. 2010-051, box 122.

76 John Corr, "A Museum Tries to Shed Its Dusty Image," *Philadelphia Inquirer*, 12 April 1977. Coordinating the Festival of Fishes was the recently hired assistant to the director of the museum, Robert Peck.

77 Approval of the new building came in two stages: On 12 January 1976, the board approved building a four-story addition; then, in response to staff pressure, on 8 March 1976, it agreed to add two floors to the new wing, bringing the estimated cost to a total of $2,300,000. By the time other renovation expenses were added to the budget, the total cost of phase 1 of the project rose to $4,200,000. See memo to board of trustees from Milton Wahl, 16 August 1976, coll. 2010.051, series 4, box 70, file 17; and "Capital Campaign Case Statement, Phase 1," 15 December 1976, coll. 2010, series 4, box 66, folder 9. The additions and renovations of the 19th Street building were carried out by the architectural firm Mirick, Pearson, Ilvonen, and Batcheler, with Richard Henry, and later Clark Van Zant, serving as the supervising architects. For more on this project, see coll. 2010-051, box 64.

78 Personal communication with T. P. Bennett, 21 December 2010.

79 Maria Gallagher and Robert Strauss, "Auction Leaves Bidders Bitter," *Philadelphia Daily News*, 16 June 1978. See also Raymond A. Berens, "Moonies Bid to Buy Building Blocked," *Evening Bulletin*, 16 June 1978. A third angered party was the Southern Christian Leadership Conference (SCLC), whose local chapter had hoped to buy the building to create an archive. Stanley Culbreth, the head of the SCLC's local branch, threatened to picket the Academy until he learned the full story behind the aborted auction.

80 A truncated auction was held during which a dilapidated storefront at 2501 Olive Street, also owned by the Academy, was sold to the Unification Church for $42,500, well above its estimate. This further infuriated the neighbors. With the Fidelity building ultimately selling to the city for $1,075,000, almost $700,000 less than the Academy had paid for it, no one was entirely pleased with the result. To try to soothe the ruffled feathers of both the Unification Church and the Neighborhood Association, and with permission from the Academy's board, Bennett offered the Unification Church the opportunity to retract its bid on the Olive Street building, but it declined the offer. Personal communication with T. P. Bennett, 21 December 2010. See also Nessa Forman, "A New Art Annex?" *Evening Bulletin*, 30 July 1978.

81 The contents of the Fairmount Building, moved up the Parkway in November 1973, were moved back in October 1978 at a cost of $50,000. See Pat McKeown, "3 Million Dead Bugs Move," *Evening Bulletin*, 28 September 1978, and Edgar Williams, "Creeping: The Priceless Bugs Move Again," *Philadelphia Inquirer*, 28 September 1978. For the planning of this operation, see memorandum from Frank Gill to Richard Henry, 23 February 1976, coll. 2010-051, box 64, folder 4.

82 Remarks made at the laying of the cornerstone in 1872 can be found in a special publication accompanying the *Proceedings of the Academy*, vol. 29, 1872.

83 For more information on this campaign, see "Toward the Future," campaign brochure, 1979, coll. 2010-218.

84 Interview with Betsy Payne, director of WINS, November 2010.

85 The 3,500-square-foot gallery was located on the upper level of what had been the museum's lecture hall (now the live animal center) and in a newly created space above the commons.

86 The Academy's showing of the exhibit was made possible by a grant from the SmithKline Foundation.

87 The U.S. Fish and Wildlife art collection was deposited with the Academy on a twenty-five-year renewable loan in 1977. The loan was renewed in 2003. Images are viewable online. See coll. 808.

88 The exhibit, organized and curated by Robert Peck, was distributed by the Smithsonian Institution Traveling Exhibition Service (SITES). The other three venues were the Dallas Museum of Natural History, the Telfair Academy of Arts and Sciences in Savannah, Georgia, and the New Britain Museum of American Art. Peck authored a biography to accompany the exhibition. See Robert McCracken Peck, *A Celebration of Birds: The Life and Art of Louis Agassiz Fuertes* (New York: Walker & Co., 1982; London: Collins, 1983).

89 "Comments from Questionnaires," Academy Evaluation Project, Pat Brodey, June 1977, Peck files. In 1986, the Exhibits Department received outside funding to asses the condition of its aging dioramas. Although the assessment revealed that all were suffering from years of neglect, the Serengeti Plains diorama was identified as the most seriously in need of restoration, which it underwent in 1988. Exhibit specialist Keith Russell was in charge of this effort.

90 "Comments from Questionnaires."

91 The contract between the Academy and Carlos Ramirez and Albert H. Woods Inc. can be found in coll. 2010-051, box 64, folder 5.

92 George Davis, personal communication to Robert Peck, July 1976.

93 See Ruth Patrick to Milton Wahl, 28 June 1976, coll. 2010-051, box 64, file 11.

94 The preliminary planning and preparation for the NEH application began in 1981. The NEH grant was received in 1982.

95 When Dennis Wint left the Academy to head up the Cranbrook Institute of Science in 1982, Sam Gubins, CFO of the Academy, took charge of the *Discovering Dinosaurs* project.

96 For more on Dodson's discovery, see Charles Smart, "Discovering Dinosaurs in the Field," ANAP: *Academy News* 8, no. 4 (Winter 1985), 11–13.

97 The larger-than-life sculpture by Kent Ullberg was donated by Henry Stewart in honor of his wife, Ewell, the Academy's Women's Committee, and the institution's 175th anniversary in 1987.

98 Carrie Rickey, "Steven Spielberg to Be Honored Not as a Filmmaker, But as a Philanthropist," *Philadelphia Inquirer*, 5 October 2009.

99 In addition to the National Endowment for the Humanities, funding for the exhibit came from the Pew Memorial Trust ($900,000) and more than six other foundations, some thirty corporations, and more than one thousand individual donors. See "Prehistoric Past Seen in a Fresh Perspective," *PACE*, Wilmington: *Wilmington News Journal*, 23 January 1986.

100 Personal communication between Hollister Knowlton and Robert Peck, January 1986.

101 *1986 Annual Report* (Philadelphia: Academy of Natural Sciences, 1987), 10. During the two years in which *Discovering Dinosaurs* was being designed and installed (1984–86), a smaller exhibit, *Dinosaurs: An Exhibit in the Making*, helped to satisfy public demand for some sort of exhibit about dinosaurs.

102 When the new exhibit, *Butterflies: Live and in Color,* opened in February 1995, it was one of the very first live butterfly exhibits inside a museum anywhere in the United States. After a short hiatus to make way for a national traveling exhibit on Lewis and Clark in 2004–5, another live butterfly exhibit was installed in the same location in 2006.

103 *Philadelphia Inquirer*, 16 December 1985.

104 The two building projects were part of the overall campaign, which also included provisions for raising endowment. For more details on the "Nature of the Future" campaign, see campaign literature in coll. 2010-218.

105 The insightful CFO, who put his objections in writing in a memo to the president and chairman of the board, was Samuel Gubins. He had left the Academy by the time the New Era fraud was exposed.

106 Some of this came from directing new donations through New Era, but some came from the Academy's existing endowment. For the first news report exposing the scam, see "A Bankruptcy Shakes World of Charities," *Philadelphia Inquirer*, 16 May 1995.

107 "Foundation for New Era Philanthropy," FBI website, 2010. See also Sharon Walsh, "Charity's Troubles Shake Up Nonprofits," *Washington Post*, 16 May 1995, coll. 2010.004, box 23, file 12.

108 The University of Pennsylvania had invested $2.1 million.

109 The Ponzi scheme was eventually discovered by Albert Meyer, an accounting professor at Spring Arbor College in Michigan (which had received matching funds from New Era), and by the auditing firm Coopers & Lybrand, working with its client, a local religious college in Los Angeles.

110 "Foundation for New Era Philanthropy."

111 Excerpts of the Academy's letter, along with those from similar letters to donors from the United Way, Settlement Music School, and Eastern College, were published in an article about the impact of the scandal in *Main Line Life*. See Winnie Atterbury, "Non-profits Facing Loss Promise to Evaluate Internal Procedures," *Main Line Life*, unknown date, coll. 2010.004, box 23, file 12.

112 Academy letter to stakeholders as quoted in *Main Line Life*.

113 Academy of Natural Sciences annual report 2007, coll. 2010-032.

114 Academy of Natural Sciences master plan, prepared by Becker & Frondorf, SaylorGregg Architects, and Ralph Appelbaum Associates, vol. 1, 29 May 2009. The planning process, underwritten with a $250,000 grant from the Barra Foundation, was coordinated internally by Barbara Ceiga, vice president for public operations.

115 William Brown, ANSP master plan, vol. 1, 5.

116 Ibid.

117 James Macaleer, personal communication to Robert Peck, September 2009.

EQUATORIAL GUINEA: Bioko Island;
Bioko Sur Prov.; 2.2km from Moka
Malabo on road to Luba; N 03°21'45.2" E
008°40"17.7"; 1500m; montane
rainforest; MV light;
5.xii.1999; Orfe, Spearman &
Weintraub
EQUATORIAL GUINEA: Bioko Island;
Bioko Norte Prov.; Mobil Abayak
Compound; N 03°45'02.2" E
008°45'30.3"; elev. ±50m; 2° lowland
forest and adj. disturbed veg./ Musa
plntn.; at light; loc.#76; leg.
Orfe, Spearman
EQUATORIAL GUINEA: Bioko Island;
Bioko Norte Prov.; along stream behind
Mobil Abayak Compound; N
03°45'01.1" E 008° 29.4"; elev. ±50m;
2° lowland forest adj. disturbed
veg./Musa light trap; 9-
10.xii.1999; Orfe, Spearman
& Weintraub

FACING PAGE Field-pinned pyralid moths collected on Bioko Island, Equatorial Guinea, by N. Orfe, L. A. Spearman, and J. D. Weintraub in 1999. ANSP Entomology

Three specimens of the silver-studded blue (*Plebejus argus*) butterflies from Spain (top: male, dorsal view; middle: female, dorsal view; bottom: male, ventral view). ANSP Entomology Department. This widespread butterfly species occurs from western Europe to temperate East Asia and has a symbiotic relationship with ants.

Chapter 17

Reaching Out: Festivals and Friends

It was such a cohesive group, all with the same motives and objectives, and we had such fun.
—Caryl Wolf, Women's Committee member, 2010

The Academy began its existence as a "gathering of gentlemen," mostly amateurs in natural science, who volunteered their time to collect and arrange specimens according to the taxonomy of the period, speak and write about their particular interests, and encourage others to join them. As time went on, women were included in Academy affairs and scientific endeavors, sometimes as professionals and often as enthusiastic volunteers. Over the years, numerous men and women have given generously of their time, energy, and ideas in many ways to support, enhance, and grow the institution. They volunteered as curators in various departments in the nineteenth and early twentieth centuries; in administrative capacities as trustees; as fund-raisers and innovative members of the Women's Committee and the Friends of the Library; as teachers, writers, artists, docents, and preparators; and in countless other roles. Along with the millions of specimens in the Academy's collections, the library's priceless holdings, and the scientific staff, administration, and support personnel, these dedicated individuals *are* the Academy of Natural Sciences.

The Women's Committee, a particularly successful example of the Academy's volunteer tradition, was founded in 1954. According to Margaret Dilks, the founder, first chair, and first woman on the Academy's board of trustees, the committee got off to a rather shaky start. The group's initial undertaking intended to raise funds—a treasure hunt through the museum—almost ended in disaster, for just as dinner was about to be served, the elevator bringing food from the basement jammed on a piece of lettuce. The elevator was repaired in time, and the affair ended up a success nevertheless. Thus encouraged, noted Dilks, "we launched a series of parties which became bigger and better as we went along."[1]

The stated purpose of the committee, aside from raising money, was to further the aims of the Academy in its services to the community by undertaking appropriate projects or programs that would stimulate public interest and help the institution fulfill its mission. Dilks called the committee members "idealists,"[2] dedicated, talented young women who assumed demanding tasks as volunteers for decades. According to Caryl Wolf, a longtime committee member and Academy trustee, "It was such a cohesive group, all with the same motives and objectives, and we had such fun. There were many lasting friendships made during those years."[3]

On a January night in 1955, the Women's Committee took over the Academy for the "Dinosaur Ball," the first dance ever held in the museum. More than five hundred members and guests banqueted and swirled to the music of Meyer Davis's orchestra (figure 17.2). Along the walls of the entrance lobby and the first-floor auditorium hung humorous and colorful dinosaur cartoons

17.1 Among the eighteen million scientific specimens in the Academy's care are some that document species that no longer exist as living animals. Such is the case with this great auk skull (photographed against an intentionally headless plate of the bird from one of the library's rare books). The skull was collected in 1795 on Funk Island, Newfoundland, Canada. When it was purchased for the Academy from a European collection by Thomas B. Wilson in the mid-nineteenth century, the bird had already been hunted to extinction. ANSP Ornithology Department #173609. Purcell photograph.

173609

painted by Barbara Tyson and May Clark, which remained in place throughout the lecture season. Josephine Henry designed and executed a striking exhibit showing a dinosaur emerging from a huge packing case. She also painted dinosaur tracks on the museum floor leading to the ballroom and made a dinosaur costume for a roller skater who entertained the guests with daredevil feats. The event benefited a fund for the installation of a new dinosaur exhibit in the Academy's Hall of Earth History.[4]

Creative ideas for a series of balls followed in the ensuing years. For the 1957 event, several members went to an Arthur Murray Dance Studio to learn the steps of the specially invented Dinosaur Dip so that they would be able to teach partygoers the dance. The particular focus of the evening was a coming-out party for "Dolores," a seventeen-foot-high, twenty-two-foot-long *Corythosaurus* (duck-billed crested dinosaur), which went on permanent display.[5]

A particularly successful endeavor was "The Great Auk," a dinner dance and "auktion" staged by the committee in 1969. Six thousand invitations brought in five hundred guests, who danced to 30th Street Station, a progressive jazz band, and bid on an eclectic list of items. Among these were an hour on the Wanamaker organ, a stuffed chimpanzee, a genuine World War II airplane ejection seat, a building lot in Ocean City, and membership in the Playboy Club. Caryl Wolf recalled that there was no water at the bars, so drinks were served straight, thus enlivening auction bids and the betting on turtle races. The net proceeds from all the fun assisted the Academy's teaching of natural science to inner-city children.[6]

In 1973, when the Academy moved the president's office and some of its science departments to its newly purchased building near the Philadelphia Museum of Art, the Women's Committee planned a glamorous party they called "The Wild Goose." Suggestions for entertainment were a golden egg contest; Mother Goose–related themes; an Emperor Goose and Gander; and other ideas in the same vein.[7] As fate would have it, this goose—the move to the new building—laid an egg, and not a golden

17.2 The "Dinosaur Ball," 1955. ANSP Archives coll. 417.

one, for the whole project proved an unfortunate financial miscalculation (see chapter 16). In 1978, the Academy moved back to its building on Logan Circle and, over time, raised millions to expand and improve the museum, its library, and its laboratories.

Women's Committee members played an important role in recruiting lecturers for the Academy. Ewell Stewart (1924–1987), also an Academy trustee, who had been to Africa on safari and was acquainted with African writers and scientists, was instrumental in bringing several to Philadelphia. The first of these was the colorful Joy Adamson (1910–1980), naturalist, artist, and author of the popular book *Born Free* (later made into a movie), which described her experiences raising a lion cub named Elsa. Well over six hundred people came to the Adamson lecture and dinner in the late fall of 1962, while a separate matinee presentation for children drew an audience of equal size.[8]

An even more remarkable triumph for Stewart was her success in persuading Louis S. B. Leakey (1903–1972) to give an Academy talk at a black-tie dinner in 1964.[9] Leakey, the famous Kenyan anthropologist, archaeologist, and naturalist, was well known for his important work in establishing human evolutionary development in Africa.[10] While at the Academy, he suggested that Jane Goodall, a young Englishwoman who was studying chimpanzees at a reserve on the shores of Lake Tanganyika, might also make a good speaker for the Academy.[11] Goodall came for the first time in February 1966,[12] again for the Academy's annual meeting in April 1979, and once more in May 2004 (figure 17.4).

Mary Leakey (1913–1996), Louis Leakey's wife and a famous anthropologist and archaeologist in her own right, delivered her lecture "The Evidence of Human Origins: The 3,600,000-Year-Old Footprints" at the Academy in February 1980. These were the footprints found preserved in volcanic ash in Laetoli, Tanzania, the earliest record of bipedal gait. "What we have discovered to date at Laetoli will cause yet another upheaval in the study of human origins," Leakey wrote in *National Geographic*. "For in the gray petrified ash of the beds—among the spoor of the extinct predecessors of today's elephants, hyenas, hares—we have found hominid footprints that are remarkably similar to those of modern man."[13] Mary Leakey assumed leadership of the famous Olduvai Gorge excavations in Africa from her husband in 1962. Under the auspices of the Women's Committee, she returned to speak at the Academy a second time in April 1981.

Another Women's Committee initiative involved offering early evening courses intended "for the serious amateur natural scientist."[14] During one spring session in 1981, Samuel Fuller of the Limnology and Ecology Department lectured on freshwater invertebrate biology; Frank B. Gill, chairman and associate curator of vertebrate biology, elucidated the importance of song in bird communication; and Ernest E. Schuyler, associate of the Botany Department, spoke on native and exotic woody plants.[15]

In 1979, the committee held its first Wildfowl Exposition at the majestic Memorial Hall in Fairmount Park. The show and sale featured exquisitely carved wooden birds and antique duck and geese decoys as well as prints and paintings. This novel fund-raiser was the brainchild of Women's Committee

17.3 Cartoon by Signe Wilkinson. In the 1970s and 1980s, a series of parties for young adults, such as the "Fossil Fling," were held at the Academy to increase membership. Private collection.

17.4 Ruth Patrick, James Baker (Academy president, 2002–6), Emily Baker, and Jane Goodall, May 2004. ANSP Archives coll. 2010-30.

chair Elizabeth Dolan. That year's issue of *Frontiers,* which also served as the event's catalog, stated:

> With its sponsorship of the first Philadelphia Wildfowl Exposition, the Academy of Natural Sciences hopes to encourage and support both an appreciation for wildlife and an interest in wildlife art. [The artists'] paintings, sculpture and carvings give witness to the beauty and wonder of a delicate natural world, and serve as reminders to our responsibility for its care and protection.[16]

Educational displays, films, and seminars on wetland habitat and on the propagation, rearing, migration patterns, and management of wildfowl provided the background for an exhibit of historical decoys and contemporary bird carvings by America's leading wood-carvers (figure 17.5).[17]

The 1979 Wildfowl Expo was so successful that it was continued for the next four years, with the exception of 1982, when Memorial Hall was closed for roof repairs.[18] A gold medal, awarded by the Academy for a lifetime contribution to the preservation of nature, and made possible by the Women's Committee, brought to the Academy and the Expo such luminaries as Roger Tory Peterson (1980), Ansel Adams (1981), Sir Peter Scott (1982, in absentia because the event was not held that year), Eliot Porter (1983), and Peter Matthiessen (1984). Expanded to include other forms of wildlife art, the Expo was held at the Academy in 1983 and 1984.[19] After that, because the event required too much Academy exhibit space, and there were fewer sponsors and patrons, greater expenses, and a smaller gate, the committee decided to retire this fund-raiser and put on a "big splash" for the opening of "Discovering Dinosaurs" (a multimillion-dollar exhibit discussed in chapter 16).[20]

Super Sunday, the most successful long-term event the Women's Committee staged to raise the Academy's profile and provide necessary funds, was first held in early October 1971 and continued for twenty-seven years. A giant block party designed to celebrate and benefit the cultural institutions on the Benjamin Franklin Parkway, it was an instant success and drew larger and larger crowds each year. Super Sunday brought together city dwellers and suburbanites to enjoy Philadelphia's treasured museums. Whether wondering at dinosaurs, marveling at physics and chemistry displays, admiring magnificent paintings, sampling restaurant delicacies or street-corner favorites, or hearing different bands and dancing in the middle of Philadelphia's great boulevard, the audience spent a joyous day—especially when the weather cooperated (figure 17.6).

Setting up Super Sunday was an enormous job. Committee members, on hands and knees, marked out spaces in chalk all along the Parkway, which they had rented to vendors for the day. These ministalls sold everything from funnel cakes to T-shirts, egg rolls, puppets, jewelry, cotton candy, tarot card readings, and raft tours. They also provided a platform for a variety of political and environmental crusaders and "free hugs" from the Unity Church of Christ. Several large stages erected at Eakins Oval and Logan Circle became settings for various bands and dance groups. There was a "crafts alley" and a "sports alley." From the proceeds of the event, the Women's Committee netted somewhere between $40,000 and $60,000 each year, depending on the weather and the size of the crowds. The highest estimate of attendance given by police was four hundred thousand people in 1980.[21]

At the first Super Sunday, according to Mary Lee Cope, no one thought of providing portable toilets for the crowds, which actually had positive results. People sought restrooms in institutions on the Parkway, which many had never been in before. Some were inspired to return by their first sight of great collections, thereby fulfilling one of the goals of the organizers.[22]

The last Super Sunday was held in 1998. There were many reasons not to continue the festival, including the occasionally dangerous atmosphere that had nearly gotten out of hand,[23] but principally because Exxon, a backer for the previous three years, withdrew its sponsorship. The city representative who handled festivals tried to help the Women's Committee find another sponsor, but without success. Ellen Anderson,

17.5 Elizabeth Dolan, Thomas Peter Bennett, and Roger Tory Peterson at the Academy's Wildlife Exposition, 1979. ANSP Archives coll. 2010-300.

Women's Committee chair at the time, noted, "Super Sunday used to have a great many vendors, which was a money-making avenue for us, and all the proceeds went to the Academy, but now there are many other events that vendors can go to for two or three days. [Our festival lasts] just six hours, and if it rains, vendors have lost."[24] "Philadelphians will miss Super Sunday for what it represented," said Joseph Callan, the special events manager for Fairmount Park. "It was the granddaddy of them all."[25]

Among the many worthwhile projects the Women's Committee supported with the money made through Super Sunday was Project VIREO (VIsual REsources in Ornithology), an entrepreneurial enterprise that was established at the Academy in 1979.[26] Dr. J. Peterson Myers, the first director of VIREO, described the project at the time: "We have the premier library of ornithological photography in the world, with contributions from the likes of [Eliot] Porter, [Olin Sewall] Pettingill, [Crawford] Greenewalt, [Roger Tory] Peterson, Leslie [Brown] (his complete collection), Helen and Allan Cruickshank (their complete collections), etc. We keep the originals in archival (cold temperature), low humidity chambers, and make a series of reproduction quality duplicates. These duplicates are what the users see."[27]

From the origin of bird photography in the nineteenth century, countless millions of images have been recorded in black and white and in color, in still photographs, motion pictures, and digital images. Many represent irreplaceable raw data for scientific research and education. Yet before the establishment of VIREO, the majority were inaccessible and unused, scattered among neglected collections, and deteriorating because chemical processes take their toll on photographic emulsions. The idea of VIREO was to bring these photographs into a centralized, professionally curated collection.[28]

Such a library assists the science of ornithology, since photographs can complement traditional methods of collection and study. The colors of dried museum specimens fade perceptibly with time, and study skins, though useful in other ways, provide no information on bird posture, behavior, or habitats. Combining the study of traditional specimens with true-color reproductions of the species greatly increases the understanding of the complex color patterns of birds.[29]

The idea for VIREO came from Frank B. Gill, then the Academy's assistant curator of vertebrate biology. Gill reasoned that VIREO would provide talented amateurs, ornithologists,

17.6 Super Sunday on the Parkway, looking toward City Hall. *Philadelphia Daily News*, 1981. ANSP Archives coll. 2010-222.

and photographers with a means of contributing to the study and protection of birds, as well as give the Academy a way of exploring applications of new visual and computer technologies in the natural sciences (figures 17.7, 17.8).[30]

Academy trustee Crawford Greenewalt (1902–1993), a chemical engineer and retired president of the DuPont Company, funded the initiative with a $200,000 challenge grant.[31] Greenewalt's long-standing interest in photography and the physics of flight had already led him to amazing innovations in filming hummingbirds in flight. He contributed his collection to the Academy as the first large group of bird photographs in the VIREO collections.[32] Eight years later, he presented his specially built camera and flash equipment to the Academy for the use of the VIREO staff.[33]

Eliot Porter (1901–1990), whose life span nearly paralleled Greenewalt's, said of his own work: "My standards of photographic excellence developed from the works of Ansel Adams and Alfred Stieglitz from whom I learned that the entire image area was important if one's goal was the production of works of art. Ornithologists were apparently satisfied with obtaining simply an image of a bird."[34]

John James Audubon's realization of this same artistic idea—that habitat background was an important context for a bird image—was an innovation in avian painting in the early nineteenth century.

Today VIREO contains 164,000 images of more than seven thousand species of birds taken by more than six hundred photographers from around the world. It is the most comprehensive bird image databank worldwide, with seventy-five thousand images online and more added daily.[35] Every type of habitat is represented—mountain, tundra, boreal forest, prairie, steppe, savanna, tropical and dry rain forest, desert, mangrove swamp, and many more (figure 17.9).

Aside from the scientific and educational purposes to which its images have been applied, the collection has been sought by commercial enterprises.[36] VIREO has kept pace with the electronic revolution by providing images for iPhone use, including identification guides and a GPS app for birders with which a user can select a bird and be guided to the spot of its most recent nearby sighting.[37]

By 2000, the number of Women's Committee members had declined to the point that it was no longer possible to meet the committee's previous financial commitments to the Academy, as they had once provided for VIREO. The principal reason was that older members had died or retired, and it was difficult to enlist younger women who were taking professional jobs outside the home. At a meeting that spring, it was resolved to include men on the committee and to change the name to Friends of the Academy of Natural Sciences (FANS).[38] But this did not solve the problem, for by that December's monthly meeting, more members had been excused (fifteen) than attended (twelve).[39] And the minutes for the first meeting in January 2001, in giving instructions for the next gathering, unwittingly provided a metaphor for the end of the endeavor: "The Women's Committee office has been moved. From the front of the building, go past the snakes and the rest rooms, there make a sharp left and go

17.7 *Frilled coquettes* (hummingbirds, male and female), courting. ANSP, VIREO collection. Crawford Greenewalt photograph.

17.8 *Frilled coquettes* (male and female), sleeping. ANSP, VIREO collection. Crawford Greenewalt photograph.

FACING PAGE 17.9 Douglas Wechsler, director of VIREO, photographing a bird in the field. ANSP, VIREO collection.

through the exit door. Our office is on the right. We are behind the gazelle diorama."[40]

The committee had been a successful and profitable adjunct to the Academy for nearly fifty years, under the direction of many able chairpersons, a skilled treasurer,[41] and a large group of hardworking, dedicated volunteers. Included in the minutes for November 1987 was a letter to the members written by Ewell Stewart, recently deceased. Her parting message summed up the organization's important role:

> The Women's Committee over two decades of giving to the larger needs of the Academy, has been such a force filling in gaps where needed, topping off drives for special exhibits, providing provocative speakers and marvelous programs. I don't know truly what the Academy would do without our sustained aid, but at the same time we have all enjoyed the challenge and the camaraderie of working together.[42]

Many members of the Women's Committee were part of the Friends of the Library, a broad-based group that continues to furnish creative ideas and support to one of the most important parts of the Academy. In the late 1980s, Academy librarian Carol Spawn invited Robert Looney (1925–1996), retired curator of prints, drawings, and photographs at the Free Library of Philadelphia, to join the Friends of the Library for the group's "stamping day" project. This was an effort to mark the valuable plates of the Academy's illustrated books with a special code, instantly recognizable to book dealers, as a guard against theft. "This invitation I accepted readily," Looney noted, "knowing what treasures the library housed and the Friends I was likely to meet."[43] After many hundreds of hours of volunteering in the library, he recalled, "I have found my work for the ANSP library to be of consuming interest."[44]

Created by Academy founders at its first meeting in 1812, the library is unique in the Western Hemisphere for its holdings of rare and historic works in every discipline of natural science, as well as for its priceless archives and manuscript collections (figure 17.10). It is a repository of the key works in the history of science, including many magnificently illustrated volumes such as the pre-Linnaean (published before 1750) classics of Gesner, Aldrovandi, Maria Sybilla Merian, and Catesby; the great bird books of Wilson, Audubon, Gould, Wolf, and Elliot; the floras of Redouté, Michaux, and Sowerby; and the 1849–50 watercolors of the American Southwest by Edward and Richard Kern (figure 17.11). William Henry Jackson's early photographs of

17.10 Merou [grouper] from *Voyage de Decouvertes de l'Astrolabe exécuté par ordre du Roi, pendant les années 1826-1827-1828-1829, sous le commandement de M.J. Dumont d'Urville*, Paris 1833, Atlas, Zoology, plate 3. ANSP Library.

17.11 *Palissade*. Maria Sibylla Merian. *Metamorphosis Insectorum Suriamensium* (1726), plate 11. ANSP Library.

Yellowstone geysers, Eadweard Muybridge's photographs of human and animal motion, and Brooke Dolan's exceptional pictures of Tibet in the early 1940s are among some of the many photographic treasures housed in the archives. Some of the manuscripts found there include letters from such luminaries as Darwin, Lyell, and von Humboldt as well as notes, drawings, and manuscripts from Say, Leidy, and hundreds of others.

The Friends of the Library, established by Ewell Sale Stewart, for whom the library was named at the Academy's annual meeting in 1992, has been a vibrant and sustaining adjunct to this most important branch of the Academy.[45] The Friends have sponsored prominent speakers and book signings, collected and held major book sales to raise funds aside from membership dues, and volunteered for a variety of duties (figure 17.12).[46]

In 1990, construction began on a $6.4 million addition to the Academy, to create a significantly expanded exhibit area, new research facilities, a cafeteria, and a much-needed new library stack building. Substantially increased space for archives, maps, and rare books, including a special room for book treasures (the Wolf Room), allowed for improved care and access to these valuable collections (figure 17.13). Four floors of the new structure, encompassing sixteen thousand square feet of air-conditioned space, provided ample space for the overcrowded collection of 200,000 volumes, which had been crammed into a 1910 building designed for one-third that number.[47]

During construction, the Academy rented a huge brick warehouse, built in 1896 as a lace factory and used for many years as a carpet-weaving establishment, in northeast Philadelphia to store the entire contents of the old stack building. As far as the eye could see in that vast space were cardboard cartons, their contents carefully labeled and keyed to a master list at the Academy.[48] Every day, the stack manager traveled to the warehouse in the library's van to bring back the books and journals requested by Academy scientists and outside readers. Carol Spawn supervised this massive undertaking.[49] A significant number of these volumes had been given to the library over the years by their authors, amateur and professional, some of great renown and others obscure, but all the texts were given with great respect for the high ideals and achievements in natural science for which the institution has been known since its founding. The celebration of the library collection's reinstallation took place in June 1991, at the Academy's 180th annual meeting.[50]

Dedicated groups such as the Women's Committee and the Friends of the Library, and numerous volunteers under the more-than-twenty-year direction of Lois Kuter, are essential to the continued growth and functioning of the Academy. These "idealists" have done much to make the Academy the institution that it is today.

PTS

17.12 Friends of the Library conserving books. ANSP Archives coll. 2011-003.

17.13 Librarian Eileen Mathias in Wolf Rare Book Room, 2011. ANSP. Kate Anderton photograph.

Notes

1. Mrs. John Hyland Dilks, "The Idealists," *Frontiers* 26, no. 3 (February 1962): 92.
2. Ibid.
3. Caryl Wolf, personal interview, 15 December 2010.
4. *Frontiers* (April 1955): 124.
5. Women's Committee minutes, ANSP archives, coll. 786, file 3.
6. ANSP coll. 386, file 11; and Caryl Wolf personal interview, 15 December 2010.
7. Women's Committee minutes, 12 October 1973, ANSP coll. 786, file 3.
8. Women's Committee minutes, 7 December 1962, ANSP coll. 786, box 1, file 9. Joy Adamson also wrote *Forever Free*. She was killed in 1980 in what first appeared to be a lion attack but turned out to be murder committed by a former employee.
9. Louis Leakey died suddenly in London in 1972 on his way to speak at the Academy.
10. Louis Leakey entry at http://www.answers.com/topic/louis-leakey.
11. ANSP archives, coll. 417 B, vol. 7.
12. Women's Committee minutes, 14 January 1966 (in anticipation of Goodall's visit), ANSP coll, 786, box 1, file 3.
13. Mary Leakey, "3–6 Million Years Old: Footprints in the Ashes of Time," *National Geographic* 155, no. 4 (April 1979): 446; ANSP coll. 786, box 1, file 26.
14. ANSP spring courses 1981, sponsored by the Women's Committee, ANSP coll. 2011-002.
15. Ibid.
16. Robert McCracken Peck, introduction to Philadelphia Wildfowl Exposition catalog (Philadelphia: Academy of Natural Sciences, 1979).
17. *Academy News*, 2, no. 1 (Spring 1979). The name was subsequently changed to Wildlife Expo.
18. Women's Committee minutes, 3 March and 8 April 1982, ANSP coll. 2010-213, box 1, file 3.
19. Women's Committee minutes, 9 December 1982; 8 December 1983, coll. 2010-213, box 2, file 2.
20. Women's Committee minutes 4 November 1984, ANSP coll. 2010-213, box 2, file 2.
21. "The Original Parkway Bash Is Off," *Philadelphia Inquirer*, 19 August 1998. ANSP coll. 2010-213, box 1.
22. Mary Lee Lowry Cope, personal communication, 3 January 2011.
23. Ibid.
24. Ibid.
25. *Philadelphia Daily News*, 20 August 1998, ANSP coll. 2010-213, box 1.
26. Women's Committee minutes, 14 October 1982, coll. 786, box 2, file 8.
27. Dr. Peterson Myers to Franz Lanting, 6 September 1982, ANSP coll. 2010-122, file Franz Lanting.
28. "Proposals," unpublished manuscript in ANSP VIREO coll. 2010-122.
29. Ibid.
30. J. P. Myers, R. F. Cardillo, and M. A. Culbertson, "VIREO: Visual Resources for Ornithology," *American Birds* 38, no. 3 (May–June 1984): 267.
31. Women's Committee minutes, 8 October 1981, ANSP coll. 2010-213, box 1.
32. See Crawford Greenewalt, *Hummingbirds* (Garden City, NY: Published for the American Museum of Natural History by Doubleday, [1960]).
33. Crawford Greenewalt to Frank Gill, Wilmington, 24 January 1989, ANSP VIREO coll. 2010-122.
34. Eliot Porter to Dr. J. P. Myers, Santa Fe, New Mexico, 25 May 1987, ANSP VIREO coll. 2010-122.
35. VIREO website, http://vireo.acnatsci.org.
36. ANSP VIREO coll. 2010-122.
37. Interview with Douglas Wechsler, director of VIREO, December 2010.
38. Women's Committee minutes, 18 May 2000, ANSP coll. 2010-213, box 1, file 12.
39. Ibid., 8 December 2000.
40. Ibid., 13 January 2001.
41. Henni Slap was treasurer of the Women's Committee for many years.
42. Typewritten copy of letter from Ewell Stewart (before the June 1987 meeting), included with Women's Committee minutes for 6 November 1987, ANSP coll. 2010-213, box 1, file 1.
43. Robert Looney to Patricia Stroud, Friends of the Library chair, 3 April 1991, ANSP archives, coll. 2011-002.
44. Ibid.
45. Friends of the Library newsletter 6, no. 2 (April 1992); ANSP coll. 2011-002.
46. Certain devoted Friends must be singled out for the time, support, and enthusiasm they have given to the library over many years: Jean Bodine, Caryl Wolf, Judy Coslett, Jo Klein, Cecily Littleton, Howard Rowland, Robert Looney, Leonard Evelev, Bill and Elizabeth McLean, Howard Wood, Donald Cresswell, and Ernest Schuyler.
47. "Academy of Natural Sciences Is Embarking on an Expansion," *Philadelphia Inquirer*, 24 May 1990, ANSP coll. 2011-002.
48. Personal tour with Carol Spawn and Keith Thomson, Academy president, 1991.
49. Friends of the Library Newsletter 5, no. 1 (April 1991); ANSP coll. 2011-002.
50. Friends of the Library Newsletter 5, no. 2 (October 1991), in anticipation of the event; ANSP coll. 2011-002.

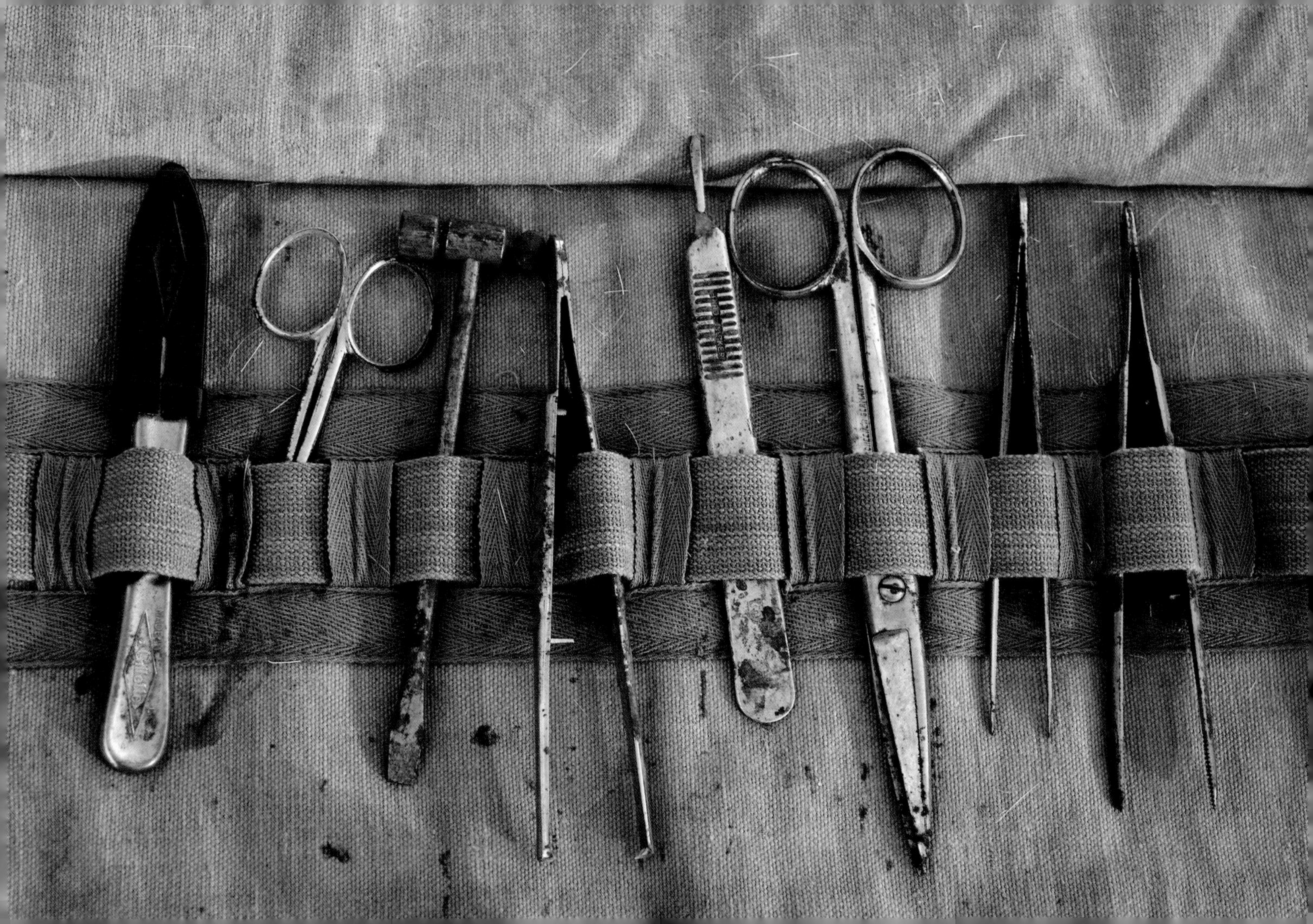

Chapter 18

The Academy's "Glorious Enterprise" Completes Its Second Century

This is not some archaic branch of the animal kingdom. This is our branch.
You're looking at your great-great-great-great-cousin!
—Neil Shubin on fossil discovery *Tiktaalik roseae*, 2006

The Academy of Natural Sciences has moved its collections no less than seven times in the past two hundred years. When the scientific holdings were still relatively small, these relocations were handled by the members themselves. In later years, professional movers were hired to help, but only after countless hundreds of hours had been invested in organizing and packing the specimens for transport. The most recent move—up the Parkway in 1973 and back again in 1978—cost more than $100,000, not counting the staff and volunteer time involved.[1] As expensive and labor-intensive as all of these moves have been, they pale into insignificance when compared to the effort and cost of acquiring the collections and the years of careful stewardship that have followed.

As their individual tags and labels reveal, every one of the more than eighteen million specimens in the Academy's care has its own story to tell. Each was found, secured, preserved, and carefully transported to Philadelphia, often passing through many hands (sometimes over many years) before arriving at the museum. At the Academy, every plant, shell, insect, fish, bird, mammal, and fossil bone undergoes a thorough examination in which it is compared to countless other specimens both in the collection and in the published literature. Since the 1970s, it has been possible to make further comparisons by examining the DNA of the organisms in question.[2] Once an identification is determined (or a new species named), the specimen is formally labeled, recorded in a database, and then integrated into the collection, where it remains available to researchers in perpetuity.

Like books in a library, every specimen is uniquely identifiable and quickly retrievable when needed for further research. Because of their importance, many of the Academy's most unusual and important specimens have been described and illustrated in the scientific literature and can be seen in printed form in libraries around the globe. In recent years, a subset of specimens from almost every department has been digitally photographed and can be examined by anyone with access to the Internet. Scientists wanting to physically examine the specimens can do so in person at the Academy or by requesting them on loan.

That a natural history museum like the Academy has such large collections of the world's flora and fauna is not surprising, but the scale of the holdings often astounds even those who know the institution well. There are a handful of museums with larger collections, but few contain the number of historically and scientifically important types (the actual specimens used to describe a new species for the first time).

As impressive as the collective whole of these holdings might be, it is through the individual specimens that the most interesting stories are told—like the tiny bird (a crossbill) that flew

18.1 More than 1,800 bird specimens collected in Australia by the British ornithologist and Academy member John Gould (1804–1881) were purchased for the Academy by Thomas B. Wilson in 1847. Revered as the starting point for the science of ornithology in Australia, they were the focus of the museum's first international traveling exhibition in 2005. Purcell photograph.

Academy of Nat. Sciences, Philadelphia.
22915 TYPE
New South Wales.
Acad. Nat. Sci. 22950 (327)
Philadelphia
Australia
Acad. Nat. Sci. 22950 (VN 327) ♂. W. A.
Philadelphia
Possible holotype of Euphema splendida Gould: looks
an older specimen than Stone's holotype 22951. However,
neither bird has a long enough tail. Status pending
C. T. Fisher Feb 1997

aboard an American troop ship one thousand miles from its eastern U.S. homeland during World War II and was collected by an Academy member on his way home from military service in Europe, or a once-common stick insect collected from a South Pacific island just two years before its population was decimated by rats and its species brought to the brink of extinction.[3] There are plants heroically carried from Greenland by the Arctic explorer Elisha Kent Kane, and others gently plucked from meadows in California's High Sierra by conservationist John Muir (figure 18.3). Here are Hemingway's fish, Audubon's birds, Jefferson's fossils, and the plants collected by Lewis and Clark. What tales these specimens tell historians, what volumes they speak to the scientists who can read their stories!

Collections like the Academy's don't build themselves, of course; they are slowly created through countless millions of hours of hard work. The process, though endlessly fascinating and enormously appealing to its participants, has never been easy. From the Prohibition era, when Rodolphe Meyer de Schauensee found the scientific contents of his barrels of preserved fish thrown into New York Harbor by thirsty stevedores eager to drink the slimy alcohol in which the fish were stored, to the Vietnam War era, when malacologist George Davis dodged

18.2 Curators, collection managers, and associates of the various systematics departments pose with symbolic representations of their respective disciplines in this 2003 photograph by Doug Wechsler. ANSP.

18.3 A yellow rockfringe (*Epilobium obcordatum*) collected by conservationist John Muir (1838–1914) in the Sierra Nevada in 1875 is just one of 1.4 million plant specimens in the Academy's herbarium. ANSP Botany Department.

18.4 Team members help guide the Academy's expedition boat on a 1987 fish-collecting expedition on the Orinoco River. Suspicious looking containers aboard led to a confrontation with the Venezuelan military. R. M. Peck photograph.

Vietcong snipers to collect tiny snails in the hope of solving one of Asia's great health problems, finding, collecting, and safely transporting scientific specimens to the Academy has proved every bit as hazardous and challenging in the twentieth century as it was in the century before. During their careers with the Academy, entomology curator Dan Otte, ichthyology curator Barry Chernoff, and ornithologist Nate Rice have each faced gun-toting soldiers in various parts of the world while carrying out the work they consider central to their research. Others have faced dangerous animals, debilitating parasites, hostile people, and all manner of medical mishaps. Poisonous snakes and ravenous piranhas notwithstanding, it is often human misunderstandings that cause the greatest hazard to the Academy's collectors.

"Young men with guns, a little authority, and time on their hands make for a dangerous combination," observes the Academy's longtime curator of entomology Dan Otte, who was taken prisoner overnight by the Botswana Defense Forces while he was conducting a cricket survey in 1983.[4] "It is almost impossible to convince a bunch of hyped-up nineteen year olds in combat mode that you are spending time in a war zone looking at insects," says Otte.[5]

In Chernoff's case, the suspicion was that he was smuggling. Because plying hundreds of miles of river along the Venezuelan-Colombian border in search of new species of fish required carrying large amounts of fuel, the long wooden boat on which his Academy expedition was traveling was loaded with dozens of metal gasoline drums. "A few had formalin [for preserving specimens], but most were full of gasoline," recalls Academy fellow Robert Peck, who served as photographer, chronicler, and historian of this and a number of other Academy expeditions in the 1980s and 1990s. "Since there was no way of knowing what was inside any of the barrels, the border patrol drew the obvious conclusion that we were either smuggling drugs from Colombia or gasoline from Venezuela [where the cost was heavily subsidized, making gas much more readily available than in the neighboring country], and probably both."[6]

As the Academy's wooden *falca* chugged slowly south, beyond the last towns of the Amazon basin and into the narrow, twisting headwaters of the Orinoco, a supercharged cigarette boat darted from a hidden side stream to apprehend its passengers and seize its cargo (figure 18.4). Before anyone onboard knew what was happening, the research boat was overrun with soldiers brandishing automatic weapons. Fortunately, expedition leader Chernoff and his Venezuelan colleagues were able to calm the jittery soldiers and talk them down from the firefight they were expecting. After a tense review of the expedition's permits to transport goods—and even to travel in this restricted region—the Academy's vessel was escorted to a nearby border post. There the rattled research team downed copious quantities of Colombian coffee while the soldiers continued to search the boat for illicit cargo. "They couldn't believe we had no contraband," recalls Peck, "for if we were not smuggling drugs or gasoline, why else would we have come to such a place?" The answer to that question lies at the very heart of the Academy's research.

With some notable exceptions, until Ruth Patrick began to apply taxonomic knowledge to detecting environmental problems in water systems in the 1930s, much of the scientific work undertaken by Academy members and staff had to do with the discovery and classification of new species. Some of the most productive and proficient taxonomists in the United States devoted their professional careers to finding, naming, describing, and classifying unknown species in the wild and in their

18.5 James A. G. Rehn (1881–1965) examines one of the thousands of insect specimens he collected for the Entomology Department during his long association with the Academy. ANSP Archives coll. 457.

laboratories at the Academy. It is estimated that Henry Weed Fowler named and personally illustrated some 1,404 species of living fish during his sixty-eight years at the Academy (1894–1962), while Henry Pilsbry, the dean of American malacologists through much of the twentieth century, may have assigned as many as six thousand new names to marine invertebrates during his equally long Academy tenure (1887–1954). James A. G. Rehn's impressive record of entomological discoveries (some 138 new genera and 592 new species), as well as Morgan Hebbard's (104 new genera and 593 new species), were exceeded only by that of their successor, Dan Otte.

For Rodolphe Meyer de Schauensee (1901–1984) and James Bond (1900–1989), who worked with and eventually succeeded Witmer Stone as the Academy's curators of ornithology, the biogeographic distribution of birds was of as much interest as their physical appearance (a topic that had occupied Stone's predecessor John Cassin and his Academy colleagues through much of the nineteenth century). While both Bond and Meyer de Schauensee penned popular field guides (Meyer de Schauensee on the birds of Venezuela, Colombia, and China, and Bond on the birds of the West Indies), much of their field research focused on how birds populated different parts of the world over time.

Among the first of a new generation of scientists to join the Academy in the late 1960s and early 1970s, Frank Gill began to move the focus of avian research at the museum away from taxonomic nomenclature and toward areas of behavioral science, biochemistry, and molecular-based classification. His studies of the feeding dynamics of sunbirds in Africa and hummingbirds in Costa Rica and his interest in the hybridization of warblers and the genetics of chickadees stimulated new and different ways of thinking in the department (figure 18.6). Gill was joined by a succession of talented colleagues, including Tom Lovejoy, who pioneered the study of habitat requirements for tropical and neotropical bird populations; J. P. Myers, who was among the first to use radio tracking to study shorebird migration; and Fred Sheldon, who studied the genetic evolution of the birds of Indonesia. Working both together and sequentially, these men and their associates kept the Academy at the cutting edge of ornithological research throughout the 1970s and 1980s.[7]

From 1983 to 1988, the department served as the center for a labor-intensive Pennsylvania Breeding Bird Atlas project, which gathered information from amateur and professional ornithologists throughout the commonwealth to create a comprehensive look at the health and distribution of Pennsylvania's birds.[8] Just three years later, Gill launched an even more ambitious publishing enterprise to provide definitive accounts of every bird species found breeding in the continental United States. Jointly funded by the Academy and the American Ornithologists Union,

The Birds of North America gathered the most recent ornithological knowledge from specialists from across the country.[9] Under the able direction of Alan Poole, individual species accounts were edited, illustrated, and distributed to subscribers around the world during the project's ten-year tenure at the Academy.[10]

The 1982 arrival of Robert S. Ridgely as associate curator of ornithology returned some of the department's focus to its historical strength in the birds of Central and South America. With support from the MacArthur Foundation, and in association with the Museo Ecuatoriano de Ciencias Naturales in Quito, Ridgely and ornithology collection manager Mark Robbins spent much of the next two decades making a comprehensive survey of the birds of Ecuador, with frequent forays into other parts of South America. Their research led to a dramatic increase in the number of expeditions conducted by the department. This, in turn, generated a large increase in the size of the Academy's bird collections. In the process, both Robbins and Ridgely were able to discover a number of previously unknown birds.[11] After many decades of work on the topic, in 2001 Ridgely published a two-volume work on the birds of Ecuador with coauthor and illustrator Paul Greenfield.[12]

The most memorable—and challenging—of Ridgely and Robbins's many South American expeditions during this period was a trip into the uncharted Cordillera de Cutucu on the Ecuadorian-Peruvian border in 1984. Prior to the Academy's expedition, the area had been a blank spot on the scientific map because of its physical inaccessibility, disputed political status, and the hostile nature of its Shuar (Jivaro) Indian inhabitants. Ridgely was confident that with the help of a Peace Corps volunteer who could speak some of the local Indian language and the logistical support of neighboring non-Indian farmers, he could penetrate the mountainous area, inventory its birds, and return with the information he needed. Unfortunately, as the six-week expedition demonstrated, his optimism was ill founded.

"Almost from the start there were serious logistical problems," recalls Frank Gill, who made up part of the ten-person expeditionary team.[13] "We enlisted the help of people living just outside the Shuar tribal lands to help us carry in our equipment on mules and horses, but they were soon fighting among themselves and could not get along with the local Shuars we employed as guides."[14] There were also serious physical obstacles. In his published account of the trip, the expedition's chronicler, Robert Peck, describes the team's entry into the Cutucu mountains from the Upano River Valley:

> The trail grew softer as we climbed; it turned, at last, into a trench of slimy goo. Leaving the solid ground of the valley, we pushed on through oozing yellowish mud that clung to our legs like peanut butter, slowing our progress to an agonizing crawl. . . .

18.6 Ornithologist Frank Gill researching the relationship between sunbirds and their flower food source, Lake Naivasha, Kenya, about 1973. ANSP Archives.

18.7 Ornithology collection manager Mark Robbins (left) and entomologist George Glenn ponder the mysteries of the Cordillera de Cutucu in Ecuador during a challenging research trip in 1984. R. M. Peck photograph.

> We had trouble staying upright. . . . So did the horses. Time and again, they lost their footing on the slippery rocks that underlay the trail. Sometimes, if the path was especially steep and narrow where they fell, they rolled back down it like misthrown bowling balls, scattering us to the sides and grinding the canvas packs of food and clothing into the mud.[15]

As the traveling conditions worsened, the mule drivers grew tired of the struggle and increasingly fearful of the Shuar, in whose lands outsiders are never welcome. Within days of entering the Cutucu, the porters abandoned the expedition, taking their surviving pack animals and a good deal of the food with them. As their helpers fled back to the "safe" side of the Shuar border, Ridgely and his team found themselves with a fraction of their original supplies and no easy way of transporting what was left.[16]

After several weeks of difficult but successful mist-netting of birds from a succession of makeshift camps, Ridgely, Gill, and several of the other expedition members retraced their muddy tracks out of the Cutucu and returned to the United States. This left four of the original team members and a small group of Shuar helpers to complete the avian inventory.[17] The already bad situation took another turn for the worse a few days later, when local Shuar leaders, inherently hostile to outsiders, concluded that the curious group of gringos who had arrived in their midst weeks before could not be enduring the hardships of the Cutucu simply to look at birds. "They thought it more likely that we were looking for gold and other valuables," writes Peck, and this they would not allow.[18] Fearing the discovery of rich resources on their land, the Shuar called for the expedition's termination. According to an Indian hunter who wandered into camp, a hit order had been issued. All of the intruders were to be killed or driven from the mountains (figure 18.8).

For the next two weeks, the expedition lived in a state of siege, moving from campsite to newly created campsite, climbing ever higher into the Cutucu in hope of escaping the Shuar.[19] All the while, the team continued to observe and gather birds and other specimens for the Academy's collections. "In some ways it was a throwback to an earlier time," recalls George "Skip" Glenn, who collected several hundred insects during the expedition. "We were an interdisciplinary group trying to get an overview of what was living in this remote and inaccessible place."[20] In the end, two new spiders, a new grasshopper, three new species of frog, and an undescribed owl species were discovered.[21]

After several weeks of continued effort, with their supplies exhausted, the group decided to make its escape from the mountains. During a harrowing encounter on one of the final days of the expedition, the team narrowly avoided disaster when Peck used his emaciated appearance to convince three machete-armed Shuar that he was afflicted with a contagious disease. The Indians, more fearful of infection than eager for blood, allowed the party to pass. "I have been on hundreds of expeditions in South America," says Ridgely, "but the Cutucu ranks far and away above the rest in difficulty. It is a wonder any of us survived."[22]

Ridgely's many years of research in neotropical ornithology, including the information gleaned in the Cordillera de Cutucu,

18.8 The Academy's extensive research on the flora and fauna of South America extends from the nineteenth century to the present day. Shown here are members of the (primarily ornithological) Cordillera de Cutucu Expedition, Ecuador, of 1984. R. M. Peck photograph.

18.9 The rampant deforestation of the Amazon River basin and other parts of South America have caused hardship for native peoples and pushed many tropical plant and animal species to the brink of extinction. This picture was taken on an Academy expedition in Amazonas Territory, Venezuela, in 1985. R. M. Peck photograph.

FACING PAGE 18.10 John Gould (1804–1881) and his wife, Elizabeth (1804–1841), spent two years studying birds in Australia before collaborating on a seven-volume book on that subject (1840–48). Among the six hundred species of Australian birds illustrated in that massive work is the now-extinct paradise parrot (*Psephotus pulcherrimus*) shown here. The original specimens that served as the models for these and the book's other illustrations were acquired by the Academy in 1847. Gould was elected a corresponding member of the Academy in 1843. ANSP Library.

PSEPHOTUS PULCHERRIMUS: *Gould.*

took tangible form in his two-volume guide to the birds of Ecuador and his multivolume work on the birds of South America, coauthored and illustrated by his longtime friend and Academy gold medalist Guy Tudor.[23] It also inspired him to turn his energy to trying to protect the birds he studies. He did this first by working with RARE, an international conservation group that had its headquarters at the Academy from 1987 to 1991.[24] Then, in 1998, Ridgely helped establish the Ecuador-based Jocotoco Foundation, to protect rare and endangered species from habitat loss (figure 18.9).[25]

Ever since 1847, when Thomas B. Wilson enabled the museum to acquire 1,800 Australian bird specimens from the British naturalist John Gould (1804–1881), the Academy has had an important place in the history of Australian ornithology. When Leo Joseph joined the Ornithology Department in 1997, that connection was revived and strengthened. In 2004, he and Academy fellow Robert Peck worked together to create an exhibition on the Australian collections of John Gould for the Australian embassy in Washington, D.C. The exhibit proved such a success that it was expanded for further exhibition at the South Australia Museum in Adelaide and at the Tasmanian Museum and Art Gallery in Hobart.[26] Thus the Academy, in its first international traveling exhibition, was able to return to Australia, at least temporarily, the type specimens of some of Australia's most iconic birds (figure 18.10).

Joseph's scientific research focused on the population genetics and distribution of Australian birds. Between 2002 and 2005, he and collection manager Nate Rice updated and expanded the historic Australian holdings of the department with several collecting trips to Australia before Joseph returned to his native country to oversee the National Wildlife Collection in Canberra.[27] Working in collaboration with Harvard, Yale, and the University of Kansas, Rice made further collecting forays to Australia and also to Africa and Vietnam, adding both skins and tissue samples to the Academy's wealth of specimens. In addition to building the collection, Rice oversaw its rehousing in new cabinets between 2001 and 2003.

The longevity and consistency of the Academy's research in so many different fields are among the institution's great strengths, for they enable scientists to document many kinds of environmental change over time. In the 1960s and early 1970s, trustee Charles C. G. Chaplin and ichthyologist James Böhlke made a comprehensive survey of the coral reefs of the Bahamas. While the immediate result of their efforts was a seminal book on the fishes of that region, they left a detailed record of the fish populations and appearance of the reefs they visited that continues to pay scientific dividends (figure 18.11).[28] In 2005, with the support of Chaplin's son Gordon, the Academy began a study of the changes that have occurred over the past sixty years on the very same reefs. "It is rare to have such a well documented baseline for any area, especially one as inaccessible and fragile as a reef," observes Katrina Ilves, who heads up the project.[29] Such studies enable Academy scientists and others to better understand the effects of water pollution, habitat disruption, and climate change on the complex ecosystems that lie at the base of the aquatic food chain.

18.11 Academy trustee Charles C. G. Chaplin (1906–1991) led a series of research expeditions to the reefs of the Bahama Islands in the 1950s, 1960s, and early 1970s. The baseline data he and ichthyology curator James Böhlke collected have proved invaluable in monitoring subsequent changes in the reef fish communities of the Caribbean. Chaplin family archives.

The Academy's long-standing interest in marine fishes has been matched, and in recent years exceeded, by a focus on freshwater fishes. Since the 1980s, under the direction of a succession of ichthyological curators—Barry Chernoff, Scott Schafer, and, most recently, John Lundberg—the Academy has been an international leader in the study of South American fishes.[30]

An ambitious multidisciplinary expedition to the Brazilian Amazon led by Ruth Patrick in 1955 represents an early attempt to study an entire aquatic system in the tropics.[31] That effort, resumed with the establishment of the Stroud Water Center's Maritza Biological Field Station in Costa Rica, has revealed fundamental differences between the freshwater ecosystems of the tropics and those of more temperate climates, not the least of which involves the diversity of fish species that live in these two very different habitats.[32]

The Academy's current curator of ichthyology, John Lundberg, and collection manager Mark Sabaj Pérez are among the handful of experts worldwide who are unraveling the complex interrelationships between tropical fish and trying to place them in an evolutionary context. Comparing living fish to their extinct relatives preserved as fossils, Lundberg has shed new light on the changing geography of South America by demonstrating how different water bodies and their fish fauna were connected before the uplift of the Andes fourteen million years ago.

In his long and productive career, Lundberg has discovered and published as new to science more than thirty-seven species, eleven genera, and two families of fish. He has led expeditions to Brazil, Venezuela, Argentina, Colombia, Panama, Suriname, Costa Rica, Mexico, Zambia, and South Africa, as well as to rivers and fossil-bearing outcrops across the United States. From 2003 to 2010, Lundberg, Sabaj Pérez, and four other ichthyologists (from Auburn, Cornell, and the Florida Museum of Natural History) co-led an international team of more than four hundred researchers from fifty-three countries making an "all catfish species inventory" with support from the National Science Foundation (figure 18.12). "Catfishes are among the most prominent fish groups with over 3,100 species already recognized and hundreds more waiting to be discovered," says Lundberg. "The All Catfish project's mission is no less than the discovery of all catfish species and their phylogenetic relationships."[33]

Dan Otte, the Academy's curator of entomology since 1975, studies an even more abundant and widely distributed group of organisms. Like his predecessors James Rehn, Morgan Hebbard, and Radclyffe Roberts, Otte has specialized in the insect group Orthoptera (grasshoppers, crickets, locusts, mantids, stick insects, earwigs, and cockroaches), and, like his predecessors, he has traveled the world in search of his subjects. Australia, South Africa, Namibia, Botswana, Kenya, Tanzania, Malawi,

18.12 Ichthyology Department collection manager Mark Sabaj Pérez takes a tissue sample for molecular analyses from a thorny catfish (*Oxydoras sifontesi*) in the Río Ventuari, a tributary of the upper Orinoco River, Amazonas Territory, Venezuela, 2010. D. Werneke photograph.

New Caledonia, Fiji, the Solomon Islands, Mexico, Canada, and many parts of the United States, from Hawaii to Pennsylvania, have been scoured by Otte, generating as many tales of adventure as there are stamps on his passport. Like Lundberg, Otte has discovered, named, and published on an astonishing number of new organisms (138 new genera and 1,608 new species). He has also had at least eight species of insects named in his honor.

Along with Titian Peale, Joseph Leidy, and Henry Fowler, Otte is numbered among the select group of Academy curators who are as widely praised for their artistic abilities as for their scientific discoveries (figure 18.13). He has used his artistic talent to complement his taxonomic knowledge by illustrating hundreds of crickets and grasshoppers in exquisite watercolors for his own and other publications.[34]

A man with a finely tuned ear for music, Otte pioneered the use of sound to identify crickets, locusts, and katydids, much as experienced ornithologists use bird calls to identify their subjects. Many insects in the order Orthoptera attract or identify their mates by generating sounds of different pitch and speed, some so high and fast as to escape detection by the human ear. Otte is able to record and play back these calls at various speeds, rendering them more clearly audible and observing the visual graphing of the songs on his computer. What might be perceived as a cacophonous blur of meaningless noise emerging from a summer field or a tropical forest is to Otte a revealing mix of distinctive calls. These help him identify and inventory the individual and family participants in a seasonal mating chorus.

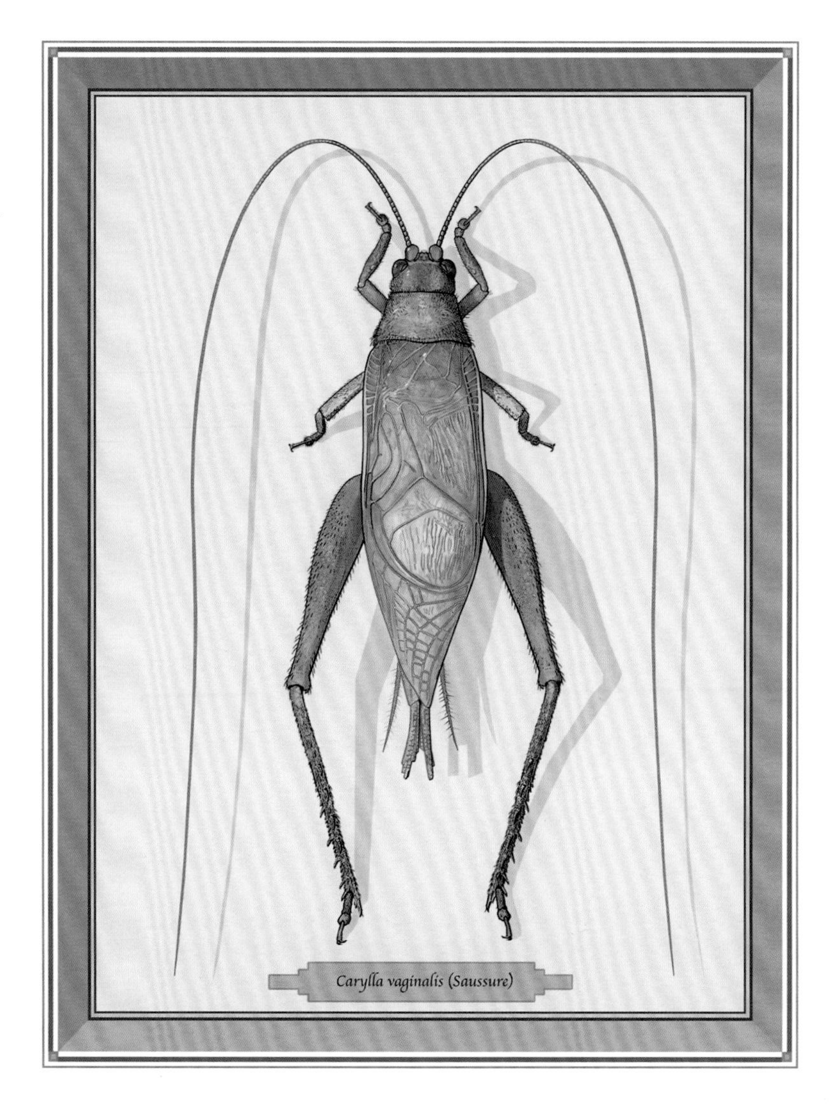

Since 1990, the Entomology Department's special focus on Orthoptera has been expanded to include other insect groups through the work of curator Jon Gelhaus, department associate Robert Allen, and collection manager Jason Weintraub. Gelhaus is one of the world's leading experts on crane flies—the long-legged, large, gossamer-winged insects that vaguely resemble mosquitoes and have almost as wide a distribution as Otte's Orthoptera. Gelhaus's investigations have taken him throughout the United States and into Mongolia, where he has explored the insect's ability to adapt to harsh climates and hostile environments. Allen, a world authority on beetles and apterygotes (primitive flightless insects), has amassed the world's largest private collection of Apterygota, contributing this collection to the Academy along with his many publications on the subject.

Weintraub's extensive experience with the insects of Southeast Asia and his personal interest in local moths and butterflies round out the Entomology Department's range of expertise and give breadth and depth to the educational programs offered by the department.

The focus on applied research that began to accelerate in the systematics departments in the 1970s was nowhere more apparent than in the Malacology Department, where George M. Davis applied his knowledge of the evolution, distribution, and life cycles of certain freshwater snails to a health issue of global concern.[35] Schistosomiasis, a disease affecting more than two hundred million people worldwide, is transmitted by a parasite that has coevolved with a number of closely related freshwater snail species that serve as its intermediate host. While many millions of dollars were being spent each year treating the symptoms of the disease, Davis focused his attention—and the financial support he received from the World Health Organization and the National Institutes of Health—on identifying its underlying causes and the mechanism through which it is distributed.

18.13 Entomology curator Daniel Otte named a new genus of cricket *Carylla* in honor of Academy trustee and Women's Committee member Caryl Wolf. He painted this illustration of it to present to her at the time he received the Academy's Joseph Leidy Award (2009). Watercolor on paper. Private collection.

FACING PAGE 18.14 Insects collected in the field are dried, pinned, and brought back to the Academy for classification and labeling before they are integrated into the Entomology Department collection. These insects and related materials were gathered by Greg Cowper, a curatorial assistant, in Texas in 2010. Purcell photograph.

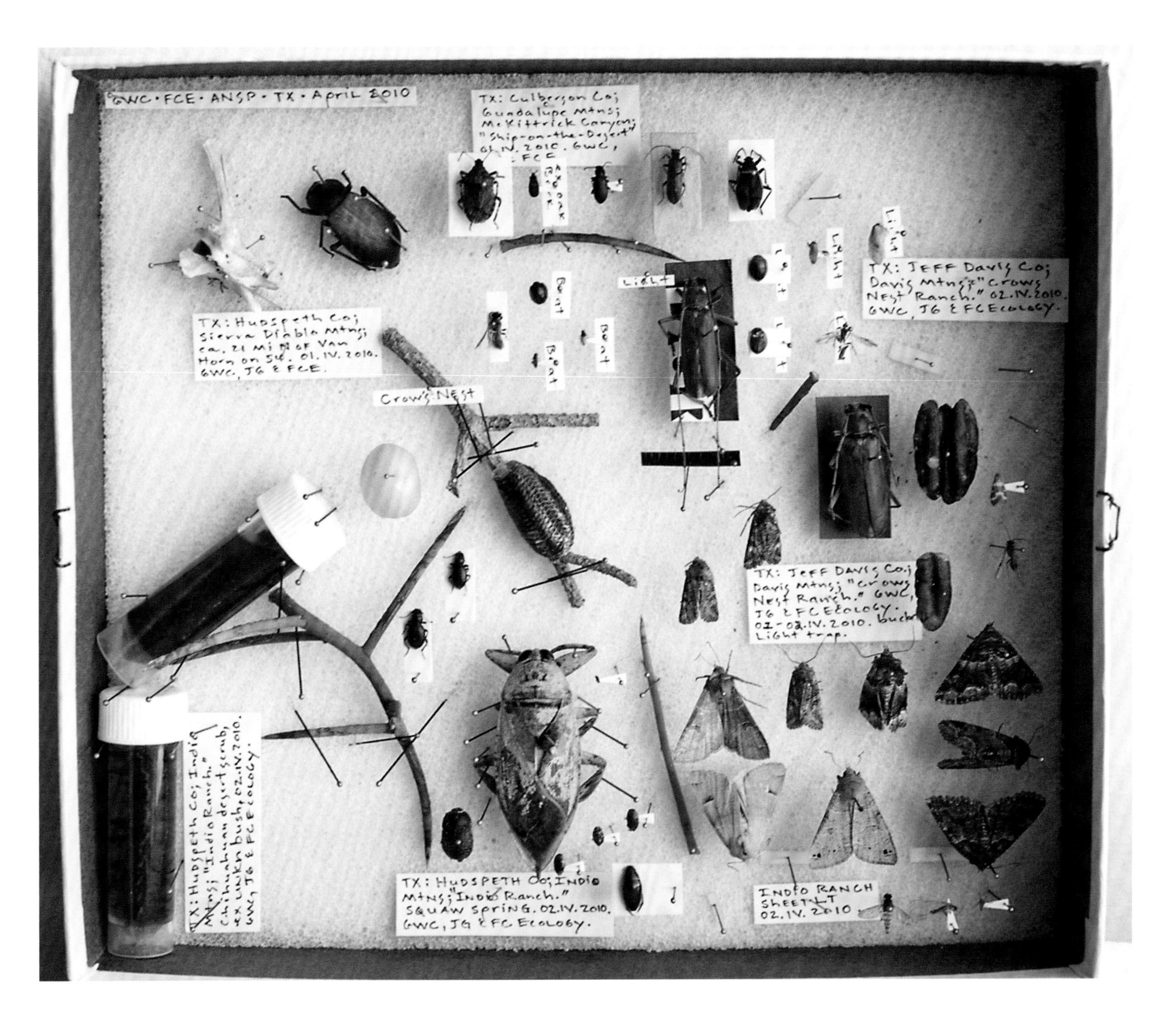
GWC • FCE • ANSP • TX • April 2010
TX: Culberson Co; Guadalupe Mtns; McKittrick Canyon; "Ship-on-the-Desert" 01.IV.2010. GWC, FCE
TX: Hudspeth Co; Sierra Diablo Mtns; ca. 21 mi N of Van Horn on 54. 01.IV.2010. GWC, JG & FCE.
TX: JEFF DAVIS Co; Davis Mtns; "Crows Nest Ranch." 02.IV.2010. GWC, JG & FCEcology.
Crow's Nest
TX: JEFF DAVIS Co; Davis Mtns; "Crows Nest Ranch." GWC, JG & FC Ecology. 01-02.IV.2010.
Light trap.
TX: Hudspeth Co; Indio Mtns; "Indio Ranch." Chihuahuan desert scrub, ex unkn bush, 02.IV.2010. GWC, JG & FC Ecology.
TX: HUDSPETH Co; Indio Mtns; "Indio Ranch." SQUAW SPRING. 02.IV.2010. GWC, JG & FC Ecology.
INDIO RANCH SHEET LT 02.IV.2010

His studies took him to the Mekong River and through combat zones in Vietnam, Laos, and Cambodia at the height of the Vietnam War. "I was given extraordinary access to rich collecting areas by the U.S. military," Davis recalls. "In exchange, I was asked to report whatever I might observe in places Americans were unlikely to go."[36] The collaboration of convenience allowed Davis to make important and potentially lifesaving discoveries. He did further research in India, Thailand, Malaysia, Japan, Taiwan, and the Philippines, as well as Brazil, Uruguay, Argentina, Panama, Costa Rica, and South Africa.

Davis's unique knowledge of the life cycles and living requirements of the schistosomiasis parasite hosts enabled him to predict distribution of the disease in parts of Southeast Asia where it had not yet been identified. This, in turn, allowed the affected countries to effectively allocate medical resources to the infected areas. Davis's expertise led to an invitation from the Chinese government to Davis to help establish and codirect a center for molluscan research at the Institute of Parasitic Diseases in Shanghai in 1997. It is unusual for any foreigner—especially an American—to be intimately involved with an issue as politically sensitive as national health in China, but Davis's expertise was recognized as unique and his reputation for identifying health risks unrivaled. "George knows the taxonomy of the water-based organisms that carry human parasites better than anyone else in the world," said Zheng Feng, director of the Chinese Academy of Preventive Medicine. "He is helping us build a baseline of information on environmental conditions all over China."[37]

A colorful figure who was never modest about his capabilities (he once pointed out to an *Inquirer* reporter that his IQ was almost the same as Albert Einstein's and proudly admitted that he was "absolutely an elitist" and a "citizen of the world"),[38] Davis oversaw a period of dynamic growth in the collection, the usefulness of which he greatly enhanced by starting a computerized database that now covers all of its ten million specimens.

While the medical applications of Davis's research enabled him to attract a great deal of external funding, he was fortunate that the Academy's Malacology Department had already acquired a semi-independent financial base. In 1953, seventeen years before Davis joined the Academy, Alfred Ostheimer Jr., a successful businessman with a passionate interest in Indo-Pacific marine mollusks, with his family and friends, established the Natural Science Foundation, an independent supporting organization for Academy malacology.[39] Its purpose was much like that of the Academy as a whole, "to encourage and promote research and education in the natural sciences."[40] While its primary interest was in funding (and participating in) field research and collecting, the group also created and helped to endow the Henry A. Pilsbry chair in malacology in 1954 in order to attract new talent to the Academy (figure 18.15).[41] When Pilsbry took an "early retirement" at the age of ninety-two, after an astonishing sixty-seven years of service to the institution (1887–1954), the fund enabled the Academy to attract the eminent malacologist R. Tucker Abbott (1919–1995) from the Smithsonian Institution to succeed him. During his fifteen years in the Pilsbry chair (1954–69), Abbott broadened the base of local support and raised the Academy's profile nationally and internationally by engaging the shell collecting community.[42] He was succeeded as chairman by Robert Robertson (1969–72), an expert on Bahamian mollusks and molluscan anatomy. Robertson was in turn succeeded by George Davis, who had served as associate curator of malacology since 1970 and assumed the chairmanship of the department in 1978.

Davis's successor (in 2000) and the current occupant of the Pilsbry chair, Gary Rosenberg, has been associated with the Academy throughout his professional career. Though broadly based in all aspects of malacology, he has specialized in the

18.15 Henry A. Pilsbry (1862–1957), author of more than one thousand scientific papers and several definitive books on mollusks, spent nearly seventy years building, identifying, and interpreting the Academy's malacology collections. The Pilsbry Chair of Malacology was established in his honor in 1954. ANSP Archives coll. 457.

marine and land snails of Jamaica and the Philippines. During his tenure, with support from the National Science Foundation, Rosenberg and collection manager Paul Callomon have completely rehoused the collection. They have also built on the tradition established by Tucker Abbott of connecting with shell-collecting enthusiasts and knowledgeable amateurs to organize and host at the Academy an annual shell show and sale that is considered the best in the eastern United States.

In recent years, the department has given a new applied use to its ten million specimens by serving as the collection of record for the U.S. Department of Agriculture. Every day, dozens of packages of preserved snails arrive at the Academy for analysis by David Robinson, a former Academy staff member who is now the Agriculture Department's chief malacologist. His office sits between Rosenberg's and Callomon's and just down the hall from the third-largest collection of mollusks in the world.[43] The snails in the incoming boxes have been found by airport and seaport inspectors on shipments of fruit and vegetables arriving in the United States from around the world. It is Robinson's job to identify the species (based on comparisons with specimens already in the Academy's collections) and determine if they pose a threat to humans, livestock, American agriculture, or the natural environment. Should he identify a mollusk that might carry a health risk or become an invasive pest (like the infamous zebra mussels that now infest many of the Great Lakes), he can have the entire shipment of produce from which it came rejected or destroyed. Millions of dollars of international trade—and billions of dollars of future U.S. agricultural production—are at stake with every identification.

Invasive species of the botanical kind are one area of expertise of the Botany Department's curator emeritus Alfred "Ernie" Schuyler, who, since his arrival at the Academy in 1962, has applied his training to studying and teaching about the plants of North America. While his colleagues in the Botany Department have worked to build the international stature of the collection through their study of plants from other parts of the world, Schuyler has focused on flora closer to home. "I have had the privilege of working with a diverse group of botanists over the years, most of them internationalists," says Schuyler. "James Mears studied the chemistry of plants from South America, Benjamin Stone and David Frodin specialized in the flora of Southeast Asia and Pacific islands, Lucinda McDade focused on the *Acanthaceae* [a family of plants found most frequently in tropical and subtropical areas], Richard McCourt is an expert on green algae and the evolution of early land plants, Tatyana Livshultz studies reproductive mechanisms in the globally distributed milkweed family, and collection manager Alina Freire-Fierro studies plants of the High Andes. All of them have added to the Academy's collections and reputation world-wide, while I have focused on more domestic subjects."[44] As an often-consulted expert on changing plant populations in wetlands and disturbed habitats, Schuyler makes frequent use of the department's historical collections, which now number more than 1.4 million specimens. "We can use the historical strengths of our collections—including some of the oldest botanical specimens in the United States—to document changes in habitat and climate over time," says Schuyler. "Our collections show us when and where an invasive species may have arrived and help document why rare and endangered native plants have been pushed to the brink of extinction."[45]

If Schuyler were forced to cite his favorite botanical subjects, they would be aquatic plants, sedges, and the globally rare plants of eastern North America, but his knowledge extends well beyond the life cycles of living species and into the history of botanical exploration, publication, and illustration. "The Academy's library is one of my favorite places because it is among the best in the world for books on botanical subjects," says Schuyler. "There is scarcely a botanist or illustrator of note from the sixteenth century to the present day who is not represented in our library [figure 18.16]. Understanding how these works contribute to our knowledge of systematic botany, and having the opportunity to see the ongoing links between art and science is an extraordinary privilege," he observes.[46]

The matchups between specimens and publications extends into every area of systematics, with historically important collections of mammals, reptiles, amphibians, insects, shells, fish, birds, plants, and minerals (those that are still at the Academy), as well as fossil vertebrates and invertebrates. The fossil collections date back to the dawn of paleontology in America and include specimens pondered by Thomas Jefferson and his contemporaries. The first dinosaur teeth ever described from an American site and the first nearly complete skeleton of a dinosaur found anywhere in the world are at the Academy. But what make collections come alive for visitors is how well they are interpreted and how they relate to the lives of the average person.

Ted Daeschler, who revitalized the Academy's once preeminent but then dormant Vertebrate Paleontology Department beginning in the late 1980s, is a master at generating excitement about his collection and his work (figure 18.17). Like former Academy president Keith Thomson, Daeschler has long

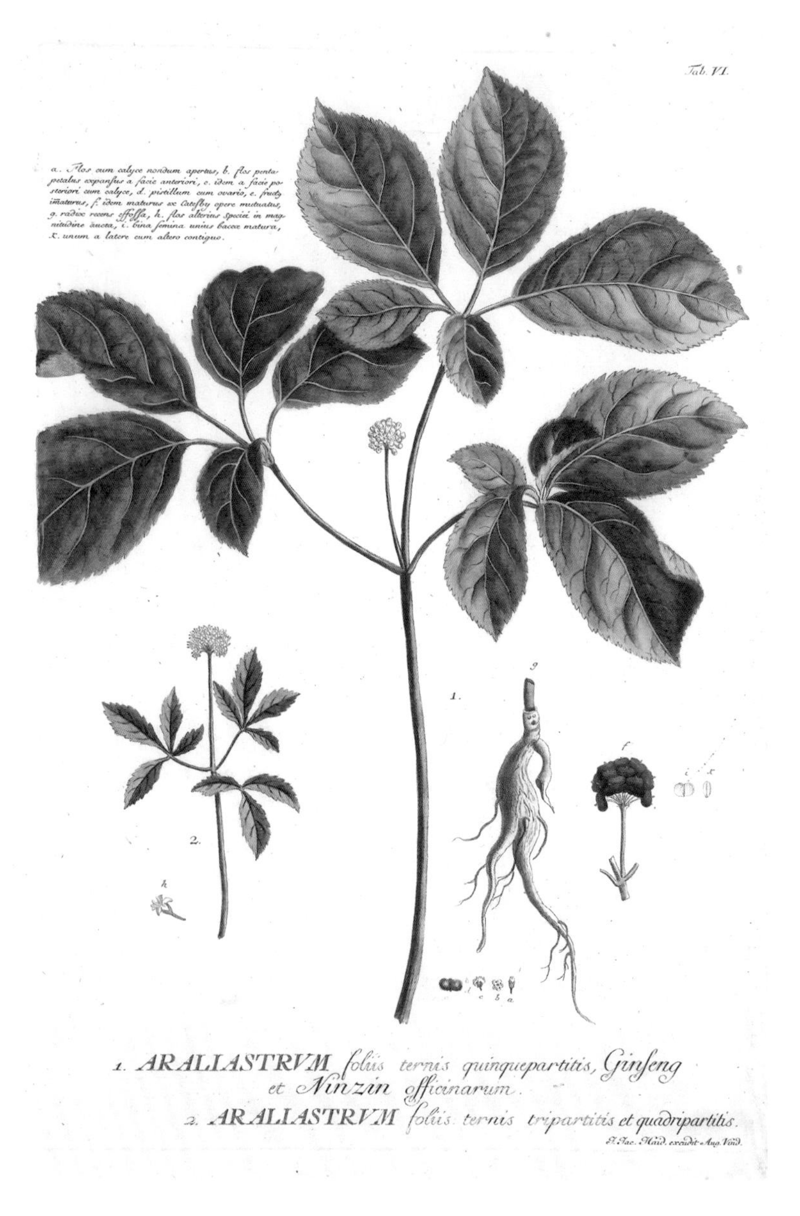

had an interest in the lobe-finned fishes that lived during the earth's Devonian period, approximately 410–360 million years ago. Among other things, Thomson studied the coelacanth, the so-called living fossil that, although believed extinct since the end of the Cretaceous period (sixty-five million years ago), was rediscovered living deep in the ocean off South Africa in 1938.[47] The fishlike creatures that formed Daeschler's favorite fossils are no longer living, but they are just as important as the coelacanth in documenting the evolution of life. They may also be as difficult to find.

After making several important discoveries in recently excavated road-cuts in central Pennsylvania in the early 1990s, Daeschler and his University of Pennsylvania colleague Neil Shubin determined that in order to explore more of the rock strata most likely to contain fossils of the particular age that interested them, they would have to mount an expedition to far northern Canada, to a forbidding, windswept area more than six hundred miles above the Arctic Circle in Nunavut.[48] The undertaking was logistically challenging and expensive, with need for transportation, equipment, and camping supplies for a team of six for five weeks.[49] Because no one had ever looked for fossils there, Daeschler and his team were certain that "whatever we found was going to be of scientific interest."[50] The discoveries made on Melville Island on their first trip in the summer of 1999 were interesting, but those made during subsequent expeditions to Ellesmere Island, some 350 miles to the east, proved even more so. Among the array of previously unknown vertebrate fossils unearthed by the Daeschler-Shubin team, the most extraordinary was a 375-million-year-old creature that documented the transition from free-swimming fish to land-walking tetrapod. The scientists named the creature *Tiktaalik roseae*, a combination of the Inuit word for "large fish in stream" and the name of a generous supporter. Unlike most fish, it had a broad skull, a flexible neck, and eyes mounted on the top of its head like a crocodile (figure 18.18). It also had a big, interlocking rib cage, suggesting that it had lungs and did at least part of its breathing through them. Most important of all, when Daeschler, Shubin, and their Harvard colleague Farish Jenkins examined the pectoral fins and skeleton, they found many features of a rudimentary forelimb, complete with primitive versions of a wrist and an elbow. "This is not some archaic branch of the animal kingdom," Shubin told a science reporter from *Time* magazine. "This is our branch. You're looking at your great-great-great-great-cousin!"[51]

When the Daeschler-Shubin discovery was published as the cover story in the international scientific journal *Nature* in April 2006, it made headlines and news reports around the world.[52] While paleontologists and historians of science greeted the discovery with applause, those of a more fundamentalist religious persuasion found this stunning evidence of evolution unsettling. "There is new and powerful evidence in *Tiktaalik* for the steps that backboned animals took to crawl out of the sea in the first place," observed Michael J. Novacek, provost of science at the American Museum of Natural History. "Many who reject evolution in favor of divine creation claim that the fossil record doesn't contain the so-called transitional species anticipated by Darwin's theory. This ancient, walking fish is yet more evidence that such an argument is simply wrong."[53]

18.16 Christoph Trew (1695–1769) and botanical artist Georgius Dionysius Ehret (1708–1770) offered this illustration of a ginseng in their landmark book *Plantae Selecta*, published in Nuremberg, Germany, in 1750. ANSP Library.

TOP 18.17 Paleontologist and vertebrate biology curator Ted Daeschler on a collecting trip to Nunavut. It was in this forbidding landscape, six hundred miles above the Arctic Circle, that Daeschler and his colleagues discovered the 375-million-year-old fossil of *Tiktaalik roseae*. A. Gillis photograph.

BOTTOM 18.18 The discovery of *Tiktaalik roseae* made the front cover of *Nature* in April 2006 and was mentioned in headlines around the world. ANSP Daeschler photograph.

Like his Academy predecessors Joseph Leidy and Edward Drinker Cope, Ted Daeschler has made groundbreaking discoveries in evolution that have helped us to better understand the history of life on earth. Work by other Academy scientists is focused on the present and future direction of our ecosystems.

Since the early 1990s, Clyde Goulden and a number of his colleagues have been studying the effects of climate change in central Asia.[54] Goulden, a student of G. Evelyn Hutchinson, headed the Academy's Limnology Department from 1972 to 1976 and has directed the Academy's Asia Center since 2007. He has spent most of his professional career researching aquatic systems, including the life forms in high-altitude lakes from the Andes to the Himalayas. When he and Louis "Sandy" Sage, then director of the Academy's Patrick Center, received an invitation from a Buddhist lama to visit Burytia (Siberia) to look at the water systems connected to Lake Baikal in 1994, his research interests began to take a new direction. "For most international scientists, especially those from the U.S., the door to doing research in Mongolia and Siberia had been closed for more than 70 years," says Goulden. "The invitation from Lama Tensing presented a wonderful opportunity for us."[55]

Lama Tensing, fearing the negative effects of unleashed capitalism on the forests and water systems of his native Burytia, realized that the best way to protect the environment there was to enlist the help of the international scientific community. When he asked friends at the United Nations which American institution knew the most about freshwater, he was directed to the Academy. "Lama Tensing wanted us to come and see for ourselves how beautiful and pristine the rivers of Burytia were," recalls Sage, who headed the first Academy expedition to the region. "He was convinced that if we once saw it we would tell others and that some day the whole area might become an eco- tourist destination, thus preventing the wholesale cutting of the forests and destruction of the rivers he loved."[56]

With seed money from the Catherwood Foundation and the National Science Foundation, Sage, Goulden, and Academy fellow Robert Peck headed off for six weeks of jeep and foot travel across the Mongolian steppe and into the Sayan Mountains with Lama Tensing as their guide.[57] When Goulden saw Lake Hovsgol near the Mongolian-Siberian border, he realized that it offered a unique opportunity to study a large freshwater lake that had been little affected by human activity (figure 18.19). "Amazingly, here was a hundred mile long by twenty mile wide lake in the heart of Asia that had not one industrial plant, not even a pleasure boat affecting its water," recalled Goulden. There were a few nomadic herdsmen and two small villages at either end, but otherwise the lake was pristine. "It was—and remains—the largest unspoiled lake in the world," said Goulden.[58]

From his first encounter with Lake Hovsgol, Goulden focused on finding out as much as he could about the lake and its tributaries. Academy entomologist Jon Gelhaus and an eager group of students soon joined him for a succession of interdisciplinary research trips. His work ultimately evolved into two five-year studies of climate change, funded by the World Bank and the National Science Foundation.[59] "When we first started our studies in Mongolia, climate change was not much on the radar," says Goulden, "but our studies over time, and a formal system of interviews with local herdsmen, has enabled us to uncover some very significant and revealing data about what is going on there."[60] With colder winters, warmer summers, and changing patterns in rainfall, the Mongolian steppes are proving an increasingly difficult place for people to live but a very fertile area for scientific study. And, as with so many Academy projects, the consistency and longevity of the undertaking gives it especially meaningful dimensions. For his efforts on behalf of Mongolia's environment over a period of almost twenty years, Goulden was presented with the nation's highest honor for foreigners—the Friendship Medal—by the country's president in 2007.

Goulden and his colleagues have produced many scientific papers, organized symposia, and even published a handbook on climate change for the use of Mongolia's herders, but what Goulden is most proud of are the many Mongolian and American students he has helped to train over his years of research and teaching (figure 18.20). "The Academy was founded by a group of men who wanted to discover new things and then share their knowledge with others," says Goulden. "That's pretty much

18.19 Lake Hovsgol in northern Mongolia became a center of research interest for Academy scientists shortly after the fall of communism in that Central Asian nation. As one of the most pristine lakes in the world, it is an important site for studies of species diversity and climate change. R. M. Peck photograph.

what we are doing in Mongolia—and in many other parts of the world. We are all part of the same system. As new generations come along it will be up to them to understand and protect our shared resources. If we can help them do that better, we will have made a meaningful contribution, not just to science, but to the world in which we live."[61]

Sharing information about life on earth, a defining mission for the Academy since its founding, has taken on new urgency as the world's natural systems have been stressed by expanding human populations. In 2004, Academy president James Baker, a former administrator of the National Oceanographic and Atmospheric Administration, with the encouragement of David Hart, director of the Patrick Center, began the Town Square program, in which Academy scientists and other authorities on a wide range of environmental issues could engage in active conversations with the public. Through these programs, the Academy has become an important forum for discussion about environmental subjects of universal concern. The Center for Environmental Policy was established in 2008 to sharpen this focus and formalize an activity with which the Academy has been involved informally since its founding: helping citizens and government entities make better-informed decisions about environmental issues.[62]

"For two hundred years, the Academy has been pushing the frontiers of scientific knowledge," observes George Gephart, the institution's twenty-ninth president. "The collections it has built and the knowledge it has gained have made important contributions in the past and will be increasingly valuable in the years to come. As we plan the Academy's future, we proudly reflect on that legacy, knowing that the foundation that has been laid will benefit the scientific and educational community for generations to come."[63]

On 26 October, 2011, Gephart and John A. Fry, the president of Drexel University, joined to announce the completion of a historic agreement between the Academy and the university. As it begins its third century, the renamed Academy of Natural Sciences of Drexel University will continue to stand at the forefront of research and education, reaffirming the position of Philadelphia, the birthplace of modern science in North America, as a city central to the advancement of the natural and environmental sciences. The glorious enterprise continues.

RMP
PTS

18.20 Clyde Goulden explains his aquatic research procedures to Lama Samayev Tensing near the Mongolia-Siberia border, August 1994. R. M. Peck photograph.

18.21 The Academy has been involved in science education since 1814, when it offered its first lectures on natural history. Its commitment to education continues through a wide range of formal and informal offerings. Here, Patrick Center staff scientist Paul Overbeck explains the ecology of a freshwater lake to a group of curious fifth-graders. ANSP Patrick Center.

Notes

1 This figure is based on the $50,000 cost of the final move in 1978; see Pat McKeown, "3 Million Dead Bugs Move," *Evening Bulletin*, 28 September 1978, and Edgar Williams, "Creeping: The Priceless Bugs Move Again," *Philadelphia Inquirer*, 28 September 1978.

2 The earliest DNA sequencing was carried out at a number of different commercial and university laboratories throughout the United States and overseas. With a gift from trustee Wistar Morris, the Academy established its own center for molecular studies in 2004.

3 Isolated on a small rocky outcrop of the coast of Lord Howe Island, the so-called Land Lobster (*Dryococelus australis*) is considered the rarest living insect in the world.

4 In 1983, the Botswana Defense Forces were involved in the neighboring Angolan civil war. Otte's experience occurred near the Botswana-Angola border.

5 Daniel Otte, personal communication to Robert Peck, January 2011.

6 Robert Peck, personal communication, January 2011.

7 Tom Lovejoy served on the Academy's scientific staff from 1971 to 1973. He served as a trustee from 1987 to 2008. J. P. Meyers joined the Ornithology Department in 1981 and remained until becoming vice president for science at the National Audubon Society in 1987. Fred Sheldon worked at the Academy from 1987 to 1993. For a brief history of the Academy's ornithology department, see Frank B. Gill, "Philadelphia: 180 Years of Ornithology at the Academy of Natural Sciences," in *Contributions to the History of North American Ornithology*, ed. William E. Davis Jr. and Jerome A. Jackson, Memoirs of the Nuttall Ornithological Club, no. 12 (Cambridge, MA: 1995).

8 This project was directed by Frank Gill and Daniel Brauning and resulted in the publication of several books, including Daniel W. Brauning, ed., *Atlas of Breeding Birds in Pennsylvania* (Pittsburgh: University of Pittsburgh Press, 1992).

9 Although the Academy continued to provide housing and logistical support for BNA, it withdrew its financial support in 1997.

10 *The Birds of North America* was based at the Academy from 1992 to 2002, when the project moved to Cornell University's Laboratory of Ornithology in Ithaca, New York, where the publication became electronic.

11 Robbins discovered a new species of cotinga (*Doliornis remseni*) in 1992; see Mark B. Robbins, Garry H. Rosenberg, and Francisco Sornoza Molina, "A New Species of Cotinga (*Doliornis remseni*)," *Auk* 111, no. 1 (1994): 1–7. Ridgely discovered a new species of antpita (*Grallaria ridgelyi*) in 1997; see Niels Krabbe, D. J. Agro, N. H. Rice, M. Jacome, L. Navarrete, and M. F. Sornoza, "A New Species of Antpita (*Formicariidae grallaria*) from the Southern Ecuadorian Andes," *Auk* 116, no. 4 (1999): 882–90.

12 Robert S. Ridgely and Paul J. Greenfield, *The Birds of Ecuador (Field Guide, and Status, Distribution, and Taxonomy)* (Ithaca, NY: Comstock Publishing Associates; Cornell University Press, 2001).

13 The Cordillera de Cutucu Expedition took place in June and July of 1984. The principal participants were Robert Ridgely, Frank Gill, Mark Robbins (Academy ornithologists); George "Skip" Glenn (Academy entomologist); Robert Peck (chronicler, photographer, and historian of the expedition); Thomas Schulenberg (ornithologist from Louisiana State University); Paul Greenfield (ornithologist and artist from Quito); Steve Greenfield; Todd Miller (Peace Corps volunteer and interpreter); and Juan-Jose Espinosa (representative from the National Museum of Natural History in Quito).

14 Frank Gill, personal communication, November 2010.

15 Robert McCracken Peck, *Headhunters and Hummingbirds: An Expedition into Ecuador* (New York: Walker and Co., 1987), 8–9.

16 Food was stolen by the deserting porters and lost along the trail.

17 The remaining team members were Mark Robbins, George Glenn, Robert Peck, and Peace Corps volunteer Todd Miller.

18 Peck, *Headhunters and Hummingbirds*, 78.

19 A tribal group that traditionally lives in the warmer climate at the base of the mountains, the Shuar were thought less likely to pursue the intruders in the colder, wetter, higher elevations of the Cutucu. Personal field notes, Robert Peck, July 1984.

20 George "Skip" Glenn, personal communication, August 2010.

21 For a description of the birds found during the Cutucu expedition (excluding the owl subsequently named *Otis petersoni* in honor of Roger Tory Peterson), see Mark B. Robbins, Robert S. Ridgely, Thomas S. Schulenberg, and Frank B. Gill, "The Avifauna of the Cordillera de Cutucu, Ecuador, with Comparisons to Other Andean Localities," *Proceedings of the Academy of Natural Sciences of Philadelphia* 139 (1987): 243–59. For a description of the frogs collected by Peck, including a new species named *Eleutherodactylus pecki* in honor of its discoverer, see William E. Duellman and John D. Lynch, "Anuran Amphibians from the Cordillera de Cutucu, Ecuador," *Proceedings of the Academy of Natural Sciences of Philadelphia* 140, no. 2 (1988): 125–42. The new grasshopper was described in Carlos S. Carbonell, *The Grasshopper Tribe Phaeopariini* (Philadelphia: The Orthopterists' Society, Academy of Natural Sciences, 2002), 91. The new spiders discovered by Peck were described by Herbert Levi at the Museum of Comparative Zoology; see *Bulletin of Museum of Comparative Zoology* 159 (1):1–144.

22 Robert Ridgely, public remarks at a dinner at the Academy of Natural Sciences, spring 2010.

23 Robert S. Ridgely and Guy Tudor, *The Birds of South America,* 2 vols. (Austin: University of Texas Press, 1989–1994).

24 The conservation group's name is derived from Rare Animal Relief Effort (RARE).

25 In addition to the *Jocotoco antpita,* the other Ridgely discovery was the El Oro Parakeet (*Pyrrhura orcesi*), discovered in 1980 and described for science in 1987. See Mark Jaffe, "Eagle-eyed 'Mr. Parrot' Finds a New Species," *Philadelphia Inquirer*, 26 June 1986, 1-B and 10-B. See also Robert S. Ridgley and M. B. Robbins, "*Pyrrhura orcesi*, a New Parakeet from S.W. Ecuador," *Willson Bulletin* 100 (1988): 173–82.

26 *John Gould's Birds of Australia* was held at the South Australia Museum in Adelaide from September through November 2004, and at the Tasmanian Museum and Art Gallery from 3 December 2004, through 10 April 2005.

27 The Joseph-Rice collecting project in Australia included five trips between 1999 and 2003. In collaboration with Harvard's Museum of Comparative Zoology, Rice made additional trips to Australia in 2004 and 2005. He collaborated with the Australian National Wildlife Collection in 2006, during which trip he procured an additional one thousand study skins for the collection.

28 James E. Böhlke and Charles C. G. Chaplin, *Fishes of the Bahamas and Adjacent Tropical Waters* (Wynnewood, PA: Livingston Press for the Academy of Natural Sciences of Philadelphia, 1968; reprinted 1992).

29 Katrina Ilves, personal communication to Robert Peck, May 2010. The project was led by Loren Kellogg from 2005 to 2006.

30 Previous curators James Böhlke and William Smith-Vaniz focused on Caribbean fishes and North American freshwater fishes respectively.

31 Joe Pancoast, "Local Scientists Return with 6000 Peruvian Fish After 3 Months' Expedition," *The Sunday Bulletin* (Philadelphia), 15 January 1956.

32 J. G. Lundberg, M. H. Sabaj-Pérez, W. M. Dahdul, and O. Aguilera, "The Amazonian Neogene Fish Fauna," in *Amazonia, Landscape and Species Evolution: A Look into the Past*, ed. Carina Hoorn and Frank Wesselingh (Oxford: Wiley-Blackwell, 2010), 281–301.

33 John Lundberg, "Angling in the Amazon and Beyond," *Natural History* (May 2007): 58. As part of this survey, Mark Sabaj-Pérez in 2004 led part of a "transcontinental catfish expedition" that started in western Peru, went over the Andes, and then down to Amazon tributaries on the frontier with Brazil—where the other team terminated; they missed meeting in real time. The expedition added significantly to the Academy's ichthyology collection and to the all-catfish survey.

34 In 1991, Otte collaborated with his friend and former Academy colleague Richard Estes and provided all of the illustrations for Estes's *The Behavior Guide to African Mammals: Including Hoofed Mammals, Carnivores, Primates* (Berkeley: University of California Press, 1991).

35 George Davis joined the Academy staff in 1970, serving first as associate curator of malacology. In 1978, he was named chairman of the department and occupant of the Pilsbry chair. He left the Academy in 2000 to join the faculty at George Washington University.

36 George M. Davis, personal communication with Robert Peck, June 1997.

37 Zheng Feng, quoted in Robert McCracken Peck, "A Tiny Terror Behind a Giant Dam," *Philadelphia Inquirer*, 1 June 1998, Health and Science section, D-1 and D-4.

38 Margaret Kirk, "Snails and the Secrets of Life," *Today: Philadelphia Inquirer Magazine*, 23 January 1983.

39 Although the initials of the two foundations are the same, Ostheimer's privately funded NSF (established 28 December 1953) should not be confused with the National Science Foundation (also NSF), which was established by Congress in 1950.

40 *Proceedings of the Third Annual Meeting of the Natural Science Foundation*, Oct. 1956, ANSP Malacology Department archives, 1.

41 The chair was established but not fully funded in 1954. On Pilsbry's death in 1957, there was a renewed effort to fully endow the chair. See *Proceedings of the Fourth Annual Meeting of the Natural Science Foundation*, Oct. 1957, ANSP Malacology Department archives.

42 Abbott went on to serve as curator of shells at the Delaware Museum of Natural History and then the founding director of the Bailey Matthews Shell Museum on Sanibel Island in Florida.

43 The Academy's twelve million mollusk specimens form a collection that ranks in size behind only the Natural History Museum in Paris and the Smithsonian Institution in Washington, D.C.

44 Alfred E. Schuyler, personal communication to Robert Peck, January 2011.

45 Schuyler, personal communication to Peck, January 2011.

46 Schuyler, personal communication to Peck.

47 Keith S. Thomson, *Living Fossil: The Story of the Living Coelacanth* (New York: W. W. Norton, 1991).

48 Nunavut is the largest and newest federal territory in Canada. Although the boundaries were established in 1993, this Inuit homeland was officially named a territory in 1999.

49 Participants in the first of Daeschler's Nunavut expeditions included Neil Shubin, Farish A. Jenkins Jr. of Harvard University, and several graduate students from Harvard and the University of Pennsylvania.

50 Ted Daeschler, as quoted in Mark Jaffe, "Expedition to the Arctic for Fossils," *Philadelphia Inquirer*, 23 August 1999.

51 J. Madeleine Nash, "Our Cousin the Fishapod," *Time* (17 April 2006), 58.

52 Edward B. Daeschler, Neil H. Shubin, and Farish A. Jenkins Jr., "A Devonian Tetrapod-like Fish and the Evolution of the Tetrapod Body Plan," and "The Pectoral Fin of *Tiktaalik roseae* and the Origin of the Tetrapod Limb," *Nature* 440, no. 6 (6 April 2006): 757–63, 764–71.

53 Michael J. Novacek, "Darwin Would Have Loved It," *Time* (17 April 2006), 60.

54 In addition to his Academy colleagues, Goulden has collaborated with colleagues from the University of Pennsylvania, the National University of Mongolia, and the Mongolian Academy of Sciences.

55 Clyde Goulden, personal communication with Robert Peck, January 2011.

56 Louis Sage, personal communication with Robert Peck, January 2011.

57 Also participating in this first Mongolian-Siberian expedition were Douglas Gill from the University of Maryland and Andrew Ingersoll, a volunteer.

58 Clyde Goulden, personal communication with Robert Peck, January 2011.

59 The NSF project ran from 2007 to 2011. It was preceded by a five-year study (2002–) supported by the Global Environment Fund of the World Bank.

60 Clyde Goulden, personal communication with Robert Peck, January 2011.

61 Ibid.

62 The Town Square program and Center for Environmental Policy, both directed by Roland Wall, were given their initial funding by Environmental Associates, the Lounsbery Foundation, and the William Penn Foundation.

63 George W. Gephart Jr., personal communication with Robert Peck, February 2011.

In addition to studying the external appearance, habitat, and behavior of fish, ichthyologists often rely on the bone structure of their subjects to aid in classification. Before the invention of the X-ray and the perfection of clearing and staining specimens, a detailed examination of fleshed-out, dry skeletons was required. In the nineteenth century, the Austrian anatomist Carl Joseph Hyrtl (1810–1894) perfected the art of preparing articulated specimens for this purpose. The European flounder shown here (originally *Pleuronectes flesus*, now *Platichthys flesus*) was collected at the mouth of the Elbe River in the North Sea and prepared as a skeleton by Hyrtl sometime before 1850. It was acquired for the Academy by Edward Drinker Cope (1840–1897) in the 1870s as part of a large collection of Hyrtl's fish skeletons. ANSP Ichthyology Department #78046. Purcell photograph.

Milestones in the Academy's History

1812 Academy founded by seven amateur naturalists: Jacob Gilliam, Camilus Macmahon Mann, Nicholas Parmentier, Thomas Say, John Shinn Jr., John Speakman, and Gerard Troost. Members elected within the first year include William Bartram and Alexander Wilson.

The Adam Seybert mineral collection, the largest and most important in America at the time (approximately two thousand specimens), is purchased on behalf of the Academy for $750.

1814 First public lecture, on botany, offered by the Academy.

1817 First issue of the Academy's *Journal* published.

First Academy-sponsored expedition (to Florida).

1818 Thomas Jefferson elected a corresponding member.

1828 Natural History Museum opens to public at 12th and Sansom Streets.

1831 John James Audubon elected a corresponding member.

1840 The Academy moves into first building built specifically to house its scientific collections and activities, at Broad and Sansom Streets.

State legislature approves perpetual exemption from taxation for the Academy.

1841 First issue of the Academy's *Proceedings* published.

1846 Duc de Rivoli collection of 12,500 bird specimens acquired by the Academy.

1847 John Gould's collection of birds from Australia purchased for £1,000.

American Medical Association (AMA) founded at the Academy.

1848 American Association for the Advancement of Science (AAAS) founded at the Academy.

1849 Thomas Jefferson's fossil collection put on long-term deposit at the Academy by the American Philosophical Society. (The deposit is made a permanent gift in 1987.)

1855 The Academy sponsors Paul Du Chaillu's first expedition to West Africa, during which he becomes the first westerner to see and collect live gorilla specimens from the wild.

1856 Lewis and Clark plant specimens (removed from Philadelphia by Frederick Pursh and acquired at auction in England by Edward Tuckerman in 1842) given to the Academy.

Ferdinand V. Hayden sends fossil teeth from Montana to Joseph Leidy, world-renowned scientist and, later, president of the Academy. Leidy recognizes them as the first dinosaur fossils from North America.

1857 The Academy's bird collection declared "the largest and richest in the world," with twenty-nine thousand specimens.

1858 Joseph Leidy identifies first dinosaur skeleton, *Hadrosaurus foulkii*, discovered by William Parker Foulke in Haddonfield, New Jersey.

1860 Charles Darwin elected a corresponding member of the Academy.

Jessup Fund for male students established by children of Augustus E. Jessup. (In 1893, his daughter established another fund for female students. These two funds are later merged.)

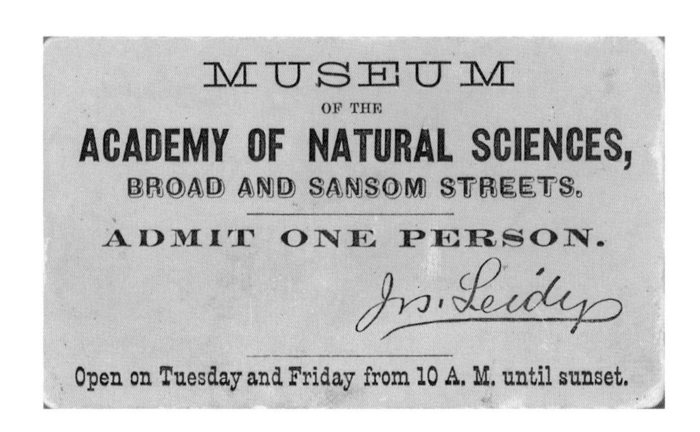

An admission ticket to the Academy's museum signed by Joseph Leidy as chief curator. ANSP Archives coll. 417b.

FACING PAGE The Academy's building at the northwest corner of Broad and Sansom Streets as it appeared before additional floors were built (in 1846 and 1855) to accommodate the institution's growing collections. The Academy occupied this building from 1840 until moving to its present location on Logan Square in 1876. ANSP Archives coll. 49.

THE ACADEMY OF NATURAL SCIENCES OF PHILADELPHIA.

1868 British artist Benjamin Waterhouse Hawkins mounts Leidy's *Hadrosaurus foulkii,* and it goes on public display in the Academy's museum, the first articulated dinosaur skeleton to be exhibited in any museum in the world.

1869 In response to the huge public interest in *Hadrosaurus foulkii,* the curators vote to impose a ten-cent admission fee.

1872 Cornerstone laid for Academy building at 19th and Race Streets.

1876 The Academy holds its first meeting at its new location on Logan Square in time to celebrate the U.S. centennial.

The American Entomological Society (founded 1859) chooses the Academy to serve as its institutional headquarters and repository for its collections and library; forms Entomological Section of the Academy.

1877 Mineralogical Section established.

1883 Collections of the State Geological Survey confided to the Academy by act of the legislature.

1888 Hayden Memorial Geological Award for best publication, exploration, discovery, or research in geology and paleontology established.

1891 Academy supports and participates in Robert Peary's exploring expedition to Greenland.

1892 Succession of buildings completes the Academy complex from Race Street to Cherry Street.

1893 Archives established as part of the Academy's library.

1896 Ludwick Institute begins its support of educational programming at the Academy.

1897 Lewis and Clark plant collection and Benjamin Smith Barton herbarium deposited at the Academy by the American Philosophical Society.

1910 A new façade is given to the Academy's building at 19th and Race Streets to match the Federal style of the new library wing on 19th Street.

A group of well-dressed naturalists, calling themselves "the Botanicals," prepare to embark on a day of collecting, May 1912. ANSP Archives coll. 2010-070.

1923 Leidy Medal for best publication, exploration, discovery, or research in the natural sciences established at international meeting.

1924 Academy charter and bylaws changed to move managerial responsibilities from council of curators to board of trustees (elected by the members).

1928 Charles M. B. Cadwalader named first managing director of the Academy.

1929 First of many habitat groups (dioramas) built to display specimens brought back for that purpose by members of the Academy.

Large portions of the Academy's ethnographic collection (including the Moore, Haldeman, Gottschall, and Peary collections) sold to George G. Heye and the Heye Foundation in New York.

1936 Education Department established.

1937 The Academy celebrates its 125th anniversary with International Symposium on Early Man.

1938 The Academy organizes the first national exhibition on the life of John James Audubon.

1939 Department of Archaeology and Anthropology dissolved.

1947 H. Radclyffe Roberts replaces Charles Cadwalader as managing director (Cadwalader continues to serve as president). Roberts remains managing director until 1963 and then continues to oversee scientific research and publications until 1972.

Limnology Department founded by Ruth Patrick to apply understanding of aquatic ecosystems to environmental problems, especially water pollution.

1948 Mineralogy office and laboratory replaced by Limnology Department; mineralogical research suspended.

1953 Admission fees once again charged for public museum.

1954 Henry A. Pilsbry chair in malacology established.

Academy's Women's Committee established by Mrs. Margaret Day Dilks.

1957 Skeleton of *Corythosaurus intermedius* dinosaur erected in Dinosaur Hall.

1960 Junior Museum, created by the Junior League of Philadelphia, opens inside the Academy.

Richard Hopper Day Memorial Medal for outstanding exploration and discovery in natural science established.

1963 John W. Bodine becomes the first paid president and CEO of the Academy.

1966 Stroud Water Research Center established in Avondale, Pennsylvania.

1967 Benedict Estuarine Research Laboratory opens in Benedict, Maryland.

1970 The Academy participates in the first national Earth Day.

1971 The Academy's Women's Committee conceives and organizes Philadelphia's first annual Super Sunday celebration.

1972 The Academy acquires the Fidelity Mutual Life Insurance Company building for $1.7 million, moving large parts of its collection there the following year.

1976 The Academy celebrates the nation's bicentennial with the international symposium "The Changing Scenes in the Natural Sciences, 1776–1976."

1977 Outside-In, a children's nature museum, established within the Academy.

1978 The Academy sells the Fidelity Mutual Life Insurance Company building and reconsolidates collections at 19th Street and the Parkway.

1979 $3.6 million research wing opens.

New auditorium completed.

George Washington Carver Science Fair begins, growing into one of the largest such fairs in the United States.

Women's Committee organizes first annual Wildlife Art Exposition.

1980 The Academy's Gold Medal for Distinction in Natural History Art established.

1981 New Hall of Changing Exhibits opens in Cherry Street building.

Widener Center for Science Education opens.

$4.6 million modernization project completed; part of $11.5 million raised for building improvements.

Entrance on 19th Street made barrier-free.

1982 *A Celebration of Birds: The Art of Louis Agassiz Fuertes* becomes the first national traveling exhibit initiated by the Academy.

A grant from the National Endowment for the Humanities funds planning for a completely redesigned Dinosaur Hall.

The Women in Natural Science (WINS) science-enrichment program for high school girls begins in collaboration with the School District of Philadelphia.

1983 The Division of Limnology rededicated as the Patrick Center for Environmental Research to commemorate Ruth Patrick's fiftieth year with the Academy.

1984 *Discovering Dinosaurs* (scheduled to open in 1986) moves closer to completion with the opening of *Dinosaurs: An Exhibit in the Making*.

Outside-In, the Academy's children's nature museum, moves to enhanced quarters on the third floor.

1986 *Discovering Dinosaurs* opens.

1987 Ethnographic collections made by T. R. Peale and others transferred permanently to the University of Pennsylvania (where they have been on loan since 1937).

1992 New library stack building opens. Archives move into new facilities.

1997 Morton skull collection formally transferred to the University of Pennsylvania (where it has been on deposit since 1966).

2004 Academy hosts national exhibition on the Lewis and Clark Expedition.

"Town Square," a forum for public discussion of environmental issues, established.

2005 *John Gould and the Birds of Australia*, the Academy's first international traveling exhibit, shown in Adelaide and Hobart, Australia.

2006 The Academy sells nineteen thousand specimens from its historic mineral collection. The original (1812) Seybert collection and the Vaux collection are retained.

2007 The Academy celebrates Ruth Patrick's one hundredth birthday with a formal dinner attended by the governor, the mayor, and 450 guests.

Asia Center established.

2008 Center for Environmental Policy established.

2009 An endowment established for a full-time archivist. The position is named in honor of explorer, collector, and Academy trustee Brooke Dolan.

Arctic explorer Lonnie Dupre carries Academy flag to the North Pole on the one hundredth anniversary of Robert Peary's claimed attainment of the pole.

2011 The Academy establishes a new, long-term affiliation with Drexel University. The Academy is to maintain its mission, identity, principal location, collections, library, endowment, and governance.

2012 In partnership with Drexel University, the Academy celebrates the beginning of its third century with the publication of *A Glorious Enterprise* and a yearlong series of exhibitions and special events.

The Academy's exploration of the natural world continues. Seen here, a team from the Ornithology Department surveys a remote valley in the Andes. Cordillera Las Lagunillas Expedition, Ecuador, 1992. R. M. Peck photograph.

Presidents of the Academy of Natural Sciences

	ELECTION/HIRE	SERVED UNTIL	SERVICE TIME
Gerard Troost**	23 Mar 1812	31 Dec 1817	5 yrs, 9 mos
William Maclure**	31 Dec 1817	23 Mar 1840*	22 yrs, 3 mos
William Hembel**	26 May 1840	Dec 1849	9 yrs, 7 mos
Samuel George Morton**	Dec 1849	15 May 1851*	2 yrs, 5 mos
George Ord**	Dec 1851	Dec 1858	7 yrs
Isaac Lea**	Dec 1858	Dec 1863	5 yrs
Thomas Bellerby Wilson**	Dec 1863	28 Jun 1864	6 mos
Robert Bridges**	Dec 1864	Dec 1865	1 yr
Isaac Hays**	Dec 1865	Dec 1869	4 yrs
William S. W. Ruschenberger**	Dec 1869	Dec 1881	12 yrs
Joseph Leidy**	Dec 1881	30 Apr 1891*	9 yrs, 4 mos
Isaac Jones Wistar**	Dec 1891	Dec 1895	4 yrs
Samuel G. Dixon**	Dec 1895	Dec 1918	23 yrs
John Cadwalader**	Dec 1918	Dec 1922	4 yrs
Richard A. F. Penrose**	Dec 1922	Feb 1926	4 yrs
T. Chalkley Palmer	Feb 1926	Nov 1928	2 yrs, 9 mos
Effingham B. Morris	Nov 1928	Jan 1937	8 yrs, 2 mos
Charles M. B. Cadwalader (President and Managing Director)	1937	Feb 1951	14 yrs
H. Radclyffe Roberts (Managing Director)	1947	1972	25 yrs
M. Albert Linton**	Feb 1951	1962	11 yrs
George R. Clark**	1962	1963	1 yr
John W. Bodine	1963	1 Nov 1970	7 yrs
William W. Marvel	15 Apr 1971	1974	3 yrs

	ELECTION/HIRE	SERVED UNTIL	SERVICE TIME
Milton Wahl (Managing Director and CEO)	10 June 1974	15 Aug 1976	2 yrs, 1 mo
Thomas Peter Bennett (President and CEO)	16 Aug 1976	16 Dec 1985	9 yrs, 4 mos
John Schmidt (Acting)	1 Jan 1986	20 Jan 1987	1 yr
Keith Stewart Thomson** (President and CEO)	21 Jan 1987	31 Dec 1995	8 yrs 11 mos
Phelan Fretz (Acting)	1 Jan 1996	25 Aug 1996	8 mos
Paul A. Hanle (President and CEO)	26 Aug 1996	2000	3 yrs, 5 mos
Seymour S. Preston III (President and CEO)	15 Apr 2000	19 May 2002	2 yrs, 1 mo
D. James Baker** (President and CEO)	20 May 2002	2006	4 yrs
Ian Davison (Acting)	2006	2007	8 mos
William Y. Brown (President and CEO)	Feb 2007	Jan 2010	3 yrs
Edward B. Daeschler (Acting)	Feb 2010	July 2010	6 mos
George W. Gephart Jr. (President and CEO)	Aug 2010		

*Died in office

**Also a member of the American Philosophical Society

Trustees of the Academy of Natural Sciences

Allmon, Warrren D.	2005–2011
Anderson, John C.	1999–2004
Asher, John L., Jr.	1999–2004
Ashurst, John	1924–29
Atkinson, Nolan, Jr.	1985–87
Austen, Peter	2011–
Baillie, Robert A.	1985–99
Baker, D. James	2002–5
Bales, John F., III	1994–2010
Barringer, Brandon	1947–90
Bartol, George E., III	1979–85
Beavers, E. M.	1979–88
Beck, William F.	1986–92
Bell, Vincent G.	1990–97
Bennett, Elizabeth Chew	1976–88
Bennett, Thomas Peter	1976–86
Besson, Michel L.	1981–83
Blaine, James G. (Mrs.) (Joannie)	1983–2000
Blanchard, Elwood P.	1990–97
Blank, Samuel A.	1968–72
Bodine, John	1962–88
Bok, Cary W.	1930–50
Bonnet, William E.	1973–97
Boothby, Willard S., Jr.	1983–97
Boyer, Francis	1971
Boyle, William F.	1968–72
Brisbon, Mrs. Delores F.	1986
Brokaw, Howard P.	1977–97
Brown, William Y.	2007–10
Burch, Robert L.	2002–8
Byron, T. Clark	2009–
Cadwalader, Charles M. B.	1926–59
Cadwalader, John	1925
Campbell, Edgar C., Jr.	1968–72
Canning, Annabelle B.	2004–8
Carpenter, R. R. M.	1929–47
Carpenter, William K.	1948–86
Catherwood, Cummins, Jr.	1979–97
Chaplin, Charles C. G.	1958–90
Chappell, Robert Earl, Jr.	1988–97
Clark, Clarence H., III	1930–37
Clark, George R.	1948–97
Clifford, Maurice C.	1965–97
Coleman, G. Dawson	1934–50
Coleman, Joseph E.	1982–91
Conklin, Edwin G.	1937–50
Connelly, Christine C.	1992–98
Constable, John	1999–2002
Cook, William R.	1994–98
Coslett, Mrs. Edward W., Jr.	1989–97
Cox, Charles K.	1971–72
Crawford, Alan, Jr.	1983–97
Crawford, Edward E.	1985–97
Crawford, Robert W.	1955–68
Croasdale, O. R.	1985–98
Cutler, Carl S.	2010–
Davies, Richard L.	1948–57
Day, Charles	1930
De Schauensee, R. M.	1933–83
Dickey, Charles D., Jr.	1962–64
Dilks, Margaret Day	1955–97
Dolan, Brooke, II	1935–44
Dolan, Thomas, IV (Mrs.)(Elizabeth)	1981–95
Donley, Edward	1981–83
Donnelly, Timothy	2005–10
Dorrance, John T., Jr.	1983–88
d'Ortona, Paul	1964–67
Driever, Lawrence	1982–97

Drinker, Henry Sandwith, Jr.	1935–60
Driscoll, Lee F., Jr.	1977–97
Dwyer, Edward	1969–72
Eagleson, William B., Jr.	1966–72
Eckert, Mildred Longstreth	1966–68
Eichler, Eric Y.	1986–2011
Emlen, Arthur Cole	1940–41
Ernsberger, Jack L.	1987–97
Fenninger, Carl W.	1952–70
Fessenden, Mrs. Samuel	1976–95
Fields, Robert J.	1994–2001
Fisher, John R. S.	1997–2002
Ford, David B.	1995–2007
Forman, Harvey J.	1996–
Forney, Robert C.	1985–97
Foster, Frank B.	1929–40
Freundlich, Richard L.	1979–80
Fritsch, Robert B.	1987–97
Frorer, Kathryn Smith	2008–9
Gadomski, Robert E.	1999–2002
Garnier, Jean-Pierre	1991–97
Garvin, Florence W.	2006–7
Gates, Thomas S.	1928–47
Gelb, Morris	1988–92
Gemmill, Elizabeth H.	2005–11
Gephart, George W., Jr.	2010–
Gill, Frank B.	2006–8
Gluckman, Stephen J.	1991–97
Godfrey, Peter	1978–97
Gowen, James E.	1930–50
Gravell, James H.	1939
Gray, Prentiss N.	1929–34
Greenewalt, Crawford H.	1952–92
Greenewalt, David	1994–2002
Griffith, Michael G.	1993–97
Groome, Harry C., III	1981–86
Haas, Janet F.	1995–2000
Halloran, Harry R., Jr.	1992–2001
Hamilton, N. Peter	2001
Hanle, Paul A.	1996–2000
Hansen, Mrs. Richard	1990–96
Harrar, J. George	1924–36
Harris, John A., IV	1971
Harris, Mary C.	2007–8
Harrison, George L.	1924–36
Hart, W. B.	1948–57
Hass, John R.	1999–2005
Hattersley, Gordon B.	1994–99
Heaps, Marvin D.	1982–98
Hearn, Gail W.	1979–2011
Heckscher, Cynthia	2009–
Henry, T. Charlton	1926–34
Hernandez-Valez, Lydia	1998–2002
Higginbotham, A. Leon, Jr.	1965–70
Horan, John J.	1968–95
Horstmann, Ignatius J., II	1965–80
Howard, Curtis D.	1994–97
Howard, Edgar B.	1935–38
Hoyle, Lawrence T., Jr.	1999–2005
Huffnagel, Frank B.	1962–66
Huggins, David	1992–95
Humphry, Arthur E.	197?–78
Hutchinson, Pemberton	1977–87
Ince, Richard W.	1990–91
Ingersoll, Robert S., Jr.	1948–53
Jackson, John T.	1978–97
Jefferson, Edward G.	1982–86
Johnson, Craig N.	1987–2006

Johnson, E. R. Fenimore 1932–79
Jones, Horace C. 1981–97
Katz, Sam 1996–2002
Kauffman, Virgil 1969–85
Keeley, Frank J. 1925–30
Kermes, Kenneth N. 1988
Koron, Ronald P. 1994–96
Krout, John E. 1981–83
Lax, Mrs. Frances R. 1986–97
Lazar, David P. 2000–
Lea, Jane Jordan 1968–70
Lee, Robert S. 1982–97
Lewis, Drew L., Jr. 1979–2007
Lewis, Hal 1976–88
Linton, M. Albert 1952–65
Lovejoy, Dr. Thomas E., III 1987–2008
Macaleer, R. James 1998–
MacDonald, William F., Jr. 1999–2005
MacElree, Mrs. Lawrence E.(Jane) 1984–98
MacMurtrie, William J. 1982–95
Mann, Frederic R. 1954
Marrazzo, William J. 1992–96
Marval, William J. 1971–7?
Masland, Frank E., Jr. 1957–72
Mattleman, Herman 1991–97
McLean, Sandra L. 2011–
McLean, William L., III 1982–2006
McMinn, William 1977–95
Meads, Donald E. 1998–2004
Merz, J. Frederick, Jr. 1999–2005
Mirick, Henry D. 1965–97
Model, Allen J. 2009–
Montgomery, Edward A., Jr. 1985–
Moore, Anthony K. 2011–
Moore, Cecil B. 1975–78
Moore, J. Percy 1944–64
Moore, Victor E. 1955–66
Morello, Valerie J. 1999–2005
Morris, Alfred 1979
Morris, Effingham B. 1924–36
Morris, I. Wistar, III 2005–
Mulroney, John P. 1981–2002
Murphy, William B. 1959–92
Myers, Frank J. 1941–47
Nalle, Horace D. 1956–68
Nalle, Richard T., Jr. 1979–97
Newbold, Arthur E., Jr. 1927–45
Newburger, Frank L., Jr. 1968–97
Oates, Patrick M. 2011–
O'Neill, Bertram L.(Mrs.)(Jane) 1970–97
Obermayer, Leon J. 1951–83
Olmstead, William W. 1979–86
Ostheimer, Alfred J., III 1955–67
Owens, Robert W. 2000–2002
Palmer, T. Chalkey 1924–27
Palumbo, Frank(Mrs.)(Kippee) 1981–97
Parrish, Morris L. 1934–37
Patrick, Ruth 1947–
Patterson, Mrs. Jefferson 1993–97
Patterson, Kent E. 1995–2004
Penrose, Charles B. 1925
Penrose, Richard A. F., Jr. 1924–26
Pew, R. Anderson(Mrs.)(Mollie) 1981–97
Pitcairn, Feodor U. 1970–72
Preston, Seymour S., III 1981–
Price, Eli Kirk 1929–30
Pudlin, Mrs. Helen P. 1986–98
Rand, Samuel 1991–97
Rauch, R. Stewart, Jr. 1950–55
Rauch, Thomas M. 1965–72
Reed, Ann L. 2011–
Reed, Michael H. 1988–
Riesenbach, Marvin S. 1988–90
Roberts, H. Radclyffe 1946–79
Robin, John P. 1963–65
Roosevelt, Theodore, III 1957–79
Rorer, Gerald B. 1993–
Rorer, Gerald F. 1964–72
Ross, George M. 1980–2005
Rybott, Charles H. 1970–71

Samuels, Janet C.	1998–2004
Scheller, Joseph B.	1997–2006
Schmidt, John P.	1981–2000
Schwartz, George X.	1972–79
Seif, James M.	2001–11
Shapiro, Irving	1979–80
Shepard, Geoffrey	1987–97
Slaughter, William A.	1989–2004
Soltz, Judith E.	1992–
Sordoni, Stephen	1994–2008
Soroko, John	2011–
Stein, Louis	1962–99
Stewart, Henry B. (Mrs.) (Ewell)	1975–86
Stewart, James M.	1995–2004
Stine, Charles M. A.	1938–53
Stokes, David E.	1984–87
Stokes, F. Joseph	1965–92
Stokes, J. Stogdell	1930–46
Street, John F.	1992–98
Stroud, Morris W., III	1986–89
Stroud, Stephen M.	1995–2002
Stroud, W. B. Dixon	1968–97
Symon, Carl	1992–94
Tatum, Charles M., Jr.	1992–2002
Taylor, Lane	1965–92
Tebo, Paul V.	1994–99
Thompson, Sheldon L.	1992–2004
Thomson, Keith S.	1987–96
Tolson, Jay H.	2001–6
Toogood, Granville	1942–46
Trexler, Harry C.	1931–32
Trotter, Spencer	1925–27
Tuffey, Thomas J.	1989–95
Tyson, Barbara	1969–97
Underwood, David R.	2002–5
Van Alen, Mrs. James L.	1975–97
Van Dusen, Lewis H., Jr.	1948–97
Vanderbilt, George	1941–47
Vaux, George, Jr.	1924–26
Wahl, Milton H.	1979–90
Walker, Douglas C.	1993–2005
Warren, Kenneth J.	2007–
White, Michael S.	2005–7
Widener, George D.	1935–47
Wilkinson, Roy	1982–86
Williams, Mrs. Carole Chew	1979–97
Wilson, S. Davis	1937–39
Wister, Owen	1924
Wolf, Caryl	1991–2004
Wolf, Elias	1962–86
Woodward, Charles H.	1959–64
Woodward, Stanley	1937–40
Wright, Minturn T., III	1959–2008
Young, Marechal-Neil E.	1976–86

Awards and Medals

THE HAYDEN MEMORIAL GEOLOGICAL AWARD

Awarded in recognition of outstanding research, exploration, discovery, or publication in the fields of geology and paleontology, this medal was established in 1888 by Emma W. Hayden in memory of her husband, Dr. Ferdinand V. Hayden (1829–1887), the distinguished American geologist and early director of the U.S. Geological Survey, who was known for his research on the stratigraphy of the American West and for facilitating the creation of Yellowstone National Park.

Recipients

1890 James Hall
1891 Edward D. Cope
1892 Edward Suess
1893 Thomas H. Huxley
1894 Gabriel Auguste Daubree
1895 Karl A. von Zittel
1896 Giovanni Capellini
1897 A. Karpinski
1898 Otto Torell
1899 Gilles Joseph Gustave Dewalque
1902 Archibald Geikie
1905 Charles Doolittle Walcott
1908 John Mason Clarke
1911 John C. Branner
1914 Henry Fairfield Osborn
1917 William Morris Davis
1920 Thomas Chrowder Chamberlin
1923 Alfred Lacroix
1926 William B. Scott
1929 Charles Schuchert
1932 Reginald A. Daly
1935 Andrew C. Lawson
1938 Sir Arthur Smith Woodward
1941 Dr. Amadeus W. Grabau
1944 Joseph A. Cushman (presented in 1945)
1947 Paul Niggli
1950 George Gaylord Simpson
1953 Norman L. Bowen
1956 Raymond C. Moore
1959 Carl O. Dunbar
1962 Alfred S. Romer
1965 Norman D. Newell
1968 Elso S. Barghoorn
1971 Wilmot H. Bra
1979 Daniel I. Axelrod
1982 Stephen Jay Gould
1986 John Ostrom
1997 Edwin Colbert
2007 Edward B. Daeschler

THE JOSEPH LEIDY AWARD

Awarded in recognition of outstanding research, exploration, discovery, and publication in any field of natural science, this medal was established in 1923 in honor of Joseph Leidy (1823–1891), anatomist, paleontologist, parasitologist, and president of the Academy of Natural Sciences, whose wide range of interests, tremendous productivity, and influence in scientific thinking made him a world leader in nineteenth-century science.

Recipients

1925 Herbert Spencer Jennings
1928 Henry Pilsbry
1931 William Morton Wheeler
1934 Gerrit Smith Miller Jr.
1937 Edwin Linton

1940 Merritt Lyndon Fernald
1943 Chancey Juday
1946 Ernst Mayr
1949 Warren P. Spencer
1952 Evelyn Hutchinson
1955 Herbert Friedmann
1958 H. B. Hungerford
1961 Robert Evans Snodgrass
1964 Carl H. Hubbs
1967 Don Eric Rosen
1970 Arthur Cronquist
1975 James Bond
1979 Edward O. Wilson
1983 Ledyard Stebbins
1985 Hampton Carson
1989 Daniel Janzen
1995 Peter R. Grant and Rosemary Grant
2006 David Wake
2009 Daniel Otte
2010 Tim Flannery

THE RICHARD HOPPER DAY MEMORIAL MEDAL

Awarded in recognition of outstanding contributions in interpreting natural science and making discoveries in natural history more accessible to the general public, this medal was established in 1960 by Margaret Day Dilks in memory of her grandfather Richard Hopper Day (1847–1924), in recognition of his great interest in natural history.

Recipients

1960 Jacques Piccard
Lawrence A. Shumaker
Don Walsh
Andreas B. Rechnitzer
1964 L. S. B. Leakey
1966 H. Bradford Washburn
1967 Charles A. Berry
1969 Ruth Patrick
1973 Harrison H. Schmitt
1979 Stanton A. Waterman
1980 Crawford H. Greenewalt
1983 David Attenborough
1985 Lewis Thomas
1988 Gerald Durrell
1991 Robert M. Peck
1997 Stephen Ambrose
2000 Thomas Lovejoy
2004 Sylvia Earle
2010 Scott Weidensaul

GOLD MEDAL FOR DISTINCTION IN NATURAL HISTORY ART

This medal is awarded to a person or group of people "whose artistic endeavors and life's work have contributed to mankind's better understanding and appreciation of living things." Artistic is interpreted in its broadest sense, and the list of recipients includes artists working in a wide range of disciplines, including painting, writing, photography, and cinematography.

Recipients

1980 Roger Tory Peterson
1981 Ansel Adams
1982 Sir Peter Scott
1983 Eliot Porter
1984 Peter Matthiessen
1987 BBC Natural History Unit
1992 William Cooper
1995 Guy Tudor
2005 John McPhee
2007 Ray Troll

THE WILLIAM MACLURE AWARD

The William Maclure Award, named for the educator and philanthropist who served as the Academy's president from 1817 to 1840, was established in 2002 to recognize vision, leadership, and philanthropy.

Recipients

2002 Gordon B. Hattersley (posthumously)
2004 Seymour S. Preston, III
2006 William L. McLean, III
2008 Minturn T. Wright, III

Acknowledgments

First and foremost, the authors want to thank, with our sincerest appreciation, the late Robert L. McNeil Jr. for the generous support that made possible the publication of this history of the Academy of Natural Sciences of Philadelphia.

At the Academy library, where most of our work has been done and where most of the documents and pictures we used are located, we benefited greatly from the tireless and good-humored assistance of librarian Eileen Mathias (now retired), who scanned and organized countless illustrations for use in this book. Also, we want to extend our thanks to archivist Clare Fleming, who helped find many elusive documents, and to Sue O'Connell, who secured illustration permissions on our behalf. Joseph Annaruma and Kira Vidumsky were also of particular assistance with our archival research.

The Academy's scientific, education, and exhibition staff were of immeasurable help throughout the research and writing of this book. Those who provided information for *A Glorious Enterprise* include: in Entomology, Dan Otte, Jon Gelhaus, Robert Allen, Jason Weintraub, and Greg Cowper; in Malacology, Gary Rosenberg and Paul Callomon; in Ornithology, Nate Rice; in VIREO, Doug Wechsler and Daniel Thomas; in Invertebrate Paleontology, John Sime; and in Vertebrate Paleontology, Ted Daeschler and Ned Gilmore. Ned deserves extra thanks because he also helped us with our research in the departments of Mammalogy, Herpetology, and Mineralogy. In the Botany Department, we were assisted by Ernie Schuyler, Tatyana Livshultz, Rick McCourt, Alina Freire-Fierro, and Elana Benamy; in Ichthyology by John Lundberg, Mark Sabaj Pérez, Katriina Ilves, and Kyle Luckenbill; and at the Patrick Center by David Velinsky, Richard Horwitz, and Roger Thomas. Clyde Goulden at the Asia Center and Bern Sweeney at the Stroud Water Research Center were most generous in sharing their memories of times past. Of those retired from the Academy who provided information, we especially want to thank John Cairns, Tom Lovejoy, Frank Gill, Louis Sage, and Robert Grant.

We are grateful to George Gephart, Amy Marvin, David Rusenko, Lois Kuter, Carolyn Belardo, Dennis Murphy, Sara Hertz, and many other members of the Academy staff for their enthusiasm and support. Several former Women's Committee members, especially Mary Lee Cope and Caryl Wolf, provided amusing and affectionate anecdotes of their time on the committee.

Institutions in Philadelphia at which we conducted research and that kindly made images available for our use are the American Philosophical Society, the Historical Society of Pennsylvania, Independence National Historical Park, the Library Company of Philadelphia, the Pennsylvania Academy of the Fine Arts, the Philadelphia Museum of Art, the University of Pennsylvania's Museum of Archaeology and Anthropology (where Janet Monge and Alessandro Pezzati were of particular help), and the Wagner Free Institute of Science.

The following national institutions provided additional illustrations for use in this book: Historic New Harmony, the Indiana State Museum, the Joslyn Art Museum, the Peabody Essex Museum, the St. Louis Art Museum, and the White House Historical Association.

The authors wish to acknowledge the important role played by Thomas Peter Bennett, former president of the Academy, whose instructive criticism of early drafts of our manuscript was invaluable.

From the beginning of the project, Jerome Singerman, our editor at the University of Pennsylvania Press, has been an enthusiastic and helpful partner, and Noreen O'Connor-Abel has overseen the production of the book with care and competence. Jeff Wincapaw of Marquand Books has done an outstanding job with the book's design.

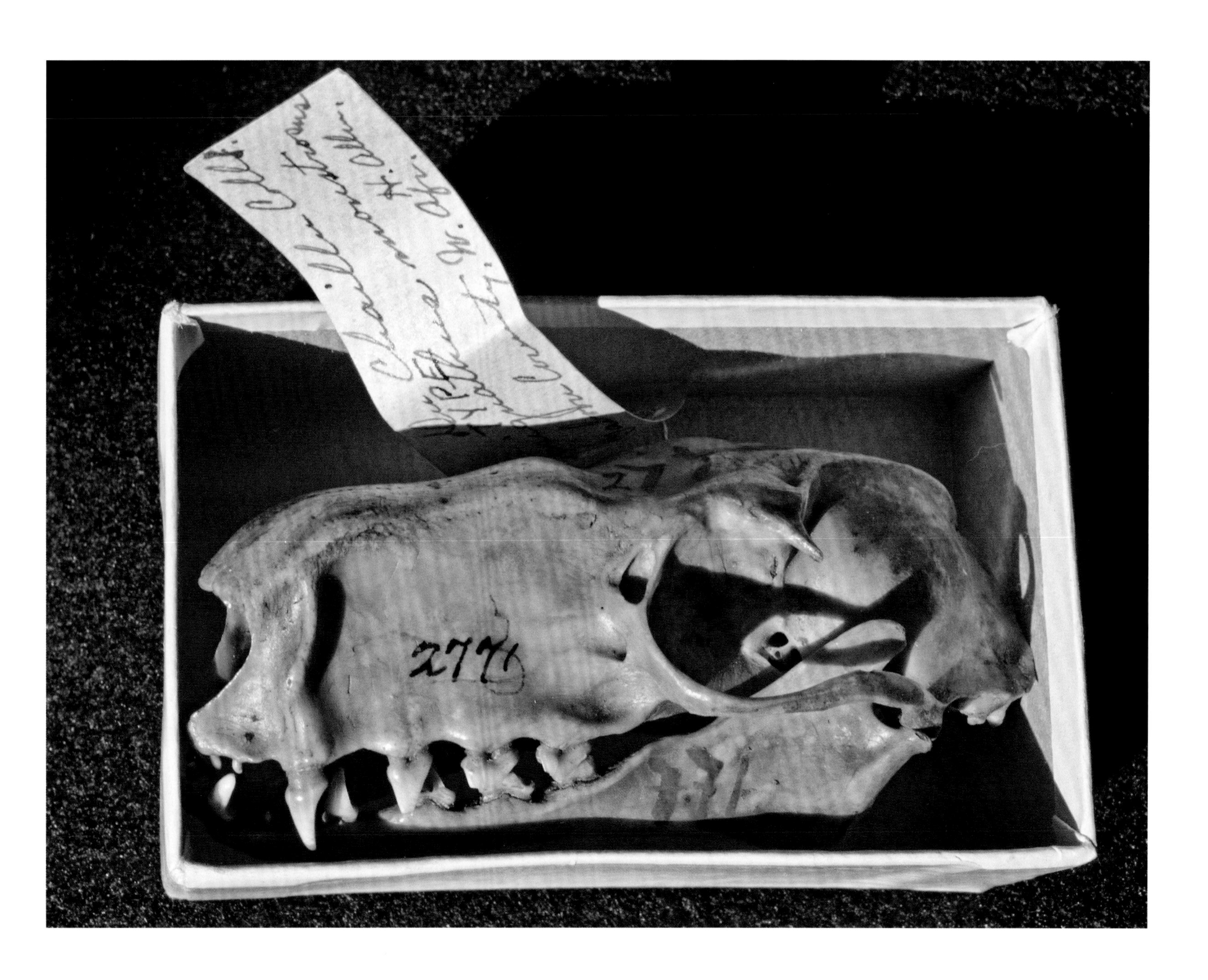

Without the stunning photographs of Rosamond Purcell, this history would not have the interest or elegance that her creative art provides. We thank Rosamond warmly for the effort she expended documenting the Academy's specimens—under skylights, in the parking lot, on the roof, or wherever there was enough natural light for her to create these magnificent images.

Most of all, we are grateful to our spouses, Susan Weld Peck and Alexander McCurdy III, for their forbearance in living with this history for what must have seemed to them like two hundred years.

RMP

PTS

Skull of a hammer-headed bat (*Hypsignathus monstrosus*) collected in Gabon, West Africa, by Paul Du Chaillu circa 1859. ANSP Mammalogy Department #2771. Purcell photograph.

Index

Page references in italics indicate illustrations and/or names in captions.

Copyright © 2012 The Academy of Natural Sciences
Photographs copyright © Rosamond Purcell.
All rights reserved. Except for brief quotations used for purposes of review or scholarly citation, none of this book may be reproduced in any form by any means without written permission from the publisher.

Published by
University of Pennsylvania Press
Philadelphia, Pennsylvania 19104-4112
www.upenn.edu/pennpress

10 9 8 7 6 5 4 3 2 1

Library of Congress Cataloging-in-Publication Data

Peck, Robert McCracken.
A glorious enterprise : the Academy of Natural Sciences of Philadelphia and the making of American science/Robert McCracken Peck and Patricia Tyson Stroud ; photographs by Rosamond Purcell.—1st ed.
p. cm.
Includes bibliographical references and index.
ISBN 978-0-8122-4380-2 (alk. paper)
1. Academy of Natural Sciences of Philadelphia—History. 2. Natural history—Research—Pennsylvania—Philadelphia. I. Stroud, Patricia Tyson. II. Purcell, Rosamond Wolff. III. Title.
QH70.U62P557 2012
508.748'11—dc23 2011034991

Produced by Marquand Books, Inc., Seattle
www.marquand.com

Edited by John Pierce
Designed by Jeff Wincapaw
Layout and typesetting by Susan E. Kelly
Typeset in Arno Pro with captions in Whitney
Proofread by Laura Iwasaki
Color Management by iocolor, Seattle
Printed and bound in China by C&C Printing Co., Ltd.

Pages ii–iii: Ammonite (*Ammonites giganteus*) from Tisbury, Wiltshire, from the collection of Etheldred Benett. ANSP Invertebrate Paleontology Department. This specimen was described and illustrated by William Buckland in *Geology and Mineralogy*, 1837, vol. 1, plate 41. Purcell photograph.

Page v: Shell with Buddhas (*Cristaria placatus*) from Min River, China, presented by S. Drinker, May 1854. ANSP Malacology Department #56518. Purcell photograph.

Title page: Henry E. Crampton, Jr. (left) and a Japanese colleague named Mr. Kono collect partulid snails in Saipan, 1921. ANSP Archives coll. 755.

Pages viii–ix: Common American wildcat or bobcat (*Lynx rufus*, now *Felis rufus*) by John James Audubon. Plate 1 from *The Viviparous Quadrapeds of North America* (1845–48). ANSP Library.

Page x: Top: Yellowstone geyser photographed by William Henry Jackson during a surveying expedition to Wyoming with Ferdinand V. Hayden. ANSP Archives coll. 34; Bottom: Ferdinand V. Hayden's surveying team. One of the 274 photographs by William Henry Jackson presented to the Academy's library by Hayden in 1876. ANSP Archives coll. 34.

Page xi: An assembly of fish from the South Pacific from *Voyage de Decouvertes de l'Astrolabe exécuté par ordre du Roi, pendant les années 1826-1827-1828-1829, sous le commandement de M.J. Dumont d'Urville*. Paris, 1833, Atlas, Zoology, plate 19. ANSP Library.

Pages xii–xiii: Wolf skull (*Canis lupus*, syntype of *Canis gigas*) collected by John Kirk Townsend in the Rocky Mountains in 1834; E. D. Cope Collection #531. ANSP Mammalogy Department #2266. Purcell photograph.

Page xiv: Common piddock (*Pholas dactylus*) in stone. ANSP Malacology Department #329635. Purcell photograph.

Page xviii: *Chiva*, on which members of the Denison-Crockett Expedition sailed to Dutch New Guinea in 1937 to collect bird specimens for the Academy. ANSP Archives coll. 113 IV.